32095.

DAUGHTY.
WRL

Engineering GNVQ: Intermediate

WALCAT

Engineering GNVQ: Intermediate

Editor:

Mike Tooley

Dean of Faculty, Brooklands College of Further and Higher Education

Authors:

John Bird

Senior Lecturer, Highbury College of Technology

Bruce Newby

Head of School of Electrical and Electronic Engineering, Brooklands College of Further and Higher Education

Roger Timings

Head of Department of Engineering (retired), Henley College, Coventry

620.0071141 TOO.
TEMP REF.

BUTTERWORTH HEINEMANN

Butterworth-Heinemann
Linacre House, Jordan Hill, Oxford, OX2 8DP
A division of Reed Educational and Professional Publishing Ltd

A member of the Reed Elsevier plc group

OXFORD BOSTON JOHANNESBURG
MELBOURNE NEW DELHI SINGAPORE

First published 1996
Reprinted 1997

© Reed Educational and Professional Publishing Ltd 1996

All rights reserved. No part of this publication may be reproduced in any material form (including photocopying or storing in any medium by electronic means and whether or not transiently or incidentally to some other use of this publication) without the written permission of the copyright holder except in accordance with the provisions of the Copyright, Designs and Patents Act 1988 or under the terms of a licence issued by the Copyright Licensing Agency Ltd, 90 Tottenham Court Road, London, England W1P 9HE. Applications for the copyright holder's written permission to reproduce any part of this publication should be addressed to the publishers

British Library Cataloguing in Publication Data
A catalogue record for this book is available from the British Library

ISBN 0 7506 2597 X

Printed and bound in Great Britain by The Bath Press, Bath

Contents

Preface

Welcome to the challenging and exciting world of engineering! This book is designed to help get you through the core units of the Intermediate General National Vocational Qualification (GNVQ) in Engineering. It contains all of the material that makes the essential underpinning knowledge required of a student who wishes to pursue a career in any branch of engineering.

The book has been written by a team of Further and Higher Education Lecturers. They each bring their own specialist knowledge coupled with a wealth of practical teaching experience. The team worked closely during the production of the book and this has helped to ensure that the book uses a common format and approach.

About GNVQ

General National Vocational Qualifications (GNVQ) are available in schools and colleges throughout England, Wales and Northern Ireland. Their main aim is to raise the status of vocational education in the UK within a new system of high quality vocational qualifications which can be taken as an alternative to the well-established General Certificate of Education (GCSE) and General Certificate of Educational Advanced Level (GCE A level) qualifications. The Government intends that GNVQs, together with National Vocational Qualifications (NVQs), will replace other vocational qualifications and become the main national provision for vocational education and training.

Although GNVQs are primarily aimed at 16 to 19-year-olds, they are also available to adults. Furthermore, credit towards GNVQs is being gained by 14 to 16-year-olds in secondary education. One of the main objectives of GNVQs is to provide a valued alternative to GCE A level qualifications for the increasing number of students staying on in full-time education beyond 16. The Advanced GNVQ is designed to be of comparable standard to that set by GCE A levels whilst the Intermediate GNVQ is broadly comparable to GCSE standard.

GNVQs provide a broad-based vocational education that continues many aspects of general secondary education. As well as acquiring the basic skills and body of knowledge that underpin a vocational area, all students have to achieve *core skills*. The attainment of both vocational *and* core skills provides a foundation from which students can progress either to further and higher education, or into employment with further training appropriate to the job concerned.

GNVQs, like NVQs, are unit-based qualifications. Each is made up of a number of units which can be assessed separately, and this allows credit accumulation throughout a course. A certificate can be obtained for each unit when necessary and credit transfer on some units can be made between qualifications. To maintain this fundamental characteristic of GNVQs, assessment is based on the unit rather than the qualification.

The mandatory units for each GNVQ have been developed jointly by the National Council for Vocational Qualifications (NCVQ) and the GNVQ Awarding Bodies – BTEC, City and Guilds and RSA Examinations Board. NCVQ has taken a lead role in the evaluation and revision of the mandatory units. Each awarding body develops its own optional and additional units, and takes responsibility for their evaluation and revision. NCVQ has developed, evaluated and revised the core skills units.

Core skills

Core skills are a vital part of any GNVQ programme. Three core skills units, Communication, Application of Number, and Information Technology are mandatory requirements for all students. Indeed, the same core skills units are taken by all students taking GNVQs at the same level (e.g. level 2 core skills units are taken by all students following any Intermediate GNVQ programme irrespective of the vocational area or the awarding body). Other core skills may be taken as 'additional units'.

How to use this book

This book covers the four mandatory units that make up the GNVQ Intermediate Engineering programme. One chapter is devoted to each unit and each chapter contains text, worked examples, 'test your knowledge' questions, activities, and multi-choice practice questions.

The worked examples will not only show you how to solve simple problems but they will also help put the subject matter into context with typical illustrative examples.

The 'test your knowledge' questions are interspersed with the text throughout the book. These questions allow you to check your understanding of the preceding text. They also provide you with an opportunity to reflect on what you have learned and consolidate this in manageable chunks.

Most 'test you knowledge' questions can be answered with only a few minutes and the necessary information, formulae, etc., can be gleaned from the surrounding text. Activities, on the other hand, require a significantly greater amount of time to complete. Furthermore, they often require additional library or resource area research coupled with access to computing and other information technology resources.

Activities make excellent vehicles for gathering the necessary evidence to demonstrate that you are competent in core skills. To help you identify the opportunities for developing core skills and acquiring evidence, we have included core skills icons in the text, as shown below:

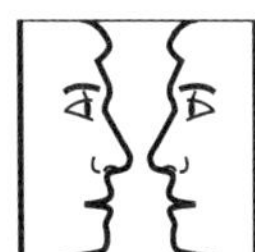
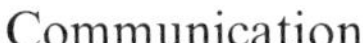
Communication

Application of number

Information technology

Look out for these icons as they will provide you with some clues as to how to go about developing and enhancing your own core skills!

Finally, here are a few general points worth noting:

- Allow regular time for reading – get into the habit of setting aside an hour, or two, at the weekend to take a second look at the topics that you have covered during the week.
- Make notes and file these away neatly for future reference – lists of facts, definitions and formulae are particularly useful for revision!
- Look out for the inter-relationship between subjects and units – you will find many ideas and a number of themes that crop up in different places and in different units. These can often help to reinforce your understanding.
- Don't expect to find all subjects and topics within the course equally interesting. There may be parts that, for a whole variety of reasons, don't immediately fire your enthusiasm. There is nothing unusual in this; however, do remember that something that may not appear particularly useful now may become crucial at some point in the future!
- However difficult things seem to get don't give up! Engineering is not, in itself, a difficult subject, rather it is a subject that *demands* logical thinking and an approach in which each new concept builds upon those that have gone before.
- Finally, don't be afraid to put new ideas into practice. Engineering is about 'doing' – get out there and do it!

Good luck with your GNVQ Engineering studies!

Mike Tooley

Unit 1 Engineering materials and processes

Summary

This unit introduces you to a selection of widely used engineering materials and components. It also introduces you to a range of workshop and manufacturing processes. Some of these processes you will eventually use in your project work. Others are more appropriate to the large scale production of materials and components. The first element starts by reviewing the selection of materials and processes used to make some common devices. We then look at these materials and processes in greater depth and breadth.

Products

Engineered products are usually assemblies of individual components. Such products can be divided into three main groups.

- Mechanical products such as gear boxes, pumps, engines and turbines.
- Electronic products such as the completed circuit boards used in computers, video recorders and television receivers.
- Electromechanical devices can range from washing machines to computer controlled machine tools. These are devices which combine electrical and mechanical components.

Electrician's screwdriver

Figure 1.1 shows an exploded view of a typical electrician's screwdriver. It has to combine the functions of a conventional screwdriver with the ability to indicate whether a mains circuit is live. It must be strong enough to tighten and undo the small brass screws found in the terminals of electrical accessories. It must be insulated to withstand the potentials (voltages) met with in domestic, industrial and commercial installations. The current through the neon indicator lamp must be limited to a safe level

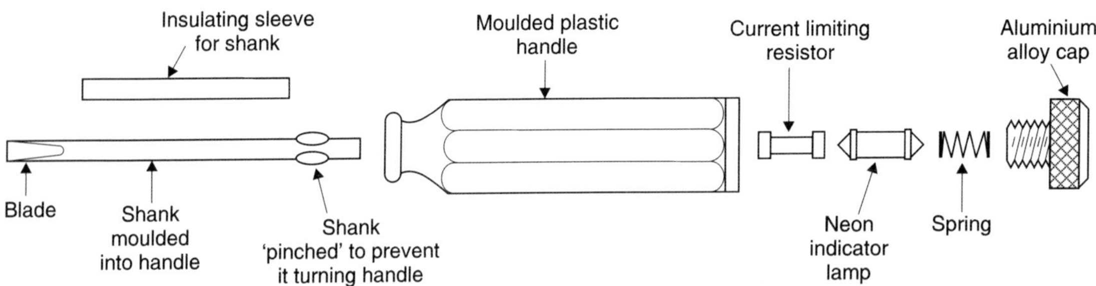

Figure 1.1 *A typical electrician's screwdriver*

under all conditions. It must be light in weight, compact and competitively priced.

These criteria can be met by careful selection of materials and manufacturing processes.

- The blade is made from a toughened medium carbon steel (0.8% carbon).
- The shank of the blade is insulated using a PVC sleeve.
- The handle is moulded from cellulose acetate. This is a tough, flame resistant plastic with good insulating properties. It is transparent so that the neon indicator lamp can be seen to light up. The blade would be moulded into the handle.
- The spring would be made from hard drawn phosphor bronze wire. This is a good conductor and corrosion resistant.
- The end cap would be made from an aluminium alloy on a computer numerically controlled (CNC) lathe. This material is light in weight, easily cut, a good conductor and corrosion resistant. The process of manufacture is suitable for large batch production.
- The neon indicator lamp and the current limiting resistor would be bought in as standard components.

All the materials chosen are readily available and relatively low in cost. They are selected for their fitness for purpose. Such screwdrivers are made and sold in large quantities and the manufacturing processes chosen lend themselves to large batch production.

Power supply

Figure 1.2 shows the circuit for a variable voltage d.c. power supply unit. It also shows some of the constructional details.

Most of the components for this project such as the transformer, rectifier pack, the sockets, switches, fuse holders, etc., would be bought in. The charger will have to have quite a high current output, so you must carefully consider how to keep the components cool. This particularly applies to the rectifier pack since solid state devices are destroyed if they are allowed to overheat. For this reason an aluminium case should be used and a heatsink

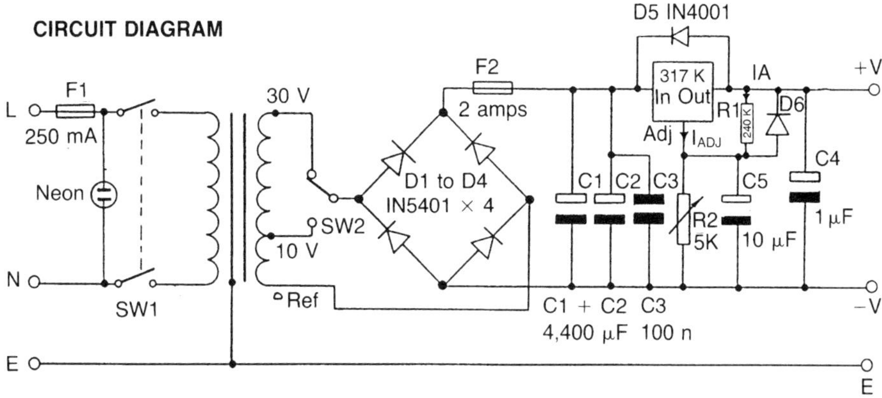

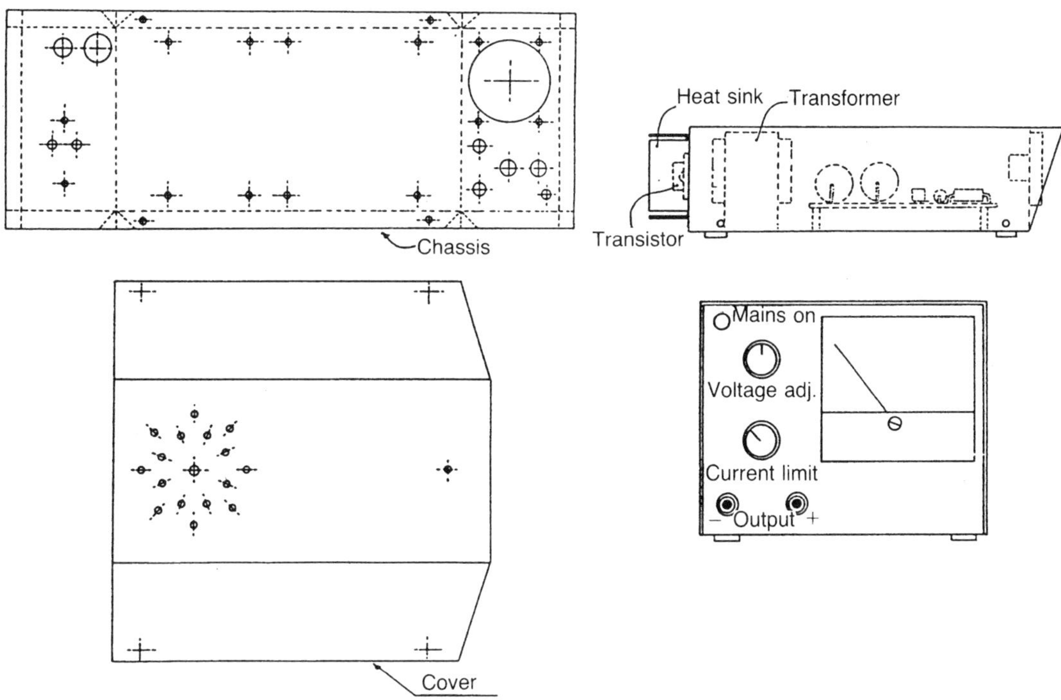

Figure 1.2

> **Test your knowledge 1.1**
>
> List THREE main factors that must be considered when choosing a material for a given component.

if recommended by the manufacturer of the rectifier. Cooling holes or louvres should be included in your design. These should be positioned so that moisture cannot enter the case. Since the equipment is connected to the mains supply and the case is of metal construction, the case must be earthed in accordance with current safety regulations.

Engineering materials

To make the engineering products that have just been described, you had to select suitable materials and components. We are now going to consider some of the main groups of materials and components available to engineers.

Figure 1.3 shows the main groups of engineering materials. The Latin name for iron is ferrum, so it is not surprising that

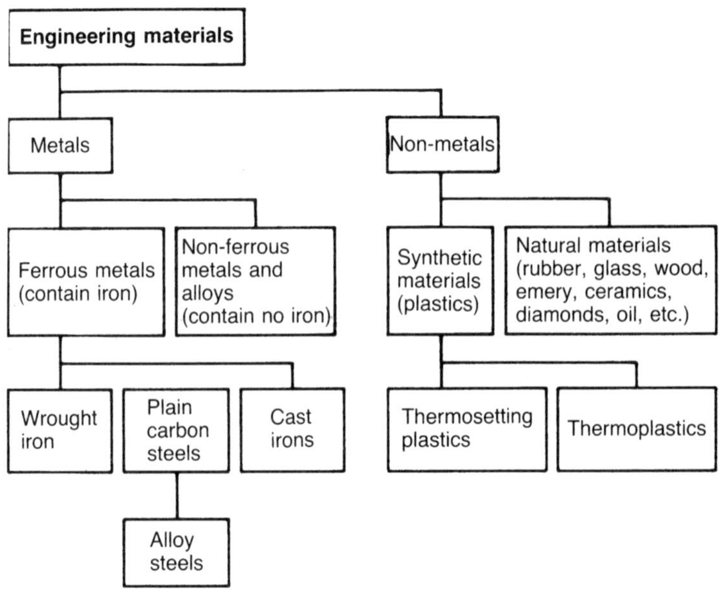

Figure 1.3 *The main groups of engineering materials*

ferrous metals and alloys are all based on the metal iron. Alloys consist of two or more metals or metals and non-metals that have been brought together as compounds or solid solutions to produce a metallic material with special properties. For example an alloy of iron, carbon, nickel and chromium is stainless steel. This is a corrosion resistant ferrous alloy. The remainder of the metallic materials available are non-ferrous metals and alloys. Non-metals can be natural, such as rubber, or they can be synthetic such as the plastic compound PVC.

Properties

When selecting a material we make certain that it has suitable properties for the job it has to do. For example we must ask ourselves the following questions.

- Will it corrode in its working environment?
- Will it weaken or melt in a hot environment?
- Will it break under normal working conditions?
- Can it be easily cast, formed or cut to shape?

We compare materials by comparing their properties. We will now consider the more important properties of engineering materials.

Chemical properties

Corrosion

This is caused by the metals and metal alloys being attacked and eaten away by chemical substances. For example the rusting of ferrous metals and alloys is caused by the action of atmospheric oxygen in the presence of water. Another example is the attack on aluminium and some of its alloys by strong alkali solutions. Take care when using degreasing agents on such metals. Copper and copper based alloys are stained and corroded by the active sulphur and chlorine products found in some heavy duty cutting lubricants. Choose your cutting lubricant with care when machining such materials.

Degradation

Non-metallic materials do not corrode but they can be attacked by chemical substances. Since this weakens or even destroys the material it is referred to as degradation. Unless specially compounded, rubber is attacked by prolonged exposure to oil. Synthetic (plastic) materials can be softened by chemical solvents. Exposure to the ultraviolet rays of sunlight can weaken (perish) rubbers and plastics unless they contain compounds that filter out such rays.

Physical properties

Electrical resistance

Materials with a very low resistance to the flow of an electric current are good electrical conductors. Materials with a very high resistance to the flow of electric current are good insulators. Generally, metals are good conductors and non-metals are good insulators (poor conductors). A notable exception is carbon which conducts electricity despite being a non-metal. You will learn about electrical resistance in your science unit. The electrical resistance of a conductor depends upon:

- its length (the longer it is the greater its resistance)
- its thickness (the thicker it is the lower its resistance)
- its temperature (the higher its temperature the greater is its resistance)
- its resistivity (this is the resistance of a material measured between the opposite faces of a metre cube of the material).

A small number of non-metallic materials, such as silicon, have atomic structures that fall between electrical conductors and insulators. These materials are called semi-conductors and are use for making solid state devices such as transistors.

Magnetic properties

All materials respond to strong magnetic fields to some extent. Only the ferro-magnetic materials respond sufficiently to be of interest. The more important of the ferro-magnetic materials are the metals iron, nickel and cobalt. Soft magnetic materials, such as soft iron, can be magnetized by placing them in a magnetic field. They cease to be magnetized as soon as the field is removed.

Hard magnetic materials, such as high-carbon steel that has been hardened by cooling it rapidly (quenching) from red heat, also become magnetized when placed in a magnetic field. Hard magnetic materials retain their magnetism when the field is removed. They become permanent magnets.

Permanent magnets can be made more powerful for a given size by adding cobalt to the steel to make an alloy. Soft magnetic materials can be made more efficient by adding silicon or nickel to the pure iron. Silicon–iron alloys are used for the rotor and stator cores of electric motors and generators. Silicon–iron alloys are also used for the cores of power transformers.

Thermal properties

These include:

- The melting temperatures of materials. Note that some plastics do not soften when heated, they only become charred and are destroyed. This will be considered later in this unit.
- Thermal conductivity. This is the ease with which materials conduct heat. Metals are good conductors of heat. Non-metals are poor conductors of heat. Therefore non-metals are heat insulators.
- Expansion. Metals expand appreciably when heated and contract again when cooled. They have high coefficients of linear expansion. Non-metals expand to a lesser extent when heated. They have low coefficients of linear expansion. Again, these properties will be considered in more detail in your science unit.

Mechanical properties

Strength

This is the ability of a material to resist an applied force (load) without fracturing (breaking). It is also the ability of a material not to yield. Yielding is when the material 'gives' suddenly under load and changes shape permanently but does not break. This is what happens when metal is bent or folded to shape. The load or force can be applied in various ways, as shown in Figure 1.4.

You must be careful when interpreting the strength data quoted for various materials. A material may appear to be strong when subjected to a static load, but will break when subjected to an impact load. Materials also show different strength characteristics when the load is applied quickly from when the load is applied slowly.

Toughness

This is the ability of a material to resist impact loads as shown in Figure 1.5. Here, the toughness of a piece of high-carbon steel in the soft (annealed) condition is compared with a piece of the same steel after it has been hardened by raising it to red-heat and cooling it quickly (quenching it in cold water). The hardened steel shows a greater strength, but it lacks toughness.

A test for toughness, called the Izod test, uses a notched specimen that is hit by a heavy pendulum. The test conditions are carefully controlled, and the energy absorbed in bending or breaking the specimen is a measure of the toughness of the material from which it was made.

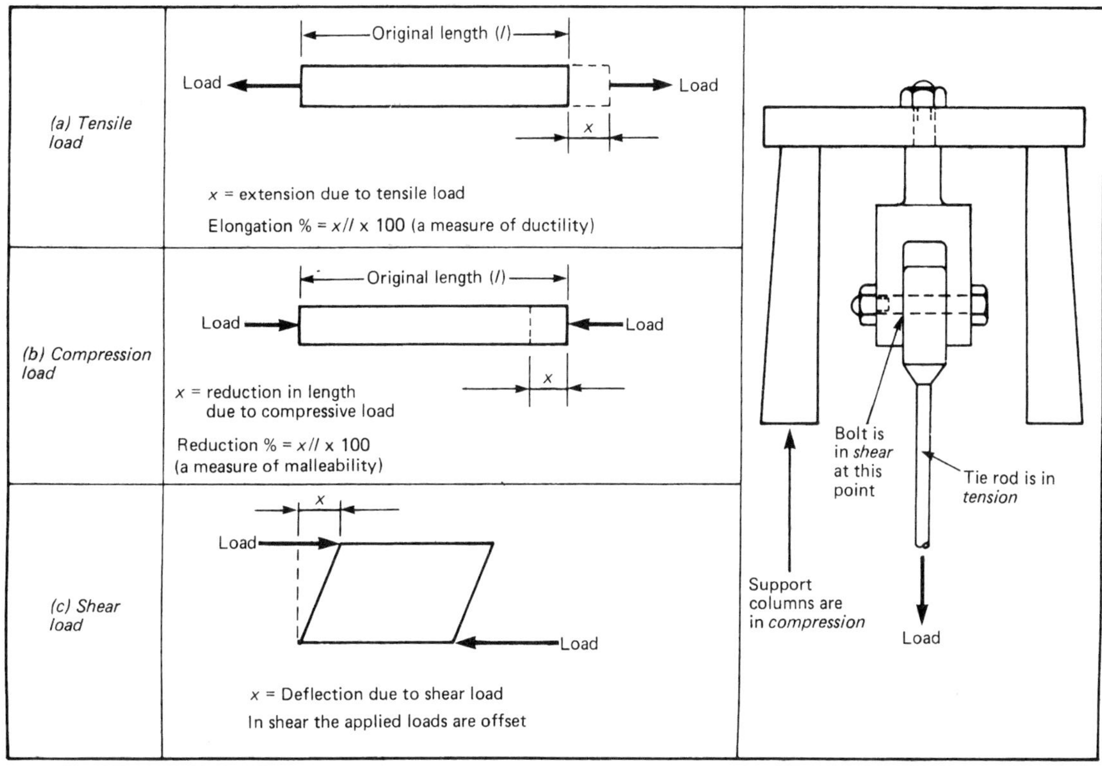

Figure 1.4 *Different ways in which a load can be applied*

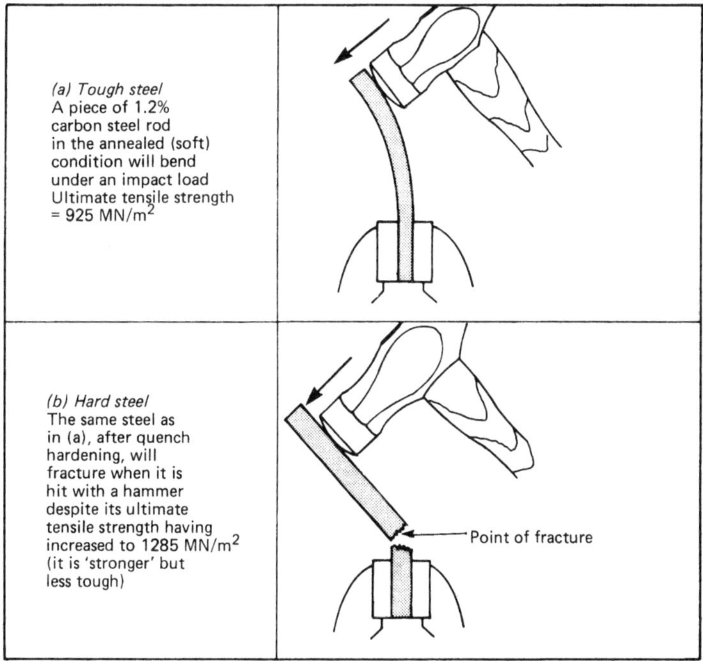

Figure 1.5 *Impact loads*

Elasticity

Materials that change shape when subjected to an applied force but spring back to their original size and shape when that force is removed are said to be elastic. They have the property of elasticity.

Plasticity

Materials that flow to a new shape when subjected to an applied force and keep that shape when the applied force is removed are said to be plastic. They have the property of plasticity.

Ductility

Materials that can change shape by plastic flow when they are subjected to a pulling (tensile) force are said to be ductile. They have the property of ductility. This is shown in Figure 1.6(a).

Malleability

Materials that can change shape by plastic flow when they are subjected to a squeezing (compressive) force are said to be malleable. They have the property of malleability. This is shown in Figure 1.6(b).

Hardness

Materials that can withstand scratching or indentation by an even harder object are said to be hard. They have the property of hardness. Figure 1.7 shows the effect of pressing a hard steel ball into two pieces of metal with the same force. The ball sinks further in to the softer of the two pieces of metal than it does into the harder metal.

There are various hardness tests available. The Brinell hardness test uses the principles set out above. A hardened steel ball is pressed into the specimen by a controlled load. The diameter of the indentation is measured using a special microscope. The

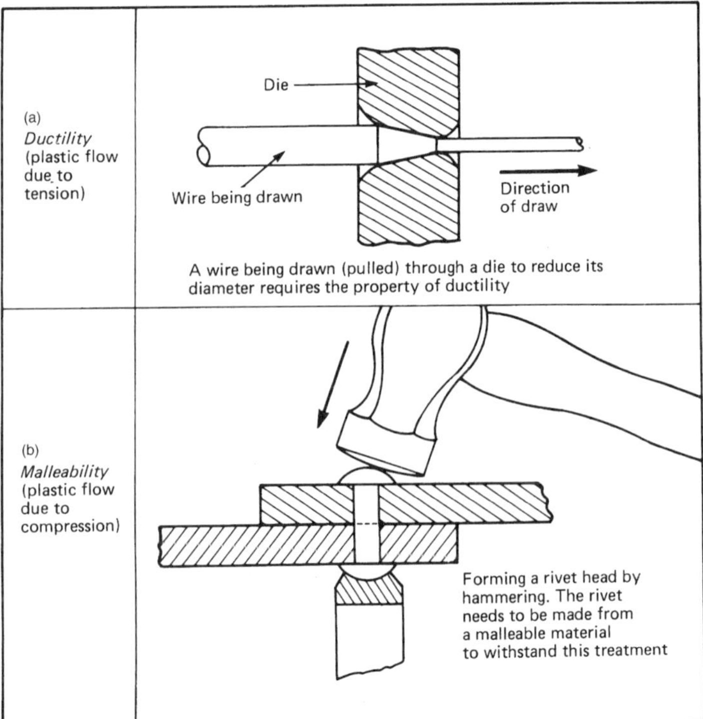

Figure 1.6 *Drawing and riveting*

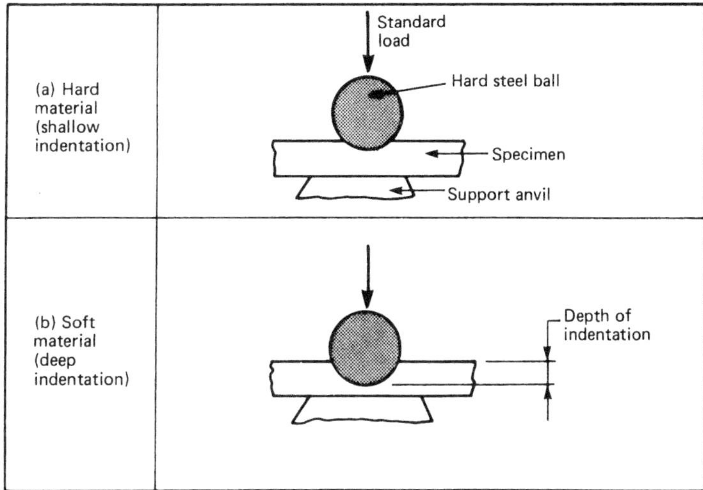

Figure 1.7 *Effect of pressing a hard steel ball into two different materials*

Test your knowledge 1.2

1. State whether the property of ductility or malleability is being exploited in the following processes. For each answer, give the reason for your choice.
   - (a) pressing out a car body panel
   - (b) forging a connecting rod
   - (c) wire drawing
   - (d) bending strip metal to make a bracket
   - (e) rolling out a cast ingot into sheet metal.
2. State the properties required by each of the following engineering products. For each answer, give the reason for your choice.
   - (a) A wire rope sling used for lifting heavy loads
   - (b) the axle of a motor vehicle
   - (c) a spring.

hardness number is obtained from the measured diameter by use of conversion tables.

The Vickers test is similar but uses a diamond pyramid instead of a hard steel ball. This enables harder materials to be tested. The diamond pyramid leaves a square indentation and the diagonal distance across the square is measured. Again, conversion tables are used to obtain the hardness number from the measured distance.

The Rockwell test uses a diamond cone. A minor load is applied and a small indentation is made. A major load is then added and the indentation increases in depth. This increase in depth of the indentation is directly converted into the hardness number and it can be read from a dial on the machine.

Rigidity

Materials that resist changing shape under load are said to be rigid. They have the property of rigidity. The opposite of rigidity is flexibility. Rigid materials are usually less strong than flexible materials. For example, cast iron is more rigid than steel but steel is the stronger and tougher. However the rigidity of cast iron makes it a useful material for machine frames and beds. If such components were made from a more flexible material the machine would lack accuracy. It would be deflected by the cutting forces.

Ferrous metals

As previously stated, ferrous metals are based upon the metal iron. For engineering purposes iron is usually associated with various amounts of the non-metal carbon. When the amount of carbon present is less than 1.8% we call the material steel. The figure of 1.8% is the theoretical maximum. In practice there is no advantage in increasing the amount of carbon present above 1.4%. We are only going to consider the plain carbon steels. Alloy steels are beyond the scope of this book. The effects of the carbon content on the properties of plain carbon steels are shown in Figure 1.8.

Cast irons are also ferrous metals. They have substantially more carbon than the plain carbon steels. Grey cast irons usually have a carbon content between 3.2% and 3.5%. Not all this carbon can be taken up by the iron and some is left over as flakes of graphite between the crystals of metal. It is these flakes of graphite that gives cast iron its particular properties and makes it a 'dirty' metal to machine. The compositions and typical uses of plain carbon steels and a grey cast iron are summarized in Table 1.1.

Low carbon steels

These are also called mild steels. They are the cheapest and most widely used group of steels. Although they are the weakest of the steels, nevertheless they are stronger than most of the non-ferrous metals and alloys. They can be hot and cold worked and machined with ease.

Medium carbon steels

These are harder, tougher, stronger and more costly than the low carbon steels. They are less ductile than the low carbon steels and cannot be bent or formed to any great extent in the cold condition without risk of cracking. Greater force is required to bend and form them. Medium carbon steels hot forge well but close temperature control is essential. Two carbon ranges are shown.

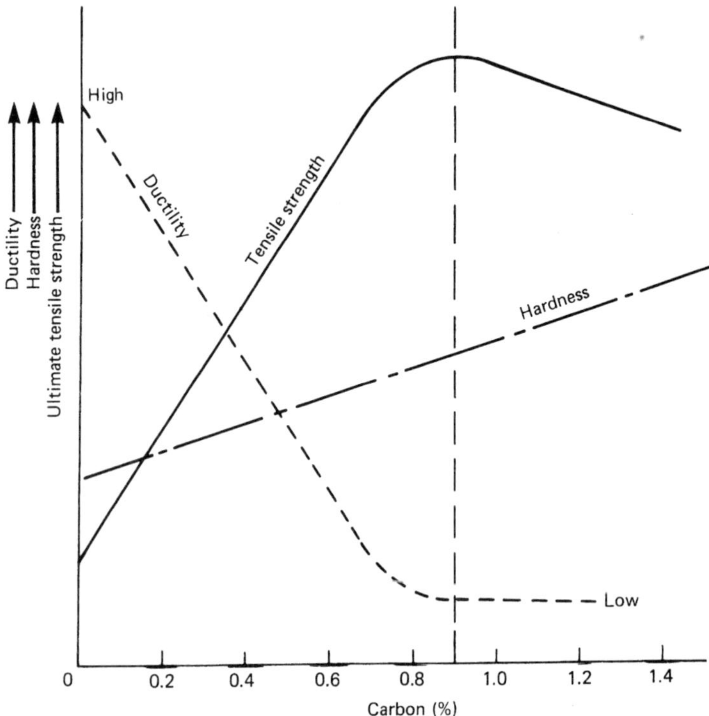

Figure 1.8 *Effect of carbon content on the properties of plain carbon steels*

Table 1.1 *Ferrous metals*

Name	*Group*	*Carbon content (%)*	*Some uses*
Dead mild steel (low carbon steel)	Plain carbon steel	0.10–0.15	Sheet for pressing out components such as motor car body panels. General sheet-metal work. Thin wire, rod, and drawn tubes
Mild steel (low carbon steel)	Plain carbon steel	0.15–0.30	General purpose workshop rod, bars and sections. Boiler plate. Rolled steel beams, joists, angles, etc.
Medium carbon steel	Plain carbon steel	0.30–0.50	Crankshafts, forgings, axles, and other stressed components.
		0.50–0.60	Leaf springs, hammer heads, cold chisels, etc.
High carbon steel	Plain carbon steel	0.8–1.0	Coil springs, wood chisels.
		1.0–1.2	Files, drills, taps and dies.
		1.2–1.4	Fine-edge tools (knives, etc.)
Grey cast iron	Cast iron	3.2–3.5	Machine castings

Test your knowledge 1.3

Select a suitable ferrous metal and state its carbon content for each of the following objects.

1 A hexagon head bolt produced on a lathe
2 a car engine cylinder block
3 a cold chisel
4 a wood-carving chisel
5 a pressed steel car body panel.

The lower carbon range can only be toughened by heating and quenching (cooling quickly by dipping in water). They cannot be hardened. The higher carbon range can be hardened and tempered by heating and quenching.

High carbon steels

These are harder, stronger and more costly than medium carbon steels. They are also less tough. High carbon steels are available as hot rolled bars and forgings. Cold drawn high carbon steel wire (piano wire) is available in a limited range of sizes. Centreless ground high carbon steel rods (silver steel) are available in a wide range of diameters (inch and metric sizes) in lengths of 333 mm, 1 m and 2 m. High carbon steels can only be bent cold to a limited extent before cracking. They are mostly used for making cutting tools such as files, knives and carpenters' tools.

Non-ferrous metals and alloys

Copper

High-conductivity copper

This is better than 99.9% pure and it is widely used for electrical conductors and switchgear components. It is second only to silver in conductivity but it is much more plentiful and very much less costly. Pure copper is too soft and ductile for most mechanical applications.

Tough-pitch copper

For general purpose applications such as roofing, chemical plant, decorative metal work and copper-smithing, tough-pitch copper is used. This contains some copper oxide which makes it stronger, more rigid and less likely to tear when being machined. Because it is not so highly refined, it is less expensive than high conductivity copper.

There are many other grades of copper for special applications. Copper is also the basis of many important alloys such as brass and bronze, and we will be considering these next. The general properties of copper are:

- Relatively high strength
- Very ductile so that it is usually cold worked. An annealed (softened) copper wire can be stretched to nearly twice its length before it snaps
- Corrosion resistant
- Second only to silver as a conductor of heat and electricity
- Easily joined by soldering and brazing. For welding, a phosphorous deoxidized grade of copper must be used.

Copper is a available as cold-drawn rods, wires and tubes. It is also available as cold-rolled sheet, strip and plate. Hot worked copper is available as extruded sections and hot stampings. It can also be cast. Copper powders are used for making sintered components. It is one of the few pure metals of use to the engineer as a structural material.

Brass

Brass is an alloy of copper and zinc. The properties of a brass alloy and the applications for which you can use it depends upon the amount of zinc present. Most brasses are attacked by sea water. The salt water eats away the zinc (dezincification) and leaves a weak, porous, spongy mass of copper. To prevent this happening, a small amount of tin is added to the alloy. There are two types of brass that can be used at sea or on land near the sea. These are Naval brass and Admiralty brass.

Brass is a difficult metal to cast and brass castings tend to be coarse grained and porous. Brass depends upon hot rolling from cast ingots, followed by cold rolling or drawing to give it its mechanical strength. It can also be hot extruded and plumbing fittings are made by hot stamping. Brass machines to a better finish than copper as it is more rigid and less ductile than that metal. Table 1.2 lists some typical brasses, together with their compositions, properties and applications.

Tin bronze

As the name implies, the tin bronzes are alloys of copper and tin. These alloys also have to have a deoxidizing element present to prevent the tin from oxidizing during casting and hot working. If

Table 1.2 *Properties and applications of brass alloys*

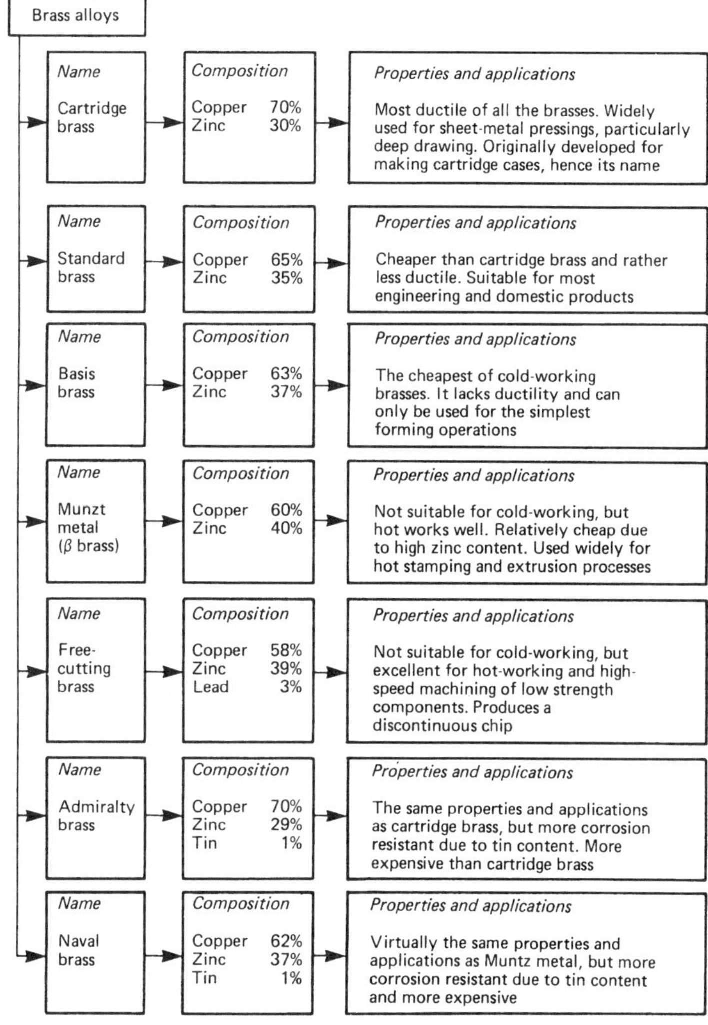

Brass alloys

Name	Composition	Properties and applications
Cartridge brass	Copper 70% Zinc 30%	Most ductile of all the brasses. Widely used for sheet-metal pressings, particularly deep drawing. Originally developed for making cartridge cases, hence its name
Standard brass	Copper 65% Zinc 35%	Cheaper than cartridge brass and rather less ductile. Suitable for most engineering and domestic products
Basis brass	Copper 63% Zinc 37%	The cheapest of cold-working brasses. It lacks ductility and can only be used for the simplest forming operations
Munzt metal (β brass)	Copper 60% Zinc 40%	Not suitable for cold-working, but hot works well. Relatively cheap due to high zinc content. Used widely for hot stamping and extrusion processes
Free-cutting brass	Copper 58% Zinc 39% Lead 3%	Not suitable for cold-working, but excellent for hot-working and high-speed machining of low strength components. Produces a discontinuous chip
Admiralty brass	Copper 70% Zinc 29% Tin 1%	The same properties and applications as cartridge brass, but more corrosion resistant due to tin content. More expensive than cartridge brass
Naval brass	Copper 62% Zinc 37% Tin 1%	Virtually the same properties and applications as Muntz metal, but more corrosion resistant due to tin content and more expensive

the tin oxidizes the metal becomes hard and 'scratchy' and is weakened. The two deoxidizing elements commonly used are:

- zinc in the gun-metal alloys.
- phosphorus in the phosphor–bronze alloys.

Unlike the brass alloys, the bronze alloys are usually used as castings. However low-tin content phosphor–bronze alloys can be extensively cold worked. Tin–bronze alloys are extremely resistant to corrosion and wear and are used for high pressure valve bodies and heavy duty bearings. Table 1.3 lists some typical bronze alloys together with their compositions, properties and applications.

Aluminium

Aluminium has a density approximately one third that of steel. However it is also very much weaker so its strength/weight ratio

Table 1.3 *Properties and applications of bronze alloys*

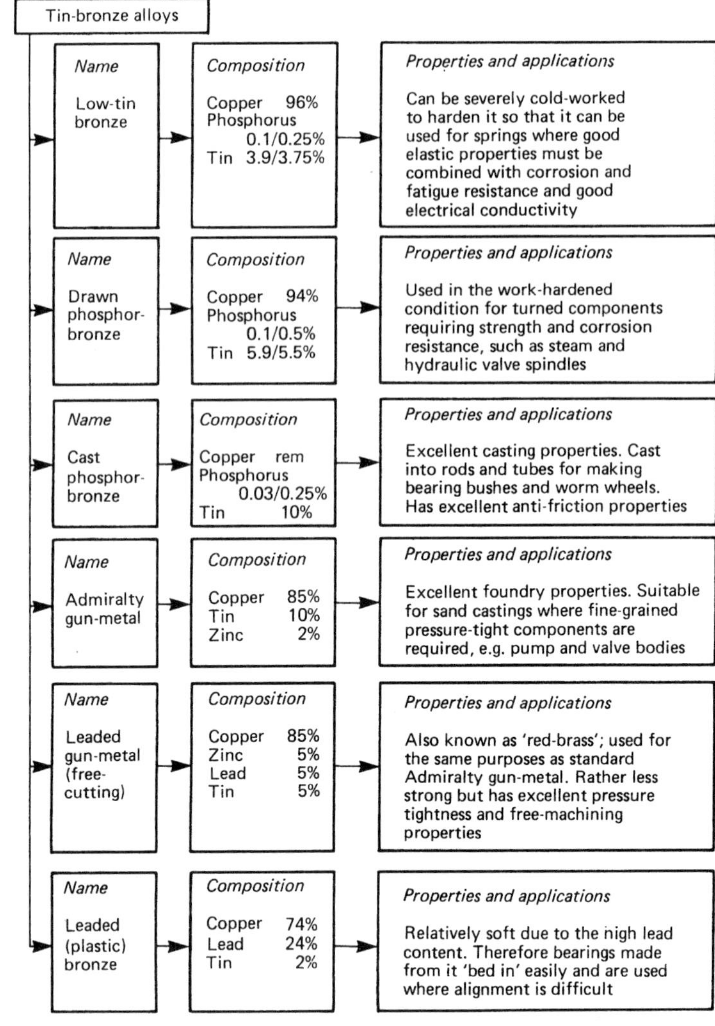

Tin-bronze alloys

Name	Composition	Properties and applications
Low-tin bronze	Copper 96% Phosphorus 0.1/0.25% Tin 3.9/3.75%	Can be severely cold-worked to harden it so that it can be used for springs where good elastic properties must be combined with corrosion and fatigue resistance and good electrical conductivity
Drawn phosphor-bronze	Copper 94% Phosphorus 0.1/0.5% Tin 5.9/5.5%	Used in the work-hardened condition for turned components requiring strength and corrosion resistance, such as steam and hydraulic valve spindles
Cast phosphor-bronze	Copper rem Phosphorus 0.03/0.25% Tin 10%	Excellent casting properties. Cast into rods and tubes for making bearing bushes and worm wheels. Has excellent anti-friction properties
Admiralty gun-metal	Copper 85% Tin 10% Zinc 2%	Excellent foundry properties. Suitable for sand castings where fine-grained pressure-tight components are required, e.g. pump and valve bodies
Leaded gun-metal (free-cutting)	Copper 85% Zinc 5% Lead 5% Tin 5%	Also known as 'red-brass'; used for the same purposes as standard Admiralty gun-metal. Rather less strong but has excellent pressure tightness and free-machining properties
Leaded (plastic) bronze	Copper 74% Lead 24% Tin 2%	Relatively soft due to the high lead content. Therefore bearings made from it 'bed in' easily and are used where alignment is difficult

is inferior. For stressed components, such as those found in aircraft, aluminium alloys have to be used. These can be as strong as steel and nearly as light as pure aluminium.

High purity aluminium

This is second only to copper as a conductor of heat and electricity. It is very difficult to join by welding or soldering. and aluminium conductors are often terminated by crimping. Despite these difficulties, it is increasingly used for electrical conductors where its light weight and low cost compared with copper is an advantage. Pure aluminium is resistant to normal atmospheric corrosion but it is unsuitable for marine environments. It is available as wire, rod, cold-rolled sheet and extruded sections for heat sinks.

Commercially pure aluminium

This is not as pure as high purity aluminium and it also contains up to 1% silicon to improve its strength and stiffness. As a result

Test your knowledge 1.4

Select a suitable non-ferrous metal for each of the following, giving reasons for your choice.

1. The body casting of a water pump
2. Screws for clamping the electric cables in the terminals of a domestic electric light switch
3. A bearing bush
4. A deep drawn, cup-shaped component for use on land
5. A ship's fitting made by hot stamping.

it is not such a good conductor of electricity nor is it so corrosion resistant. It is available as wire, rod, cold-rolled sheet and extruded sections. It is also available as castings and forgings. Being stiffer than high purity aluminium it machines better with less tendency to tear. It forms non-toxic oxides on its surface which makes it suitable for food processing plant and utensils. It is also used for forged and die-cast small machine parts. Because of their range and complexity, the light alloys based upon aluminium are beyond the scope of this book.

Non-metallic materials

Non-metallic materials can be grouped under the headings shown in Figure 1.9. In addition, wood is also used for making the patterns which, in turn, are used in producing moulds for castings. We are only going to consider some ceramics, thermosets and thermoplastics.

Ceramics

The word ceramic comes from a Greek word meaning potter's clay. Originally, ceramics referred to objects made from potter's clay. Nowadays, ceramic technology has developed a range of materials far beyond the traditional concepts of the potter's art. These include:

- glass products
- abrasive and cutting tool materials
- construction industry materials
- electrical insulators
- cements and plasters for investment moulding
- refractory (heat resistant) lining for furnaces
- refractory coatings for metals

The four main groups of ceramics and some typical applications are summarized in Table 1.4.

The common properties of ceramic materials can be summarized as follows:

Strength

Ceramic materials are reasonably strong in compression, but tend to be weak in tension and shear. They are brittle and lack ductility. They also suffer from micro-cracks which occur during

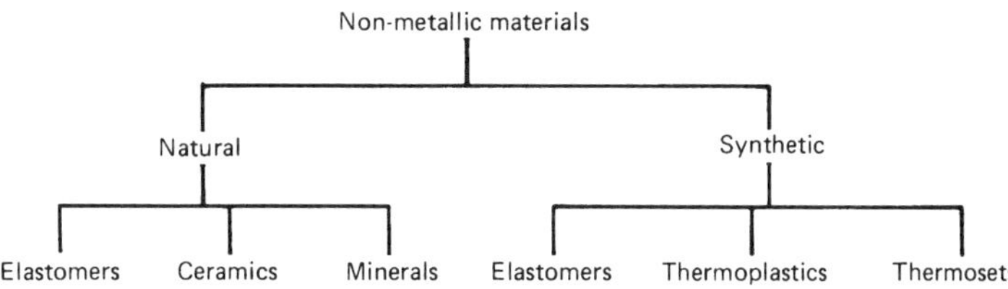

Figure 1.9 *Non-metallic materials*

Table 1.4 *Applications of ceramic materials*

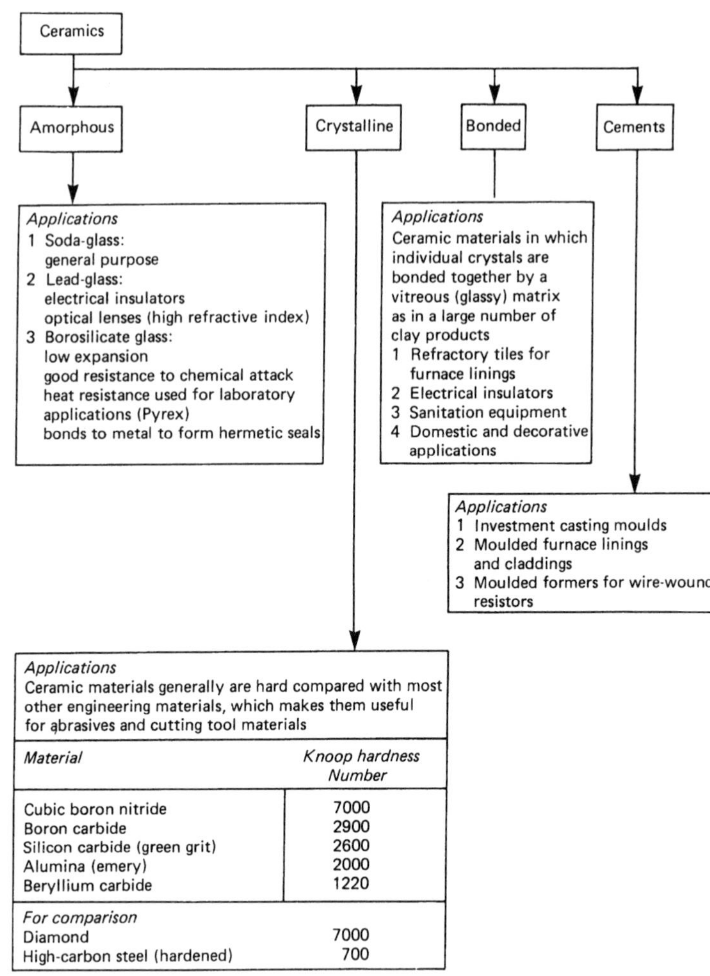

Material	*Knoop hardness Number*
Cubic boron nitride	7000
Boron carbide	2900
Silicon carbide (green grit)	2600
Alumina (emery)	2000
Beryllium carbide	1220
For comparison	
Diamond	7000
High-carbon steel (hardened)	700

the firing process. These lead to fatigue failure. Many ceramics retain their high compressive strength at very high temperatures.

Hardness

Most ceramic materials are harder than other engineering materials, as shown in Table 1.4. They are widely used for cutting tool tips and abrasives. They retain their hardness at very high temperatures that would destroy high carbon and high speed steels. However they have to be handled carefully because of their brittleness.

Refractoriness

This is the ability of a material to withstand high temperatures without softening and deforming under normal service conditions. Some refractories such as high-alumina brick and fireclays tend to soften gradually and may collapse at temperatures well below their fusion (melting) temperatures. Refractories made from clays containing a high proportion of silica to alumina are most widely used for furnace linings.

Electrical properties

As well as being used for weather resistant high-voltage insulators for overhead cables and sub-station equipment, ceramics are now being used for low-loss high-frequency insulators. For example they are being used for the dielectric in silvered ceramic capacitors for high frequency applications.

In all the above examples the ceramic material is polycrystalline. That is, the material is made up of a lot of very tiny crystals. For solid state electronic devices single crystals of silicon are grown under very carefully controlled conditions. The single crystal can range from 50 mm diameter to 150 mm diameter with a length ranging from 500 mm to 2500 mm. These crystals are without impurities. They are then cut up into thin wafers and made into such devices as thermistors, diodes, transistors and integrated circuits. This is done by doping the pure silicon wafers with small, controlled amounts of carefully selected impurities. Some impurities give the silicon n-type characteristics. That is they make the silicon electrically negative by increasing the number of electrons present. Some impurities give the silicon p-type characteristics. That is they make the silicon electrically positive by reducing the number of electrons present.

Thermosetting plastics

Themosetting plastics are also known as thermosets. These materials are available in powder or granular form and consist of a synthetic resin mixed with a 'filler'. The filler reduces the cost and modifies the properties of the material. A colouring agent and a lubricant are also added. The lubricant helps the plasticized moulding material to flow into the fine detail of the mould.

The moulding material is subjected to heat and pressure in the moulds during the moulding process. The hot moulds not only plasticize the moulding material so that it flows into all the detail of moulds, the heat also causes a chemical change in the material. This chemical change is called polymerization or, more simply, 'curing'. Once cured, the moulding is hard and rigid. It can never again be softened by heating. If made hot enough it will just burn. Some thermosets and typical applications are summarized in Table 1.5.

Thermoplastics

Unlike the thermosets we have just considered, thermoplastics soften every time they are heated. In fact, any material trimmed from the mouldings can be ground up and recycled. They tend to be less rigid but tougher and more 'rubbery' than the thermosetting materials. Some thermoplastics and typical applications are summarized in Table 1.6.

Table 1.5 *Thermo-setting plastics*

Type	*Applications*
Phenolic resins and powders	The original 'Bakelite' type of plastic materials, hard, strong and rigid. Moulded easily and heat 'cured' in the mould. Unfortunately, they darken during processing and are available only in the darker and stronger colours. Phenolic resins are used as the 'adhesive' in making plywoods and laminated plastic materials (Tufnol)
Amino (containing nitrogen) resins and powders	The basic resin is colourless and can be coloured as required. Can be strengthened by paper-pulp fillers and are suitable for thin sections. Used widely in domestic electrical switchgear
Polyester resins	Polyester chains can be cross-linked by adding a monomer such as styrene, when the polyester ceases to behave as a thermoplastic and becomes a thermoset. Curing takes place by internal heating due to chemical reaction and not by heating the mould. Used largely as the bond in the production of glass fibre mouldings.
Epoxy resins	The strongest of the plastic materials used widely as adhesives, can be 'cold cast' to form electrical insulators and used also for potting and encapsulating electrical components. Tooling epoxies are mixed with a metal powder filler and are used to make forming dies and low-cost tooling generally. Epoxy resins are self-curing by chemical reaction. No water or volatiles are produced during curing so shrinkage is negligible.

Reinforced plastics

The strength of plastics can be increased by reinforcing them with fibrous materials.

- *Laminated plastics (Tufnol).* Fibrous material such as paper, woven cloth, woven glass fibre, etc., is impregnated with a thermosetting resin. The sheets of impregnated material are laid up in powerful hydraulic presses and they are heated and squeezed until they become solid and rigid sheets, rods, tubes, etc. This material has a high strength and good electrical properties. It can be machined with ordinary metal working tools and machines. Tufnol is used for making insulators, gears and bearing bushes.
- *Glass reinforced plastics (GRP).* Woven glass fibre and chopped strand mat can be bonded togther by polyester or by epoxy resins to form mouldings. These may range from simple objects such as crash helmets to complex hulls for ocean-going racing yachts. The thermosetting plastics used are cured by chemical action at room temperature and a press is not required. The glass fibre is laid up over plaster or wooden moulds and coated with the resin which is well worked into the reinforcing material. Several layers or 'plies' may be built up according to the strength required. When

Table 1.6 *Thermoplastic materials*

Type	*Material*	*Characteristics*
Acrylics	Polymethyl-methacrylate	Materials of the 'Perspex' and 'Plexiglass' types. Excellent light transmission and optical properties, tough, non-splintering and can be easily heat-bent and shaped. Excellent high-frequency electrical insulators.
Cellulose plastics	Nitro-cellulose	Materials of the 'celluloid' type. Tough, waterproof, and available preformed as sections, sheets and films. Difficult to mould because of their high flammability. In powder form nitro-cellulose is explosive.
	Cellulose acetate	Far less flammable than nitro-cellulose and the basis of photographic 'safety' film. Frequently used for moulded handles for tools and electrical insulators.
Fluorine plastics (Teflon)	Polytetrafluoro-ethylene (PTFE)	A very expensive plastic material, more heat resistant than any other plastic. Also has the lowest known coefficient of friction (see Chapter 13), and is used for heat-resistant and anti-friction coatings. Can be moulded with difficulty to give components a waxy appearance.
Nylon	Polyamide	Used as a fibre or as a wax-like moulding material. Tough, with a low coefficient of friction. Cheaper than PTFE but loses its strength rapidly when raised above ambient temperatures. Absorbes moisture readily, making it dimensionally unstable and a poor electrical insulator.
Polyesters (Terylene)	Polyethylene-terephthalate	Available as a film or in fibre form. Ropes made from polyesters are very light and strong and have more 'give' than nylon ropes. The film is an electrical insulator.
Vinylplastics	Polythene	A simple material, relatively weak but easy to mould, and a good electrical insulator. Used also as a waterproof membrane in the building industry.
	Polypropylene	A more complicated material than polythene. Can be moulded easily and is similar to nylon in many respects. Its strength lies between polythene and nylon. Chapter than nylon and does not absorb water.
	Polystyrene	Cheap and can be moulded easily. Good strength but tends to be rigid and brittle. Good electrical insulation properties, but tends to craze and yellow with age.
	Polyvinylchloride (PVC)	Tough, rubbery, practially non-flammable, cheap and easily manipulated. Good electrical properties and used widely as an insulator for flexible and semi-flexible cables.

cured the moulding is removed from the mould. The mould can be used again. Note that the mould is coated with a release agent before moulding commences.

Although the properties of plastic materials can vary widely, they all have some general properties in common.

General properties of plastics

Strength/weight ratio

Plastic materials vary considerably in strength and some of the stronger (such as nylon) compare favourably with the metals. All plastics have a lower density than metals and, therefore, chosen with care and proportioned correctly their strength/weight ratio compares favourably with the light alloys.

Test your knowledge 1.5

Choose suitable plastic materials for the following applications, giving reasons for your choice:

1 The moulded cockpit cover for a light aircraft
2 a moulded bearing for an office machine
3 a rope for rock-climbing
4 a domestic light switch body
5 encapsulating (potting) a small transformer.

Corrosion resistance

Plastic materials are inert to most inorganic chemicals and some are inert to all solvents. Thus they can be used in environments that are hostile to the most corrosion resistant metals and many naturally occurring non-metals.

Electrical resistance

All plastic materials are good electrical insulators, particularly at high frequencies. However their usefulness is limited by their softness and low heat resistance compared with ceramics. Flexible plastics such as PVC are useful for the insulation and sheathing of electric cables.

Activity 1.1 

Examine and dismantle a typical garden lawn sprinkler of the rotary type. Sketch the components for identification purposes (detail not required) and choose a suitable material for each component giving reasons for your choice.

Engineering components

Wherever possible you should use standard commercially available components. Because these are mass produced in very large quantities, their cost is kept to a minimum. If they are made to an internationally acceptable standard, their quality is guaranteed. Some of the more widely used mechanical and electrical components will now be considered.

Mechanical components

Screwed fastenings

Screwed fastenings refer to nuts, bolts, screws and studs. These come in a wide variety of sizes and types of screw thread. When selecting a screwed fastening for any particular purpose, you should ask yourself the following questions.

- Is the fastening strong enough for the application?
- Is the material from which the fastening is made corrosion resistant under service conditions and is it compatible with the metals being joined?
- Is the screw thread chosen, suitable for the job? Coarse threads are stronger than fine threads, particularly in soft metals such as aluminium. Fine threads are less likely to work loose.

Figure 1.10 shows some typical screwed fastenings and it also shows how they are used.

There are a large variety of heads for screwed fastenings, and the selection is usually a compromise between strength, appearance and ease of tightening. The hexagon head is usually selected for general engineering applications. The more expensive cap-head screw is widely used in the manufacture of machine tools, jigs and fixtures, and other highly stressed applications. These fastenings are forged from high-tensile alloy steels, thread rolled and heat treated. By recessing the cap-head, a flush surface is provided for safety and easy cleaning. Figure 1.11 shows some alternative screw heads.

Riveted joints

Figure 1.12 shows some typical riveted joints. Riveted joints are very strong providing they are correctly designed and assembled.

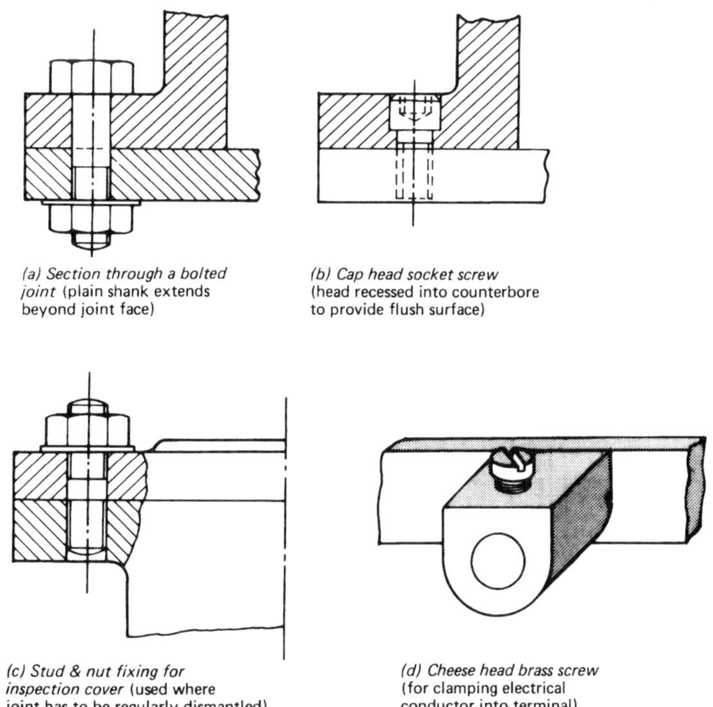

(a) Section through a bolted joint (plain shank extends beyond joint face)

(b) Cap head socket screw (head recessed into counterbore to provide flush surface)

(c) Stud & nut fixing for inspection cover (used where joint has to be regularly dismantled)

(d) Cheese head brass screw (for clamping electrical conductor into terminal)

Figure 1.10 *Some typical screwed fastenings*

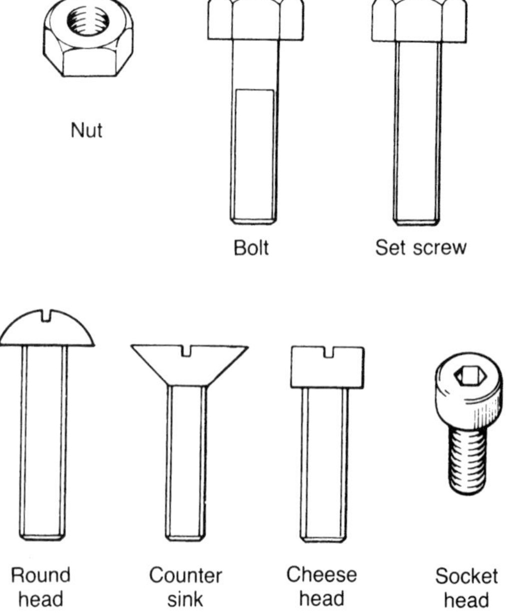

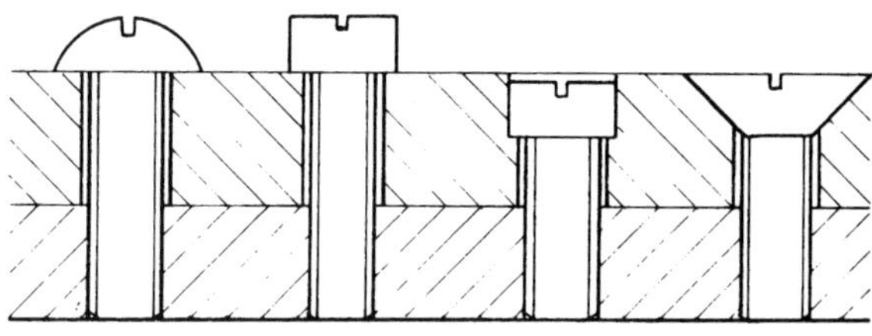

Applications of screw heads

Figure 1.11 *Various nuts, bolts and screw threads*

Test your knowledge 1.6

1 The cylinder head of a motor cycle engine is held on by studs. The studs are screwed into the cylinder casting at one end and the cylinder head is secured by nuts at the other end. The nuts must not vibrate loose. Which end of the stud should have a coarse thread? Which end of the stud should have a fine thread? Give the reason for your choice.

2 State whether a rivet should be loaded in shear or in tension giving the reason for your choice.

3 State the main advantage of a screwed joint over a riveted joint.

4 (a) Name THREE factors you would need to consider when selecting a screwed fastening for a particular application.
(b) Name THREE factors you would need to consider when selecting a rivet for a particular application.

The joint must be designed so that the rivet is in shear and not in tension. Consider the head of the rivet as only being strong enough to keep the rivet in place.

You must consider the following factors when selecting a rivet and making a riveted joint.

Material

The material from which the rivet is made must not react with the components being joined as this will cause corrosion and weakening. Also the rivet must be strong enough to resist the loads imposed upon it.

Choice of head

The rivet head chosen is always a compromise between strength and appearance. In the case of aircraft components, wind resistance must also be taken into account. Figure 1.13 shows some typical rivet heads and rivet types.

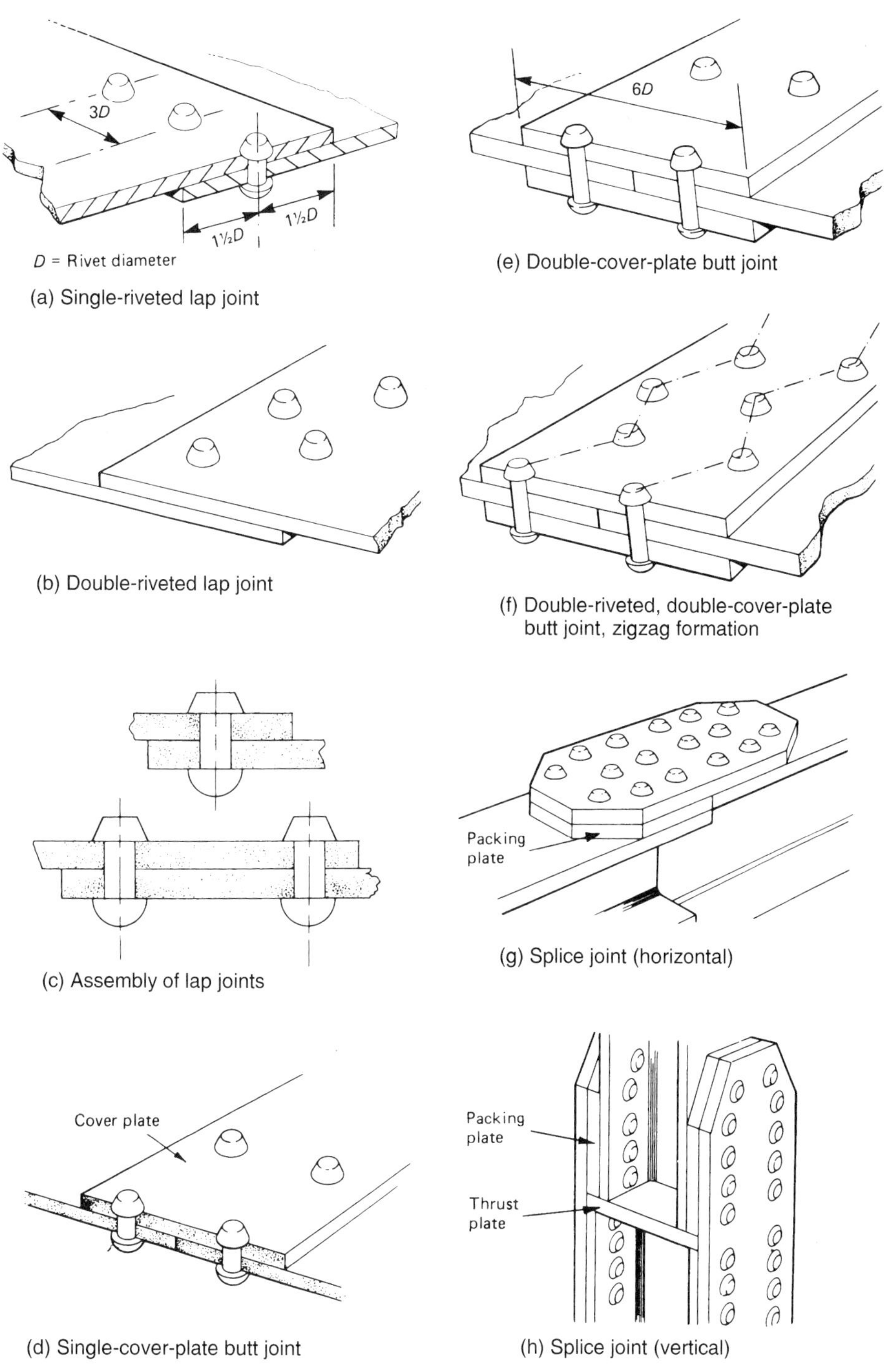

(a) Single-riveted lap joint

(b) Double-riveted lap joint

(c) Assembly of lap joints

(d) Single-cover-plate butt joint

(e) Double-cover-plate butt joint

(f) Double-riveted, double-cover-plate butt joint, zigzag formation

(g) Splice joint (horizontal)

(h) Splice joint (vertical)

Figure 1.12 *Typical riveted joints*

Electrical and electronic components

When selecting electrical and electronic components, you have to consider the following factors.

- Has the component got the correct circuit value? For example has it got the correct resistance or capacitance

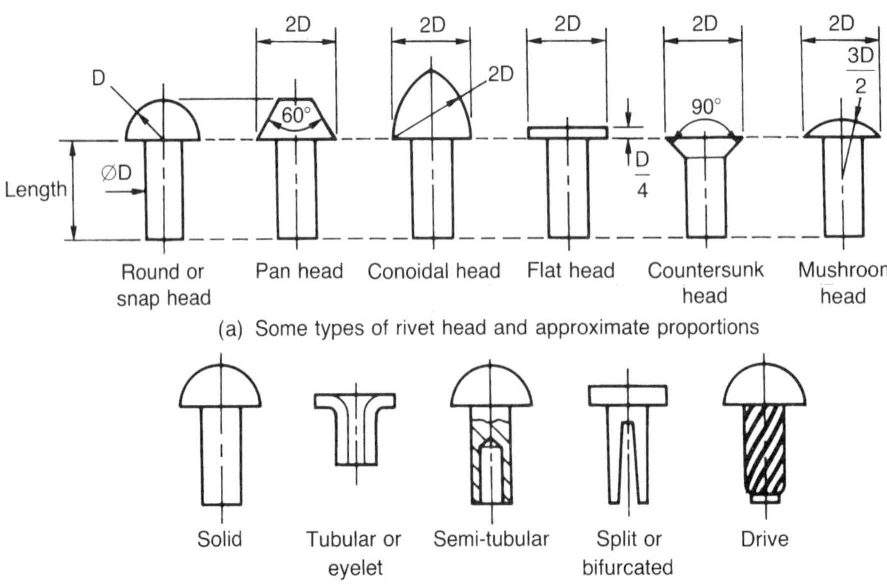

Figure 1.13 *Typical rivet heads and types*

value. Also has it got the correct tolerance grade? The better the tolerance, the more expensive the component.

- Is it insulated to withstand the potential across the component and between the component and earth?
- If it is not enclosed, is it insulated against accidental contact (electric shock)?
- Can it pass the required current without overheating?
- Has it the correct power rating in watts?
- Will it fit onto the circuit board or chassis and is it suitable for connecting to the circuit board or associated components?
- If it is being used for telecommunications or data processing, is it suitable for the high frequencies used in these applications?

You have only to look through any electronics catalogue to see how many different types of electrical and electronic components are available. We only have room to look at a few items in this unit. Large scale manufacturers of electronic equipment would buy their components direct from the makers in bulk. For small scale batch production and for prototype work it is usual to buy from a wholesale or retail supplier. This enables you to obtain all your requirements on a 'one-stop' purchasing basis.

Figure 1.14 shows some typical cables and hardware for electronic equipment.

(a) This shows examples of matrix board, strip board and a printed circuit board. The matrix board is a panel of laminated plastic perforated with a grid of holes. Pins can be fixed in the holes at convenient places for the attachment of such components as resistors and

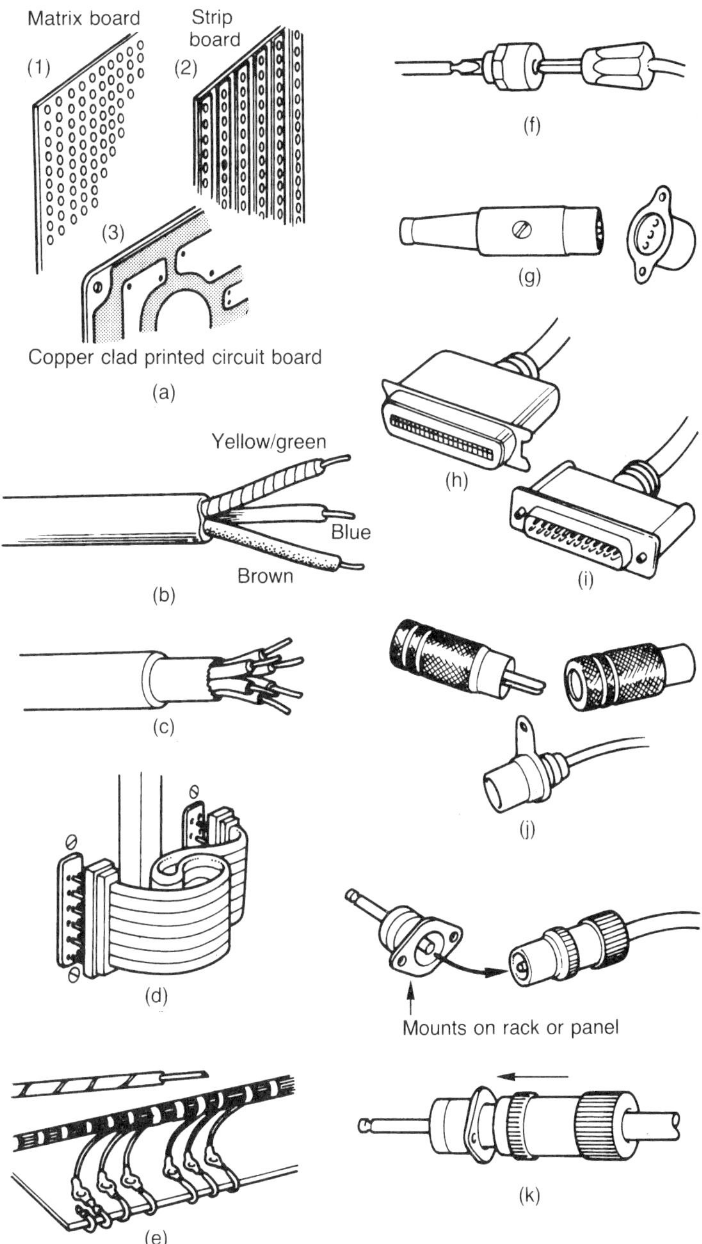

Figure 1.14 *Electronic cables and connections*

capacitors. They are merely attachment points and do not form part of the circuit.

The strip board is like a matrix board but is copper faced in strips on one side. The holes are the same pitch as the pins of integrated circuits which can be soldered into position.

The components are placed on the insulated side of the board, with the copper strips underneath. The wire leads pass through the holes in the board and are soldered on the underside. The copper tracks have to be cut wherever a break in the circuit is required.

Printed circuit boards (p.c.b.) are custom made for a particular circuit and are designed to give the most efficient layout for the circuit. The components are installed in the same way as for the strip board. Assembly will be considered in greater detail later in this unit.

(b) This is a typical flexible mains lead with PVC sheathing and colour coded PVC insulation. In selecting such a cable, the only factors you have to consider is its current handling capacity and its colour, providing the insulation is rated for mains use.

(c) This is a signal cable suitable for audio frequency analogue signals and for data processing signals. The conductors only need to have a limited current handling capacity, so many conductors can be carried in one cable. The conductors are surrounded by an earthed metal braid to prevent pick-up of external interference and corruption of the signal.

(d) This is a ribbon cable widely used in the manufacture of computers and the interconnection of computers and printers. Unlike the signal cable described earlier it is not screened against interference. However, it is cheaper and more easily terminated.

(e) Single cored PVC insulated wire is useful for making up wiring harnesses and for flying leads on PVC boards.

(f) This shows a banana plug and socket. Thes are used with single cored, flexible conductors for low-voltage power supply connections.

(g) This shows a DIN type plug and socket. These are used in conjunction with multi-cored screened signal cables. They are available for 3-way to 8-way connections inclusive and are designed so that the plug can only be inserted into the socket in one position.

(h) This show a 36-way Centronics plug as widely used for making connections to the parallel port of a computer printer.

(i) This shows a 25-way D-type plug as widely used for making connections to the parallel output port of a computer.

(j) This shows a selection of phono type plugs and sockets. These are widely used for making signal lead connections to audio amplifiers.

(k) This shows a typical plug and socket. These are used at radio frequencies for connecting aerial leads to television sets.

Figure 1.15 shows a selection of electronic components that are widely used. Resistors are used to limit the flow of an electric current in a circuit. Carbon type resistors can be used on both direct current and alternating current circuits at any frequency

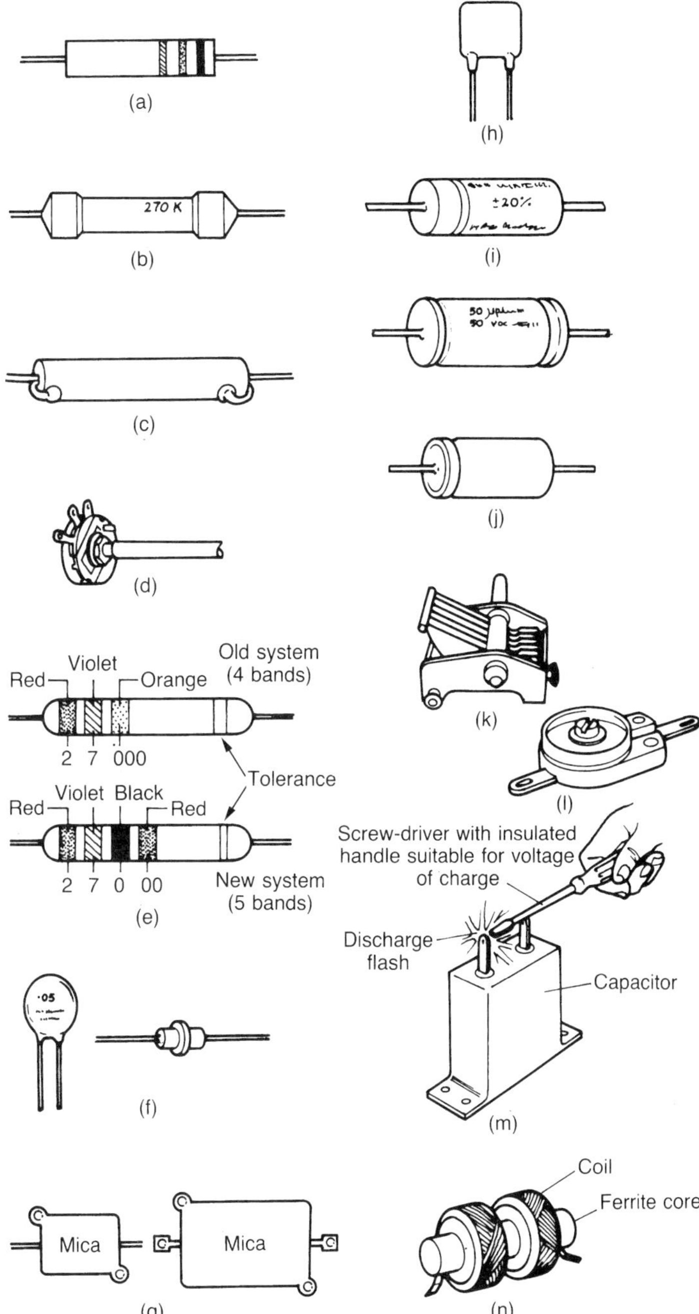

Figure 1.15 *A selection of typical electronic components*

since they are non-inductive. Wire wound resistors are inductive. They can only be used for direct current and mains frequency alternating current.

Capacitors are used to store electrical charges. Unlike batteries, they can be charged and discharged almost instantaneously. On the other hand, compared with batteries, they can only handle relatively small charges. Inductive devices such as chokes offer an impedance to alternating current. This increases as the frequency of the current gets higher. On direct current, inductive devices only offer the resistance of the wire from which they are wound.

(a) This shows a carbon rod resistor. They come in a wide range of resistance values and power ratings. These have quite wide tolerances but are suitable for general usage.

(b) This shows a high-stability resistor. These are precision resistors made from carbon film, metal film or oxide film. They are much more expensive than the ordinary carbon rod resistors. They are made to closer tolerances and are less susceptible to changes in resistance with changes in temperature. They are widely used in measuring instruments, computer applications and high stability radio frequency oscillators.

(c) This shows a wire-wound, vitreous enamelled resistor. These are inductive and only suitable for direct current, and mains frequency, applications. They are used where high power ratings are required.

(d) This shows a typical carbon track variable resistor. These are used for volume control and tone control circuits.

(e) Carbon rod and high-stability resistors are too small for much information to be marked on them so they are usually coded in some way. The colour code is as follows.

		Tolerance	
0 Black	5 Green		
1 Brown	6 Blue	±5%	Gold
2 Red	7 Violet	±10%	Silver
3 Orange	8 Grey	±20%	No colour
4 Yellow	9 White	High stability	Pink

There are two ways of applying the colour code. Let's consider the old way first. This system had four colour bands. Three represented the resistance of the resistor and one represented the tolerance. Reading from the end of the resistor, the example in Figure 1.15(e) shows bands coloured red, violet, orange.

- The first band is the first number. Red = 2
- The second band is the second number. Violet = 7
- The third band is the number of zeros. Orange = 3 = 000 (or a multiplying factor of 10^3)

So the resistance of our resistor is 27 000 Ω (or 27 kΩ).

If the resistor has a silver band as its fourth band, its resistance could range from 10% below its nominal value to 10% above its nominal value. That is, from 24 300 Ω to 29 700 Ω. Such a wide range of values would be unacceptable for many applications, so additional close tolerance bands have been added; these are:

0.1%	Violet
0.25%	Blue
0.5%	Green
1%	Brown
2%	Red

The new system uses five bands, four for the resistance value and one for the tolerance. The example in Figure 1.15(e) shows bands of red, purple, black, red.

- The first band is the first number. Red = 2
- The second band is the second number. Violet = 7
- The third band is the third number Black = 0
- The fourth band is the number of zeros. Red = 2 = 00 (or a multiplying factor of 10^2)

So the resistance of our resistor is once again 27 000 Ω.

If the resistor has a gold band as its fifth band, its resistance could range from 5% below its nominal value to 5% above its nominal value. That is, from 25 650 Ω to 28 350 Ω. The additional band provides for intermediate values of resistance. For example if we had required 27 200 Ω b1% the colours would have been red, violet, red, red, and a tolerance band coloured brown. This value could not have been achieved with the older system. Think about it.

A number and letter code is also used on circuit diagrams and often found printed on high stability resistors. This is best explained by some examples.

0.47 Ω would be marked R47
4.7 Ω would be marked 4R7
47 Ω would be marked 47R
100 Ω would be marked R100
1k Ω would be marked 1k0
10 kΩ would be marked 10k
47 MΩ would be marked 47M

Note that k = kilo = ×1000 and that M = mega ×1 000 000

(f) This shows examples of metallized ceramic capacitors. These are widely used in telecommunications equipment and in computers where high stability and compact size are required.

(g) This shows examples of silvered mica capacitors. These are also high stability capacitors suitable for radio frequency tuned circuits and for pulse operation.

(h) This shows a moulded polyester capacitor. These are self-healing and are widely used on printed circuit boards. They offer high values of capacitance in a small case size. They have a low inductance and low loss characteristics.

(i) This shows a polystyrene foil capacitor. These have ousted the foil and waxed paper capacitors found in some old equipment. They have low self-inductance, low high frequency losses and a long life. They are used for signal coupling and filter circuits.

(j) This shows some electrolytic capacitors. These are used where very high values of capacitance are required: for example smoothing capacitors in power packs. Normally these are polarized and they can only be used in direct current circuits. They must always be connected into the circuit in the correct direction as indicated on the case. Bi-polar electrolytic capacitors are available for use with low-voltage alternating currents and as signal coupling capacitors in audio amplifiers.

(k) This shows a capacitor whose capacitance can be varied. Such a variable capacitor is wired in series or in parallel with an inductance (coil) to form a resonant (tuned) circuit. This is the way the tuning circuit of your radio works.

(l) This shows a preset pressure capacitor. The capacitance increases when the screw is rotated in a clockwise direction to tighten it. Such devices are used in pretuned circuits that do not have to be varied once they have been set.

(m) SAFETY: Large capacitors can store substantial charges of electricity at high voltages. Before handling such capacitors always discharge them as shown, either with a suitable screwdriver or with a length of insulated wire.

(n) This shows a typical inductor or coil. These may be air cored or they may be wound on a ferrite core to increase their inductance for a given size. Chokes have a single winding. Transformers may have a single winding with tappings (auto-transformer) or they may have two or more windings. They can only be used on alternating current circuits. Either power circuits or signal circuits.

Figure 1.16 shows some solid state devices.

(a) and (b) show some thermistors. The resistance of these devices falls off rapidly as the temperature increases. They can be used as sensors to activate thermal protection devices. A common application is as sensors in car engines to activate the temperature gauge on the instrument cluster. As the water temperature rises, the resistance of the sensor falls and the current through the circuit increases. This causes the temperature gauge to show a higher reading. Although measuring current, the scale of the instrument is calibrated in degrees of temperature.

(c) This shows some diodes. These are electronic switching devices that only allow the current to flow in one direction. To ensure that they are connected in the circuit the correct way round, the positive end is either chamfered, or it has a red band. On larger, metal-cased diodes the diode symbol is printed on the case with the arrow head pointing to the positive pole of the diode.

(d) This shows a selection of transistors. These are switching devices that allow a small current to control a larger current. Small changes in the applied current can cause corresponding and amplified changes in the larger current. Power transistors have metal cases that can be bolted to a heat-sink. If a transistor (or any other solid state device) overheats it is destroyed. For this reason heat-sinks should

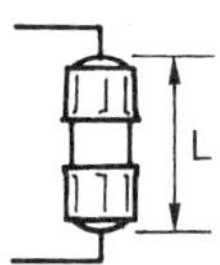

(a) **Rod type**
These special resistance elements have a very high negative temperature coefficient of resistance, making them suitable as protective elements in a wide range of circuits

Type	*Resistance Cold*	*Resistance Hot*	*Dimensions in mm*
TH-1A	650 Ω	37 Ω at 0.3 A	L. 38 Dia. 11
TH-2A	3.8 kΩ	44 Ω at 0.3 A	L. 32 Dia. 8
TH-3	370 Ω	28 Ω at 0.3 A	L. 22 Dia. 12
TH-5	4 Ω	0.4 Ω at 1 W (max)	Dia. 10 H. 4-5

(a)

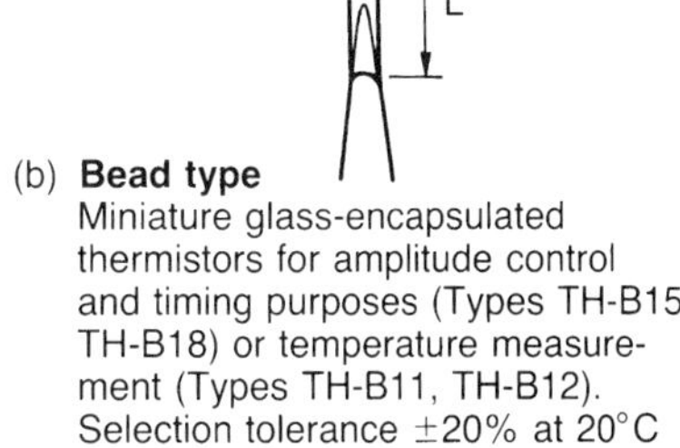

(b) **Bead type**
Miniature glass-encapsulated thermistors for amplitude control and timing purposes (Types TH-B15, TH-B18) or temperature measurement (Types TH-B11, TH-B12).
Selection tolerance ±20% at 20°C

Type	*Resistance at 20 C*	*Minimum resistance*	*Dimensions in mm*
TH-B11	1 MΩ	170 Ω	L10 Dia 2.5
TH-B12	2 kΩ	115 Ω	
TH-B15	100 kΩ	320 Ω	L25 Dia. 4
TH-B18	5 kΩ	100 Ω	L.38 Dia. 10

(b)

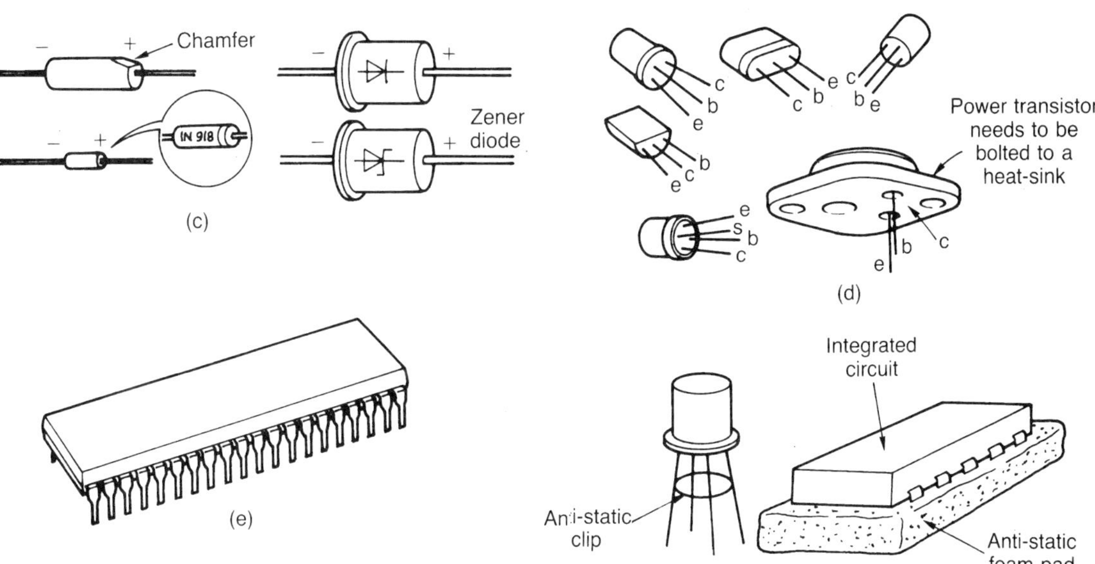

Figure 1.16 *Thermistors and solid-state devices*

be clamped to the legs of solid state devices whilst they are being soldered into the circuit. The shape of the transistor case gives an indication as to the correct way to connect it into the circuit. Consult manufacturers' tables for the different shapes and the corresponding connections.

(e) This shows a typical integrated circuit. These contain many components on one chip. Complete amplifiers, radio receivers, timers and many other devices can come ready packaged in a single case ready for installation on a circuit board.

(f) I have already mentioned the care that must be taken to prevent a solid state device being destroyed by overheating. Equally important is the care that must be taken with some devices to prevent them being destroyed by electrostatic charges. Two ways of protecting such devices are shown in Figure 1.16(f). The anti-static clip must not be removed until the device has been installed in the circuit. Whilst installing such devices, the circuit board, the soldering iron and the installer must all be bonded to the same earth point. The installer wears a wrist clamp connected to earth via a flexible copper braid.

Test your knowledge 1.7

1 Write down the colour bands for the following resistance values using the old four band system; 56 Ω, 470 Ω, 33 kΩ, 5.6 MΩ.

2 Write down the colour bands for the above values using the new five band system. Also write down the colour bands for the following values: 562 Ω, 675 kΩ.

3 A resistor has the following bands, ORANGE, WHITE, BROWN, GOLD. Write down its nominal value, and its upper and lower limits of resistance.

4 Select suitable capacitor types for the following applications.
(a) A high stability radio frequency filter network.
(b) A smoothing capacitor for a power supply unit.
(c) A coupling capacitor in the signal circuit of an amplifier of a bass guitar.

5 From manufacturers' literature or from component suppliers' catalogues sketch the following transistor bases; T03, T05, T012 and T092. Some of these bases identify the connections by letters. State what the letters stand for.

Safety

Health and safety legislation

The Health and Safety at Work, etc., Act of 1974 makes both the employer and the employee (you) equally responsible for safety. Both are equally liable to be prosecuted for violations of safety regulations and procedures. It is your legal responsibility to take reasonable care for your own health and safety. The law expects you to act in a responsible manner so as not to endanger yourself or other workers or the general public. It is an offence under the act to misuse or interfere with equipment provided for your health and safety or the health and safety of others.

Causes of accidents

Human carelessness

Most accidents are caused by human carelessness. This can range from 'couldn't care less' and 'macho' attitudes, to the deliberate disregard of safety regulations and codes of practice. Carelessness can also result from fatigue and ill-health resulting from a poor working environment.

Personal habits

Personal habits such as alcohol and drug abuse can render workers a hazard not only to themselves but also to other workers. Fatigue due to a second job (moonlighting) can also be a considerable hazard, particularly when operating machines. Smoking in prohibited areas where flammable substances are used and stored can cause fatal accidents involving explosions and fire.

Supervision and training

Another cause of accidents is lack of training or poor quality training. Lack of supervision can also lead to accidents if it leads to safety procedures being disregarded.

Environment

Unguarded and badly maintained plant and equipment are obvious causes of injury. However, the most common causes of accidents are falls on slippery floors, poorly maintained stairways, scaffolding and obstructed passageways in overcrowded workplaces. Noise, bad lighting, and inadequate ventilation can lead to fatigue, ill-health and carelessness. Dirty surroundings and inadequate toilet and washing facilities can lead to a lowering of personal hygiene standards.

Accident prevention

Elimination of hazards

The work place should be tidy with clearly defined passageways. It should be well lit and ventilated. It should have well maintained non-slip flooring. Noise should be kept down to acceptable levels. Hazardous processes should be replaced with less dangerous and more environmentally acceptable alternatives. For example asbestos clutch and brake linings should be replaced with safer materials.

Guards

Rotating machinery, drive belts and rotating cutters must be securely fenced to prevent accidental contact. Some machines have interlocked guards. These are guards coupled to the machine drive in such a way that the machine cannot be operated when the guard is open for loading and unloading the work. All guards must be set, checked and maintained by qualified and certificated staff. They must not be removed or tampered with by operators. Some examples of guards are shown in Figure 1.17.

Maintenance

Machines and equipment must be regularly serviced and maintained by trained fitters. This not only reduces the chance of a major breakdown leading to loss of production, it lessens the chance of a major accident caused by a plant failure. Equally important is attention to such details as regularly checking the stocking and siting of first-aid cabinets and regularly checking the condition and siting of fire extinguishers. All these checks must be logged.

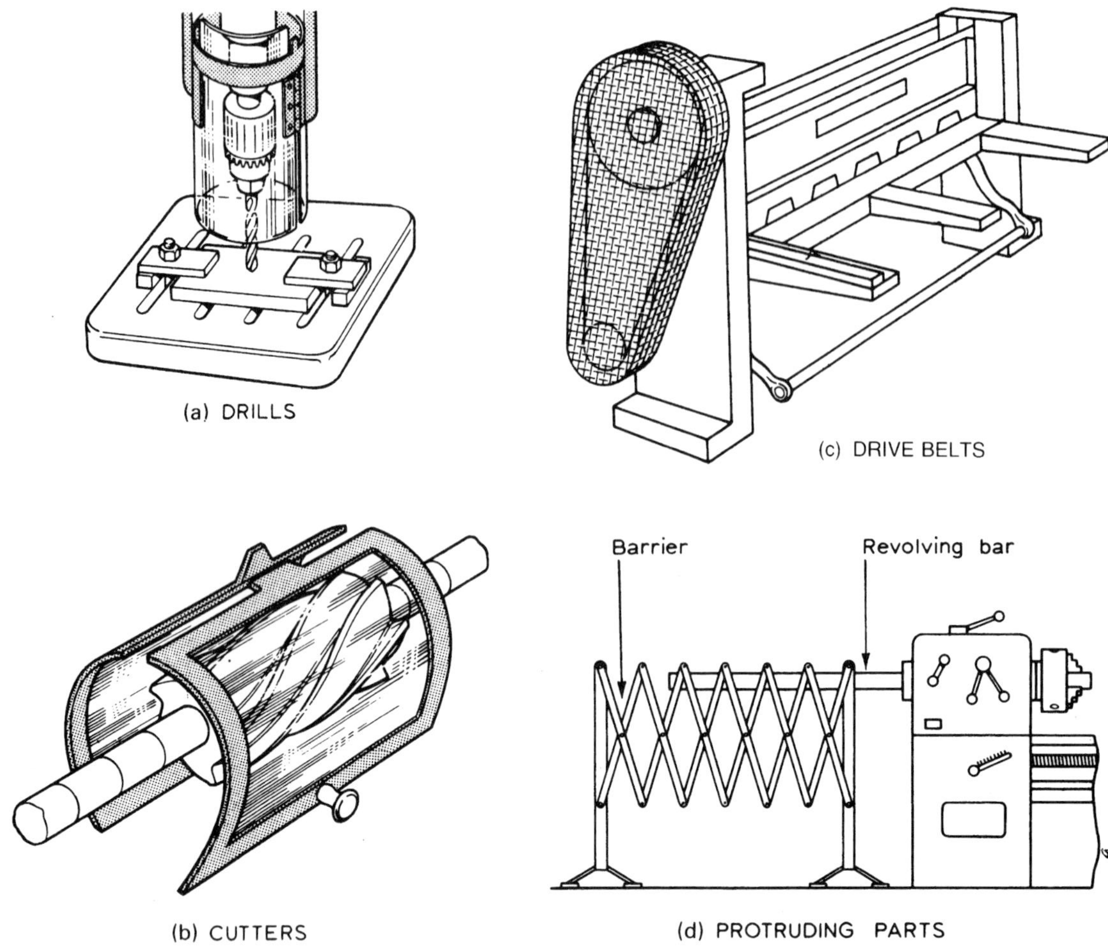

Figure 1.17 *Examples of the use of guards*

Personal protection

Suitable and unsuitable working clothing is shown in Figure 1.18. Some processes and working conditions demand even greater protection, such as safety helmets, earmuffs, respirators and eye protection worn singly or in combination. Such protective clothing must be provided by the employer when a process demands its use.

Employees must, by law, make use of such equipment. Some examples are shown in Figure 1.19.

Safety education

This is important in producing positive attitudes towards safe working practices and habits. Warning notices and instructional posters should be displayed in prominent positions and in as many ethnic languages as necessary. Information, education and training should be provided in all aspects of health and safety: for example, process training, personal hygiene, first aid and fire procedures. Regular fire drills must be carried out to ensure that the premises can be evacuated quickly, safely and without panic.

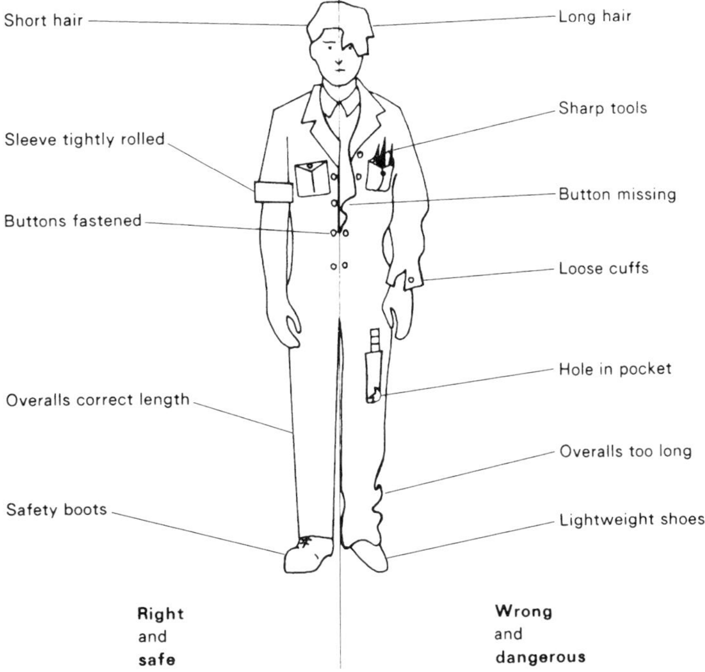

Figure 1.18 *Suitable and unsuitable clothing*

Personal attitudes

It is important that everyone adopts a positive attitude towards safety. Not only your own safety but the safety of your workmates and the general public. Skylarking and throwing things about in the workplace or on site cannot be allowed. Any distraction that causes lack of concentration can lead to serious and even fatal accidents.

Test your knowledge 1.8

1. Who is equally responsible for Health and Safety in the workplace under the terms of the Act?
2. List SIX precautions you can take that will help to prevent an accident happening to you or other workers in your workplace.

Housekeeping

A sign of a good worker is a clean and tidy working area. Only the minimum of tools for the job should be laid out at any one time. These tools should be laid out in a tidy and logical manner so that they immediately fall to hand. Tools not immediately required should be cleaned and properly stored away. All hand tools should be regularly checked and kept in good condition. Spillages, either on the workbench or on the floor should be cleaned up immediately.

Electrical hazards

Electrical equipment is potentially dangerous. The main hazards can be summarized as follows:

- electric shock
- fire due to the overheating of cables and equipment
- explosions set off by sparks when using unsuitable equipment when flammable vapours and gases are present.

(a) WEAR THE CORRECT TYPE OF PROTECTIVE CLOTHING

(b) PROTECT THE HEAD

(c) WEAR SAFETY FOOTWEAR

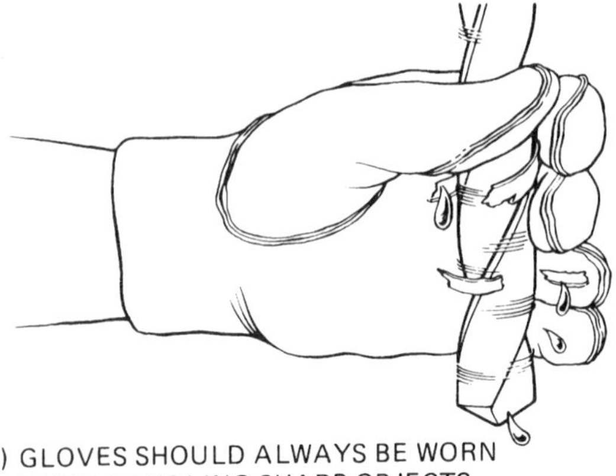

(d) GLOVES SHOULD ALWAYS BE WORN WHEN HANDLING SHARP OBJECTS, BUT NEVER WHEN OPERATING MACHINE TOOLS

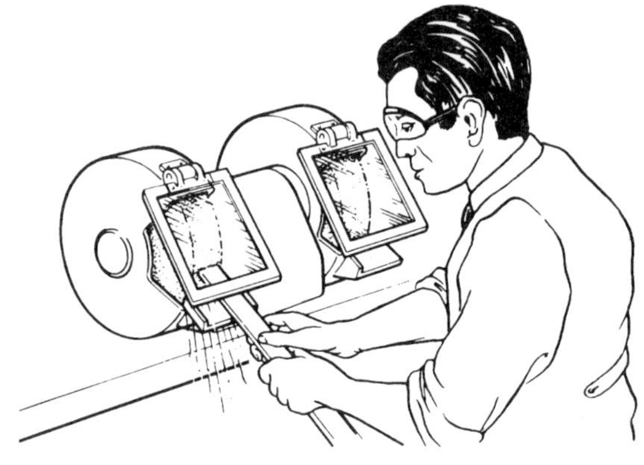

(e) ALWAYS PROTECT THE EYES WHEN USING MACHINERY

(f) WEAR A SUITABLE RESPIRATOR WHEN DUST AND FUMES ARE PRESENT

Figure 1.19 *Safety equipment and clothing*

Personal safety

Before using any electrical equipment it is advisable to carry out a number of visual checks as shown in Figure 1.20(a).

- Check that the cable is not damaged or frayed.
- Check that both ends of the cable are secured in the cord grips of the plug or appliance and that none of the conductors is visible.
- Check that the plug is in good condition and not cracked.
- Check that the voltage and power rating of the equipment is suitable for the supply available.
- If low voltage equipment is being used check that a suitable transformer is available.
- Check that whatever the voltage rating of the equipment, it is connected to the supply through a circuitbreaker containing a residual current detector (RCD).
- Check that all metal clad electrical equipment has a properly connected earth lead and is fitted with a properly connected three-pin plug as shown in Figure 1.20(b).

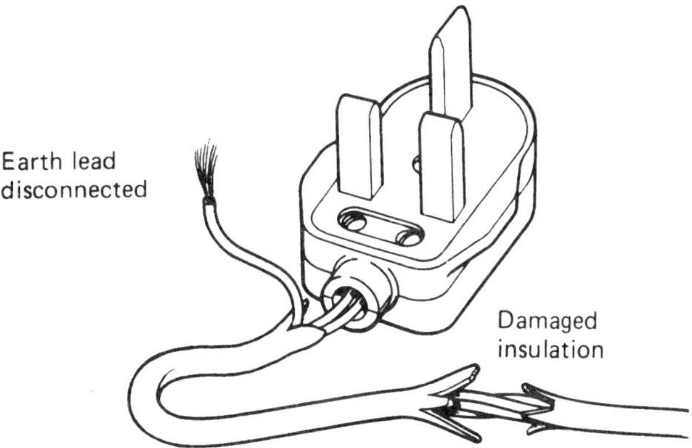

(A) EXAMINE PLUGS DAILY

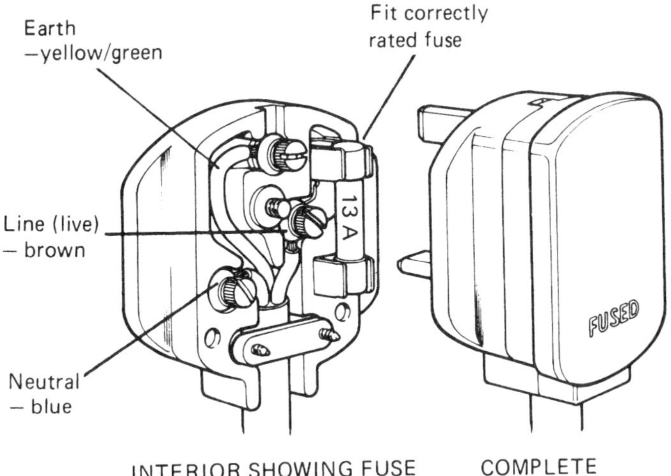

(B) Correctly connected plug with 13 A fuse

Figure 1.20 *Visual checks on electric plugs and cables*

Earthing

All exposed metalwork of electrically powered or operated equipment must be earthed to prevent electric shocks (see also battery charger project). Figure 1.21 shows the two ways in which a person may receive an electric shock. In Figure 1.21(a) the person is receiving a shock by holding both the live and neutral conductors so that the electric current can flow through his or her body. The neutral conductor is connected to earth, so it is equally possible to receive a shock via an earth path when holding a live conductor, as shown in Figure 1.21(b). It is unlikely you would be so foolish as to deliberately touch a live conductor, but you might come into contact with one accidentally.

For example, the portable electric drill shown in Figure 1.22(a) has a metal casing but no earth. The live conductor has fractured within the machine and is touching the metal casing. The operator cannot see this and would receive a serious or even fatal shock. The fault current would flow through the body of the user via the earth path to neutral.

Figure 1.22(b) shows the effect of the same fault in a properly earthed machine. The fault current would take the path of least resistance and would flow to earth via the earth wire. The operator would be unharmed. Electrical equipment must be regularly inspected, tested, repaired and maintained by qualified electricians. Note that 'double insulated' equipment does not need to be earthed.

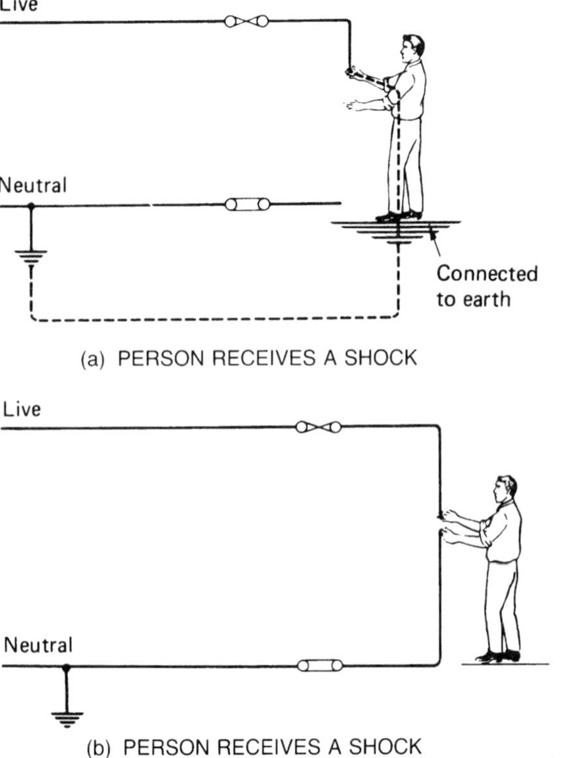

Figure 1.21 *Shock hazard*

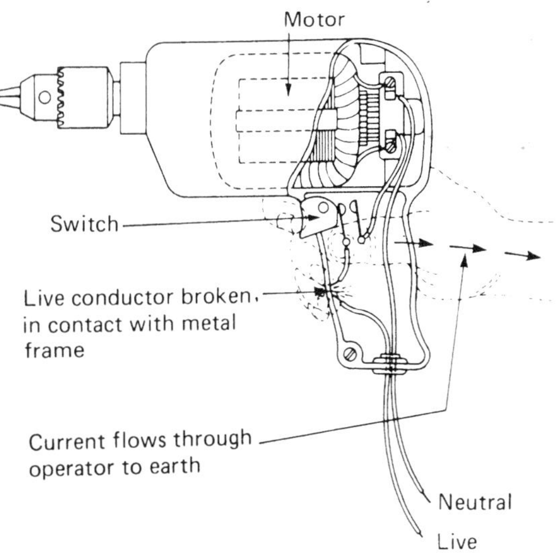

(a) ELECTRIC DRILL NOT EARTHED

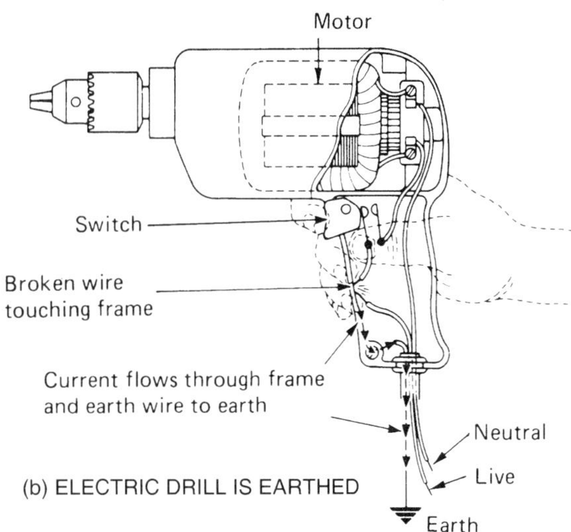

(b) ELECTRIC DRILL IS EARTHED

Figure 1.22 *Electric drill earthing to prevent shock hazard*

Procedure in the event of electric shock

- Switch off the supply of current if this can be done quickly.
- If you cannot switch off the supply, do not touch the person's body with your bare hands. Human flesh is a conductor and you would also receive a severe shock. Drag the affected person clear using insulating material such as dry clothing, a dry sack, or any plastic material that may be handy.
- If the affected person has stopped breathing commence artifical respiration immediately. Don't wait for help to come or go to seek for help. If the affected person's pulse has stopped, heart massage will also be required. Use whatever method of resuscitation with which you are familiar.

Test your knowledge 1.9

1. State THREE causes of electrical accidents and suggest how they may be prevented.
2. List the checks you would make before using a portable electric power tool.

Fire

Fire is the rapid oxidation (burning) of flammable materials. For a fire to start, the following are required:

- a supply of flammable materials
- a supply of air (oxygen)
- a heat source.

Once the fire has started, the removal of one or more of the above will result in the fire going out.

Fire prevention

Fire prevention is largely 'good housekeeping'. The workplace should be kept clean and tidy. Rubbish should not be allowed to accumulate in passages and disused storerooms. Oily rags and waste materials should be put in metal bins fitted with airtight lids. Plant, machinery and heating equipment should be regularly inspected, as should fire alarm and smoke detector systems. You should know how to give the alarm.

Electrical installations, alterations and repairs must only be carried out by qualified electricians and must comply with the current IEE Regulations. Smoking must be banned wherever flammable substances are used or stored. The advice of the fire prevention officer of the local brigade should be sought before flammable substances, bottled gases, cylinders of compressed gases, solvents and other flammable substances are brought on site.

Fire procedures

In the event of you discovering a fire, you should:

- Raise the alarm and call the fire service.
- Evacuate the premises. Regular fire drills must be held. Personnel must be familiar with normal and alternative escape routes. There must be assembly points and a roll call of personnel. A designated person must be allocated to each department or floor to ensure that evacuation is complete. There must be a central reporting point.
- Keep fire doors closed to prevent the spead of smoke. Smoke is the biggest cause of panic and accidents, particularly on staircases. Emergency exits must be kept unlocked and free from obstruction whenever the premises are in use. Lifts must not be used in the event of fire.
- Only attempt to contain the fire until the professional brigade arrives if there is no danger to yourself or others. Always make sure you have an unrestricted means of escape. Saving lives is more important than saving property.

Extinguishers

Figure 1.23 shows a fire hose and a range of pressurized water extinguishers. These can be identified by their shape and colour which is RED. They are for use on burning solids such as wood, paper, cloth, etc. They are UNSAFE on electrical equipment at all voltages.

Figure 1.24 shows two types of foam extinguisher. These can be identified by their shape and colour which is CREAM. They are for use on burning flammable liquids. They are UNSAFE on electrical equipment at all voltages.

Figure 1.25 shows a variety of extinguishers that can be used on most fires and are safe for use on electrical equipment. Again, they can be identified by their shapes and colours:

- Dry powder extinguishers are coloured BLUE and are safe up to 1000 V.
- Carbon dioxide (CO_2) extinguishers are coloured BLACK and are safe at high voltages.
- Vaporizing liquid extinguishers are coloured GREEN and are safe at high voltages.

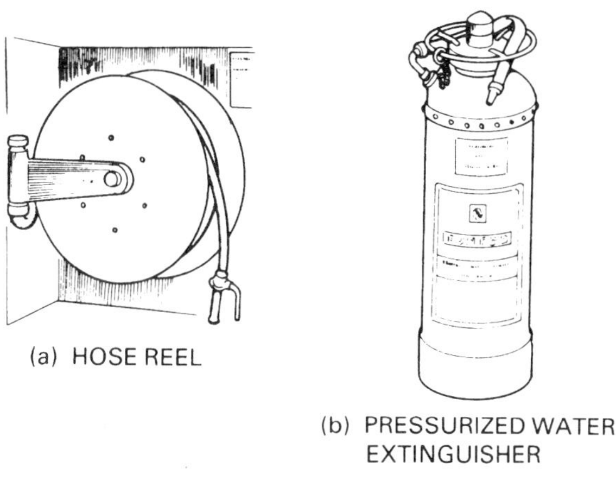

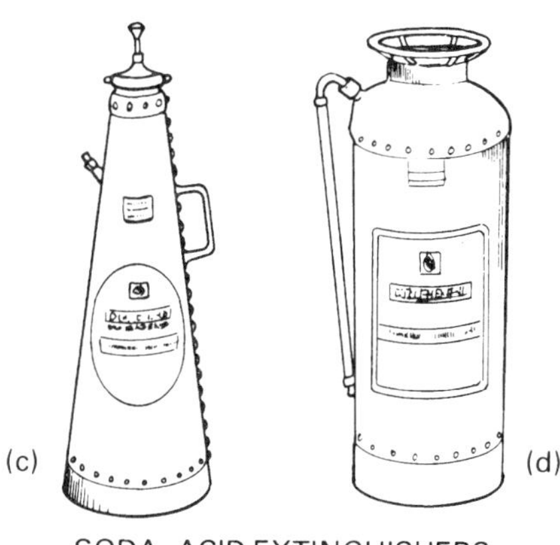

Figure 1.23 *Various types of fire extinguisher*

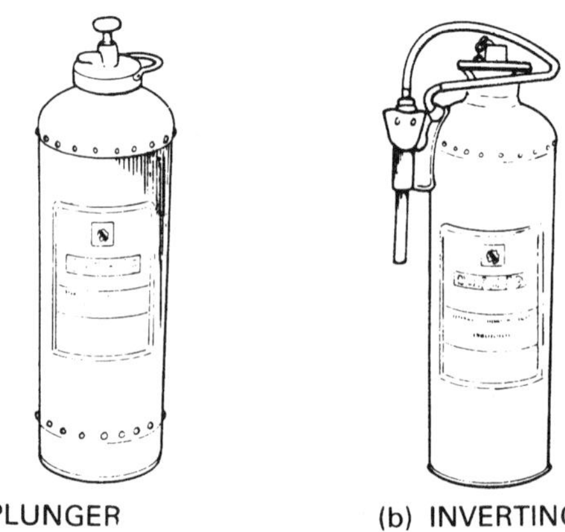

(a) PLUNGER (b) INVERTING

Figure 1.24 *Foam extinguishers*

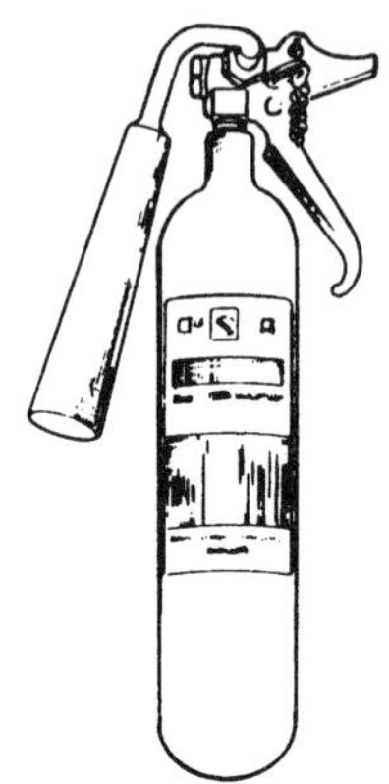

(a) Dry powder extinguisher (b) Carbon dioxide (CO_2) extinguisher (c) Vaporizing liquid extinguisher

Figure 1.25 *Extinguishers for use on electrical equipment*

These latter two extinguishers act by replacing the air with an atmosphere free from oxygen. They are no good in draughts which would blow the vapour or gas away. Remember, if the fire can't breathe neither can any living creature. Evacuate all living creatures before using one of these types of extinguisher. When using this type of extinguisher, keep backing away from the gas towards fresh air, otherwise it will put you out as well as the fire!

Figure 1.26 shows a fire blanket. Fire blankets are woven from fire-resistant synthetic fibres and are used to smother fires. The

Figure 1.26 *A fire blanket*

Test your knowledge 1.10

1 State which sort of fire extinguisher you would use in each of the following instances:
   (a) Paper burning in an office waste bin
   (b) A pan of fat burning in the kitchen of the works canteen
   (c) A fire in a mains voltage electrical machine.

2 A fire breaks out near to a store for paints, paint thinners and bottled gases. What action should be taken and in what order?

old-fashioned blankets made from asbestos must NOT be used. The blanket is pulled from its container and spread over the fire to exclude the air necessary to keep the fire burning. They are suitable for use in kitchens, in workshops, and in laboratories. They are also used where a person's clothing is on fire, by rolling the person and the burning clothing up in the blanket to smother the fire. Do not cover the person's face.

Activity 1.2

Consult the Health and Safety at Work Act and answer the following questions.

(a) What is an improvement notice?

(b) What is a prohibition notice?

(c) Who issues such notices?

(d) Who can be prosecuted under the Act?

Present your findings in the form of a brief word-processed report.

Activity 1.3

Investigate the construction of one of the lathes in your workshop and describe how the guard over the end-train gears (change wheels) is interlocked so that the lathe cannot be operated with the guard open. Illustrate your findings with a sketch.

Activity 1.4

Design a poster that can be placed in a workshop that will provide information on the types of fire extinguisher that can safely be used on different types of fire. You poster should be designed so that it can be understood by even those who may only have a very limited command of the English language.

Measuring

Before you can mark out a component or check it during manufacture you need to know about engineering measurement. All engineering measurements are comparative processes. You compare the size of the feature to be measured with a known standard.

Linear measurement

Figure 1.27 shows a steel rule and how to use it. The distance between the lines or the width of the work is being compared with the rule. In this instance the rule is our standard of length. The rule should be made from spring steel and the markings should be engraved into the surface of the rule. The edges of

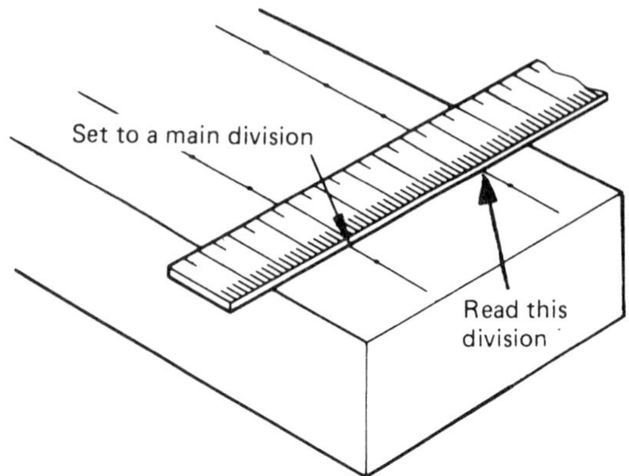

(a) MEASURING THE DISTANCE BETWEEN TWO SCRIBED LINES

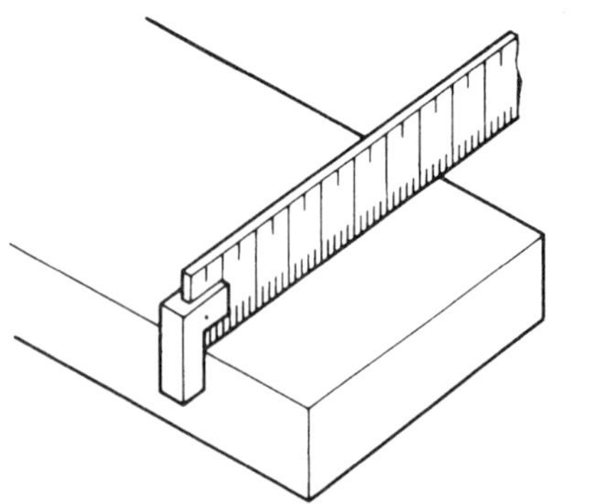

(b) MEASURING THE DISTANCE BETWEEN TWO FACES USING A HOOK RULE

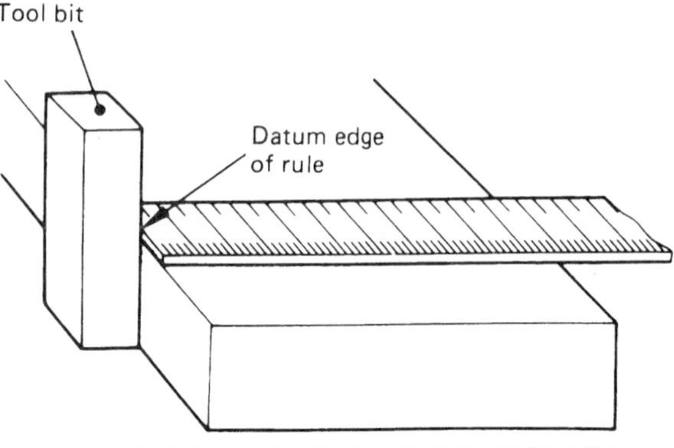

(c) MEASURING THE DISTANCE BETWEEN TWO FACES USING A STEEL RULE AND ABUTMENT

Figure 1.27 *Using a steel rule*

the rule should be ground so that it can be used as a straight edge, and the datum end of the rule should be protected from damage so that the accuracy of the rule is not lost. NEVER use a rule as a screwdriver or for cleaning out the T-slots on machine tools.

To increase the usefulness of the steel rule and to improve the accuracy of taking measurements, accessories called calipers are used. These are used to transfer the distances between the faces of the work to the distances between the lines engraved on the rule. Figure 1.28 shows some different types of inside and outside calipers. It also shows how to use calipers.

A steel rule can only be read to an accuracy of about ± 0.5 mm. This is rarely accurate enough for precision engineering purposes. Figure 1.29(a) shows a vernier caliper and how it can take inside and outside measurements. Typical vernier scales are shown in Figure 1.29(b). Some verniers have scales that are different to the ones shown in Figure 1.29(b). Always check the scales before taking a reading. Like all measuring instruments, a vernier caliper must be treated carefully and it must be cleaned and returned to its case whenever it is not in use.

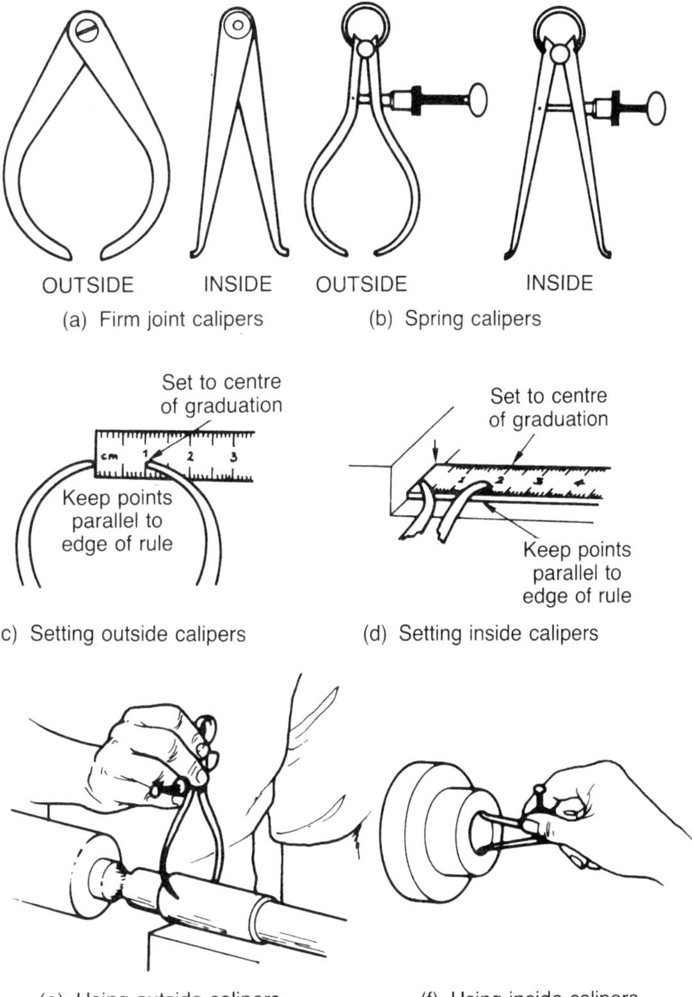

Figure 1.28 *Inside and outside calipers*

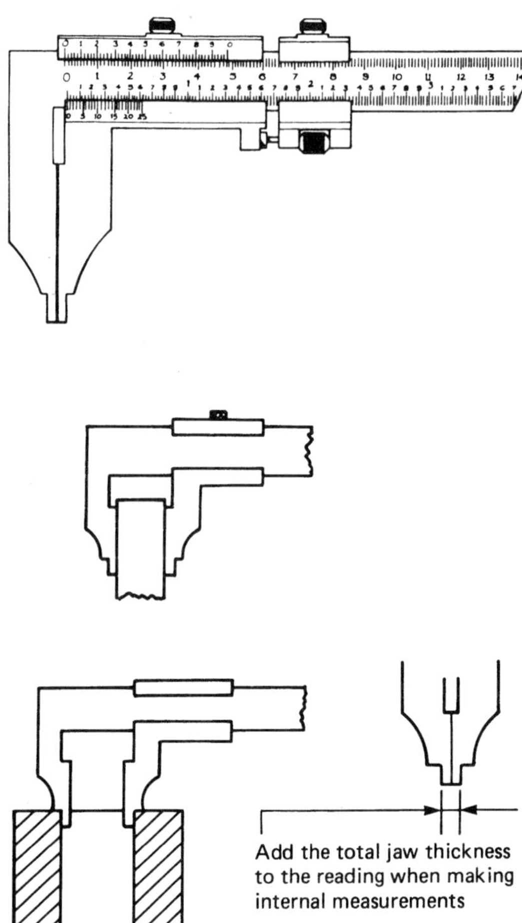

(a) Vernier caliper

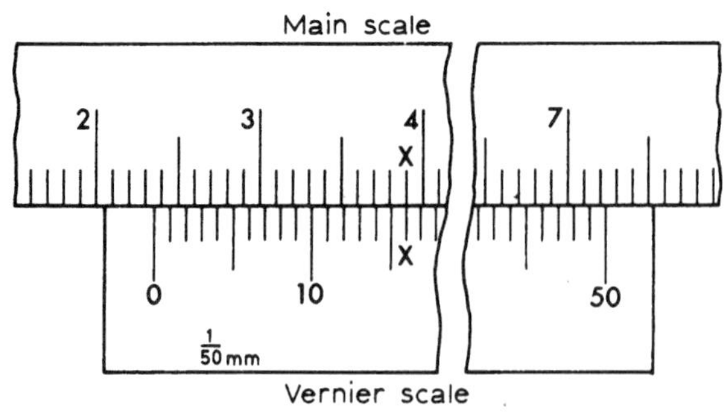

Reading
23mm on main scale
$\frac{16}{50}$mm on vernier scale
23·32mm

(b) Reading the metric vernier scale: 23 mm on main scale plus 16 × 0.02 mm on vernier scale gives total 23.32 mm

Figure 1.29 *Vernier calipers*

Test your knowledge 1.11

1 State THREE precautions that you should take to keep a steel rule in good condition.

2 Explain how you would obtain a constant measuring pressure when using a micrometer caliper.

3 Write down the reading of the micrometer caliper scales shown in Figure 1.31(a).

4 Write down the reading of the vernier caliper scales shown in Figure 1.31(b).

Vernier calipers are difficult to read accurately even if you have good eyesight. A magnifying glass is helpful. The larger sizes of vernier caliper are quite heavy and it is difficult to get a correct and consistent 'feel' between the instrument and the work. An alternative instrument is the micrometer caliper. Figure 1.30 shows a micrometer caliper and the method of reading its scales.

Micrometer calipers are more compact than vernier calipers. They are also easier to use. The ratchet on the end of the thimble ensures that the contact pressure is kept constant and at the correct value. Unfortunately micrometer calipers have only a limited measuring range (25 mm), so you need a range of micrometers moving up in size in 25 mm steps. (0–25 mm, 25–50 mm, 50–75 mm and so on to the largest size). You also need a range of inside micrometers, and depth micrometers.

Angular measurements

The most frequently measured angle is 90°. This is a right-angle and surfaces at right-angles to each other are said to be perpendicular. Right angles are checked with some sort of try square. Figure 1.32 shows a typical engineer's try square and two ways in which it can be used.

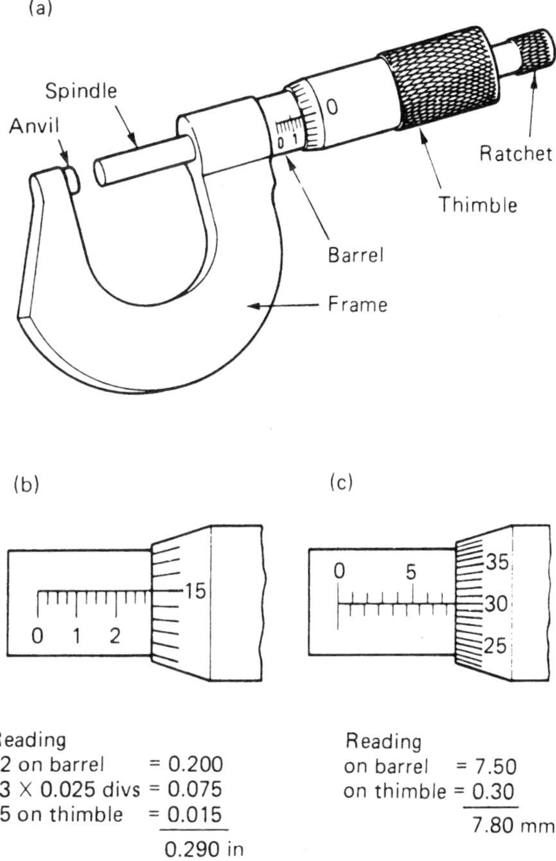

Figure 1.30 *Micrometer caliper*

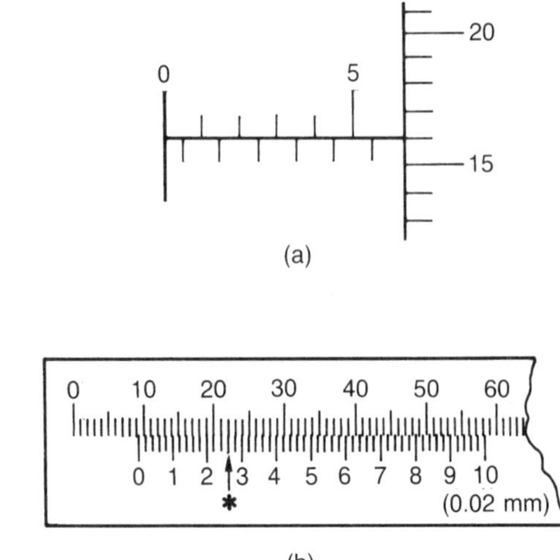

Figure 1.31 *See Test your knowledge 1.11*

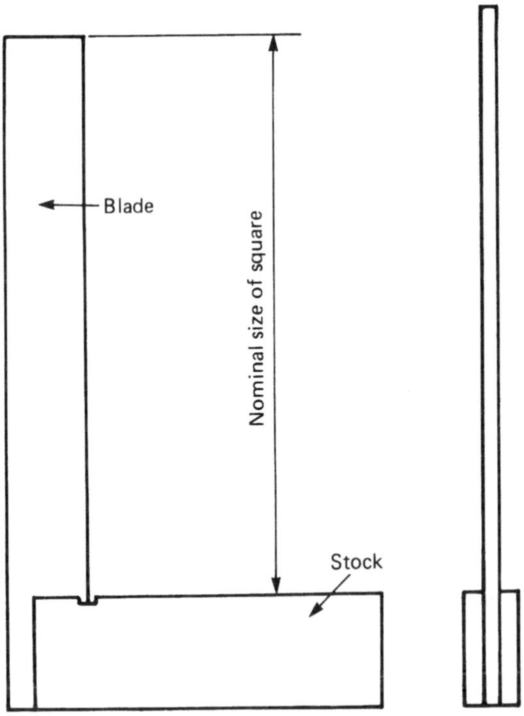

(a) Engineer's try-square

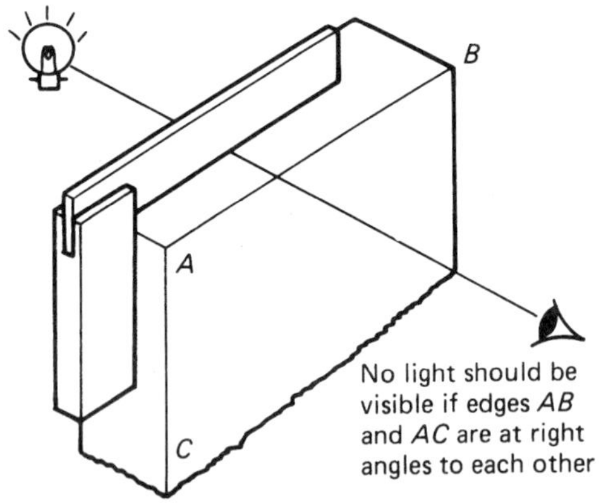

(b) CHECKING A SMALL COMPONENT FOR SQUARENESS

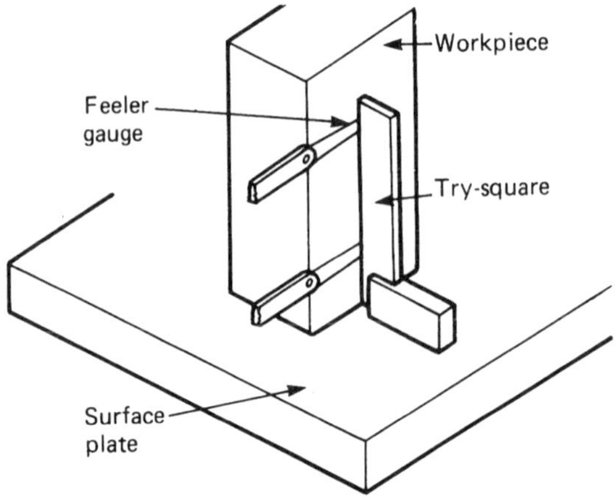

(c) CHECKING A LARGE COMPONENT FOR SQUARENESS

Figure 1.32 *Engineer's try-square*

Test your knowledge 1.12

1 Surfaces at 90° to each other are said to be:
(a) mutually parallel
(b) mutually perpendicular
(c) at a mutually acute angle
(d) at a mutually obtuse angle.

2 Name the instrument that is used to check that two surfaces are at right-angles to each other.

3 State the reading accuracy of the vernier bevel protractor shown in Figure 1.33(b).

4 Explain how the reading stated in Figure 1.33(b) is obtained.

For angles other than a right-angle, a protractor is used. This may be a simple plain protractor as shown in Figure 1.33(a) or it may have a vernier scale (vernier protractor) as shown in Figure 1.33(b).

Tolerances and gauging

So far we have only considered measurement of size. This is usual when making a small number of components. However for quantity production it requires too high a skill level, is too time consuming and, therefore, too expensive. Since no product can be made to an exact size, nor can it be measured exactly, the designer usually gives each dimension an upper and lower size. This is shown in Figure 1.34. If the component lies

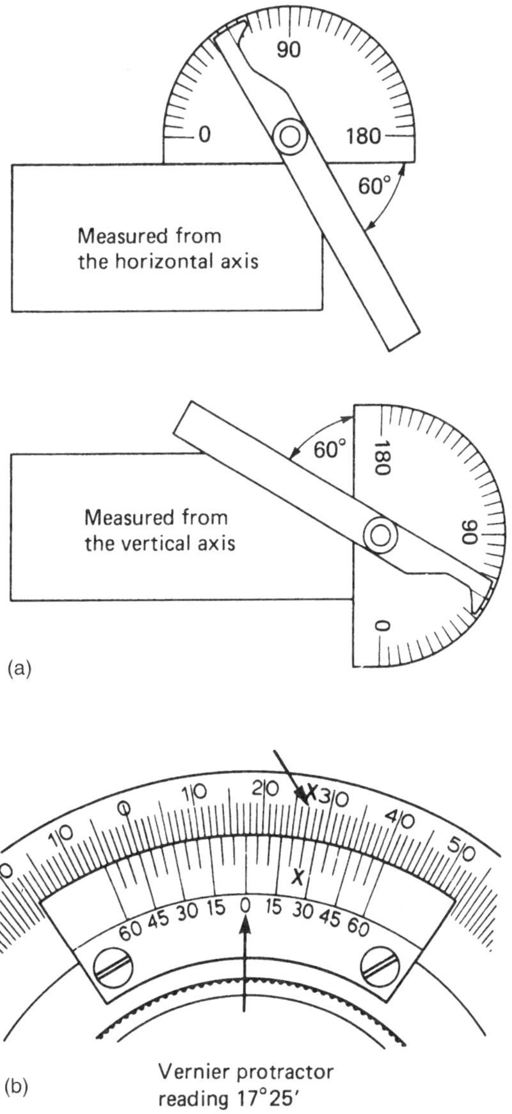

Figure 1.33 *Using a protractor*

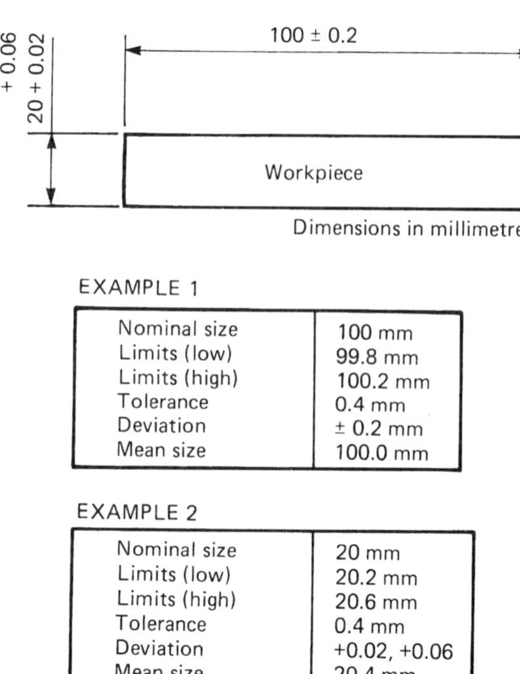

EXAMPLE 1

Nominal size	100 mm
Limits (low)	99.8 mm
Limits (high)	100.2 mm
Tolerance	0.4 mm
Deviation	± 0.2 mm
Mean size	100.0 mm

EXAMPLE 2

Nominal size	20 mm
Limits (low)	20.2 mm
Limits (high)	20.6 mm
Tolerance	0.4 mm
Deviation	+0.02, +0.06
Mean size	20.4 mm

Figure 1.34 *Use of tolerances*

anywhere between the upper and lower limits of size it will function correctly. The closer the limits, the more accurately the component will work, but the more expensive it will be to make.

A major advantage of using tolerance dimensions is that they can be checked without having to be measured. Gauges can be used instead of measuring instruments. This is easier, quicker and much cheaper. Figure 1.35 shows how a caliper gauge can be used to check the thickness of a component. Plug gauges are used in a similar manner to check hole sizes.

Some more gauges are shown in Figure 1.36. Radius gauges are used to check the corner radii of components. Feeler gauges are used to check the gap between components: for example the valve tappet clearances in a motor vehicle engine. Thread gauges are used to check the pitch of screw threads.

Test your knowledge 1.13

1 With reference to Figure 1.37, state the following:
   (a) the nominal size
   (b) the upper limit
   (c) the lower limit
   (d) the tolerance
   (e) the deviation
   (f) the mean size.

2 With the aid of sketches, explain how you would check the dimensions for the hole shown in Figure 1.37 with a plug gauge.

Activity 1.5

(a) Select suitable measuring instruments and explain how you would check the dimensions for the component shown in Figure 1.38.

(b) Write down the readings for:

   (a) the micrometer shown in Figure 1.39(a)

   (b) the vernier shown in Figure 1.39(b).

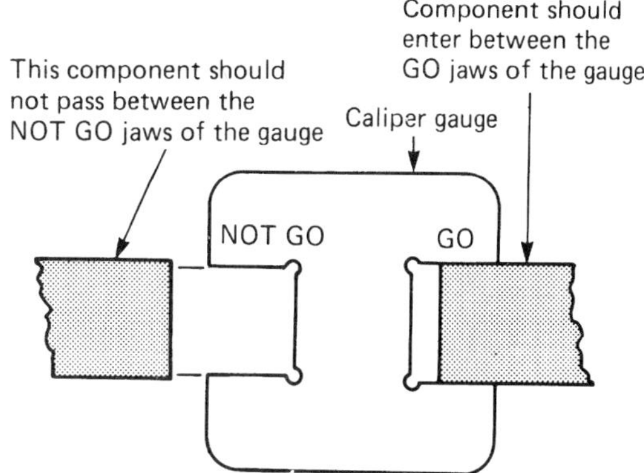

(a) CORRECTLY SIZED COMPONENT ENTERS 'GO' JAWS BUT NOT 'NOT GO' JAWS

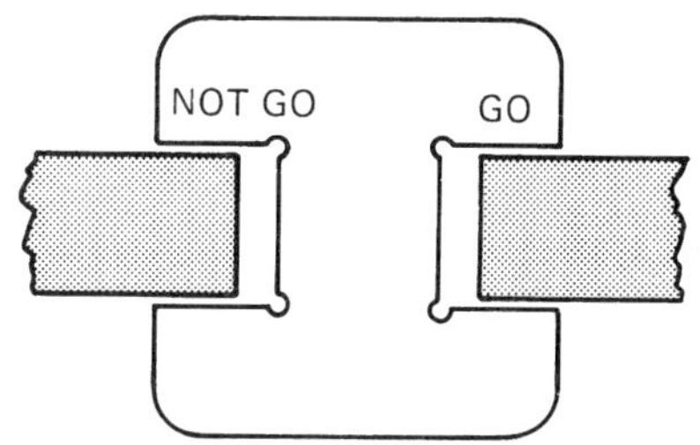

(b) UNDERSIZE COMPONENT ENTERS 'GO' AND 'NOT GO'

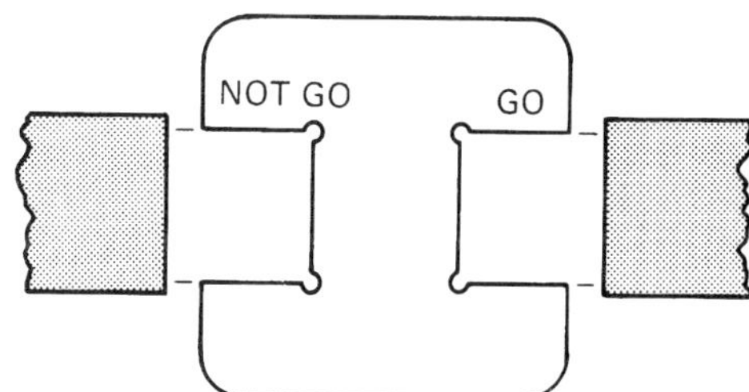

(c) OVERSIZE COMPONENT DOES NOT ENTER 'GO' OR 'NOT GO'

Figure 1.35 *Using a caliper gauge to check the thickness of a component*

Marking out

The reasons why you mark out components before making them are as follows.

- To provide you with guide lines to work to. Where only limited accuracy is required, marking out also controls the size and shape of the workpiece and the position of any holes.
- As a guide to a machinist when setting up and cutting. In this instance the final dimensional control comes from the use of precision measuring instruments together with the micrometer dials on the machine controls.
- Marking out also indicates if sufficient machining allowance has been left on cast or forged components. It also indicates whether or not such features as webs, flanges and cored holes have been correctly positioned.

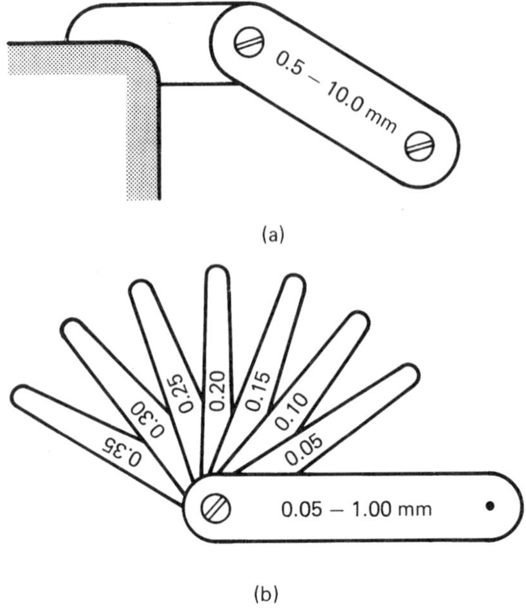

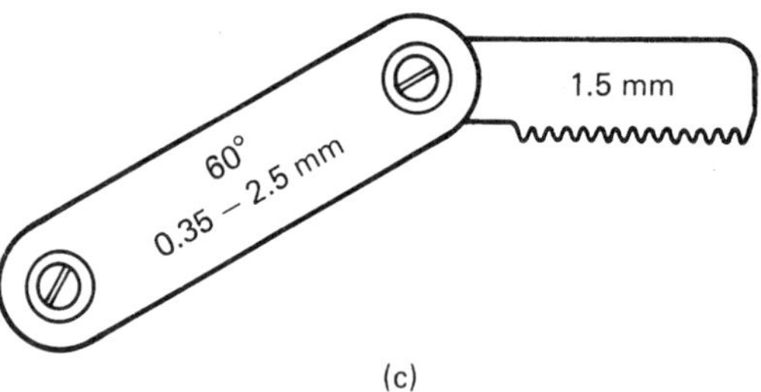

Figure 1.36 *Radius and feeler guages*

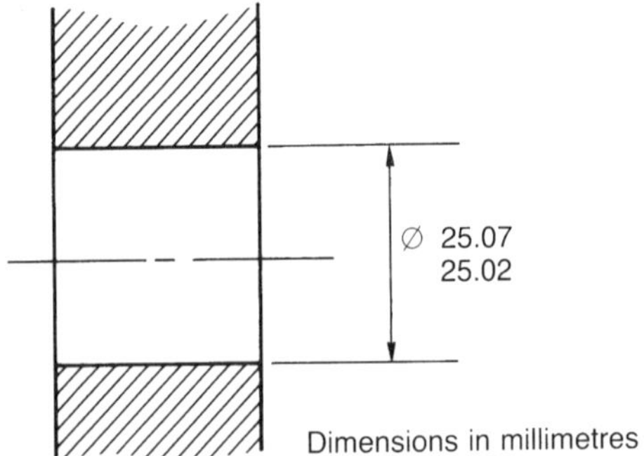

Figure 1.37 *See Test your knowledge 1.13*

Scribed lines and centre marks

Scribed lines are fine lines cut into the surface of the material being marked out by the point of a scribing tool (scriber). An example of a scriber is shown in Figure 1.40(a). To ensure that the scribed line shows up clearly, the surface of the material to be marked out is coated with a thin film of a contrasting colour. For

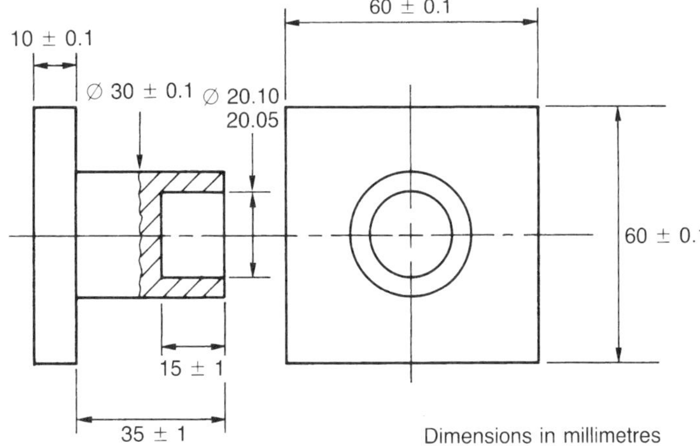

Figure 1.38 *See Activity 1.5*

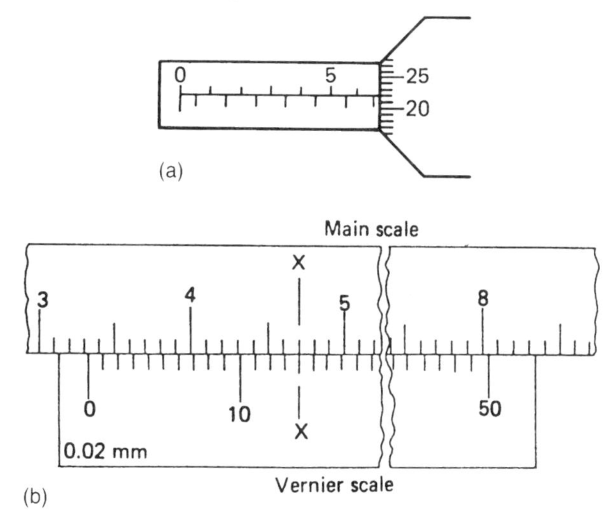

Figure 1.39 *See Activity 1.5*

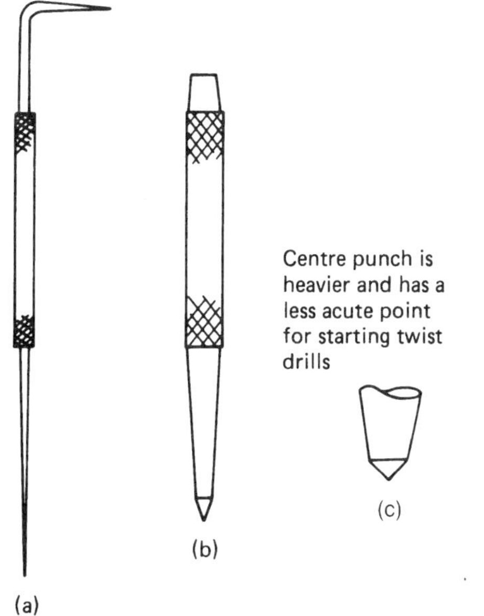

Figure 1.40 *Scriber and centre punch*

example, the surfaces of the casting to be machined are often whitewashed. Bright metal surfaces can be treated with a marking out 'ink'. Plain carbon steels can be treated with copper sulphate solution which copper plates the surface of the metal. This has the advantages of permanence. Care must be taken in its use as it will attack any marking out and measuring instruments into which the copper sulphate comes into contact.

Centre marks are made with a dot punch as shown in Figure 1.40(b) or with a centre punch, as shown in Figure 1.40(c). A dot punch has a fine conical point with an included angle of about 60°. A centre punch is heavier and has a less acute point angle of about 90°. It is used for making a centre mark for locating the point of a twist drill and preventing the point from wandering at the start of a cut.

The dot punch is used for two purposes when marking out.

- A scribed line can be protected by a series of centre marks made along the line, as shown in Figure 1.41(a). If the line is accidentally removed, it can be replaced by joining up the centre marks. Further, when machining to a line as shown in Figure 1.41(b), the half-marks left behind are a witness that the machinist has 'split the line'.
- Secondly, dot punch marks are used to prevent the centre point of dividers from slipping when scribing circles and arcs of circles, as shown in Figure 1.41(c). The correct way to set divider points is shown in Figure 1.41(d).

When a centre punch is driven into the work, distortion can occur. This can be a burr raised around the punch mark, swelling of the edge of a component, or the buckling of thin material.

Equipment for marking out

From what we have already seen, your basic requirements for marking out are:

- A scriber to produce a line
- A rule to measure distances and act as a straight edge to guide the point of the scriber
- Dividers to scribe circles and arcs of circles as shown in Figure 1.41(c).

In addition, you require hermaphrodite (odd-leg) calipers, as shown in Figure 1.42(a) and a try square. Odd-leg calipers are used to scribe lines parallel to a datum edge. A try square and scriber are used to scribe a line at right-angles to a datum edge, as shown in Figure 1.42(b).

Alternatively, lines can be scribed parallel to a datum edge using a surface table and scribing block, as shown in Figure 1.44(a). In this example the scribing point is set to a steel rule, so the accuracy is limited. Alternatively, the line can be scribed using a vernier height gauge, as shown in Figure 1.44(b). This is very much more accurate. Where extreme accuracy is required, slip gauges and slip gauge accessories can be used, as shown in Figure 1.44(c).

Test your knowledge 1.14

Describe with the aid of sketches how you would mark out the link shown in Figure 1.43 on a 6 mm thick low carbon steel blank.

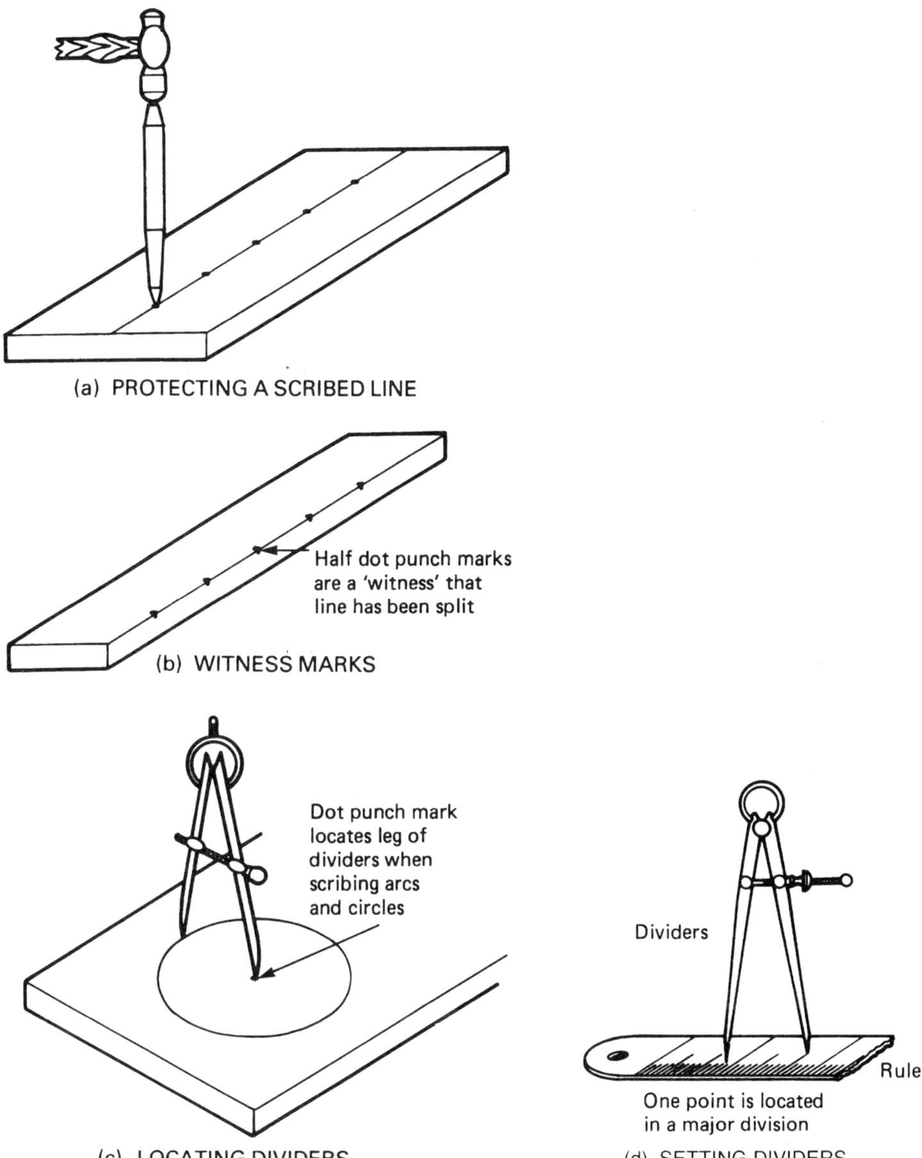

Figure 1.41 *Using dot punch and dividers*

Test your knowledge 1.15

1. Refer to Figure 1.43. Explain with the aid of diagrams how you would mark out the centre lines for the holes and the radii using a surface table and scribing block, together with whatever additional equipment is required.
2. With the aid of sketches explain how you would mark out a keyway 8 mm wide by 4 mm deep and 50 mm long on the end of a 38 mm diameter shaft.

Cylindrical components are difficult to mark out since they tend to roll about. To prevent this they can be supported on vee-blocks, as shown in Figure 1.45(a). Vee-blocks are always made and sold in boxed sets of two. In order that the axis of the work is parallel to the surface plate or table, you must always use such a matched pair of vee-blocks, and make sure that, after use, they are always put away as a pair.

Another useful device for use on cylindrical work is the box square shown in Figure 1.45(b). This is used for scribing lines along cylindrical work parallel to the axis. A centre finder is shown in Figure 1.45(c). This is used to scribe lines that pass through the centre of circular blanks or the ends of cylindrical components.

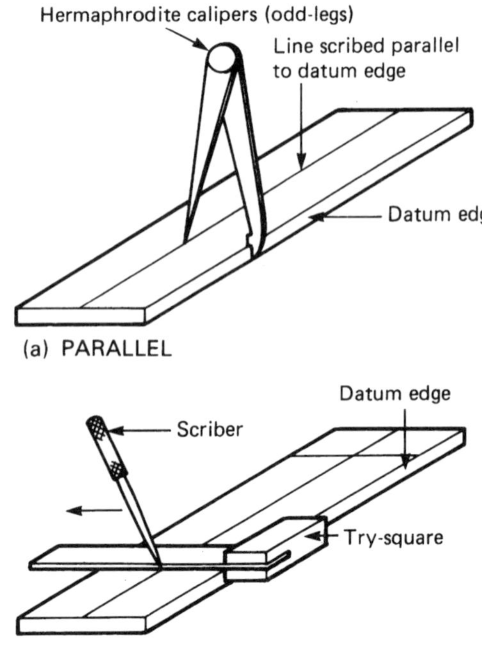

Figure 1.42 *Using odd-leg calipers, scriber and try-square*

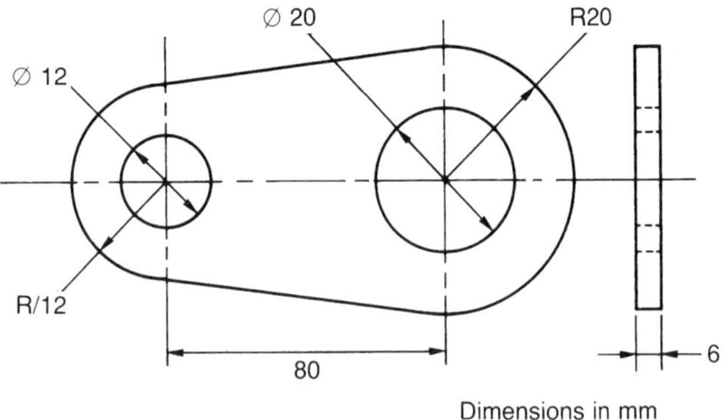

Figure 1.43 *See Test your knowledge 1.14*

Datum points, lines and surfaces

First we had better revise what we know about rectangular and polar coordinates. Examples of these are shown in Figure 1.46.

Rectangular coordinates

The point A in Figure 1.46(a) is positioned by a pair of ordinates (coordinates) lying at right-angles to each other. They also lie at right-angles to the datum edges from which they are measured. This system of measurement requires the production of two datum surfaces or edges at right-angles to each other. That is, two datum edges that are mutually perpendicular.

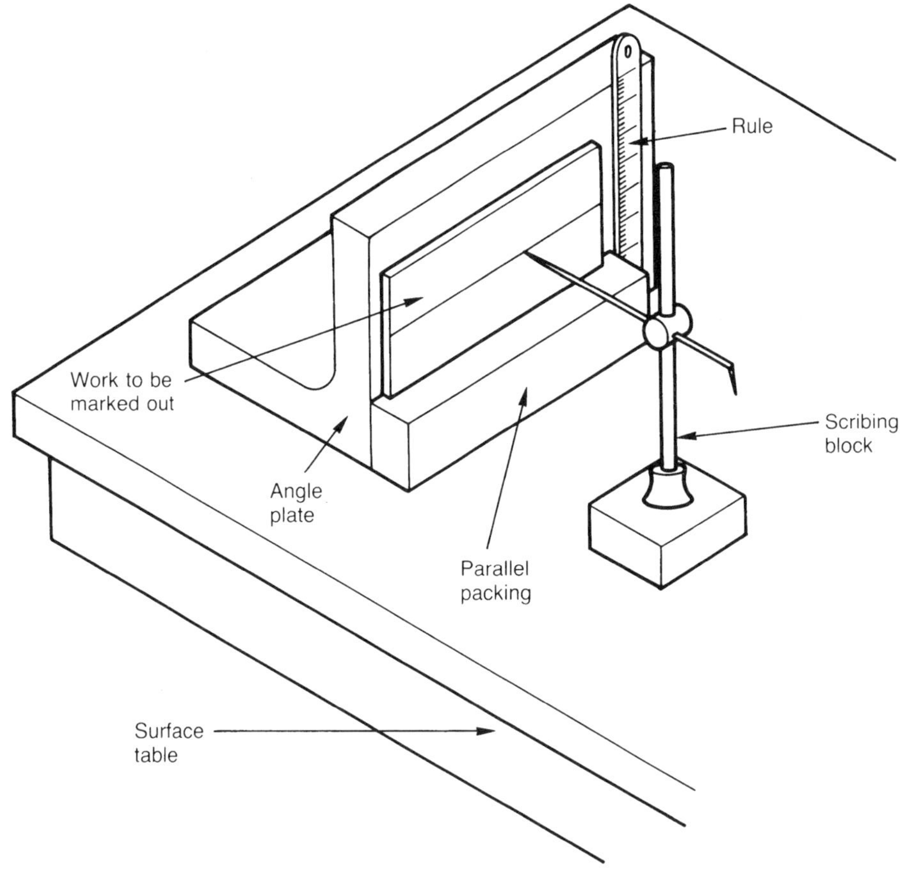

(a) marking out on the surface table.

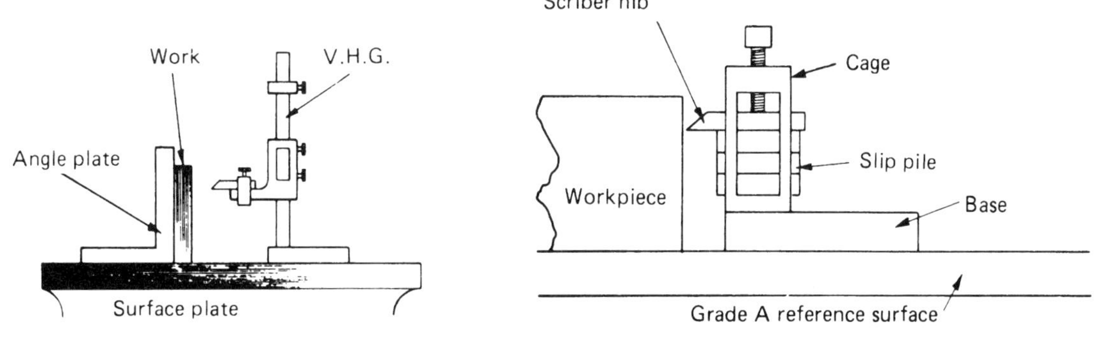

(b) Marking out with the vernier height

(c) Marking out using slip gauges and accessories

Figure 1.44 *Marking out using a surface table*

Polar coordinates

Polar coordinates consist of one linear distance and an angle, as shown in Figure 1.46(b). Dimensioning in this way is useful when the work is to be machined on a rotary table. It is widely used when setting out hole centres round a pitch circle. Quite frequently both systems are used at the same time, as shown in Figure 1.46(c)

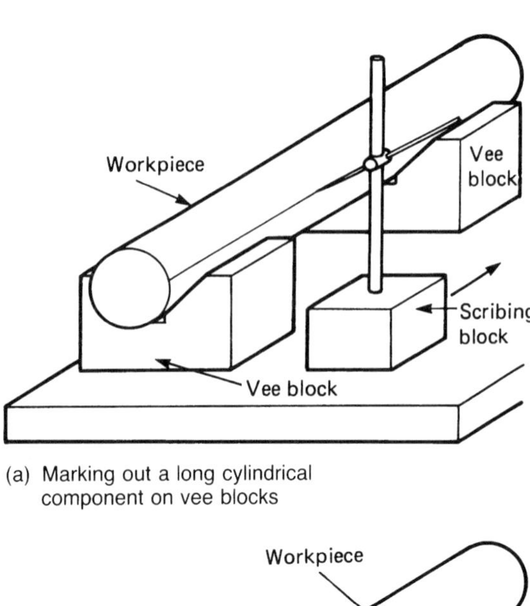

(a) Marking out a long cylindrical component on vee blocks

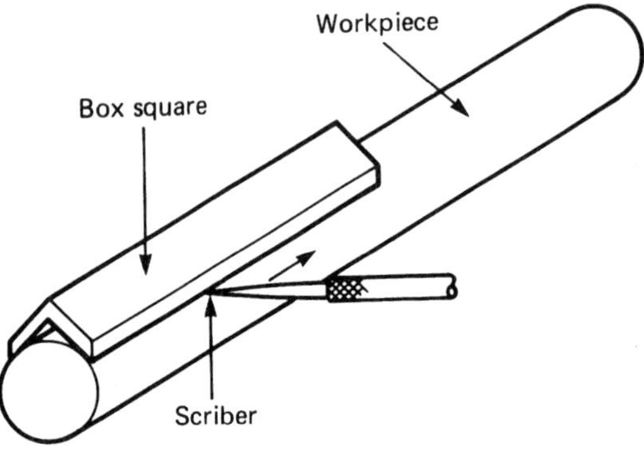

(b) Use of the box square

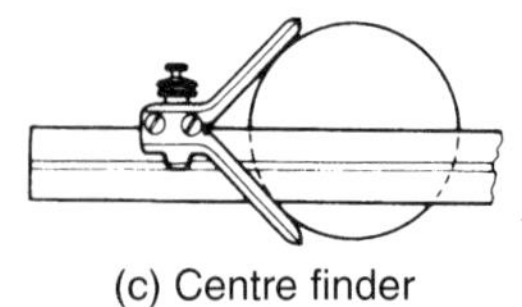

(c) Centre finder

Figure 1.45 *Marking cylindrical components*

During our discussion on marking out, I have kept referring to datum points, datum lines and datum edges. A datum is any point, line or surface that can be used as a basis for measurement. When you go for a medical check-up, your height is measured from the floor on which you are standing. In this example the floor is the basis of measurement, it is the datum surface from which your height is measured.

Point datum

This is a single point from which a number of features are marked out. For example, Figure 1.47(a) shows two concentric circles representing the inside and ouside diameter of a pipe-flange ring. It also shows the pitch circle around which the bolt hole centres are marked off. All these are marked out using dividers or trammels (beam compasses) from a single-point datum.

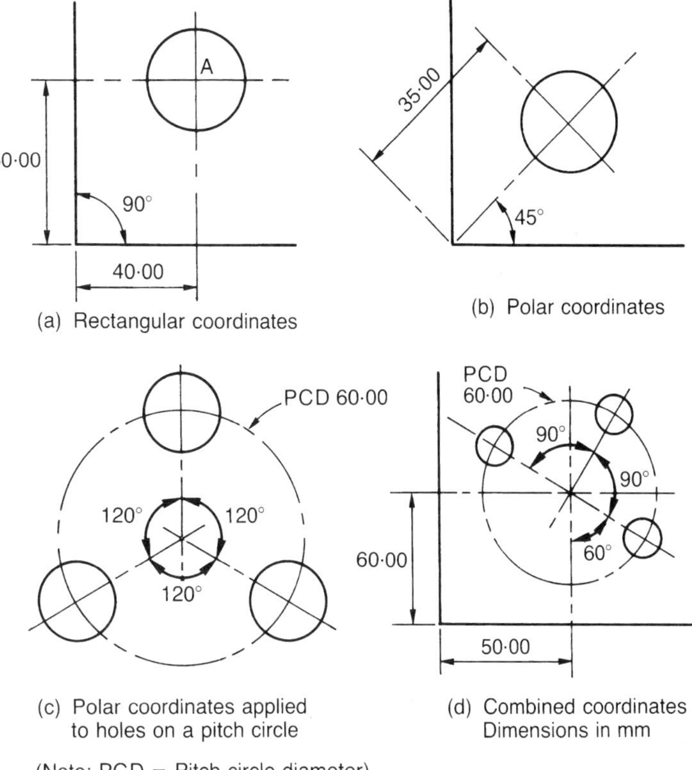

Figure 1.46 *Datum points and coordinates*

Line datum
Any line from which, or along which a number of features are marked out. An example of the use of line datums is shown in Figure 1.47(b).

Surface datum
This is also known as an edge datum and a service edge. It is the most widely used datum for marking out solid objects. Two edges are accurately machined at right angles to each other and all the dimensions are taken from these edges. Figure 1.47(c) shows how dimensions are taken from surface datums. It shows both rectangular and polar coordinates. Alternatively, the work can be clamped to an angle-plate. The mutually perpendicular edges of the angle-plate provide the surface datums. In this instance there is no need to machine the edges of the workpiece at right-angles.

Test your knowledge 1.16

Draw and dimension 8 holes each of 10 mm diameter on a 75 mm diameter pitch circle. The pitch circle to be central on a plate measuring 100 mm by 125 mm.

Activity 1.6 

With the aid of sketches explain how you would mark out the holes and the square on the conponent shown in Figure 1.48. Present your answer in the form of an instruction sheet that is suitable for use by someone with no previous experience of this type of task.

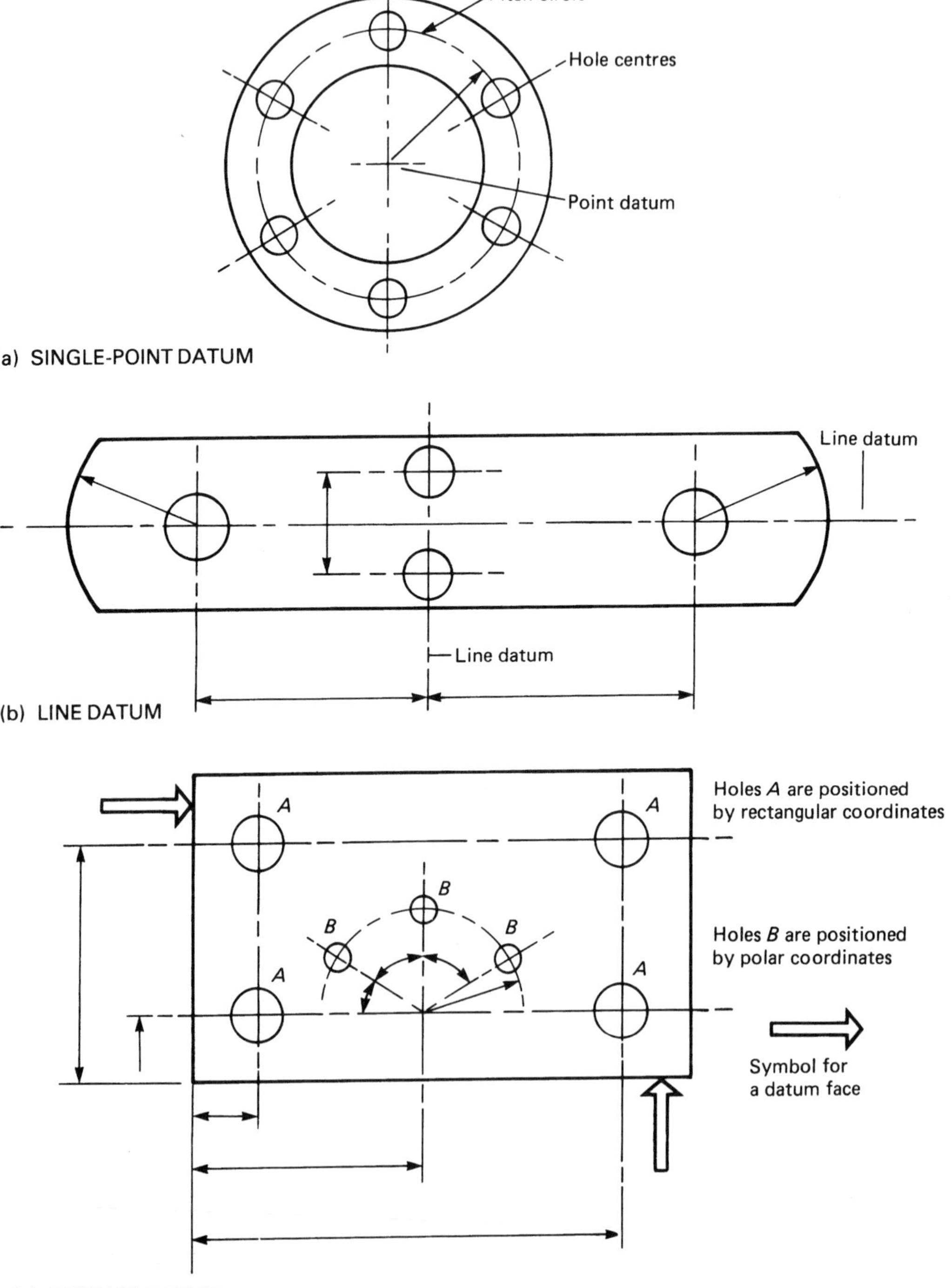

Figure 1.47 *Point, line and surface datums*

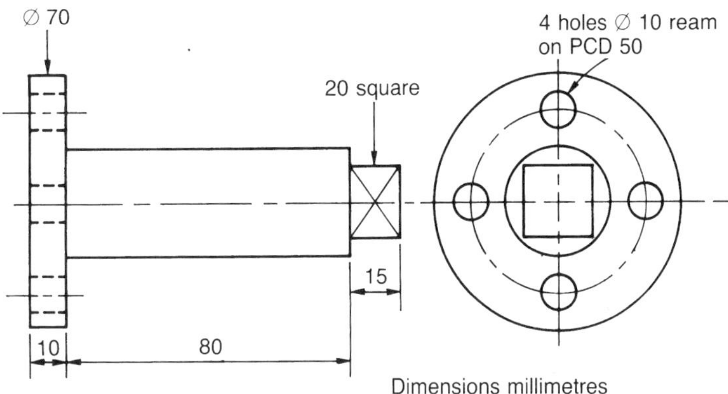

Figure 1.48 *See Activity 1.6*

Hand tools and benchwork

Fitter's bench and vice

A fitter's bench should be substantial and rigid. This is essential if accurate work is to be performed on it. It should be positioned so that it is well lit by both natural and artificial light without glare or shadows. It should be equipped with a fitter's vice. A plain screw vice is shown in Figure 1.49(a) and a quick action vice is shown in Figure 1.49(b). In the latter type of vice, the jaws can be quickly pulled apart when the lever at the side of the screw

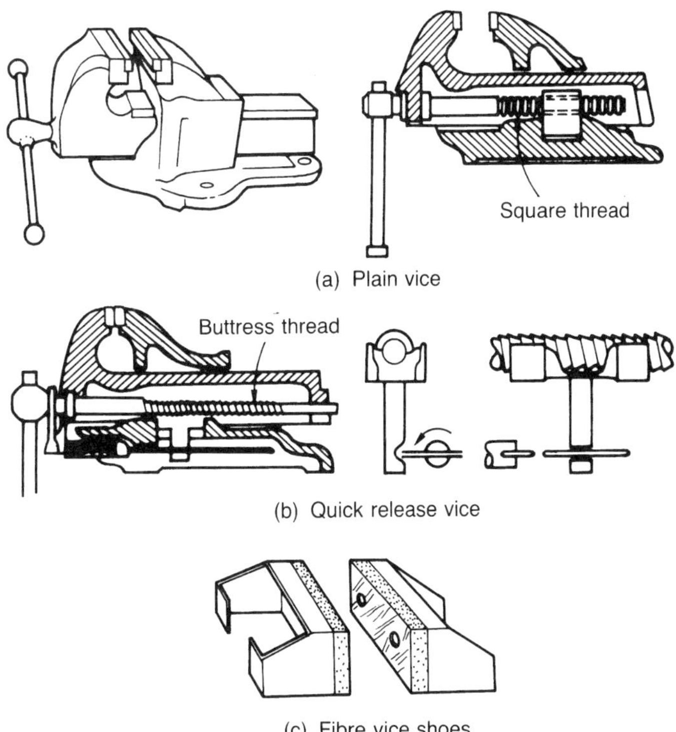

Figure 1.49 *A fitter's vice*

handle is released. The screw is used for closing the jaws and clamping the work in the usual way.

The jaws of a fitter's vice are serrated and hardened to prevent the work from slipping. This also marks the surfaces of the work. For fine work with finished surfaces, the serrated jaws should be replaced with hardened and ground smooth jaws. Alternatively vice shoes can be used. These are faced with a fibre compound and can be slipped over the serrated jaws when required. A pair of typical vice shoes are shown in Figure 1.49(c).

Cutting tools

Before we can discuss the cutting tools we use for bench fitting we need to look at the way metal is cut. Here are the basic facts.

Wedge angle

If you look at a hacksaw blade, as shown in Figure 1.50(a), you can see that the teeth are wedge shaped. Figure 1.50(b) shows how the wedge angle increases as the material gets harder. This strengthens the cutting edge and increases the life of the tool. At the same time it reduces its ability to cut. Try cutting a slice of bread with a cold chisel!

Clearance angle

If you look at the hacksaw blade in Figure 1.50(a), you can see that there is a clearance angle behind the cutting edge of the tooth. This is to enable the tooth to cut into the work.

Rake angle

This angle controls the cutting action of the tool. It is shown in Figure 1.51. I hope you can see that the wedge angle, clearance angle and rake angle always add up to 90°. This is true even when the rake angle is zero or negative as shown in Figure 1.51(b). The greater the rake angle the more easily the tool will cut. Unfortunately, the greater the rake angle, the smaller the

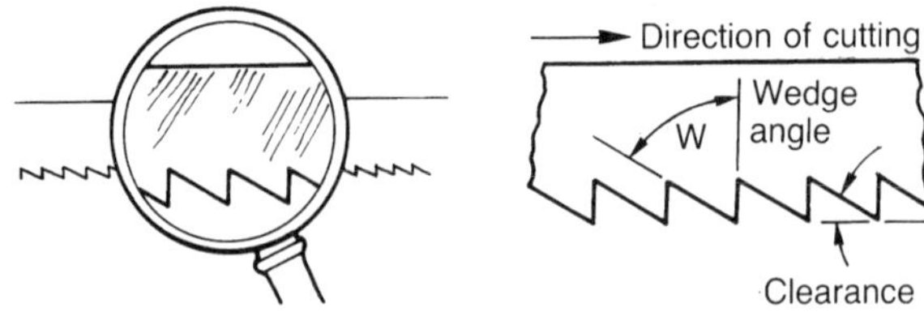

(a) Hacksaw blade showing the wedge angle

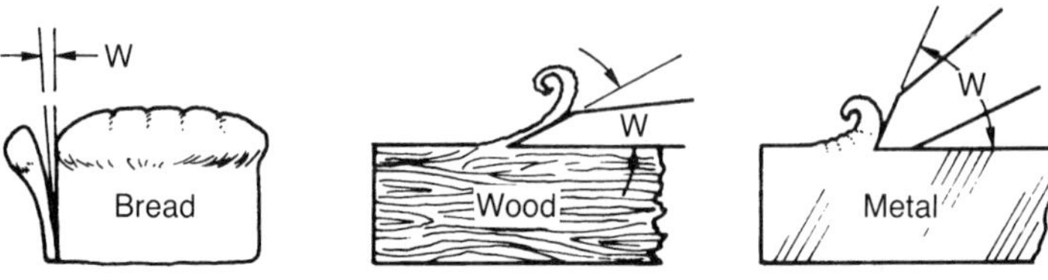

(b) Wedge angles for various materials

Figure 1.50 *Hacksaw blade and wedge angles*

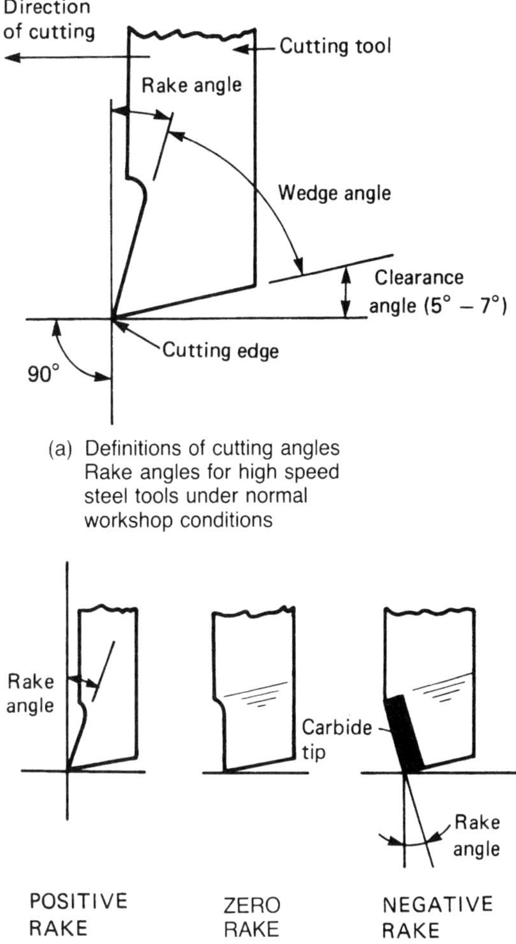

(a) Definitions of cutting angles
Rake angles for high speed steel tools under normal workshop conditions

(b) Comparison of rake angles

Figure 1.51 *Rake angle*

wedge angle will be and the weaker the tool will be. Therefore the wedge and rake angles have to be a compromise between ease of cutting and tool strength and life. The clearance angle remains constant at between 5° and 7°.

Table 1.7

Material	*Rake angle*
Aluminium alloy	30°
Brass (ductile)	14°
Brass (free-cutting)	0°
Cast iron	0°
Copper	20°
Phosphor bronze	8°
Mild steel	25°
Medium carbon steel	15°

For high speed steel tools under normal workshop conditions

Test your knowledge 1.17

1. Calculate the wedge angle for a tool if the clearance angle is 5° and the rake angle is 17°.
2. Explain why cutting fluids (suds) do not have to be used when filing, but are used when turning.
3. Explain why metal cutting tools have a larger wedge angle than wood cutting tools.
4. Explain why metal cutting tools need a clearance angle.
5. Sketch an example of any lathe tool that cuts orthogonally and any lathe tool that cuts obliquely.

Orthogonal and oblique cutting

Figure 1.52(a) is a pictorial representation of the single point cutting tool shown in Figure 1.51. Notice how the cutting edge is at right-angles to the direction in which the tool is travelling along the work. This is called orthogonal cutting. Now look at Figure 1.52(b). Notice how the cutting edge is inclined at an angle to the direction of cut. This is called oblique cutting. Oblique cutting results in a better finish than orthogonal cutting, mainly because the chip is thinner for a given rate of metal removal. This reduced thickness and the geometry of the tool allows the chip to coil up easily in a spiral.

Apart from threading operations, it is very rare to use a coolant or lubricant when using hand tools. However the conditions are very different when using machine tools. Large amounts of metal are removed quickly, considerable energy is used to do this, and this energy is largely converted into heat at the cutting zone. The rapid temperature rise of the work and the cutting tool can lead to inaccurcy and short tool life. A coolant is required to prevent this. The chip flowing over the rake face of the tool results in wear. A lubricant is required to prevent this. Usually coolants are poor lubricants, and lubricants are poor coolants.

For general machining an emulsion of cutting oil (which also contains an emulsifier) and water is used. This has a milky-white appearance and is commonly known as 'suds'. On no account try to use a mineral lubricating oil. This cannot stand up to the temperatures and pressures found in the cutting zone. It is completely useless as a cutting lubricant or as a coolant. It gives off clouds of noxious fumes and it is a fire risk.

Cold chisels

Let's now see how the cutting angles discussed in the previous section can be applied to the basic bench tools.

Figure 1.53(a) shows a typical cold chisel and names its more important features. Constantly hitting the head of a chisel causes it to mushroom, as shown in Figure 1.53(b). Never use a chisel with a mushroom head because bits of metal can fly off it when it

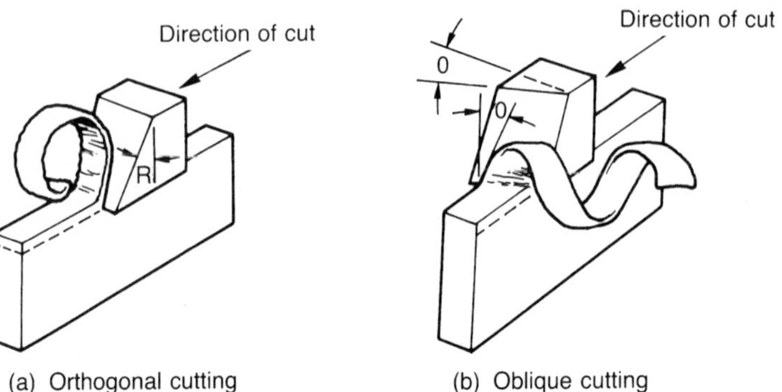

Figure 1.52 *Orthogonal and oblique cutting*

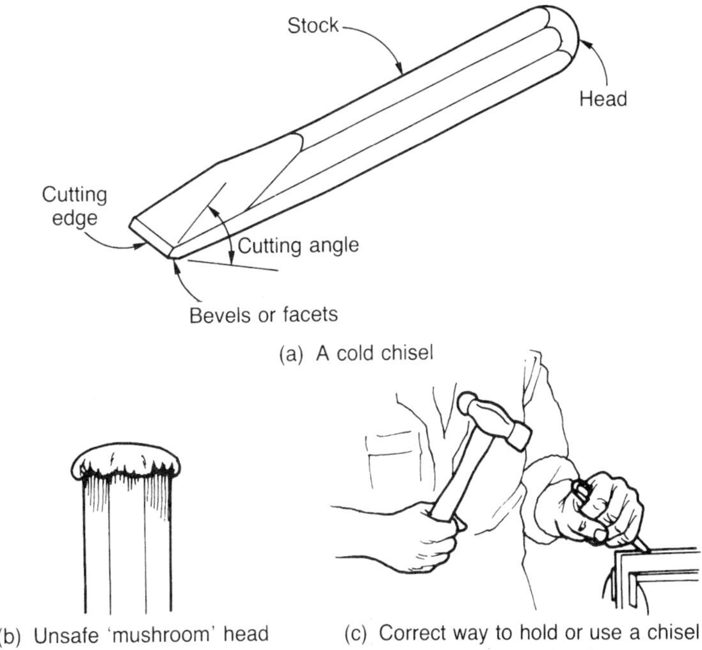

Figure 1.53 *Using a cold chisel*

is hit. These can cause an accident. When the mushroom head starts to form, it must be trimmed off on a grinding machine. Figure 1.53(c) shows the correct way to hold and use a cold chisel.

Safety when chipping

- Never chip towards another person
- Always chip towards a chipping screen
- Always wear goggles when chipping
- Always grind the mushroom head off a chisel before using it and make sure the cutting edge is sharp and in good condition. Regrind if necessary.

The chisel shown in Figure 1.53 is only one of many different types of chisel. Some further examples and their applications are shown in Figure 1.54. In the course of a working lifetime most fitters will make up many small chisels for special jobs. These are often made from hardened and tempered silver steel rod. In addition there are the fine engraving chisels or 'gravers' used by die-sinkers and engravers.

The application of the cutting angles we discussed earlier can be applied to a chisel as shown in Figure 1.55. The point angle (wedge angle) and the angle of inclination are only a guide. In practice, a fitter does not work out the angle of inclination or the rake and clearance angle, but uses experience and the feel of the chisel as it cuts through the metal to present the chisel to the work at the correct angle.

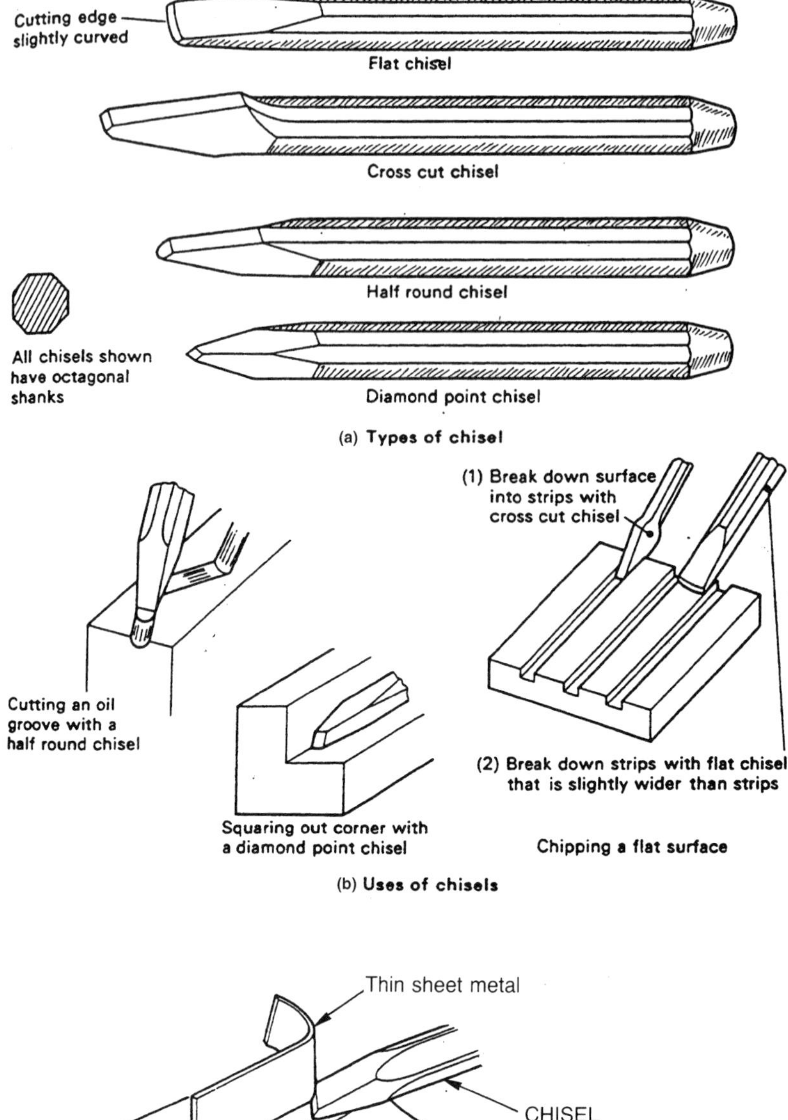

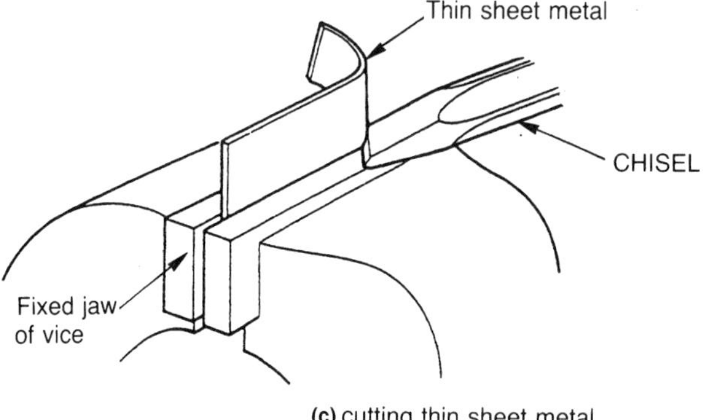

Figure 1.54 *Types of chisel and applications*

Files and filing

Files are the most widely used and important tools for the fitter. The main parts of a file are named in Figure 1.56(a). Files are forged to shape from 1.2% plain carbon steel. After forging, the teeth are machine cut by a chisel shaped tool, as shown in Figure 1.56(b). The teeth of a single-cut are wedge shaped with the rake and clearance angles essential for metal cutting. Most files used

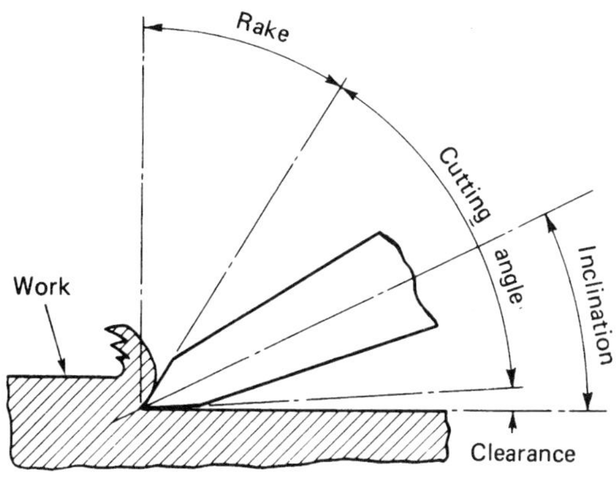

(a) CUTTING ACTION OF CHISEL

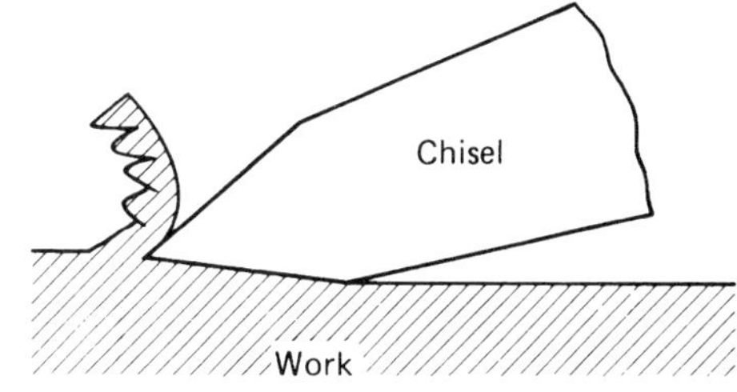

(b) ANGLE OF INCLINATION TOO SMALL CHISEL POINT RISES

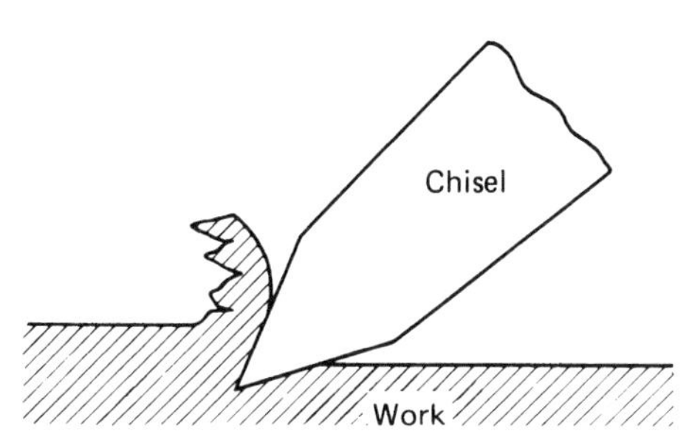

(c) ANGLE OF INCLINATION TOO GREAT CHISEL POINT DIGS IN

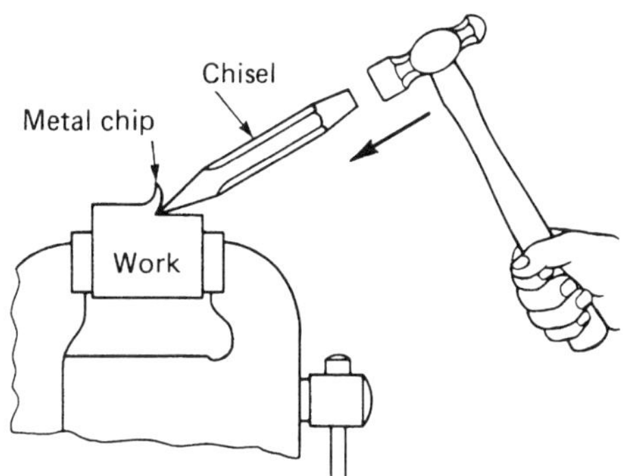

(d) CUT TOWARDS FIXED JAW OF VICE

Typical cutting and inclination angles (clearance angle constant at 7°):

Material	*Point Angle*	*Angle of inclination*
Cast iron	60°	37°
Mild steel	55°	34.5°
Medium carbon steel	65°	39.5°
Brass	50°	32°
Copper	45°	29.5°
Aluminium	30°	22°

Figure 1.55 *Chisel cutting angles*

in general engineering are double-cut. That is they have two rows of cuts at an angle to each other, as shown in Figure 1.56(c).

Files are classified by the following features:

- length
- kind of cut
- grade of cut (roughness)
- profile
- cross-sectional shape or most common use.

The grades of cut are: rough, bastard, second, smooth and dead smooth. These cuts vary with the length of a file. For example a short, second cut file will be smoother than a longer, smooth file. For further information on file cuts, see Table 1.8. The profiles and cross-sectional shapes of some typical files are shown in Figure 1.57.

Figure 1.58 shows how a file should be held and used. To file flat is very difficult and the skill only comes with years of

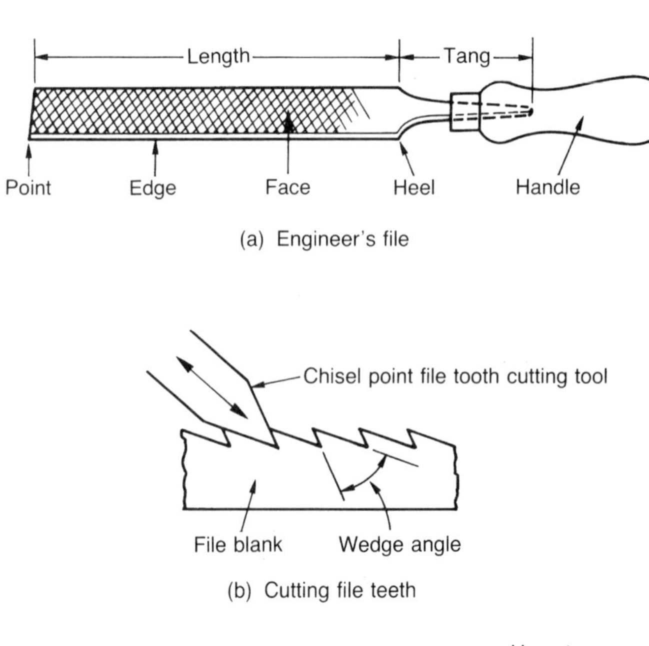

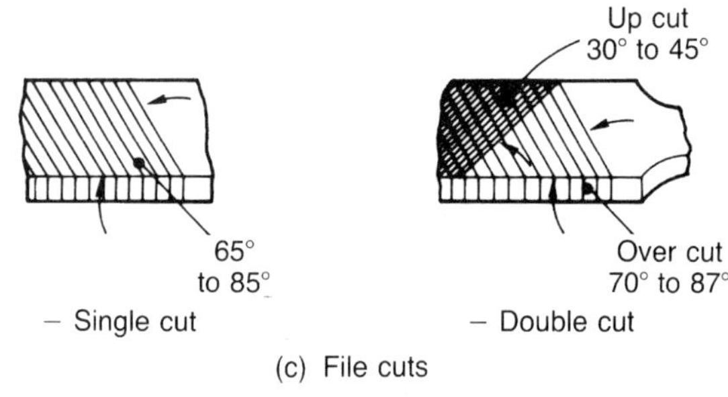

Figure 1.56 *An engineer's file*

Table 1.8

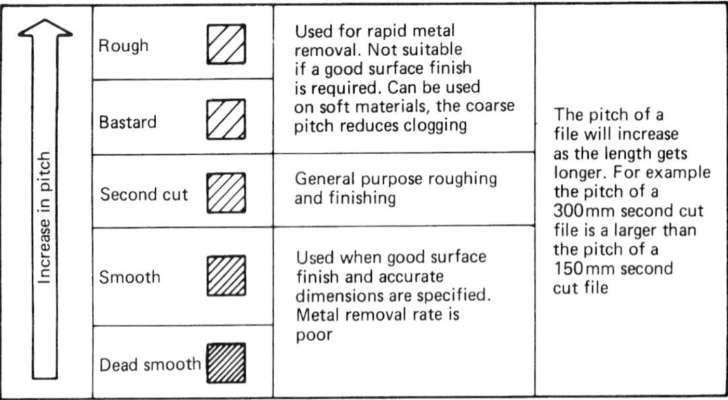

Increase in pitch (↑)	Grade	Use	Note
	Rough	Used for rapid metal removal. Not suitable if a good surface finish is required. Can be used on soft materials, the coarse pitch reduces clogging	The pitch of a file will increase as the length gets longer. For example the pitch of a 300 mm second cut file is a larger than the pitch of a 150 mm second cut file
	Bastard		
	Second cut	General purpose roughing and finishing	
	Smooth	Used when good surface finish and accurate dimensions are specified. Metal removal rate is poor	
	Dead smooth		

continual practice. Cross filing is used for rapid material removal. Draw filing is only a finishing operation to improve the surface finish. It removes less metal per stroke than cross filing and can produce a hollow surface, unless care is taken.

The spaces between the teeth of a file tend to become clogged with bits of metal. This happens mostly when filing soft metals. It is called 'pinning'. The clogged teeth tend to leave heavy score

Test your knowledge 1.18

1. Name FOUR important safety precautions you should take when using a cold chisel.
2. Explain why a cold chisel must be held at the correct angle of inclination to the work.
3. Explain what is meant by a file having a 'safe edge' and name the more common types of file that have such a 'safe edge'.
4. State how the cut of a file varies with its length.
5. Describe how you would produce a datum edge on a sawn blank by filing. The metal is 10 mm wide by 150 mm long.

Type of file	Applications
Square	Filing of keyways and slots
Three square	Filing of angled surfaces
Knife	Filing of acute angles
Hand	These two files are the general-purpose tools for filing flat surfaces and convex profiles
Flat	
Round	Used for enlarging or elongating holes
Half round	Filing of concave profiles

Figure 1.57 *Types of file and applications*

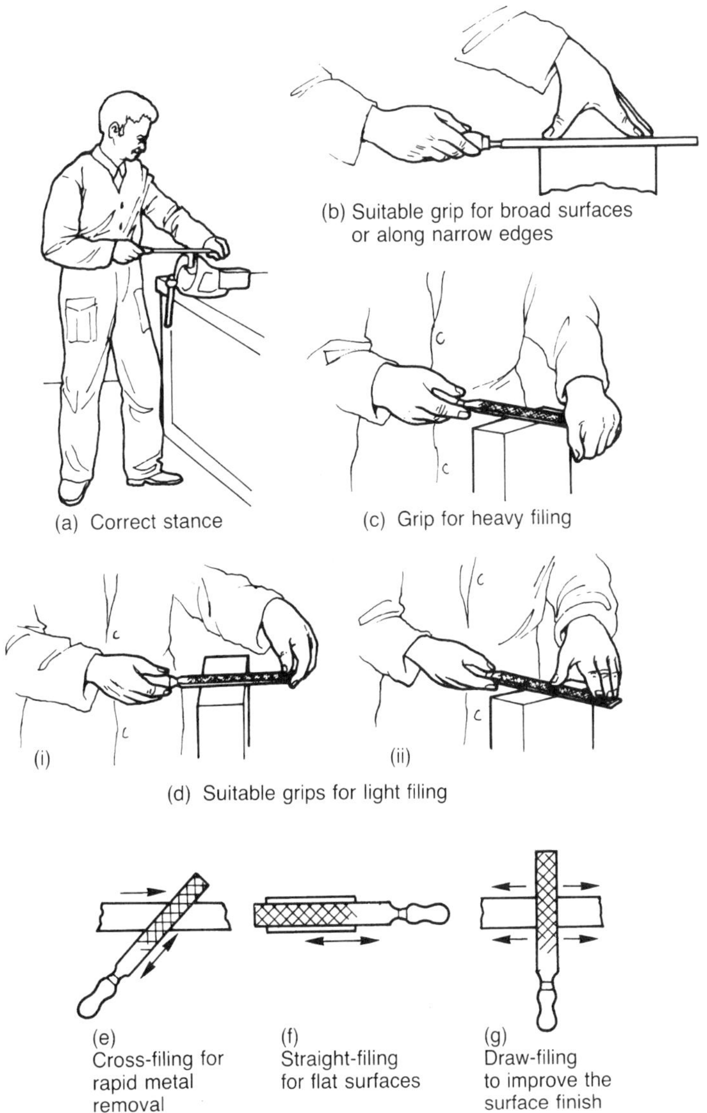

Figure 1.58 *Correct use of a file*

marks in the surface of the work. These marks are difficult to remove. The file should be kept clean and a little chalk should be rubbed into the teeth to prevent pinning. Files are cleaned with a file brush called a 'file card'.

Hacksaws and sawing

A typical hacksaw frame and blade is shown in Figure 1.59(a). The frame is adjustable so that it can be used with blades of various lengths. It is also designed to hold the blade in tension when the wing nut is tightened. The blade is put into the frame so that the teeth cut on the forward stroke. Figure 1.59(b) shows how a hacksaw should be held when being used.

There are a variety of blade types available:

- High speed steel 'all hard' blades are the most rigid and give the most accurate cut. However they are brittle and easily broken when used by an inexperienced person.
- High speed steel 'soft back' blades have a good life and, being more flexible, are less easily broken.
- Carbon steel flexible blades are satisfactory for occasional use on soft non-ferrous metals. They are cheap and not easily broken. Unfortunately they only have a limited life when cutting steels.

To prevent the blade from jamming in the slot it makes as it cuts, all saw blades are given a 'set'. This is shown in Figure 1.60. Coarse pitch hacksaw blades and power-saw blades have the individual teeth set to the left and to the right with either the

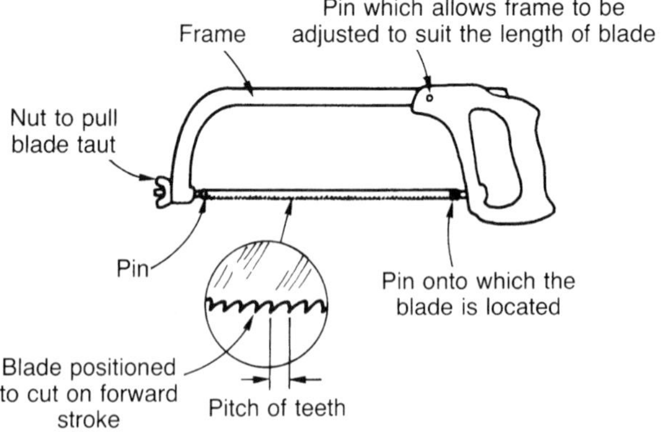

(a) Metal cutting hacksaw

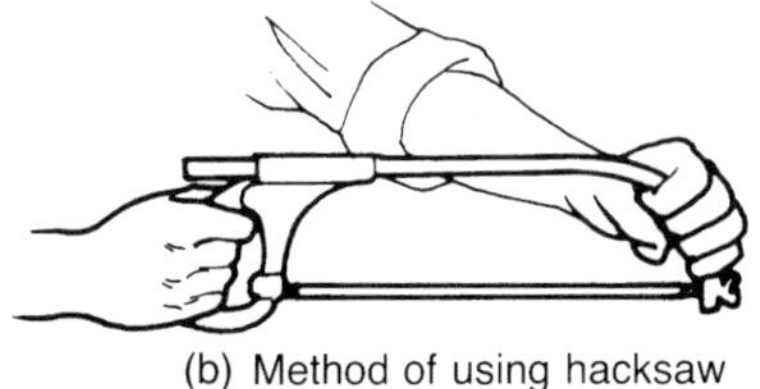

(b) Method of using hacksaw

Figure 1.59 *A typical hacksaw frame and blade*

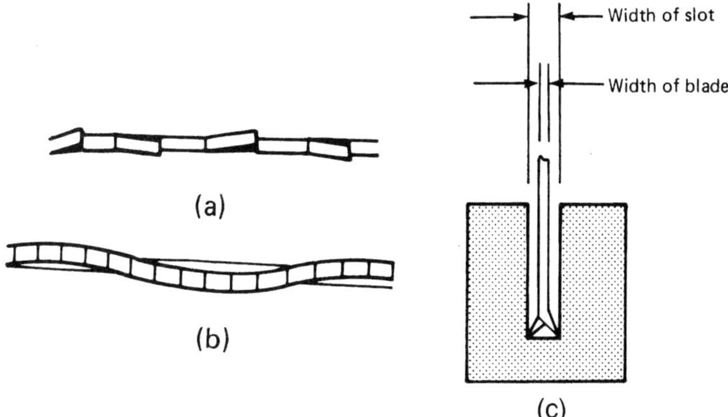

Figure 1.60 *The 'set' of a hacksaw blade*

intermediate teeth or every third tooth left straight to clear the slot. This as shown in Figure 1.60(a). This is not possible with fine pitch blades, and the blade as a whole is given a 'wave' set as shown in Figure 1.60(b). The effect of set on the cut being made is shown in Figure 1.60(c). If you have to change a blade part way through a cut, never continue in the old slot. Because the set of the old blade will have worn, the new blade will jam in the old cut and break. Always start a new cut to the side of the failed cut.

The sizes of hacksaw blades are now given in metric sizes. The length (between the fixing hole centres), the width and the thickness. However, the pitch of the teeth is still given as so many teeth per inch. The fewer teeth per inch the coarser will be the cut, the more quickly will metal will be removed, and the greater will be the set so that there is less chance of the blade jamming. However, there should always be at least three teeth in contact with the work at any one time. Therefore, the thinner the metal being cut the finer the pitch of the blade that should be used. Some typical examples are given in Table 1.9.

Test your knowledge 1.19

1 Explain why:
   (a) If you change a hacksaw blade during a cut you must start a new cut.
   (b) You should use a fine pitch blade when cutting thin metal.

2 With the aid of a sketch explain why it is sometimes necessary to turn a hacksaw blade on its side.

3 Select a suitable hacksaw blade for a craftsperson to use when cutting through a 25 mm square bar.

Table 1.9

Teeth per inch	*Material to be cut*	*Blade application*
32	up to 3 mm	Thin sheets and tubes Hard and soft materials (thin sections)
24	3 mm to 6 mm	Thicker sheets and tubes Hard and soft materials (thicker sections)
18	6 mm to 12 mm	Heavier sections such as mild steel, cast iron, aluminium, brass, copper, bronze
14	12 mm plus	Soft materials of heavy section such as aluminium, brass, copper, bronze

Screw thread cutting

Internal screw threads are cut with taps. A set of straight fluted hand taps are shown in Figure 1.61. The difference between them is the length of the lead. The taper tap should be used first to start the thread. Great care must be taken to ensure that the tap is upright in the hole and it should be checked with a try square. The second tap is used to increase the length of thread and can be used for finishing if the tap passes through the work. The third tap is used for 'bottoming' in blind holes.

The hole to be threaded is called a 'tapping size' hole and it is the same size or only very slightly larger than the core diameter of the thread. Drill diameters for drilling tapping size holes for different screw threads can be found in sets of workshop tables. For example, the tapping size drill for an M10 × 1.5 thread is 8.5 mm diameter.

A tap wrench is used to rotate the taps. There are a variety of different styles available depending upon the size of the taps. An example of a suitable wrench for small taps is shown in Figure 1.62. Taps are very fragile and are easily broken, particularly in the small sizes. Once a tap has been broken into a hole, it is virtually impossible to get it out without damaging or destroying the workpiece.

Taps are relatively expensive and should be looked after carefully. High speed steel ground thread taps are the most expensive. However, they cut very accurate threads and, with careful use, last a long time. Carbon steel cut thread taps are less accurate and less expensive and have a reasonable life when cutting the softer non-ferrous metals. Whichever sort of taps are used, they should always be well lubricated. Traditionally tallow was used, but nowadays proprietary screw-cutting lubricants are available that are more effective.

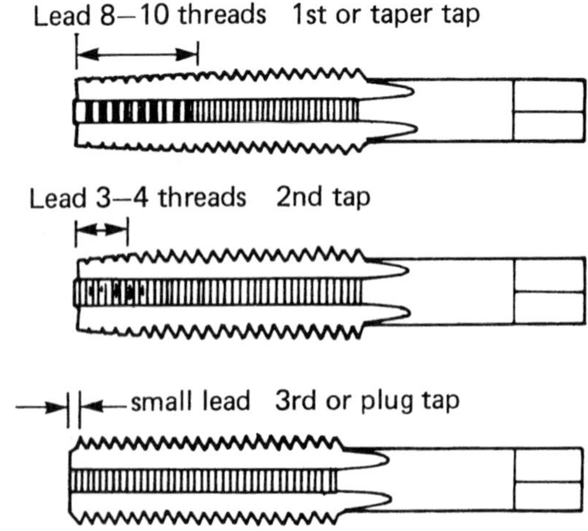

Figure 1.61 *Hand taps*

Figure 1.62 *A tap wrench*

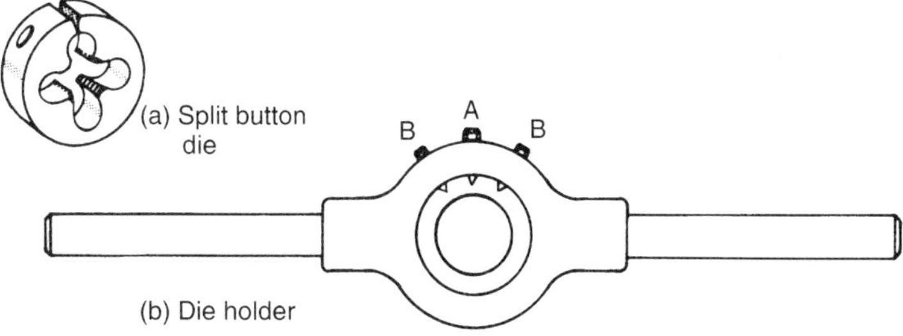

Figure 1.63 *A die holder*

External threads are cut using split button dies in a die holder, as shown in Figure 1.63. One face of the die is always marked up with details of the thread and the maker's logo. This should be visible when the die is in the die-holder. Then the lead is on the correct side for starting the cut. Screw A is used to spread the die for the first cut. The screws marked B are used to close the die until it gives the correct finishing cut. This is judged by using a standard nut or a screw thread gauge. The nut or gauge should run up the thread without binding or without undue looseness.

Again, the die must be started square with the workpiece or a 'drunken' thread will result. Also, a thread cutting lubricant should be used. Like thread cutting taps, dies are available in carbon steel cut thread and high speed steel ground thread types. For both taps and dies, each set only cuts one size and pitch of thread and one thread form.

Spanners and keys

In addition to cutting tools a fitter should also have a selection of spanners and keys available for dismantling and assembly purposes. Figure 1.64 shows a selection of spanners and keys. These are carefully proportioned so that a person of average strength will be able to tighten a screwed fastening correctly.

Use of a piece of tubing to extend a spanner or key is very bad practice. It strains the jaws of the spanner so that it becomes loose and may slip. It may even crack the jaws of the spanner so that they break. In both cases this can lead to nasty injuries to your hands and even a serious fall if you are working on a ladder. Also it over stresses the fastening which will be weakened

Test your knowledge 1.20

1 Give TWO reasons why a tap must be started square in its hole.

2 Explain how a button die is adjusted to cut the required diameter of thread.

3 With reference to workshop tables, select tapping size drills for the following threads:
(a M8 × 1.0,
(b) M12 × 1.75
(c) 4 BA.

4 Explain why important fastenings are tightened with a torque spanner.

5 Sketch and name the spanners and keys shown in Figure 1.64.

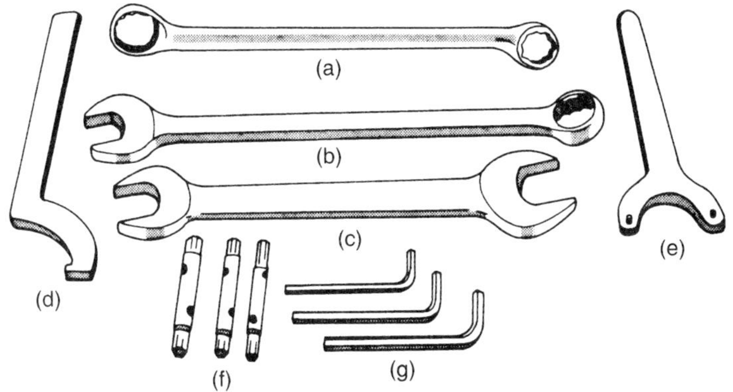

Figure 1.64 *Selection of spanners and keys*

or even broken. Always check a spanner for damage and correct fit before using it. A torque spanner should be used to tighten important fastenings.

Activity 1.7

Select suitable hand tools (giving reasons for your choice), and draw up an operation schedule for making the component shown in Figure 1.65.

Drilling

This is a process for producing holes. The holes may be cut from the solid or existing holes may be enlarged. The purpose of the drilling machine is to:

- Rotate the drill at a suitable speed for the material being cut and the diameter of the drill.
- Feed the drill into the workpiece.
- Support the workpiece being drilled; usually at right-angles to the axis of the drill. On some machines the table may be tilted to allow holes to be drilled at a pre-set angle.

Drilling machines

Drilling machines come in a variety of types and sizes. Figure 1.66 shows a hand held, electrically driven, power drill. It

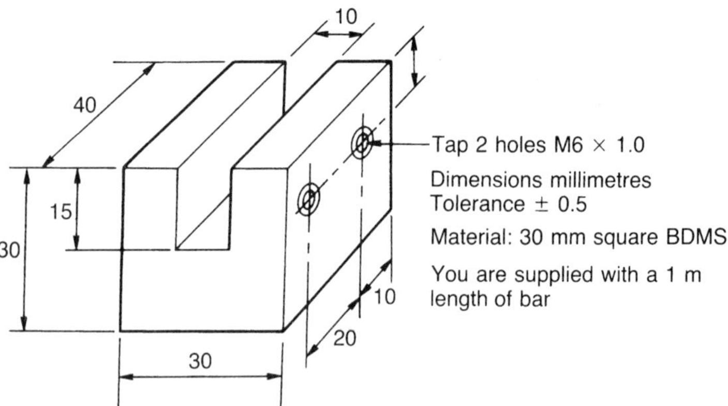

Figure 1.65 *See Activity 1.7*

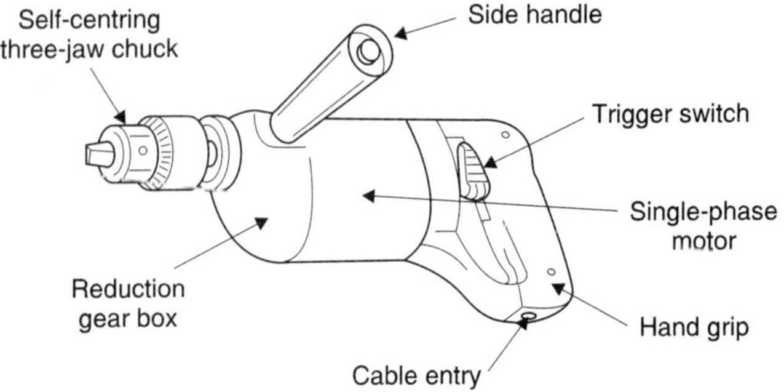

Figure 1.66 *An electric power drill*

depends upon the skill of the operator to ensure that the drill cuts at right-angles to the workpiece. The feed force is also limited to the muscular strength of the user. Figure 1.67 shows a more powerful, floor mounted machine. The spindle rotates the drill. It can also move up and down in order to feed the drill into the workpiece and withdraw the drill at the end of the cut. Holes

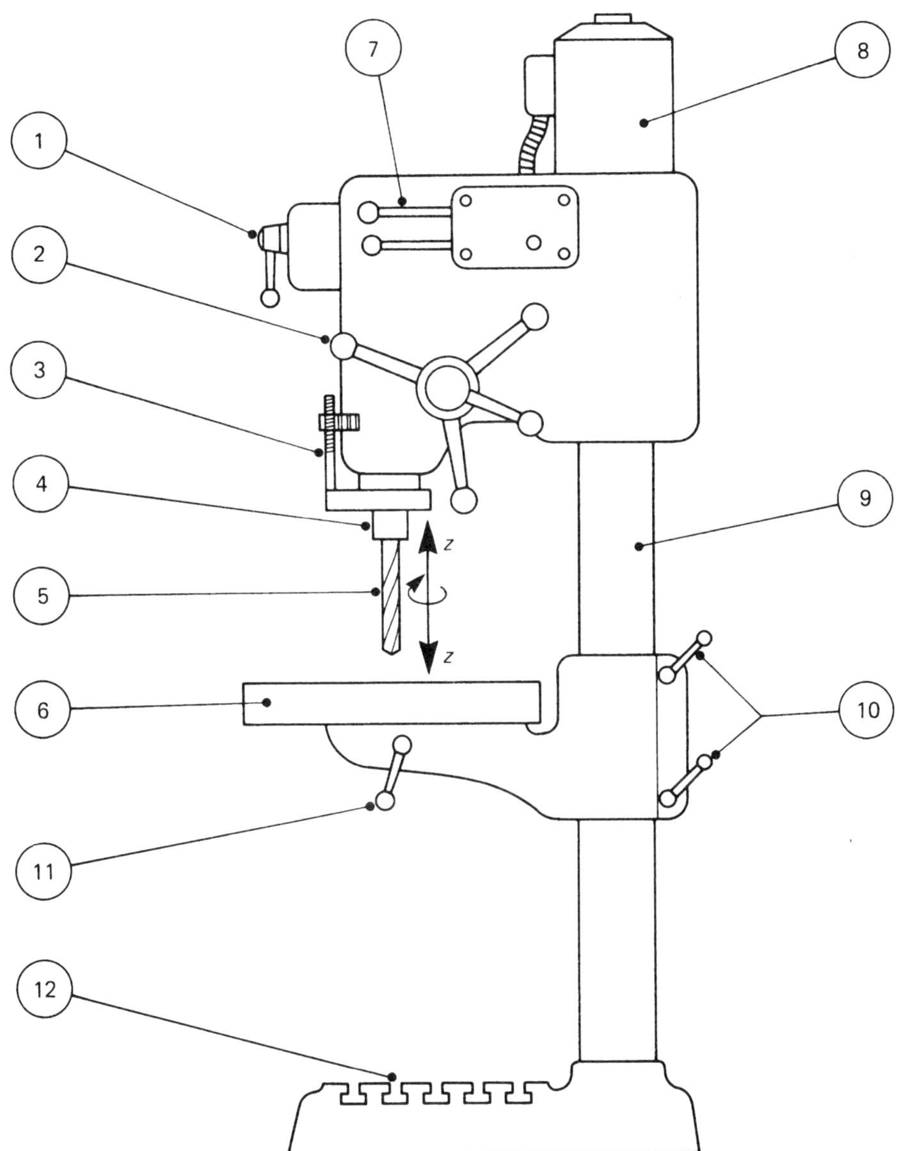

Parts of the Pillar Type Drilling Machine

1 Stop/start switch (electrics).
2 Hand or automatic feed lever.
3 Drill depth stop.
4 Spindle.
5 Drill.
6 Table.
7 Speed change levers.
8 Motor.
9 Pillar.
10 Vertical table lock.
11 Table lock.
12 Base.

Figure 1.67 *A pillar drill*

are generally produced with twist drills. Figure 1.68 shows a typical straight shank drill and a typical taper shank drill and names their more important features.

Large drills have taper shanks and are inserted directly into the spindle of the machine, as shown in Figure 1.69(a). They are located and driven by a taper. The tang of the drill is for extraction purposes only. It does not drive the drill. The use of a drift to remove the drill is shown in Figure 1.69(b).

Small drills have straight (parallel) shanks and are usually held in a self-centring chuck. Such a chuck is shown in Figure 1.69(c). The chuck is tightened with the chuck key shown. SAFETY: The chuck key must be removed before starting the machine. The drill chuck has a taper shank which is located in, and driven by, the taper bore of the drilling machine spindle.

The cutting edge of a twist drill is wedge-shaped, like all the tools we have considered so far. This is shown in Figure 1.70.

When regrinding a drill it is essential that the point angles are correct. The angles for general purpose drilling are shown in Figure 1.71(a). After grinding, the angles and lip lengths must be checked as shown in Figure 1.71(b). The point must be symmetrical. The effects of incorrect grinding are shown in Figure 1.71(c).

If the lip lengths are unequal, an oversize hole will be drilled when cutting from the solid. If the angles are unequal, then only one lip will cut and undue wear will result. The unbalanced

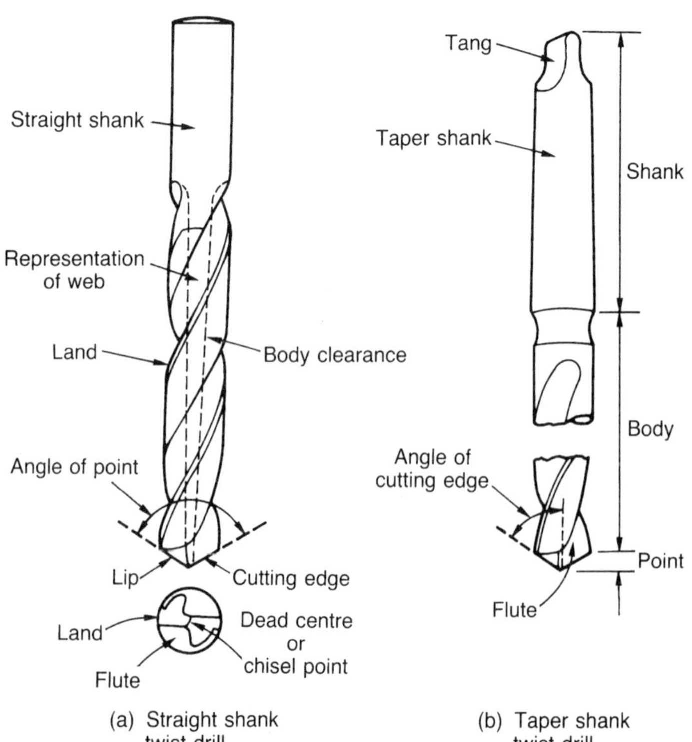

Figure 1.68 *Straight shank and taper shank twist drills*

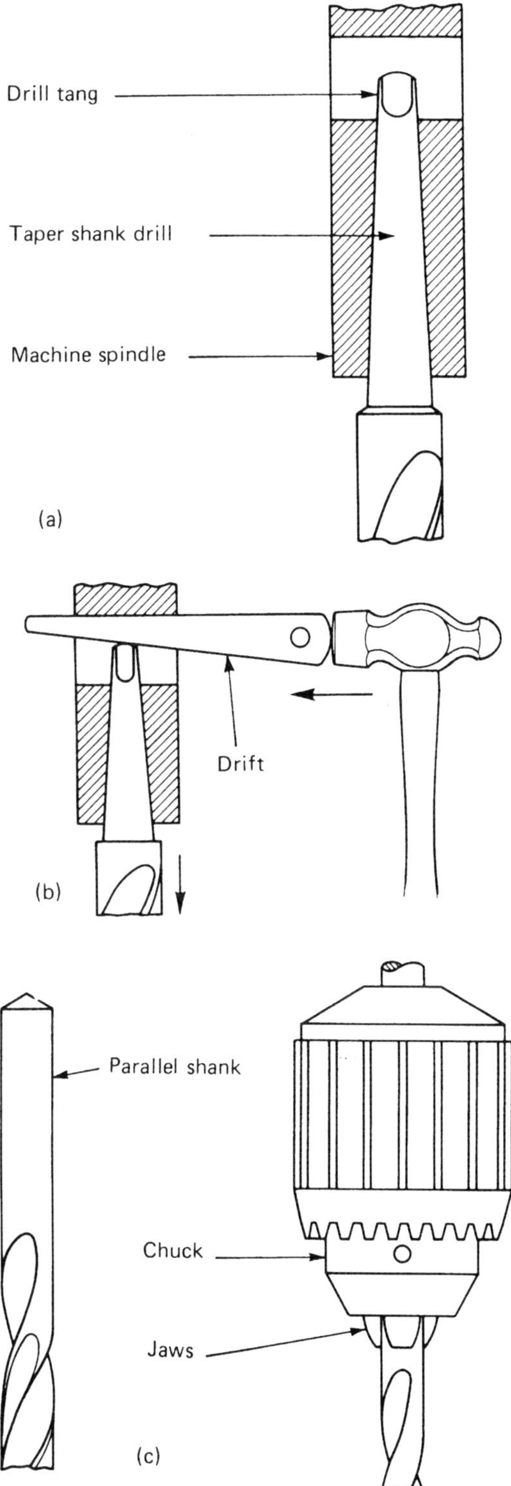

Figure 1.69 *Methods of holding a twist drill*

forces will cause the drill to flex and 'wander'. The axis of the hole will become displaced as drilling proceeds. If both these faults are present at the same time, both sets of faults will be present and an inaccurate and ragged hole will result.

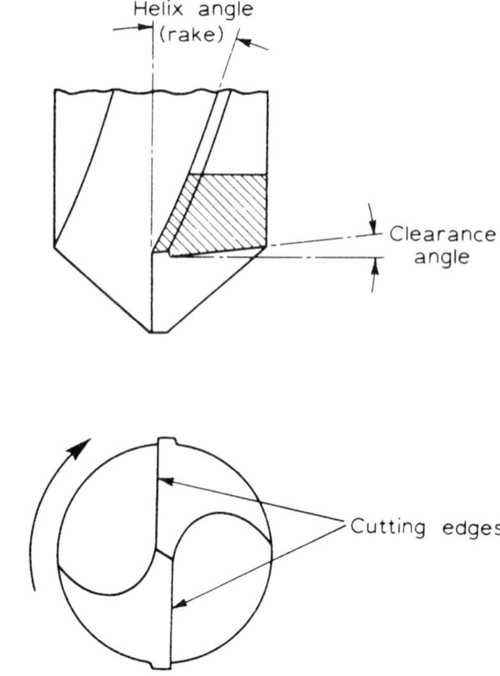

Figure 1.70 *Cutting edges of a twist drill*

Work-holding when drilling

It is dangerous to hold work being drilled by hand. There is always a tendency for the drill to grab the work and spin it round. Also the rapidly spinning swarf can produce some nasty cuts to the back of your hand. Therefore the work should always be securely fastened to the machine table.

Nevertheless, small holes in relatively large components are sometime drilled with the work handheld. In this case a stop bolted to the machine table should be used to prevent rotation.

Small work is usually held in a machine vice which, in turn, is securely bolted to the machine table. This is shown in Figure 1.72(a).

Larger work can be clamped directly to the machine table, as shown in Figure 1.72(b). In both these latter two examples the work is supported on parallel blocks. You mount the work in this way so that when the drill 'breaks through' the workpiece it does not damage the vice or the machine table.

Figure 1.72(c) shows how an angle plate can be used when the hole axis has to be parallel to the datum surface of the work. Finally, Figures 1.73(a) and 1.73(b) show how cylindrical work is located and supported using vee-blocks.

Miscellaneous drilling operations

Figure 1.74 shows some miscellaneous operations that are frequently carried out on drilling machines.

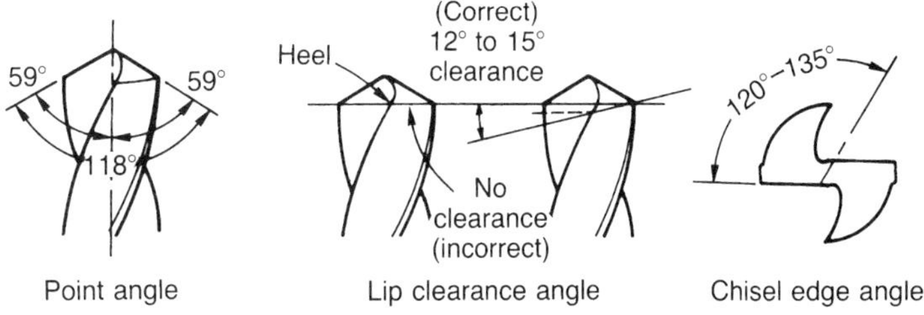

(a) Drill angles for general purpose drilling

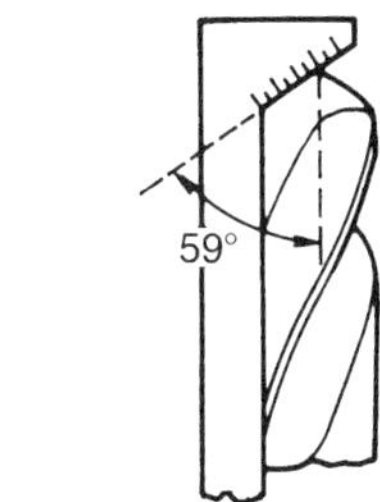

(b) Checking for correct point angle and equal lip lengths

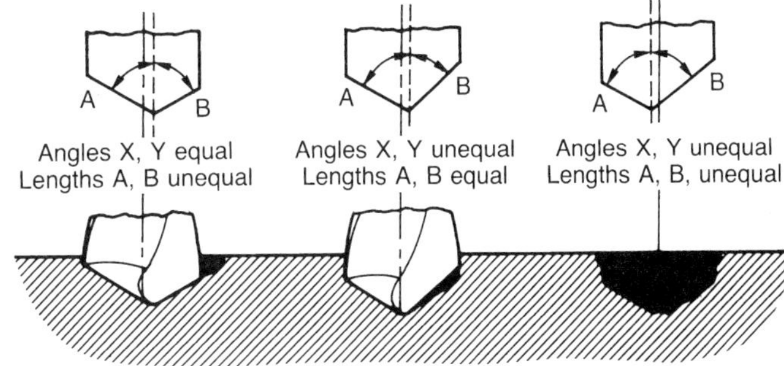

(c) Effects of incorrect grinding

Figure 1.71 *Point angles for a twist drill*

Test your knowledge 1.21

1. For clarity, no drill guards have been shown in any of the previous examples:
   (a) Sketch a typical drilling machine spindle/chuck/drill guard
   (b) Describe briefly the most common type of accident resulting from the use of an unguarded drilling machine spindle and drill.
2. Describe ONE method of locating the work under a drill so that the hole will be cut in the required position.
3. When drilling explain why:
   (a) the packing pieces must be parallel and the same size
   (b) the work should not be held by hand when drilling large diameter holes.

Countersinking

Figure 1.74(a) shows a countersink bit being used to countersink a hole to receive the heads of rivets or screws. For this reason the included angle is 90°. Lathe centre drills are unsuitable for this operation as their angle is 60°.

Counterboring

Figure 1.74(b) shows a piloted counterbore being used to counterbore a hole so that the head of a capscrew or a cheese-head screw can lie below the surface of the work. Unlike a countersink cutter, a counterbore is not self-centring. It has to have a pilot which runs in the previously drilled bolt or screw hole. This keeps the counterbore cutting concentrically with the original hole.

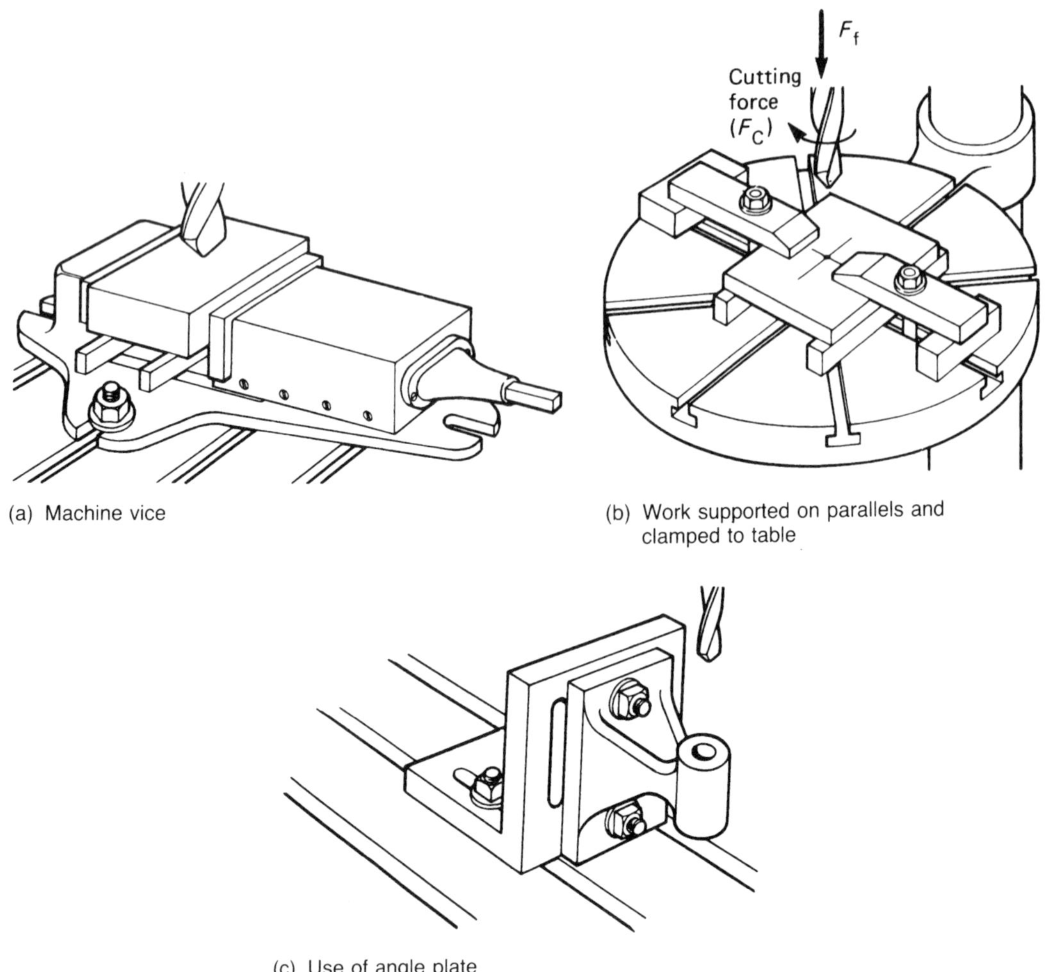

(a) Machine vice

(b) Work supported on parallels and clamped to table

(c) Use of angle plate

Figure 1.72 *Work-holding*

Spot-facing

This is similar to counterboring but the cut is not as deep. It is used to provide a flat surface on a casting or a forging for a nut and washer to seat on. Sometimes, as shown in Figure 1.74(c), it is used to machine a boss (raised seating) to provide a flat surface for a nut and washer to seat on.

Centre lathe turning

The main purpose of a centre lathe is to produce external and internal cylindrical and conical (tapered) surfaces. It can also produce plain surfaces and screw threads.

The centre lathe

Figure 1.75(a) shows a typical centre lathe and identifies its more important parts.

- The bed is the base of the machine to which all the other sub-assemblies are attached. Slideways accurately machined on its top surface provide guidance for the saddle and the

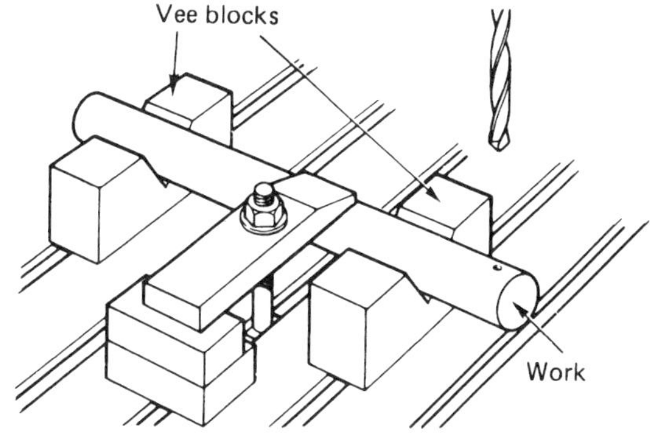

(a) HORIZONTAL

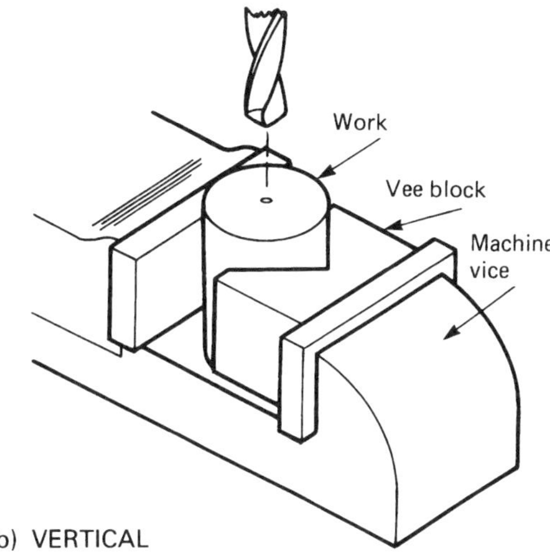

(b) VERTICAL

Figure 1.73 *Work-holding cylindrical components*

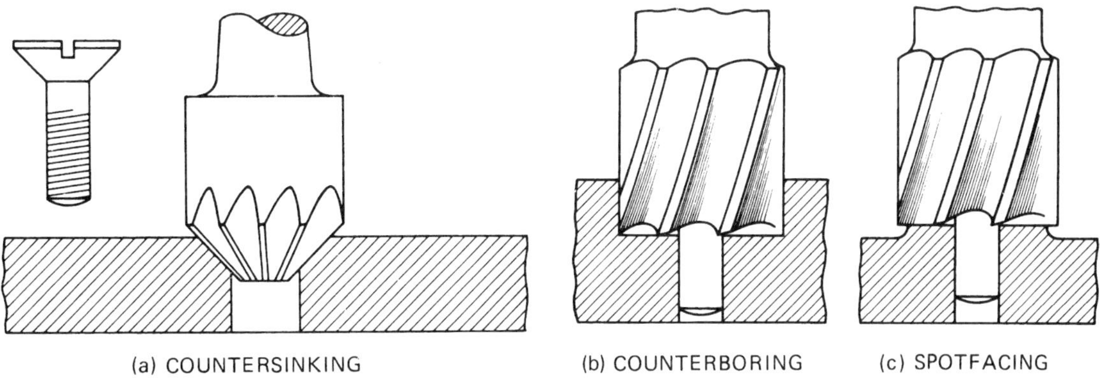

(a) COUNTERSINKING (b) COUNTERBORING (c) SPOTFACING

Figure 1.74 *Countersinking, counterboring and spot facing*

tail stock. These slideways also locate the headstock so that the axis of the spindle is parallel with the movement of the saddle and the tailstock. The saddle or carriage of the lathe moves parallel to the spindle axis as shown in Figure 1.75(b).

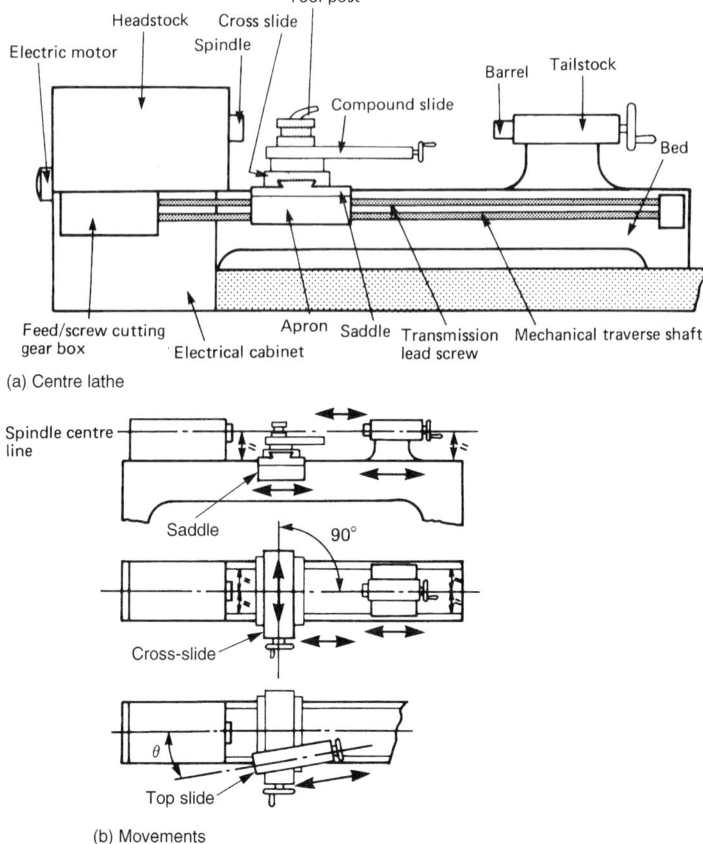

Figure 1.75 *A centre lathe*

- The cross-slide is mounted on the saddle of the lathe. It moves at 90° to the axis of the spindle, as shown in Figure 1.75(c). It provides in-feed for the cutting tool when cylindrically turning. It is also used to produce a plain surface when facing across the end of a bar or component.
- The top-slide (compound-slide) is used to provide in-feed for the tool when facing. It can also be set at an angle to the spindle axis for turning tapers, as shown in Figure 1.75(d).

Table 1.10

Cutting movement	*Hand or power tranverse*	*Means by which movement is achieved*	*Turned feature*
Tool parallel to the spindle centre	Both	The saddle moves along the bed slideways	A parallel cylinder
Tool at 90° to the spindle centre line	Both	The cross slide moves along a slideway machined on the top of the saddle	A flat face square to the spindle centre line
Tool at an angle relative to the spindle centre line	Hand	The compound slide is rotated and set at the desired angle relative to the centre line	A tapered cone

Work-holding in the lathe

The work to be turned can be held in various ways. We will now consider the more important of these.

Between centres

The centre lathe derives its name from this basic method of work-holding. The general layout is shown in Figure 1.76(a). Centre holes are drilled in the ends of the bar and these locate on centres in the headstock spindle and the tailstock barrel. A section through a correctly centred component is shown in Figure 1.76(b). The centre-hole is cut with a standard centre-

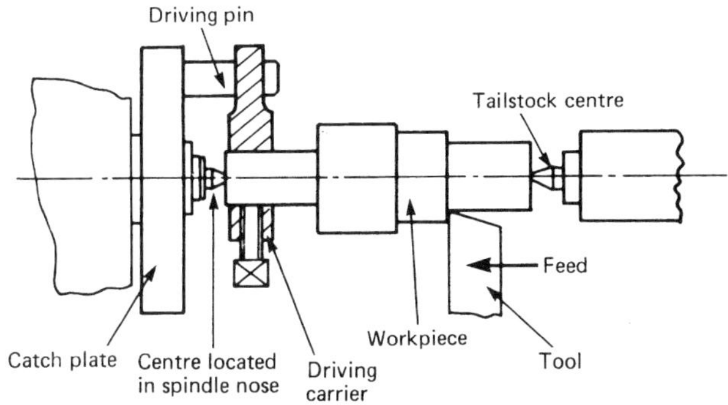

(a) Turning between centres

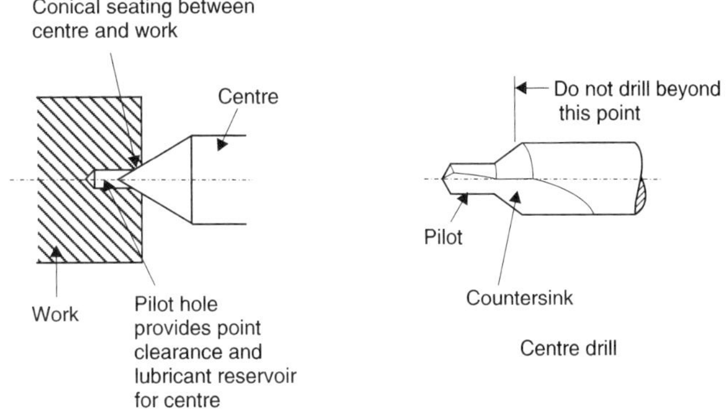

(b) The centre-hole

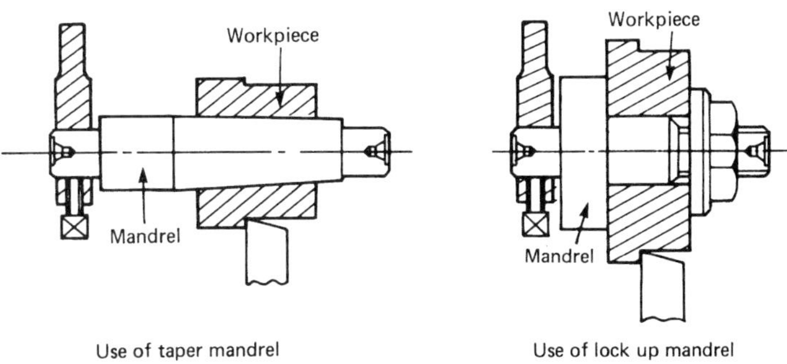

(c) Use of mandrels

Figure 1.76 *Work-holding in a centre lathe*

drill. The main disadvantage of this method of workholding is that no work can be performed on the end of the component. Work that has been previously bored can be finish turned between centres on a taper mandrel as shown in Figure 1.76(c).

Four-jaw chuck

Chucks are mounted directly onto the spindle nose and hold the work securely without the need for a back centre. This allows the end of the work to be faced flat. It also allows for the work to have holes bored into it or through it.

In the four-jaw chuck, the jaws can be moved independently by means of jack-screws. As shown in Figure 1.77(a), the jaws can also be reversed and the work held in various ways, as shown in Figure 1.77(b). As well as cylindrical work, rectangular work can also be held, as shown in Figure 1.77(c). Because the jaws can be moved independently, the work can be set to run concentrically with the spindle axis to a high degree of accuracy. Alternatively the work can be deliberately set off-centre to produce eccentric components as shown in Figure 1.77(d).

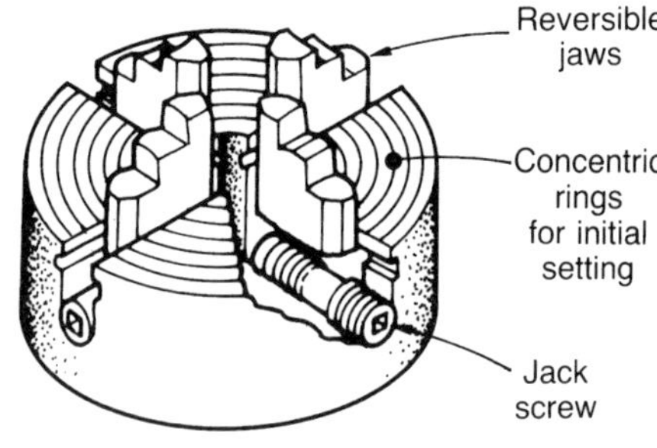

(a) Independent four-jaw chuck

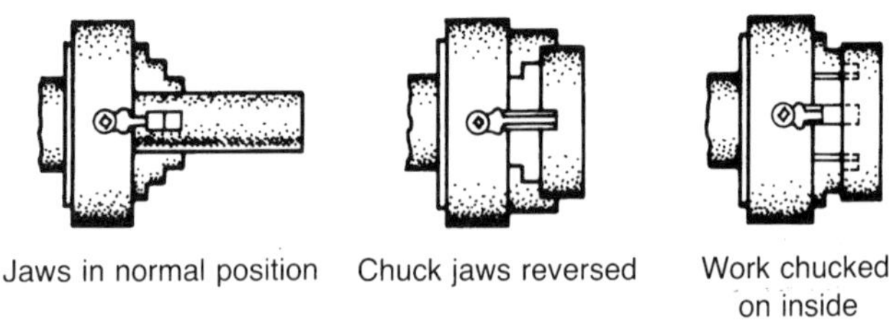

(b) Methods of holding work in a chuck

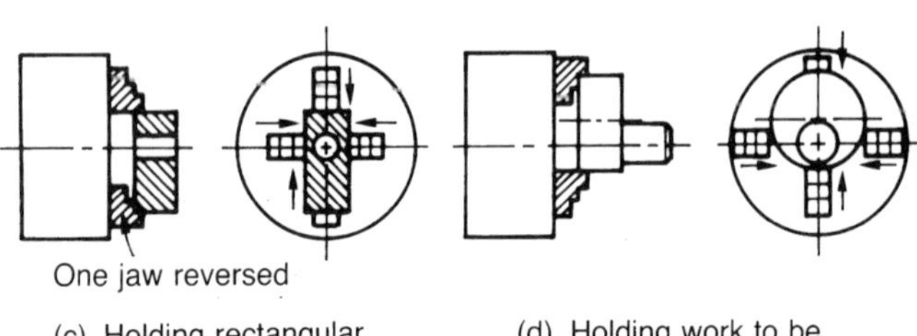

(c) Holding rectangular work

(d) Holding work to be turned eccentrically

Figure 1.77 *Four-jaw chuck*

Three-jaw chuck

The self-centring, three-jaw chuck is shown in Figure 1.78(a). The jaws are set at 120° and are moved in or out simultaneously (at the same time) by a scroll when the key is turned. SAFETY: This key must be removed before starting the lathe or a serious accident can occur. When new and in good condition this type of chuck can hold cylindrical and hexagonal work concentric with the spindle axis to a high degree of accuracy. In this case the jaws are not reversible, so it is provided with separate internal and external jaws. In Figure 1.78(a) the internal jaws are shown in the chuck, and the external jaws are shown at the side of the chuck. Again the chuck is mounted directly on the spindle nose of the lathe.

Work to be turned between centres is usually held in a three-jaw chuck whilst the ends of the bar are faced flat and then centre drilled, as shown in Figure 1.78(b).

Face-plate

Figure 1.79 shows a component held on a face-plate so that the hole can be bored perpendicularly to the datum surface. This

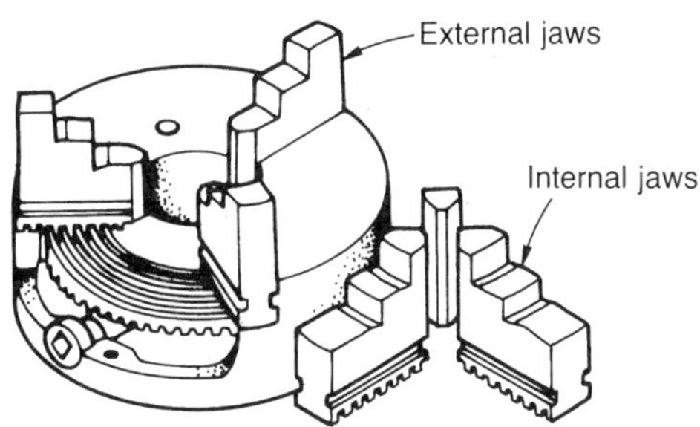

(a) Three jaw self-centring chuck

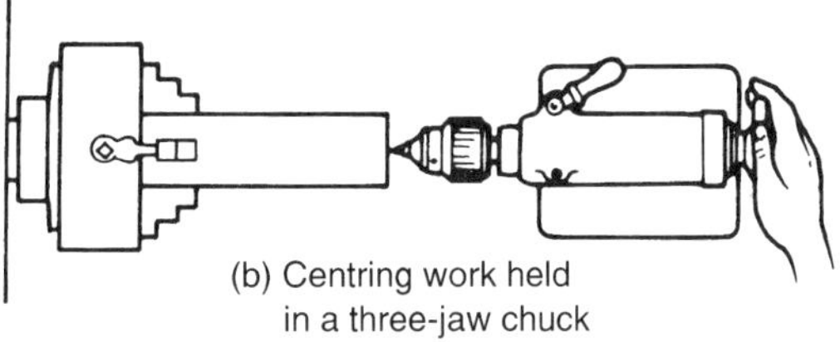

(b) Centring work held in a three-jaw chuck

Figure 1.78

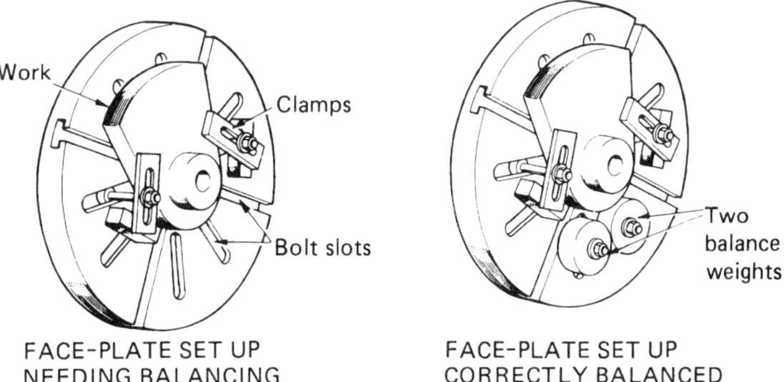

FACE-PLATE SET UP NEEDING BALANCING

FACE-PLATE SET UP CORRECTLY BALANCED

Figure 1.79 *Three-jaw chuck*

datum surface is in contact with the face-plate. Note that the face-plate has to be balanced to ensure smooth running. Care must be taken to check that the clamps will hold the work securely and do not foul the machine. The clamps must not only resist the cutting forces, but they must also prevent the rapidly rotating work from spinning out of the lathe.

Turning tools

Figure 1.80(a) shows a range of turning tools and some typical applications. Figure 1.80(b) shows how the metal-cutting wedge also applies to turning tools. Turning tools are fastened into a tool-post which is mounted on the top slide of the lathe. There are many different types of tool-post, and a four-way turret tool-post is shown in Figure 1.80(c). This allows four tools to be mounted at any one time, each of which can be quickly swung into position ready for cutting.

Parallel turning

Figure 1.81(a) shows a long bar held between centres. To ensure that the work is truly cylindrical with no taper, the axis of the tailstock centre must be in line with the axis of the headstock spindle. The saddle traverse provides movement of the tool parallel with the workpiece axis. You take a test cut and measure the diameter of the bar at both ends. If all is well, the diameter should be constant all along the bar. If not, the lateral movement of the tailstock needs to be adjusted until a constant measurement is obtained. The depth of cut is controlled by micrometer adjustment of the cross-slide.

Whilst facing and centre drilling the end of a long bar, a fixed steady is used. This supports the end of the bar remote from the chuck. A fixed steady is shown in Figure 1.81(b). If the work is long and slender it sometimes tries to kick away from the turning tool or even climb over the tool. To prevent this happening a travelling steady is used. This is bolted to the saddle opposite to the tool, as shown in Figure 1.81(c).

Surfacing

A surfacing (facing or perpendicular-turning) operation on a workpiece held in a chuck is shown in Figure 1.82. The saddle is clamped to the bed of the lathe and the tool motion is controlled by the cross-slide. This ensures that the tool moves in a path at right-angles to the workpiece axis and produces a plain surface. In-feed of the cutting tool is controlled by micrometer adjustment of the top-slide.

Boring

Figure 1.83 shows how a drilled hole can be opened up using a boring tool. The workpiece is held in a chuck and the tool

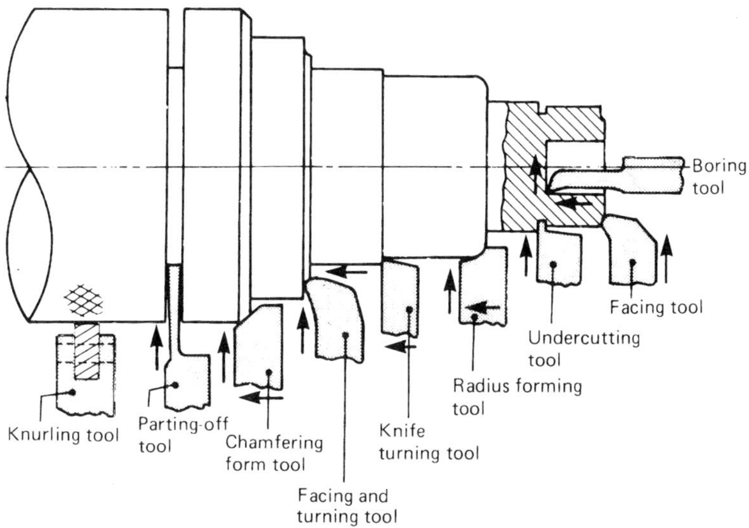

(a) Turning tools

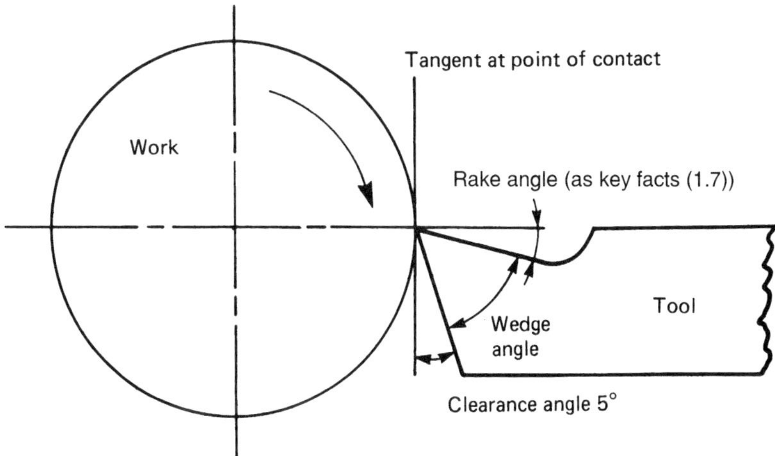

(b) Turning tool angles

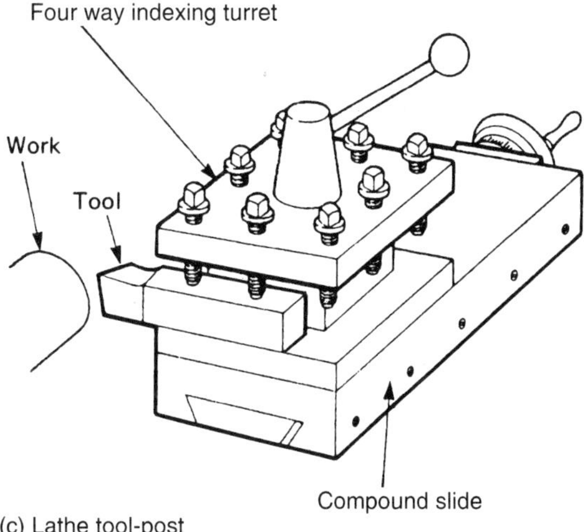

(c) Lathe tool-post

Figure 1.80 *Turning tools*

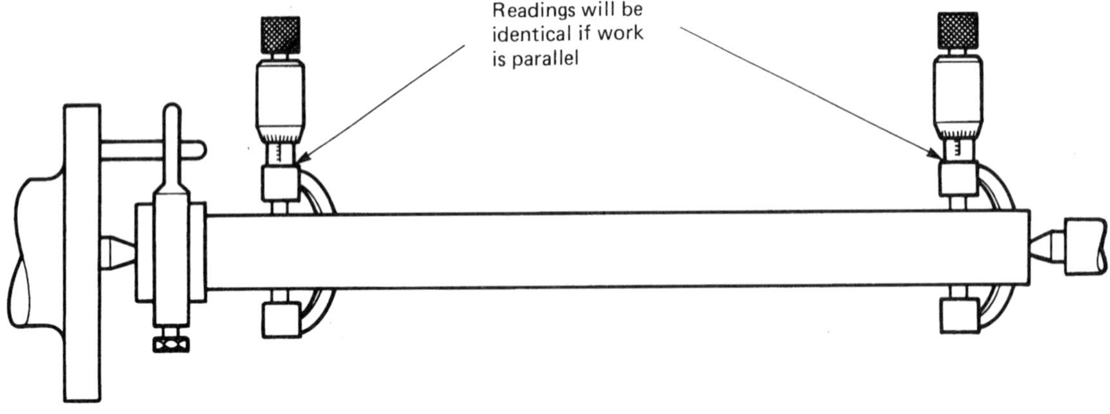

a) Checking for parallelism

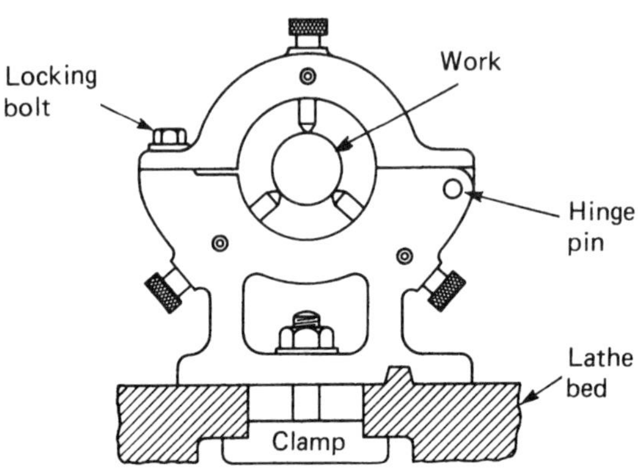

(b) Fixed steady

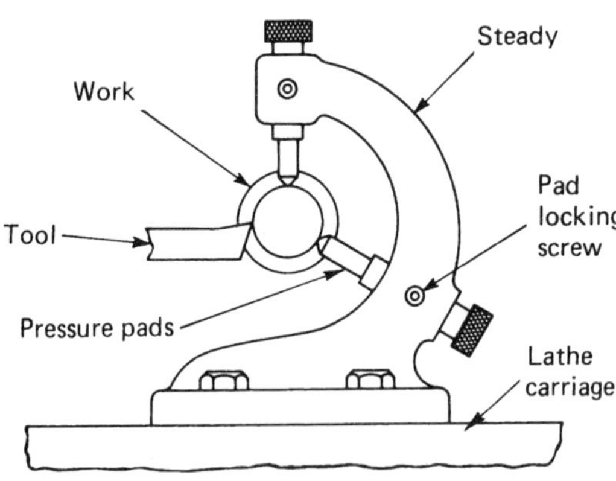

(c) Travelling steady

Figure 1.81 *Parallel turning*

movement is controlled by the saddle of the lathe. The in-feed of the tool is controlled by micrometer adjustment of the cross-slide. The pilot hole is produced either by a taper shank drill mounted directly into the tailstock barrel (poppet), or by a parallel shank drill held in a drill chuck. The taper mandrel of the drill chuck is inserted into the tailstock barrel.

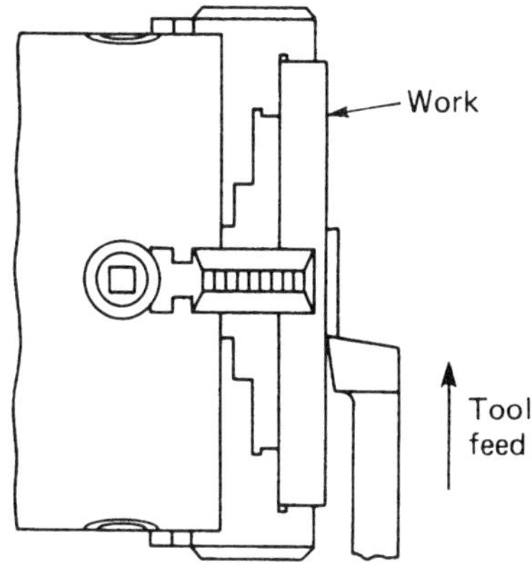

Figure 1.82 *Surfacing*

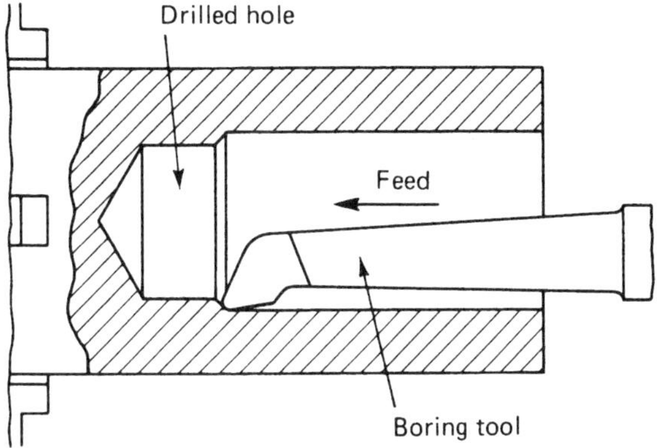

Figure 1.83 *Boring*

Conical surfaces

Chamfers on the corners of a turned component are short conical surfaces. These are usually produced by using a chamfering tool, as shown in Figure 1.80(a). Longer tapers can be produced by use of the top-slide. Use of the top (compound) slide is shown in Figure 1.84. The slide is mounted on a swivel base and it is fitted with a protractor scale. It can be swung round to the required angle and clamped in position. The taper is then cut as shown.

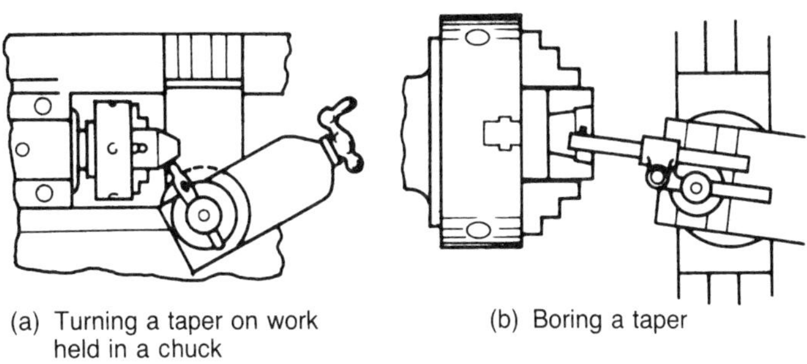

(a) Turning a taper on work held in a chuck

(b) Boring a taper

Figure 1.84 *Producing a chamfer*

Miscellaneous turning operations

Reamers are sizing tools. They remove very little metal. Since they follow the existing hole, they cannot correct the positional errors. Hand reamers have a square on the end of their shanks so that they can be rotated by a tap wrench. Machine reamers have a standard morse taper shank.

Figure 1.85(a) shows a hole being reamed in a lathe. A machine reamer is being used and it is held in the barrel of the tailstock. Because a drilled hole invariably runs out slightly, the pilot hole should be single-point bored in order to correct the position and geometry of the hole. It is finally sized using the reamer. Only the minimum amount of metal for the surface of the hole to clean up should be left in for the reamer to remove.

Standard, non-standard and large diameter screw threads can be cut in a centre lathe by use of the lead-screw to control the saddle movement. This is a highly skilled operation.

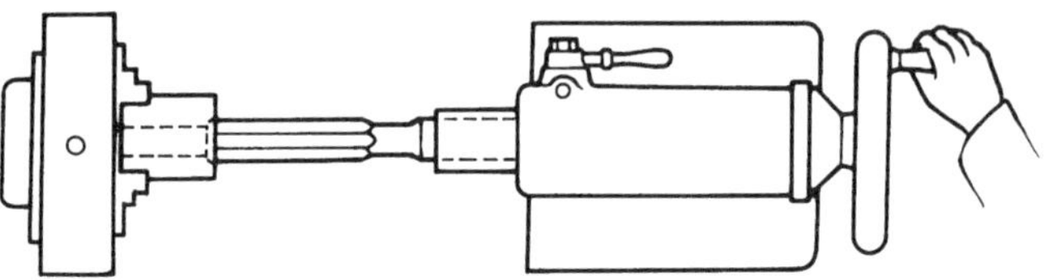

(a) Machine reamer supported in the tailstock

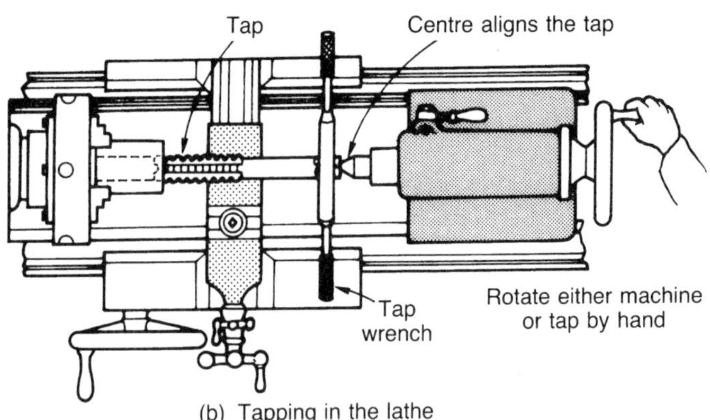

(b) Tapping in the lathe

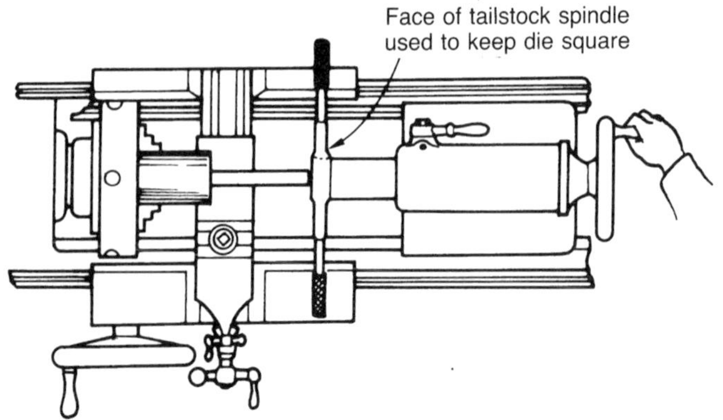

(c) Threading in the lathe with hand dies

Figure 1.85 *Miscellaneous turning operations*

Test your knowledge 1.22

1 List FIVE main operations that can be performed on a centre lathe.

2 Select suitable tools and measuring equipment (giving reasons for your choice) and draw up an operation schedule for turning the component shown in Figure 1.86(a).

3 With the aid of a sketch explain how you would hold the component shown in Figure 1.86(b) and machine the 50 mm diameter hole.

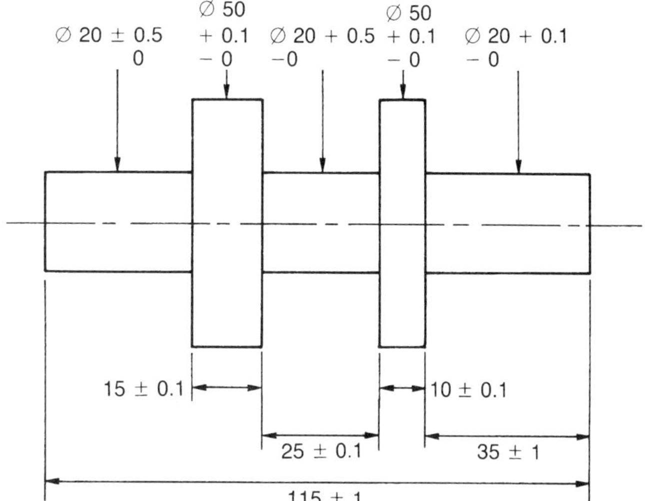

Dimensions in millimetres
Material: Phosphor bronze: Blank size ∅ 60 × 125
Hint: Turn between centres, after facing and centring

(a)

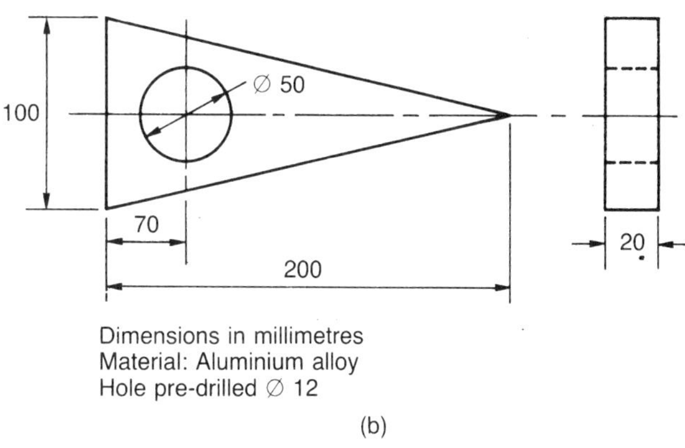

Dimensions in millimetres
Material: Aluminium alloy
Hole pre-drilled ∅ 12

(b)

Figure 1.86 *See Test your knowledge 1.22*

However, standard screw threads of limited diameter can be cut using hand threading tools as shown in Figures 1.85(b) and 1.85(c). Taps are very fragile and the workpiece should be rotated by hand with the lathe switched off and the gears disengaged.

Milling

Milling machines are mainly used to produce plain surfaces parallel and perpendicular to the machine table using rotating multi-tooth cutters.

The horizontal milling machine

This machine gets its name from the fact that the axis of the spindle lies in the horizontal plane. An example of such a machine is shown in Figure 1.87.

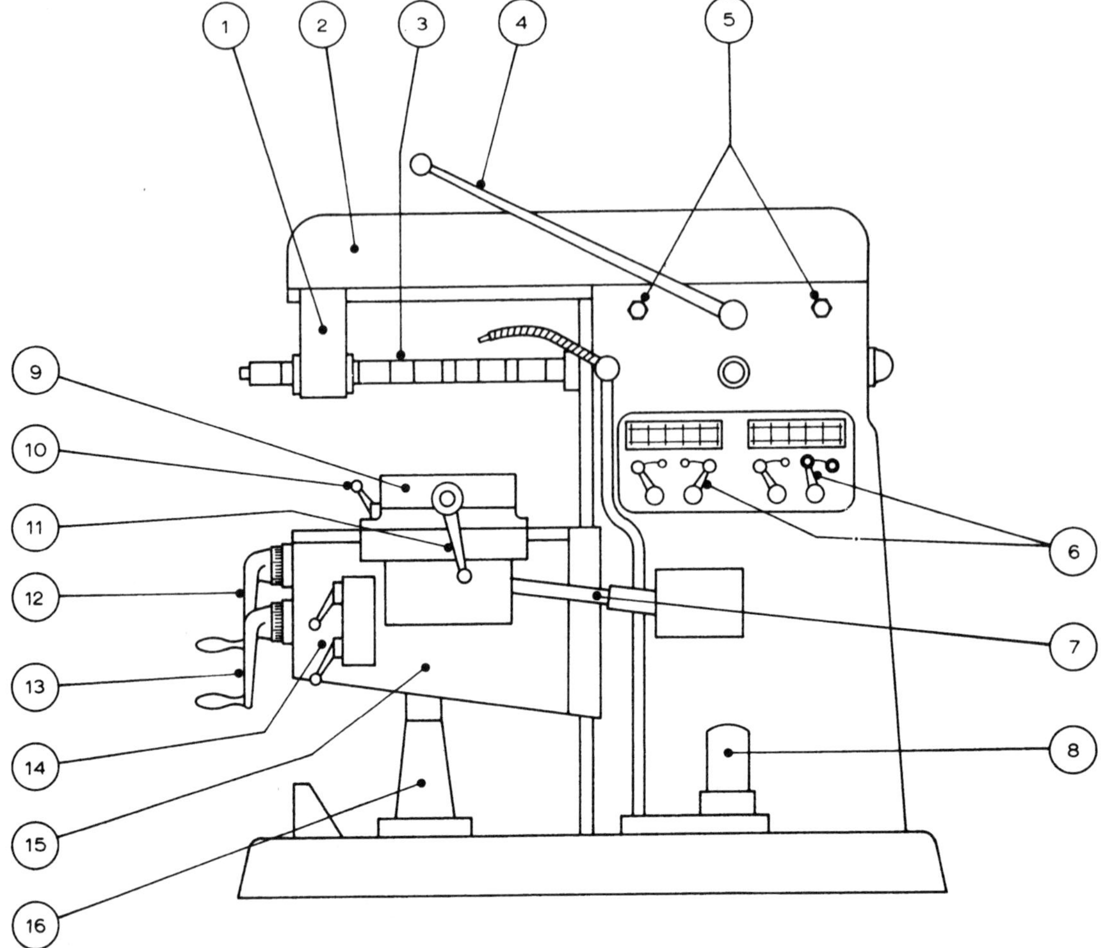

1 Arbor support or yoke (which slides on overarm).
2 Overarm.
3 Arbor (on which the cutter is mounted).
4 Clutch lever.
5 Overarm clamping screws.
6 Speed and feed selector levers.
7 Telescopic feedshaft.
8 Coolant pump.
9 Table which contains tee slots for clamping work or a vice.
10 Automatic table feed (longitudinal).
11 Longitudinal feed (hand).
12 Vertical feed (hand) which raises or lowers knee assembly.
13 Cross or transverse handle (hand).
14 Vertical and horizontal automatic-feed levers.
15 Knee which supports table and moves up and down on dovetail slides.
16 Knee elevating screw.

Figure 1.87 *A horizontal mill*

The cutter is mounted on an arbor. An arbor is shown in Figure 1.88(a). One end of the arbor is located in a taper in the spindle nose of the machine and is retained by a draw-bolt passing through the spindle of the machine, as shown in Figure 1.88(b). Drive to the arbor is positive. Blocks of metal called dogs fastened to the spindle nose engage in slots in the end of the arbor. The opposite end of the arbor is supported and located in an outrigger bearing supported by the overarm.

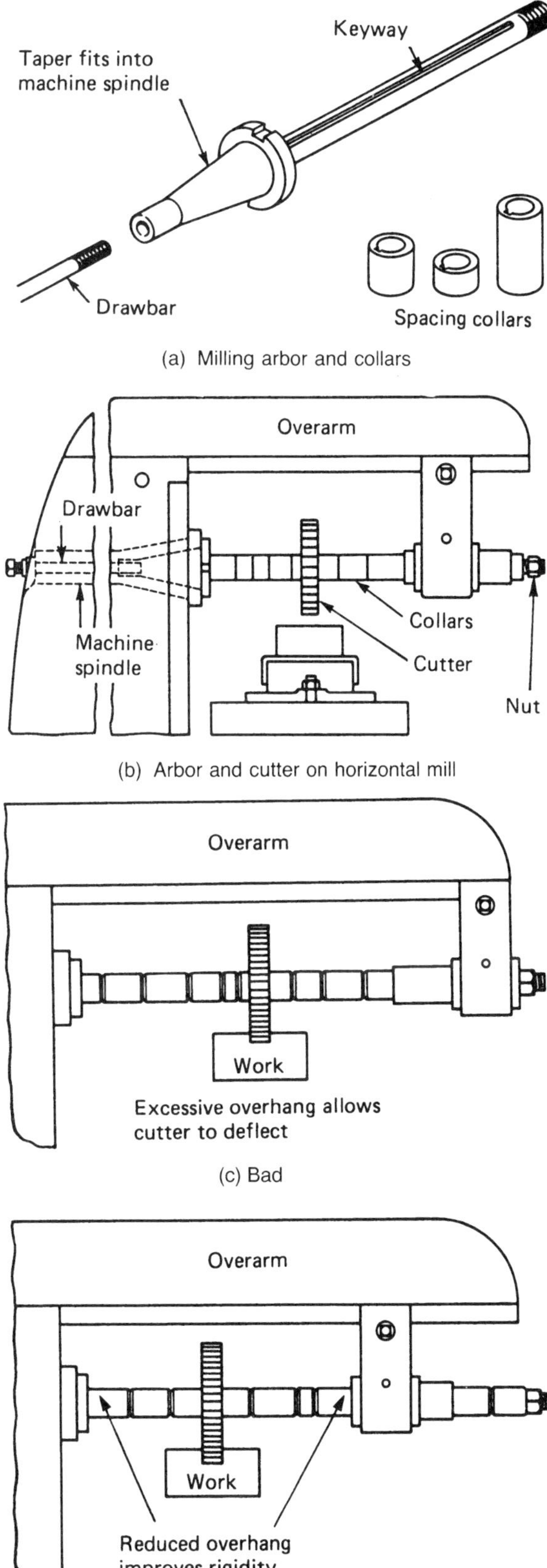

Figure 1.88 *Milling arbor and cutter positioning*

The cutter is positioned along the arbor by means of spacing collars. These should be arranged so that the overhang between the cutter and its nearest support bearing is as small as possible. Figure 1.88(c) shows the cutter badly positioned with excessive overhang. Figure 1.88(d) shows the cutter correctly positioned for improved rigidity.

For light cuts the cutter may be driven by friction alone. However, for heavy cuts the cutter should be keyed to the arbor to provide a positive drive. Cutters intended for use on horizontal milling machines have standard keyways in their bores. Arbors for horizontal milling machines have a matching keyway. Some typical cutters for horizontal milling machines are shown in Figure 1.89.

The metal cutting wedge also applies to the teeth of milling cutters, as shown in Figure 1.90(a). In addition, milling cutter teeth also have secondary clearance to prevent the 'heel' of the tooth from fouling the work, and tertiary (chip) clearance. This latter clearance angle provides a storage space for the chips of metal cut by the tooth until the tooth clears the work. Plenty of coolant helps to flush these chips away. The form of a milling cutter tooth is shown in Figure 1.90(b).

The cutting action of a milling cutter is shown in Figure 1.91. This shows that the action depends upon the direction of feed of the work relative to the direction of rotation of the cutter.

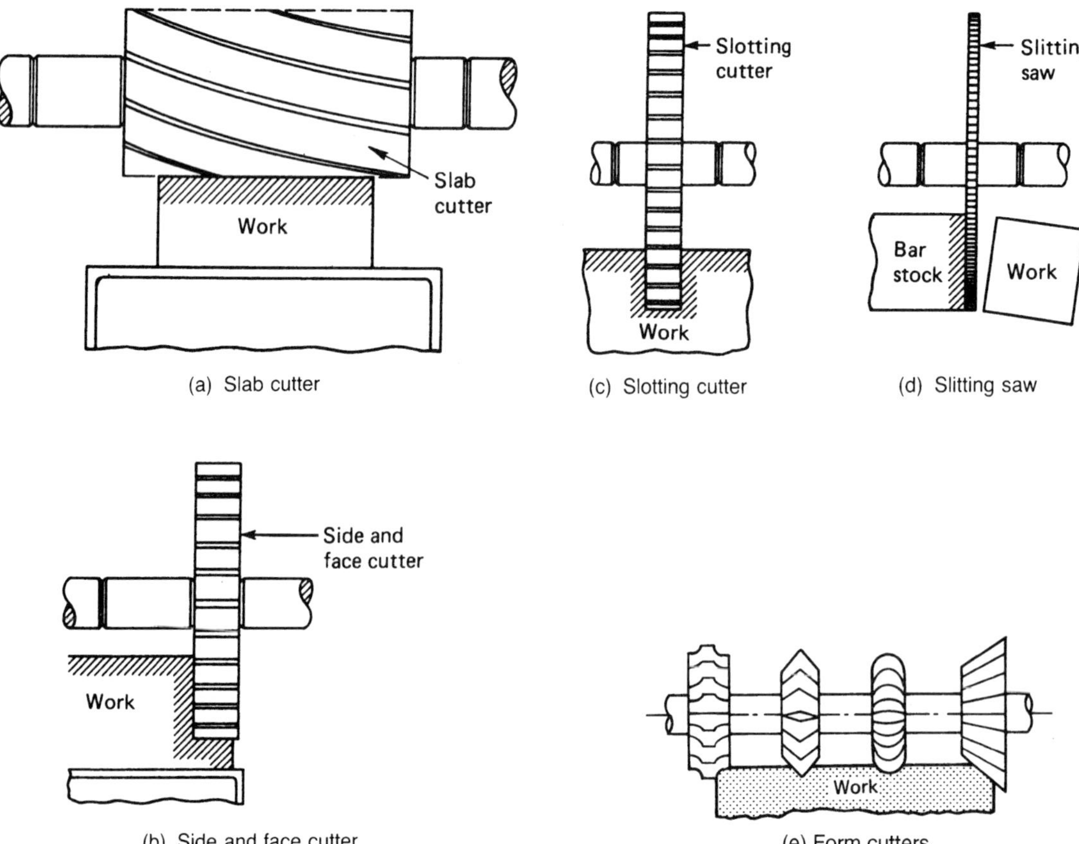

Figure 1.89 *Typical cutters for horizontal milling*

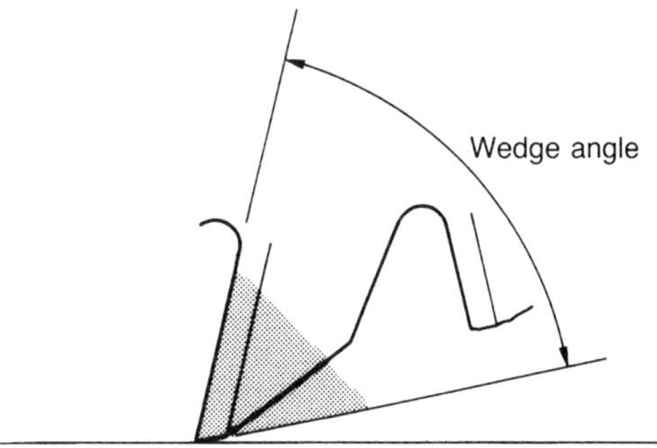

(a) The metal cutting wedge applied to a milling cutter tooth

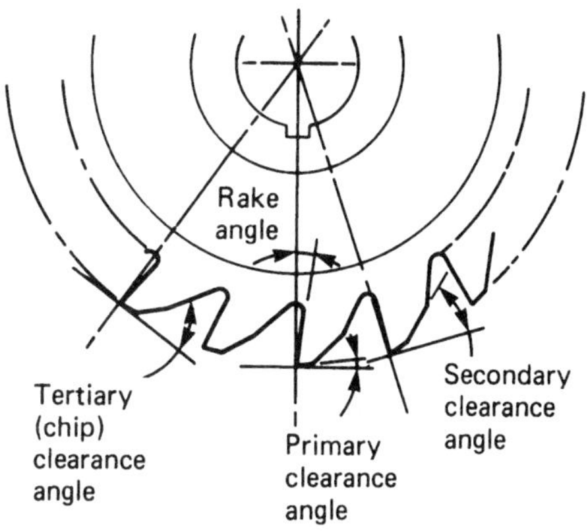

(b) Milling cutter tooth angles

Figure 1.90 *Wedge and tooth angles*

Up-cut milling

This is also known as 'conventional' milling because it is the most widely used method. Because the cutter is trying to push the work away, it is very safe to use. It does have some disadvantages. For instance the feed mechanism has to drive against the main cutting force and this can lead to wear of the feed mechanism. Also the cutter tends to rub before it commences to cut and this leads to tooth wear and blunting of the cutter.

Down-cut milling

This is also known as 'climb' milling since the cutter tries to climb over the work. Normally this would be a very dangerous method to use, resulting in serious damage to the machine, the cutter, and the work. However some machines are specially designed to work in this manner and, when such machines are well maintained and operated by skilled craftspersons, this method has several advantages. For instance the feed mechanism only has to act as a brake and to hold back the work from being drawn under the cutter too quickly. Secondly the teeth of the cutter do not rub as they commence to cut. Thirdly, the cutting action forces the work firmly down onto the worktable of the machine.

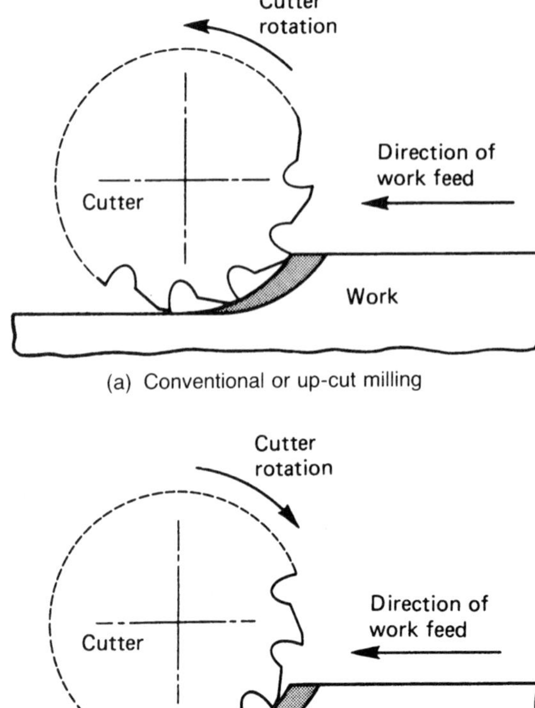

Figure 1.91 *Up-cut and down-cut milling*

The vertical milling machine

This machine gets its name from the fact that the axis of its spindle lies in the vertical plane. An example of a vertical milling machine is shown in Figure 1.92.

The cutters for use with a vertical milling machine can be mounted in various ways. Large face mills are mounted directly on the spindle nose as shown in Figure 1.93(a). The body of a face mill is fitted with a stub-arbor that is located in the taper of the spindle nose. It is retained in position by a draw-bolt passing through the spindle. The face mill is driven positively by the spindle nose dogs engaging in slots in its shank. Figure 1.93(b) shows an end mill and its chuck. Slot drills can also be held in the same chuck. The chuck has a taper shank and is mounted in the spindle nose of the milling machine. It is retained by the draw bolt passing through the spindle and is driven positively by the dogs on the spindle nose. Most end mills and slot drills have parallel shanks with screwed ends so that they are positively secured in the chuck and cannot be drawn out of the chuck by the cutting forces.

Figure 1.94 shows a typical end mill, a typical slot drill and the type of work such cutters are used for. End mills and shell end

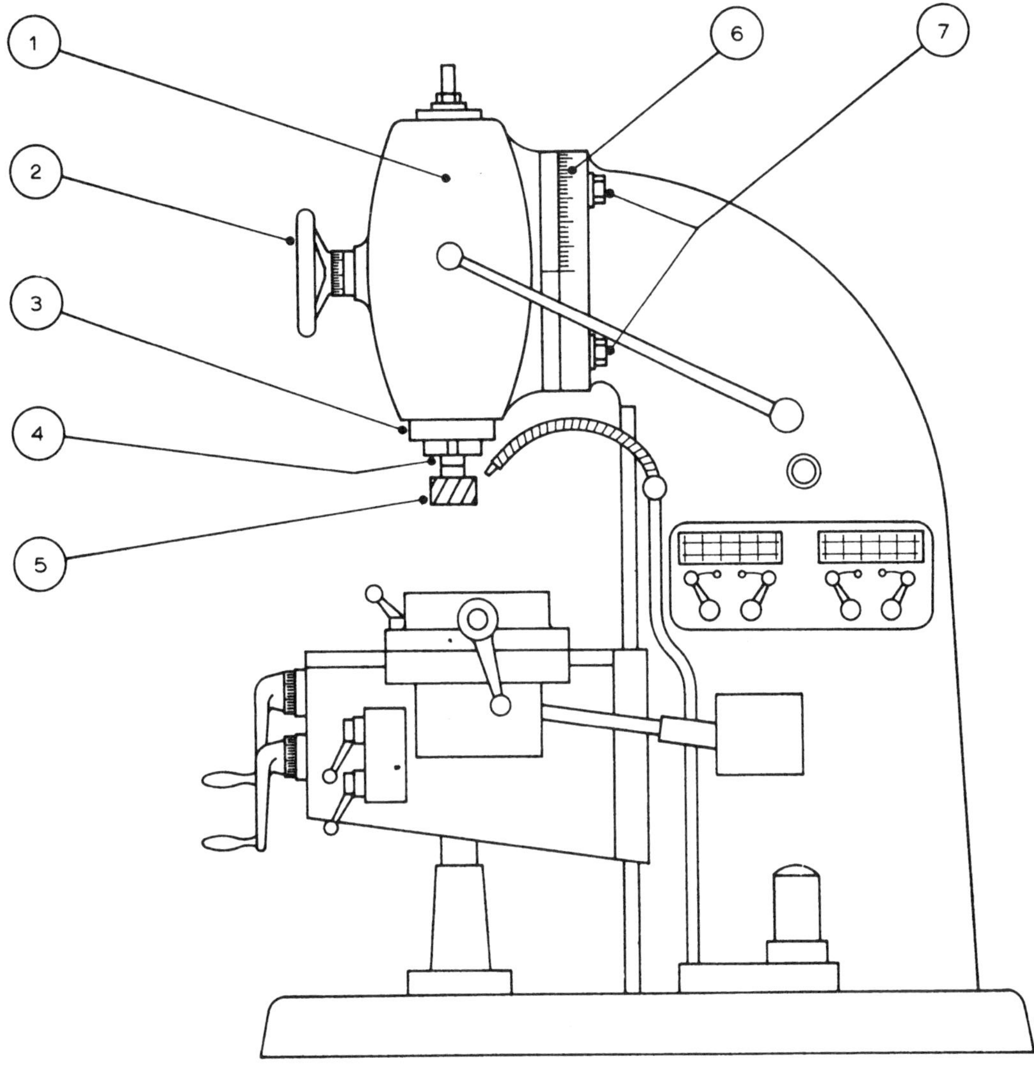

1 Vertical head which tilts.
2 Vertical feed handwheel.
3 Quill.
4 Spindle.
5 Cutter.
6 Head tilts here.
7 Head locking nuts.

Note: Remaining parts are similar to the horizontal-type milling machine (see p. 92).

Figure 1.92 *A vertical mill*

mills cannot be used for making pocket cuts from the solid. However, provided they can be fed into the work, as shown in Figure 1.94(a), end mills can remove metal more quickly and with a better surface finish than slot drills because of their larger number of teeth. For making pocket cuts from the solid, a slot drill is used, as shown in Figure 1.94(b).

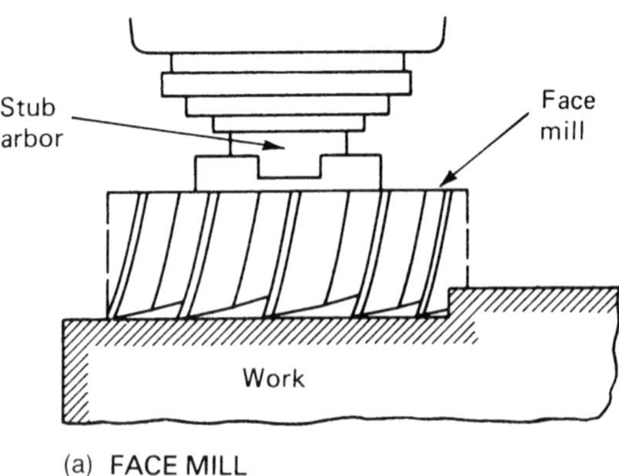

(a) FACE MILL

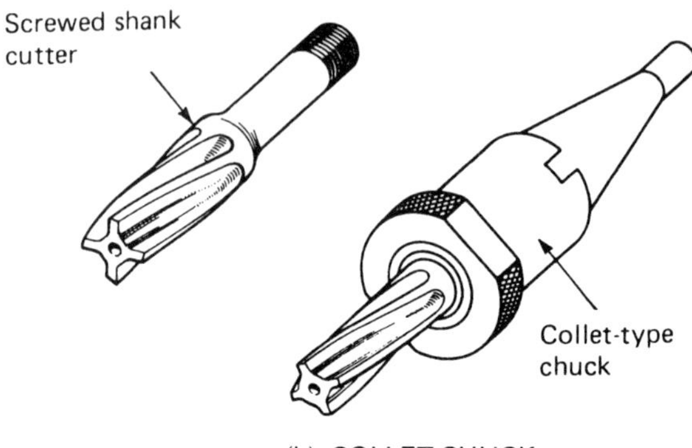

(b) COLLET CHUCK

Figure 1.93 *Face mill and end mills*

Test your knowledge 1.23

1. Describe how you would square up the blank shown in Figure 1.96(a) from a grey iron casting 85 mm × 60 mm × 35 mm.
2. With reference to Figure 1.96(b):
   (a) describe how you would hold the work whilst milling the 6 slots and the 4 flats,
   (b) calculate the number of holes to be moved for each indexing if the index plate has an 8-hole circle and a 12-hole circle,
   (c) name the most convenient type of machine and cutters for this job.
3. Figure 1.96(c) shows a keyway to be milled from the solid.
   (a) State the type of milling machine you would use.
   (b) State the type of cutter you would use, giving reasons for your choice.
   (c) Describe the method of work-holding.

Work-holding on milling machines

Work may be held in a machine vice bolted to the machine table or it may be clamped directly to the machine table. The work-holding techniques used are similar to those used for drilling but are rather more substantial to withstand the larger cutting forces. For batch and continuous production purposes a milling fixture is used. This is a custom built work-holding device which ensures that the components are located in the machine in the same place every time so that identical cuts are taken. It also ensures that all the components are correctly and securely clamped in the same manner.

Sometimes it is necessary to make a series of cuts around a circular component. Such an operation is called indexing. Figure 1.95 shows two methods of indexing. The dividing head can be used on both horizontal and vertical milling machines. The rotary table is normally used only on a vertical milling machine. The dividing head shown is called a 'simple' dividing head because the index plate locates the spindle directly. In the more sophisticated 'universal' dividing head the index plate locates the spindle via a worm and worm wheel with a standard 40:1 ratio. This, together with a range of index plates, enables an almost unlimited number of divisions to be made.

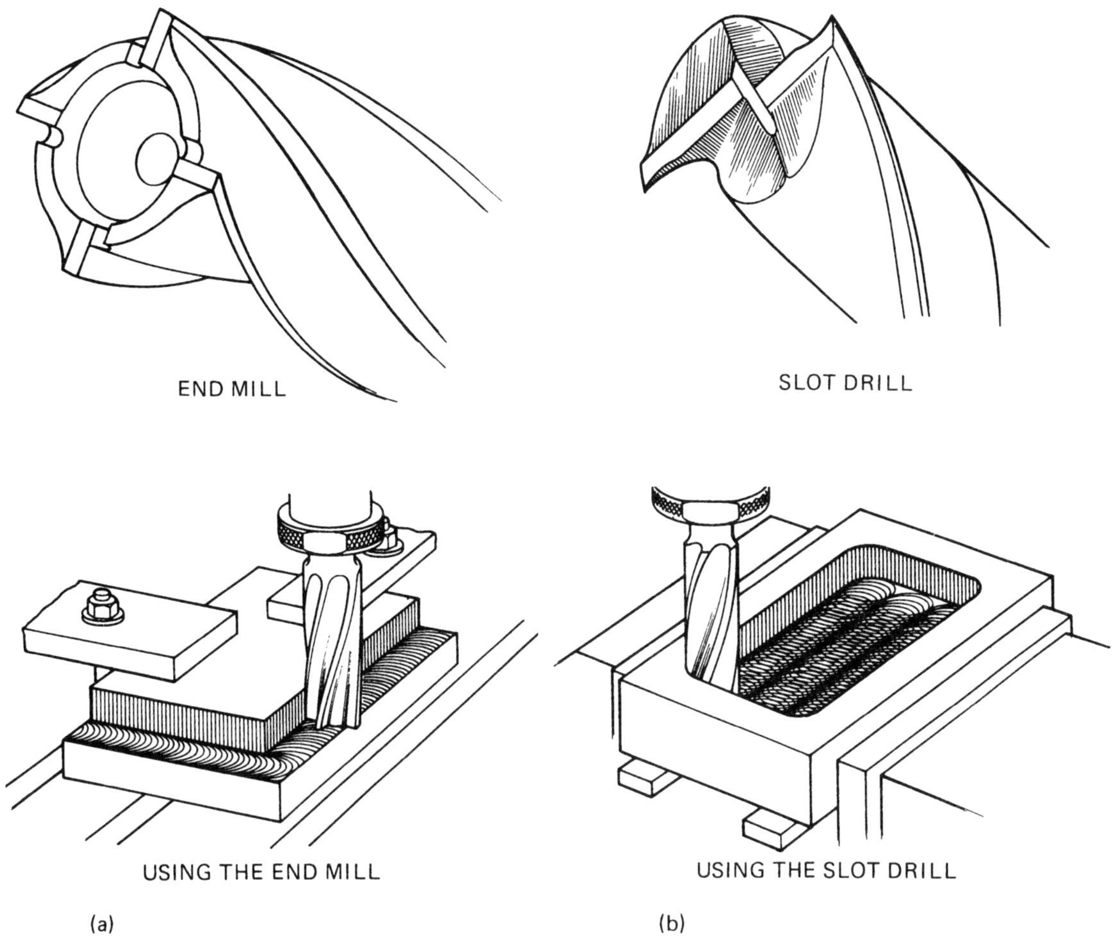

Figure 1.94 *Use of end mill and slot drill*

Activity 1.8

Figure 1.97 shows an assembly made up from three components. You have to decide how to machine each of these components. For each component:

(a) state the type of machine to be used

(b) list the tooling to be used with reasons for your choice

(c) with the aid of sketches describe the methods of workholding used

(d) draw up an operation planning sheet for each component.

Present your findings in the form of a brief word-processed report.

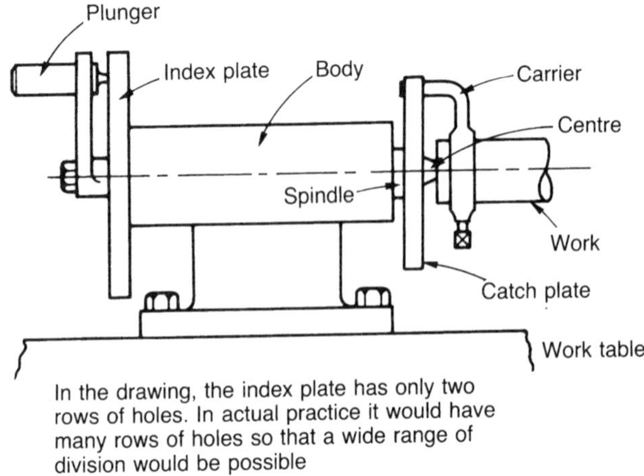

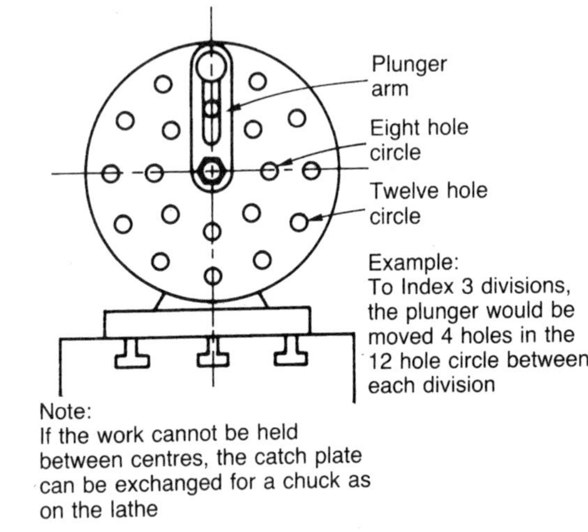

(a) Simple indexing with a direct dividing head

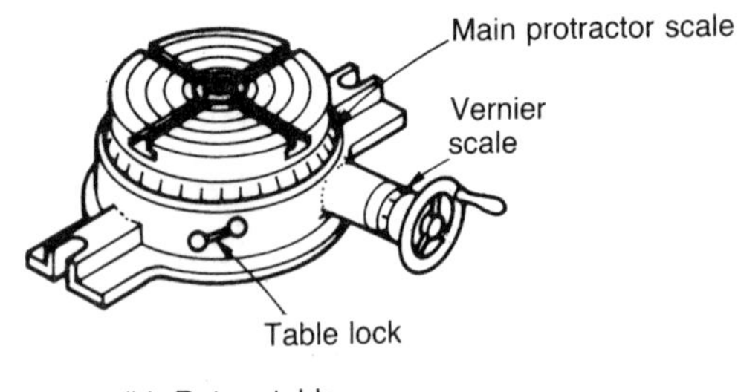

(b) Rotary table

Figure 1.95 *Indexing*

Joining and assembly

Assembly

The purpose of assembly is to put together a number of individual components to build up a whole device, structure or system. To achieve this aim attention must be paid to the following key factors.

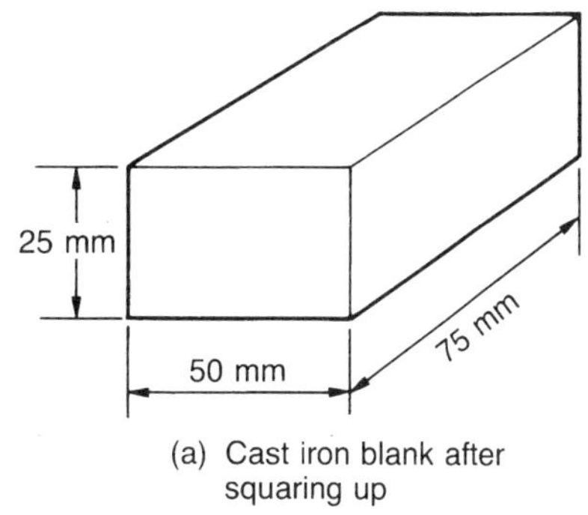

(a) Cast iron blank after squaring up

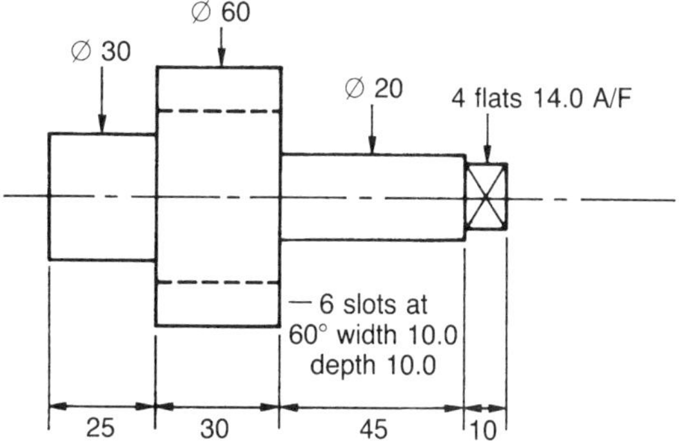

(b) Turned blank to be milled. (The ends are centred)

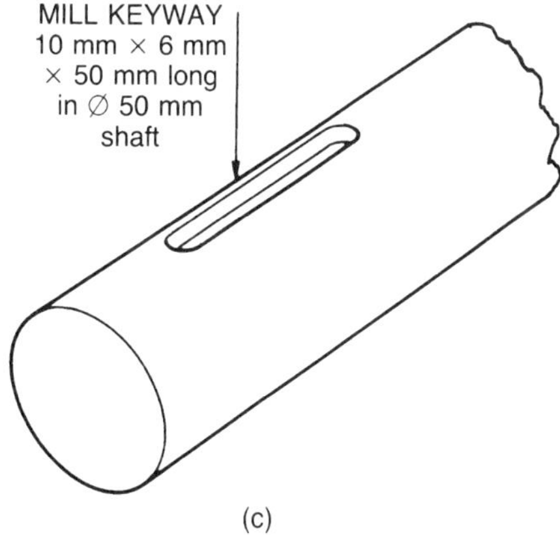

(c)

Figure 1.96 *See Test your knowledge 1.23*

Sequence of assembly

This must be planned so that as each component is added, its position in the assembly and the position of its fastenings are accessible. Also the sequence of assembly must be planned so that the installation of one component does not prevent access for fitting the next component or some later component.

Technique of joining

These must be selected to suit the components being joined, the materials from which they are made, and what they do in service. If the joining technique involves heating, then care must be taken that adjacent components are not heat sensitive or flammable.

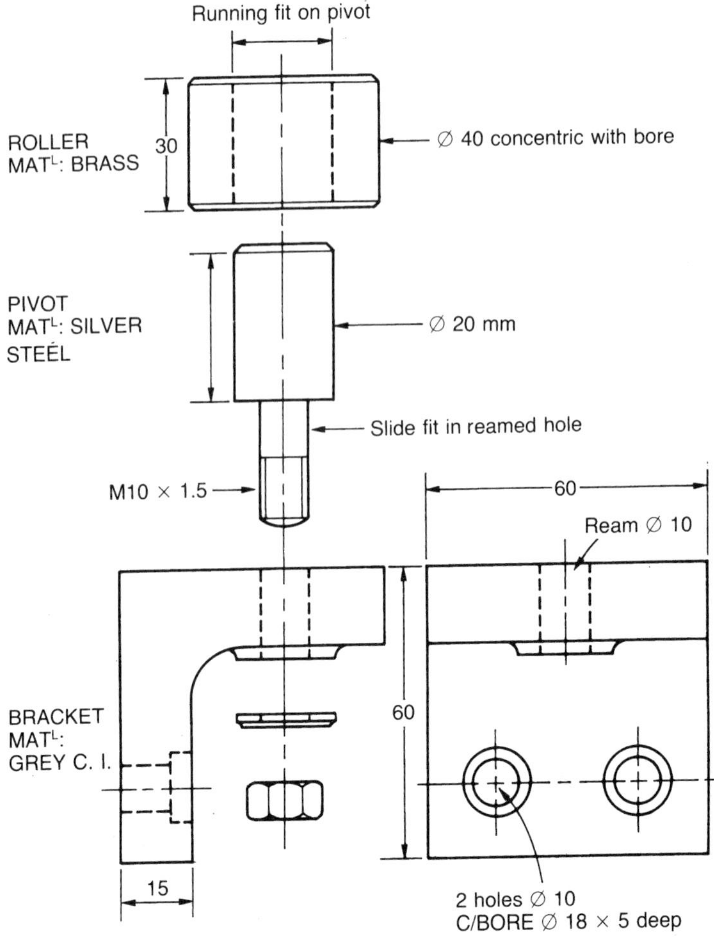

Figure 1.97 *See Activity 1.8*

Position of joints

Joints must not only be accessible for initial assembly, they must also be accessible for maintenance. You don't want to dismantle half a machine to make a small adjustment, or replace a part that wears out regularly.

Interrelationship and identification of parts

Identification of parts and their position in an assembly can usually be determined from assembly drawings or exploded view drawings. Inter-relationship markings are often included on components. For example, the various members and joints of structural steelwork are given number and letter codes to help identification on site. Printed circuit boards usually have the outline of the various components printed on them as well as the part number.

Tolerances

The assembly technique must take into account the accuracy and finish of the components being assembled. Much greater care has to be taken when assembling a precision machine tool or an artificial satellite, than when assembling structural steel work.

Protection of parts

Components awaiting assembly require protection against accidental damage and corrosion. In the case of structural steelwork this may merely consist of painting with red oxide primer and careful stacking. Precision components will require treating with an anti-corrosion lanolin based compound that can be easily removed at the time of assembly. Bores must be sealed with plastic plugs and screw threads with plastic caps. Precision ground surfaces must also be protected from damage. Heavy components must be provided with eye-bolts for lifting. Vulnerable sub-assemblies such as aircraft engines must be supported in suitable cradles.

Joining (mechanical)

The joints used in engineering assemblies may be divided into the following categories.

Permanent joints

These are joints in which one or more of the components and/or the joining medium has to be destroyed or damaged in order to dismantle the assembly; for example, a riveted joint.

Temporary joints

These are joints that can be dismantled without damage to the components. It should be possible to re-assemble the components using the original or new fastenings; for example, a bolted joint.

Flexible joints

These are joints in which one component can be moved relative to another component in an assembly in a controlled manner; for example, the use of a hinge.

Screwed fastenings

These are used to make temporary joints than can be dismantled and re-assembled at will. They are required where maintenance is necessary. We considered different types of screwed fastenings in the section on component selection. We also considered the different types of head found on screwed fastenings. Figure 1.98 shows the correct way to use some typical screwed fastenings.

Screwed fastenings must always pull down onto prepared seatings that are flat and at right-angles to the axis of the fastening. This prevents the bolt or screw being bent as it is tightened up. To protect the seating, a soft washer is placed between the seating and the nut. Taper washers are used when erecting steel girders to prevent the draught angle of the flanges from bending the bolt.

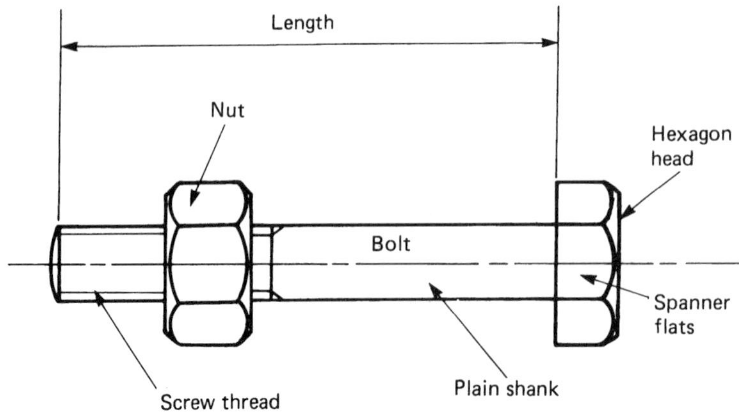

(a) HEXAGON HEAD BOLT AND NUT

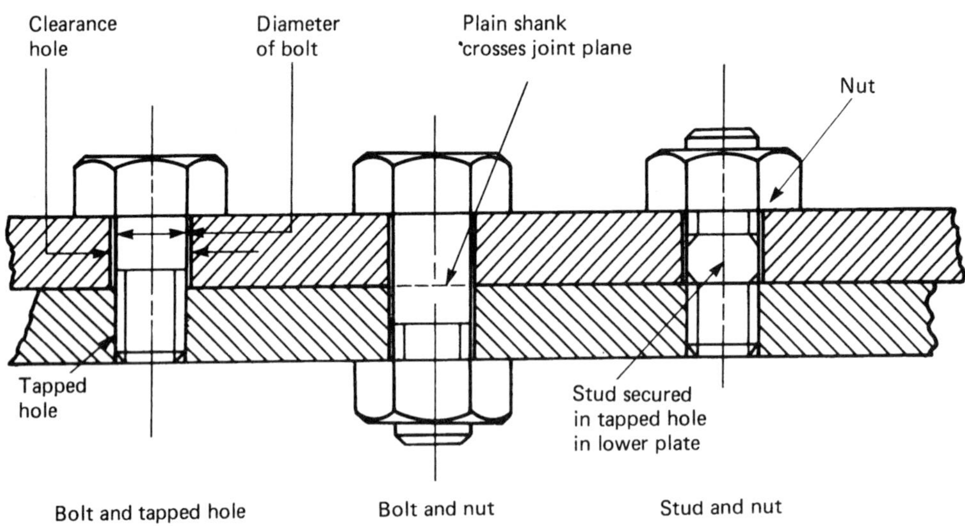

(b) TYPES OF SCREWED JOINT

Figure 1.98 *Screwed fastenings*

Locking devices are used to prevent screwed fastenings from slackening off due to vibration. Locking devices may be frictional or positive. A selection of plain washers, taper washers and locking devices is shown in Figure 1.99.

Riveting

The selection of rivets and rivet heads was considered in the section on component selection. To make a satisfactory riveted joint the following points must be observed.

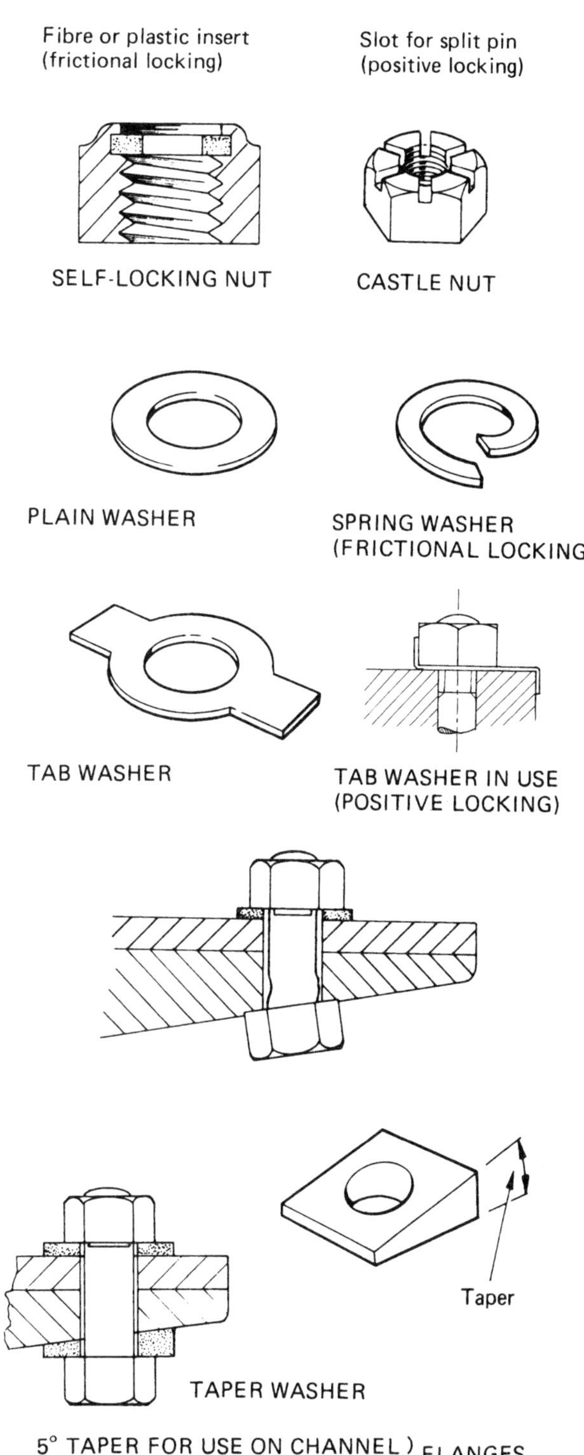

Figure 1.99 *Nuts and washers*

Hole clearance

If the clearance is too small there will be difficulty in inserting the rivet and drawing up the joint. If the hole is too large, the rivet will buckle and a weak joint will result.

Rivet length

If the rivet is too long, the rivet bends over during heading. If the rivet is too short the head cannot be properly formed. In either case a weak joint will result. Figure 1.100(a) shows the correct proportions for a riveted joint and Figure 1.100(b) shows some typical riveting faults.

The correct procedure for heading (closing) a rivet is shown in Figure 1.101(a). The drawing up tool ensures that the components to be drawn are brought into close contact and that the head of the rivet is drawn up tightly against the lower component. The hammer blows with the flat face of the hammer swells the rivet so that it is a tight fit in the hole and starts to form the head. The ball pein of the hammer head is then used to rough form the rivet head. The head is finally finished and made smooth by using a rivet snap. Where large rivets and large quantities of rivets are to be closed a portable pneumatic riveting tool is used.

'Pop' riveting is often used for joining thin sheet metal components, particularly when building up box sections. When building up box sections it is not possible to get inside the closed box to use a hold-up dolly; for example, when riveting the skin to an aircraft wing. Figure 1.101(b) shows the principle of 'pop' riveting.

Fusion welding

Welding has largely taken over from riveting for many purposes such as ship and bridge building and for structural steelwork. Welded joints are continuous and, therefore, transmit the stresses across the joint uniformly. In riveted joints the stresses are concentrated at each rivet. Also the rivet holes reduce the cross-sectional areas of the members being joined and weaken them. However, welding is a more skilled assembly technique and the equipment required is more costly. The components being joined are melted at their edges and additional filler metal is melted into the joint. The filler metal is of similar composition to that of the

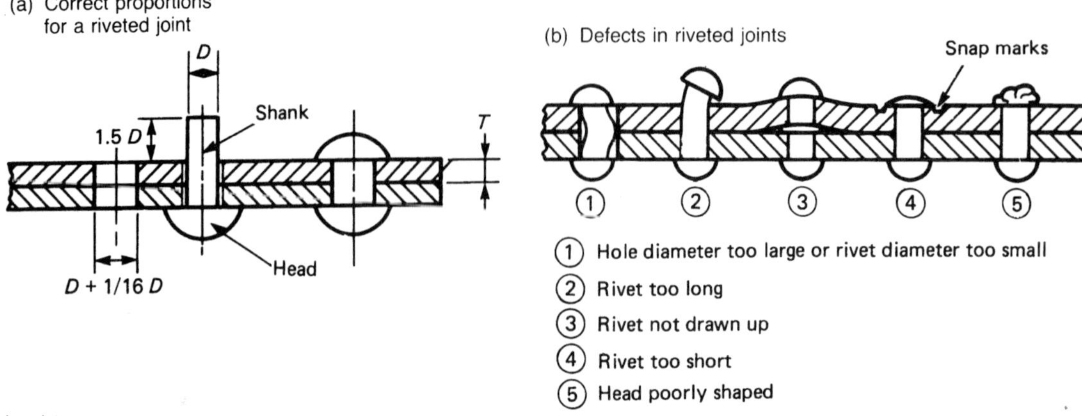

Figure 1.100 *Correct and incorrect riveted joints*

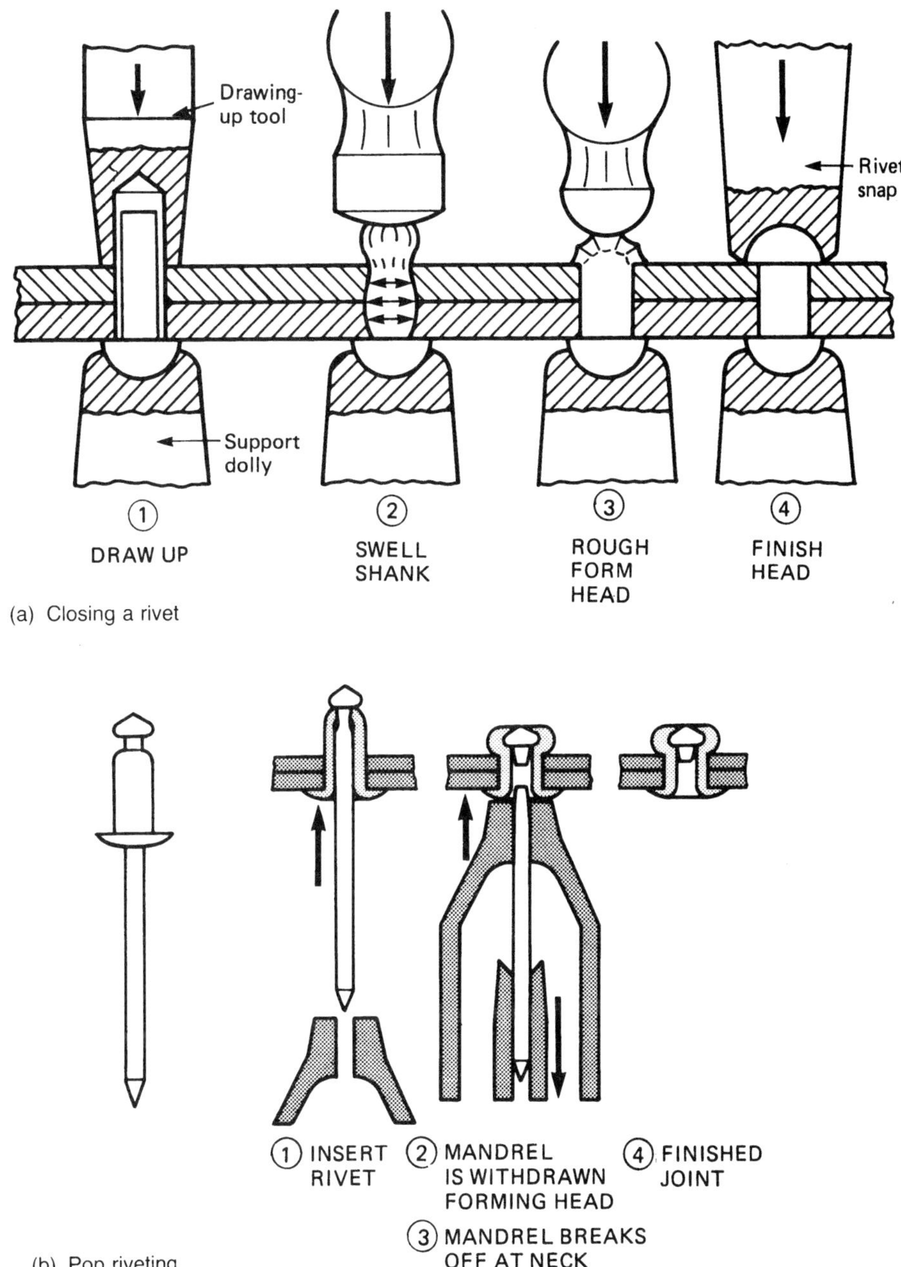

(a) Closing a rivet

(b) Pop riveting

Figure 1.101 *Correct procedure for riveting*

components being joined. Figure 1.102(a) shows the principle of fusion welding.

High temperatures are involved to melt the metal of the components being joined. These can be achieved by using the flame of an oxy-acetylene blowpipe, as shown in Figure 1.102(b), or an electric arc, as shown in Figure 1.102(c). When oxy-acetylene welding (gas welding), a separate filler rod is used. When arc welding, the electrode is also the filler rod and is melted as welding proceeds.

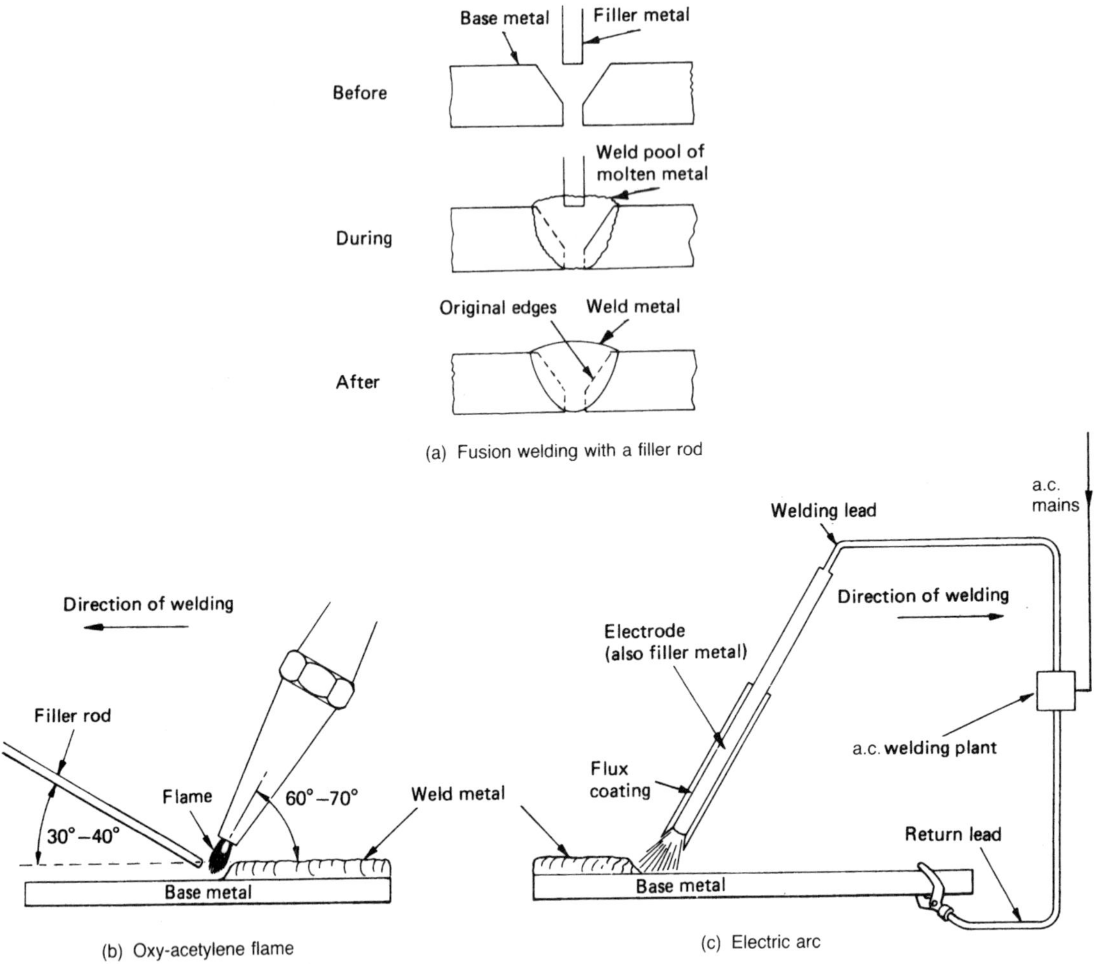

Figure 1.102 *Fusion welding*

No flux is required when oxy-acetylene welding as the molten metal is protected from atmospheric oxygen by the burnt gases (products of combustion). When arc welding, a flux is required. This is in the form of a coating surrounding the electrode. This flux coating is not only deposited on the weld to protect it, it also stabilizes the arc and makes the process easier. The hot flux gives off fumes and adequate ventilation is required.

Protective clothing must be worn when welding and goggles or a face mask (visor) appropriate for the process must be used. These have optical filters that protect the user's eyes from the harmful radiations produced during welding. The optical filters must match the process.

The compressed gases used in welding are very dangerous and welding equipment must only be used by skilled persons or under close supervision. Acetylene gas bottles must only be stored and used in an upright position.

The heated area of the weld is called the weld zone. Because of the high temperatures involved, the heat affected area can spread back into the parent metal of the component for some distance from the actual weld zone. This can alter the structure and

Test your knowledge 1.24

State THREE advantages and THREE disadvantages of welding compared with riveting.

properties of the material so as to weaken it and make it more brittle. If the joint fails in service, failure usually occurs at the side of the weld in this heat affected zone. The joint itself rarely fails.

Soft soldering

Soft soldering is also a thermal jointing process. Unlike fusion welding, the parent metal is not melted and the filler metal is an alloy of tin and lead that melts at relatively low temperatures. Soft soldering is mainly used for making mechanical joints in copper and brass components (plumbing). It is also used to make permanent electrical connections. Low carbon steels can also be soldered providing the metal is first cleaned and then 'tinned' using a suitable flux. The tin in the solder reacts chemically with the surface of the component to form a bond.

Figure 1.103 shows how to make a soft soldered joint. The surfaces to be joined are first degreased and physically cleaned to remove any dust and dirt. Fine abrasive cloth or steel wool can be used. A flux is used to render the joint surfaces chemically clean and to make the solder spread evenly through the joint. Some soft soldering fluxes and their typical applications are listed in Table 1.11.

- The copper 'bit' of the soldering iron is then heated. For small components and fine electrical work an electrically heated iron can be used. For joints requiring a soldering iron with a larger bit, a gas heated soldering stove can be used to heat the bit.
- The heated bit is then cleaned, fluxed and coated with solder. This is called 'tinning' the bit.
- The heated and tinned bit is drawn slowly along the fluxed surfaces of the components to be joined. This transfers solder to the surfaces of the components. Additional solder can be added if required. The work should be supported on

Table 1.11

Flux	*Metals*	*Characteristics*
Killed spirits (acidulated zinc chloride solution)	Steel, tin plate, brass and copper	Powerful cleansing action but leaves corrosive residue
Dilute hydrochloric acid	Zinc and galvanized iron	As above, wash off after use
Resin paste or 'cored' solders	Electrical conductor and terminal materials	Only moderate cleansing action (passive flux), but non-corrosive
Tallow	Lead and pewter	As above
Olive oil	Tin plate	Non-toxic, passive flux for food containers, non-corrosive

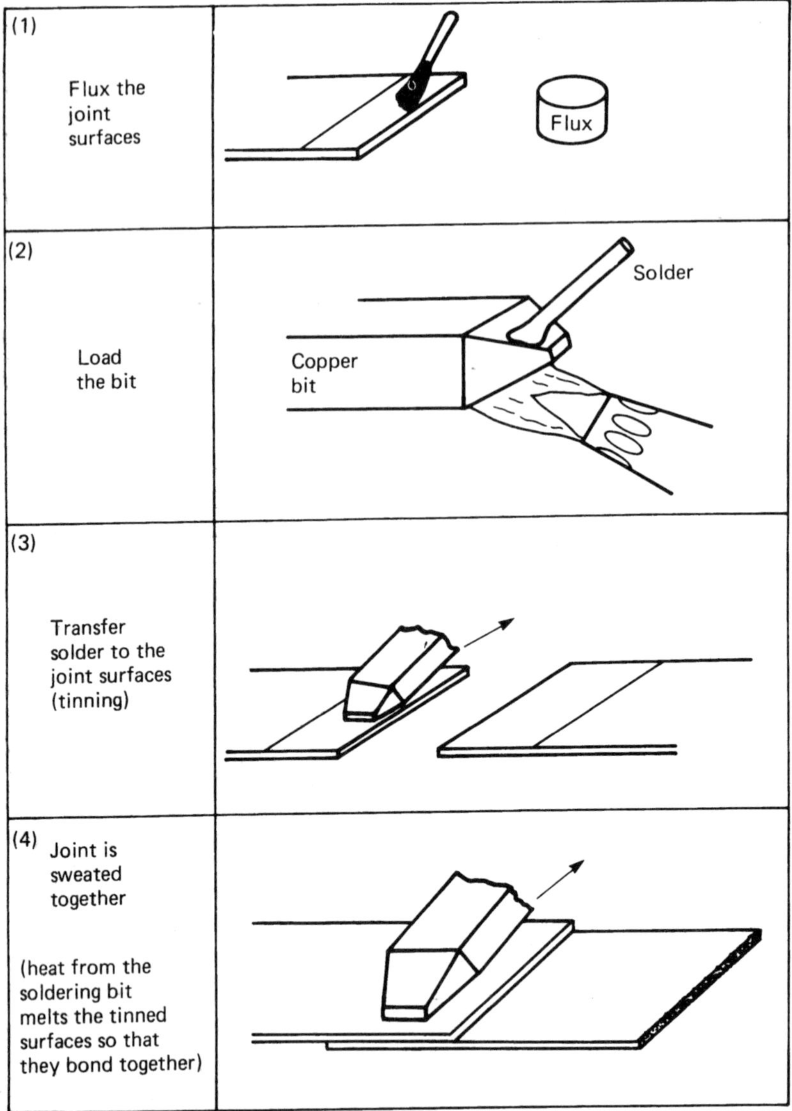

Figure 1.103 *Procedure for making a soft soldered joint*

wood to prevent heat loss. The solder does not just 'stick' to the surface of the metal being tinned. The solder reacts chemically with the surface to form an amalgam that penetrates into the surface of the metal. This forms a permanent bond.

- Finally the surfaces are overlapped and 'sweated' together. That is, the soldering iron is re-heated and drawn along the joint as shown. Downward pressure is applied at the same time. The solder in the joint melts. When it solidifies it forms a bond between the two components.

Figure 1.104 shows how a copper pipe is sweated to a fitting. The pipe and the fitting are cleaned, fluxed and assembled. The joint is heated with a propane gas torch and solder is added. This is usually a resin-flux cored solder. The solder is drawn into the close fitting joint by capillary attraction.

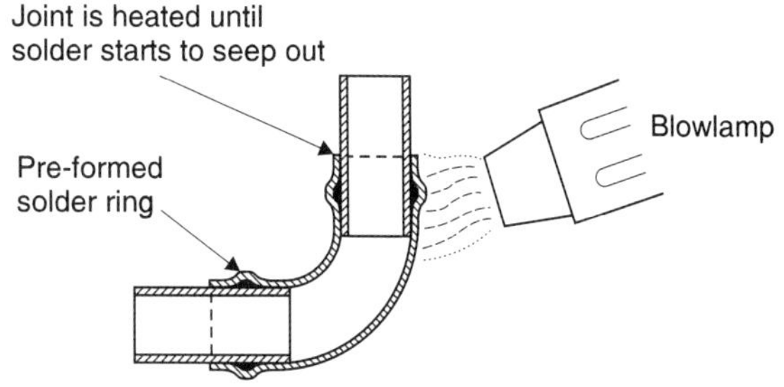

Figure 1.104 *'Sweating' a copper pipe fitting*

Hard soldering

Hard soldering uses a solder whose main alloying elements are copper and silver. Hard soldering alloys have a much higher melting temperature range than soft solders. The melting range for a typical soft solder is 183–212°C. The melting range for a typical hard solder is 620–680°C. Hard soldering produces joints that are stronger and more ductile. The melting range for hard solders is very much lower than the melting point of copper and steel, but it is only just below the melting point of brass. Therefore great care is required when hard soldering brass to copper. Because the hard solder contains silver it is often referred to as 'silver solder'. A special flux is required based on borax.

A soldering iron cannot be used because of the high temperatures involved. Heating is by a blow pipe. Figure 1.105 shows you how to make a typical hard soldered joint. Again cleanliness and careful surface preparation is essential for a successful joint. The joint must be close fitting and free from voids. The silver solder is drawn into the joint by capillary attraction.

Even stronger joints can be made using a brass alloy instead of a silver–copper alloy. This is called 'brazing'. The temperatures involved are higher than those for silver soldering. Therefore, brass cannot be brazed. The process of brazing is widely used for joining the steel tubes and malleable cast iron fittings of bicycle frames.

Test your knowledge 1.25

1. Consult solder manufacturers' literature and draw up a table showing the composition of the standard soft solders and their typical applications.
2. List the advantages and limitations of soft soldering compared with silver soldering.

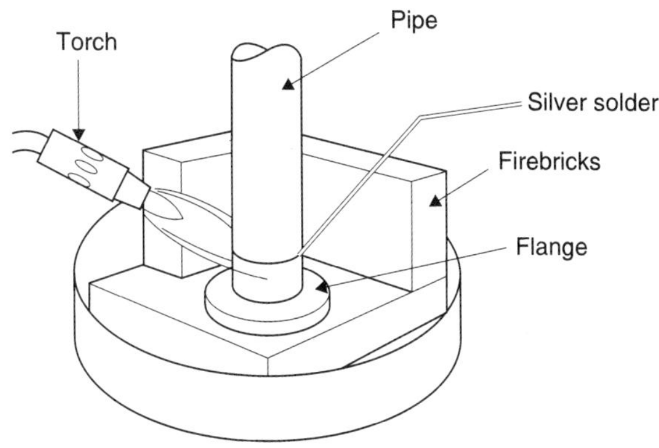

- The work is up to heat when the silver solder melts on contact with the work with the flame momentarily withdrawn.
- Add solder as required until joint is complete.

Figure 1.105 *Procedure for making a hard soldered joint*

Adhesive bonding

The advantages of adhesive bonding can be summarized as follows.

- The temperature rise from the curing of the adhesive is negligible compared with that of welding. Therefore the properties of the materials being joined are unaffected.
- Similar and dissimilar materials can be joined.
- Adhesives are electrical insulators. Therefore they reduce or prevent electrolytic corrosion when dissimilar metals are joined together.
- Joints are sealed and fluid tight.
- Stresses are transmitted across the joint uniformly.
- Depending upon the type of adhesive used, some bonded joints tend to damp out vibrations.

Bonded joints have to be specially designed to exploit the properties of the adhesive being used. You cannot just substitute an adhesive in a joint designed for welding, brazing or soldering. Figure 1.106(a) shows some typical bonded joint designs that provide a large contact area. A correctly designed bonded joint is very strong. Major structural members in modern high performance airliners and military aircraft are adhesive bonded. Figure 1.106(b) defines some of the jargon used when talking about bonded joints.

The strength of a bonded joint depends upon two factors.

Adhesion

This is the ability of the adhesive to 'stick' to the materials being joined (the adherends). This can result from physical keying or interlocking, as shown in Figure 1.107(a). Alternatively specific bonding can take place. Here, the adhesive reacts chemically with the surface of the adherends, as shown in Figure 1.107(b). Bonding occurs through intermolecular attraction.

Cohesion

This is the internal strength of the adhesive. It is the ability of the adhesive to withstand forces within itself. Figure 1.107(c) shows the failure of a joint made from an adhesive that is strong in adhesion but weak in cohesion. Figure 1.107(d) shows the failure of a joint that is strong in cohesion but weak in adhesion.

As well as the design of the joint, the following factors affect the strength of a bonded joint:

- The joint must be physically clean and free from dust, dirt, moisture, oil and grease.
- The joint must be chemically clean. The materials being joined must be free from scale or oxide films.
- The environment in which bonding takes place must have the correct humidity and be at the correct temperature.

Bonded joints may fail in four ways. These are shown in Figure 1.108. Bonded joints are least likely to fail in tension and shear. They are most likely to fail in cleavage and peel.

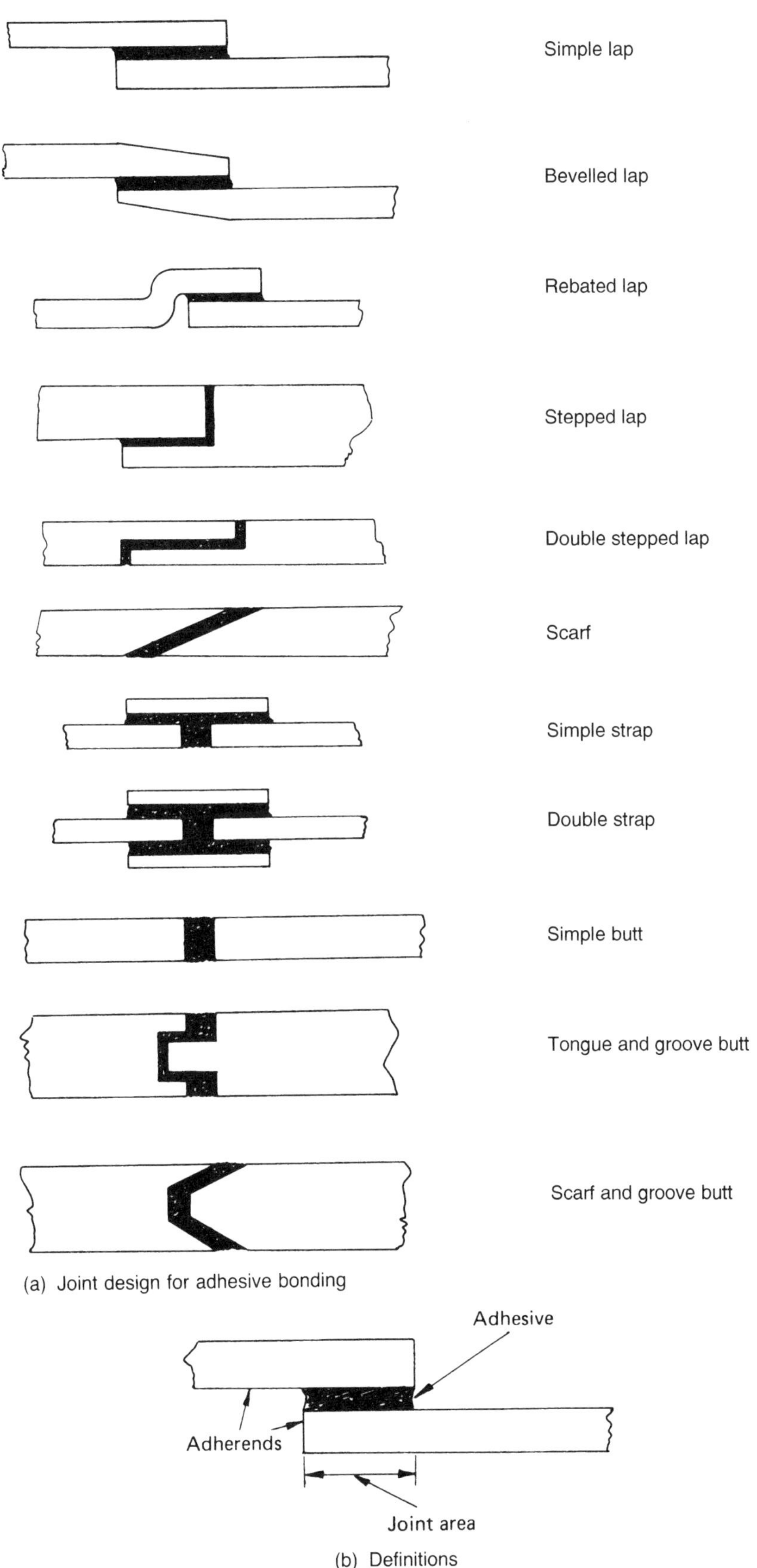

Figure 1.106 *Typical bonded joints*

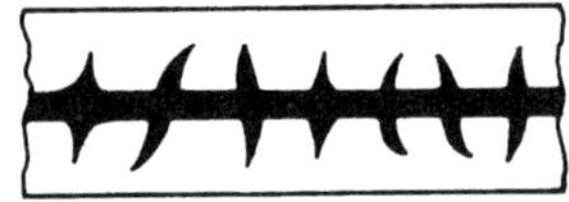

(a) Mechanical interlocking in porous materials

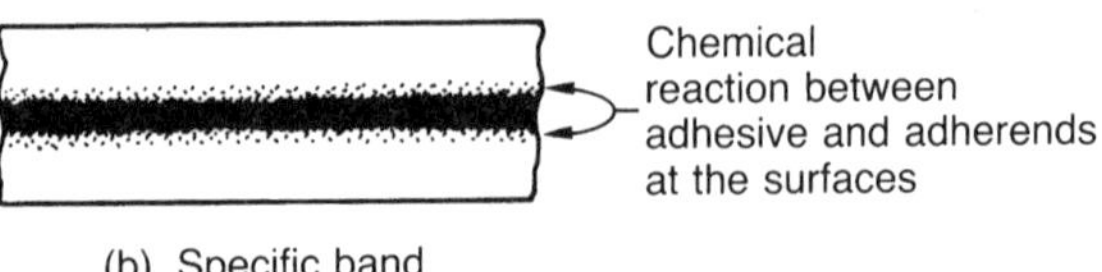

(b) Specific band

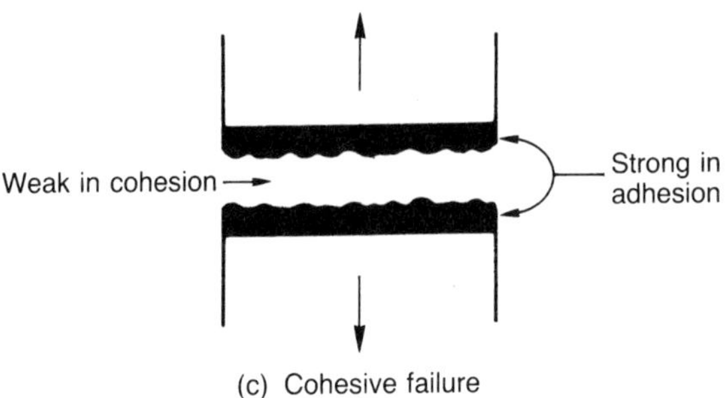

(c) Cohesive failure

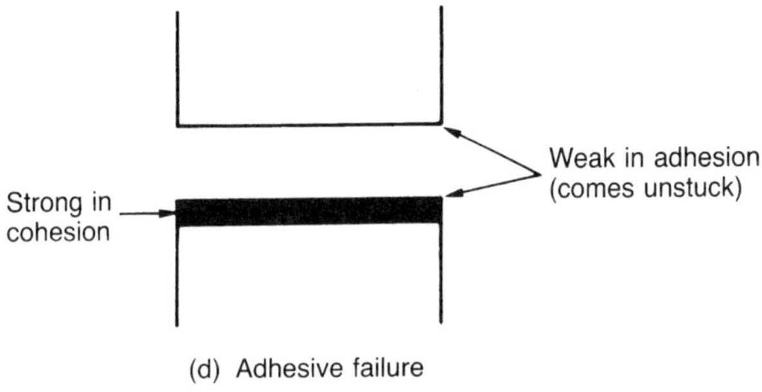

(d) Adhesive failure

Figure 1.107 *Adhesion*

Test your knowledge 1.26

1. Name THREE important SAFETY precautions that must be taken when using adhesives.
2. State TWO main requirements for a bonded joint design.
3. State TWO precautions that must be taken to produce a sound joint.

The most efficient way to apply adhesives is by an adhesive gun. This enables the correct amount of adhesive to be applied to the correct place without wastage or mess. It also prevents the evaporation of highly flammable and toxic solvents whilst the adhesive is waiting to be used.

Joining (electrical and electronic)

Again, joints may be permanent or temporary. Permanent joints are soldered or crimped. Temporary joints are bolted, clamped or plugged in.

Soldered joints

When soldering electrical and electronic components it is important not to overheat them. Overheating can soften thermoplastic

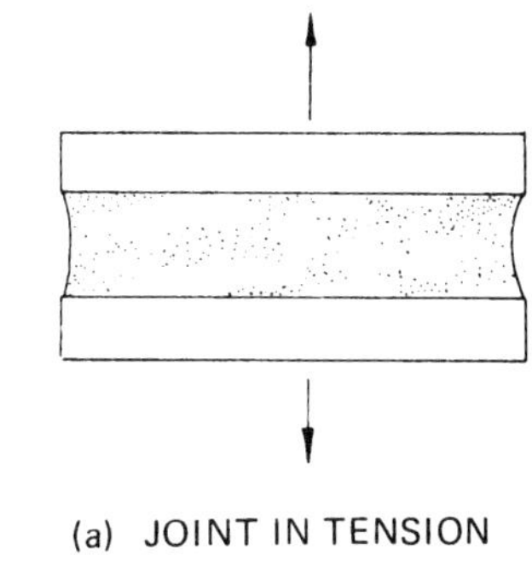

(a) JOINT IN TENSION

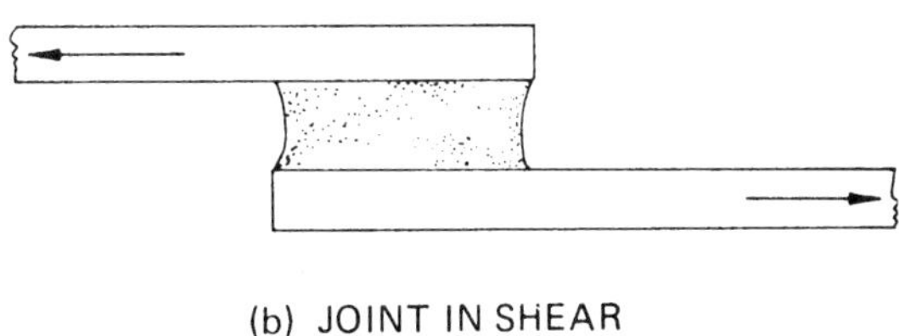

(b) JOINT IN SHEAR

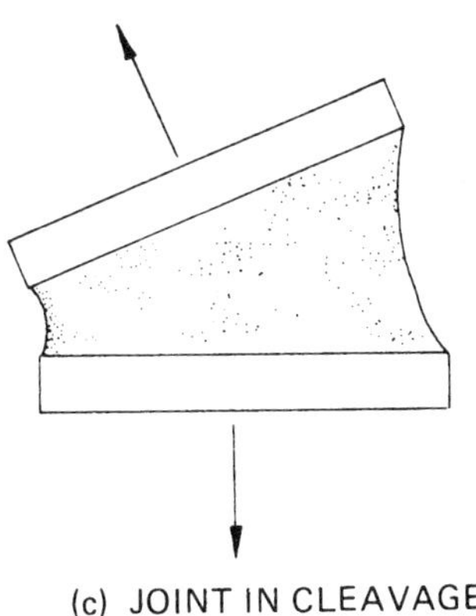

(c) JOINT IN CLEAVAGE

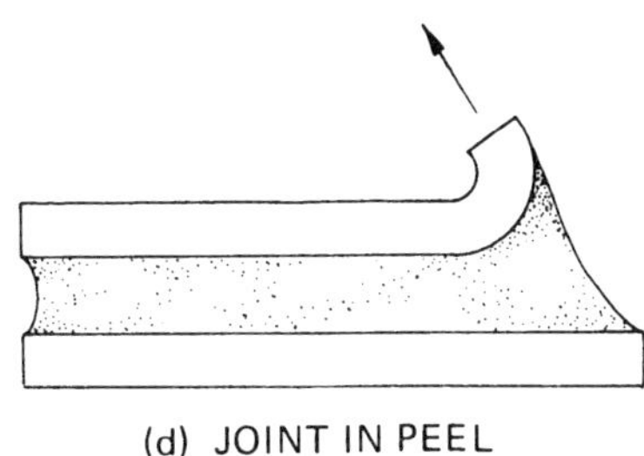

(d) JOINT IN PEEL

Figure 1.108 *Ways in which bonded joints can fail*

insulation and completely destroy solid state devices such as diodes and transistors. Very often some form of heat sink is required when soldering solid state devices.

A high tin content low melting temperature solder with a resin flux core should be used. This is a passive flux. It only protects the joint. It contains no active, corrosive chemicals to clean the joint. Therefore the joint must be kept clean whilst soldering. Even the natural grease from your fingers is sufficient to cause a high resistance 'dry' joint.

Figure 1.109(a) shows how a soldered connection is made to a solder tag. Note how the lead from the resistor is secured around the tag before soldering. This gives mechanical strength to the connection. Soldering provides the electrical continuity.

Figure 1.109(b) shows a prototype electronic circuit assembled on a matrix board. The board is made from laminated plastic and is pierced with a matrix of equally spaced holes. Pin tags are fastened into these holes in convenient places and the components are soldered to these pin tags.

Figure 1.109(c) shows the same circuit built up on a strip board. This is a laminated plastic board with copper tracks on the underside. The wire tails from the components pass through the holes in the board and are soldered to the tracks on the underside. The copper tracks are cut wherever a break in the circuit is required.

Figure 1.109(d) shows the underside of a printed circuit board (p.c.b.). This is built up as shown in Figure 1.109(c), except that the tracks do not need to be cut since they are customized for the circuit.

Large volume assembly of printed circuit boards involves the use of pick and place robots to install the components. The assembled boards are then carried over a flow soldering tank on a conveyor. A roller rotates in the molten solder creating a 'hump' in the surface of the solder. As the assembled and fluxed board passes over this 'hump' of molten solder the components tags are soldered into place.

Wire wrapping

Wire wrapping is widely used in telecommunications where large numbers of fine conductors have to be terminated quickly and in close proximity to each other. Soldering would be inconvenient and the heat could damage the insulation of adjoining conductors. Also soldered joints would be difficult to disconnect. A special tool is used that automatically strips the insulation from the wire and binds the wire tightly around the terminal pins. The terminal pins are square in section with sharp corners. The corners cut into the conductor and prevents it from unwinding. The number of turns round the terminal is specified by the supervising engineer.

Crimped joints

For power circuits, particularly in the automotive industry, cable lugs and plugs are crimped onto the cables. The sleeve of the lug or the plug is slipped over the cable and then indented by a small pneumatic or hydraulic press. This is quicker than soldering and, as no heat is involved, there is no danger of damaging the insulation. Portable equipment is also available for making crimped

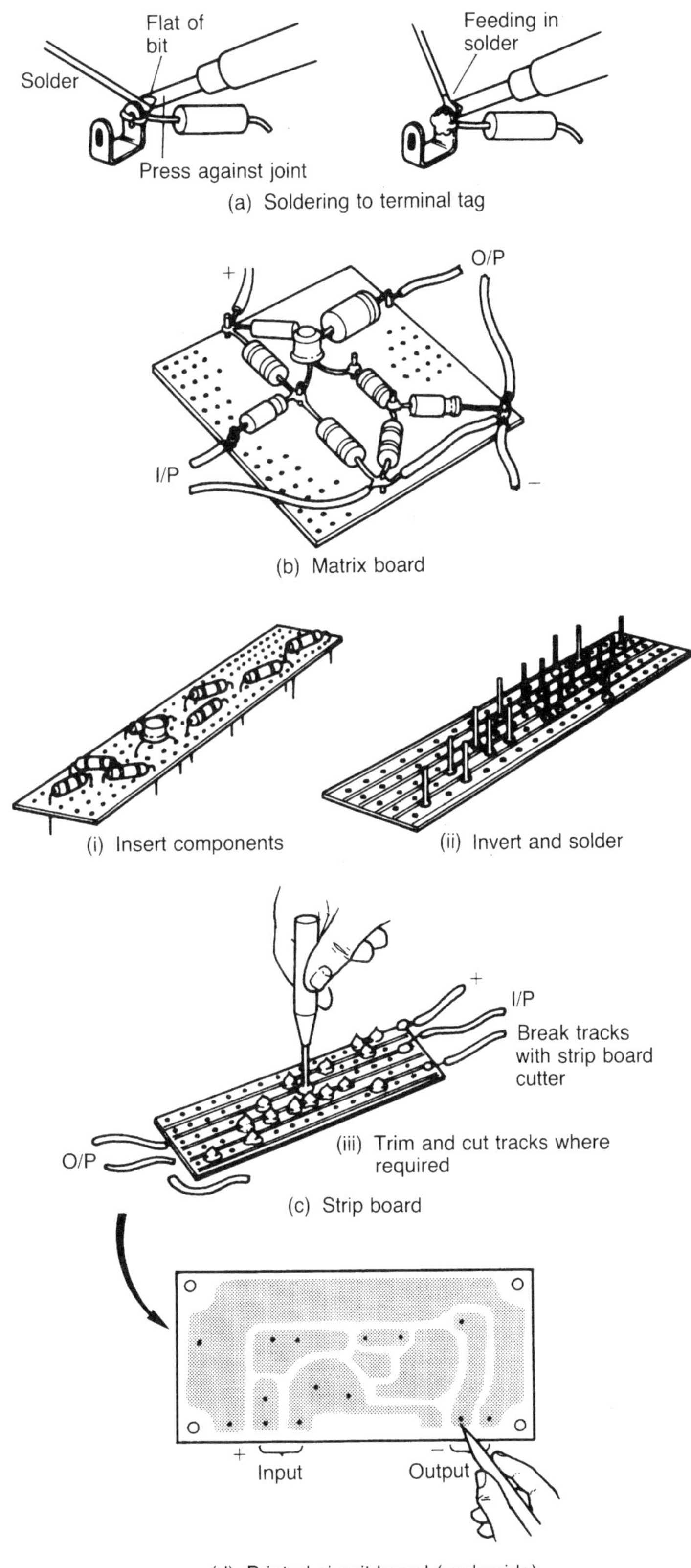

Figure 1.109 *Electronic circuit assembly*

joints on site. Hand operated equipment can be used to fasten lugs to small cables by crimping, as shown in Figure 1.110.

Clamped connections

Finally, we come to clamped connections using screwed fastenings. You will have seen many of these in domestic plugs, switches and lamp-holders. For heavier power installations, cable lugs are bolted to solid copper bus-bars using brass or bronze bolts, as shown in Figure 1.111.

Activity 1.7

1 (a) Describe the essential differences between welding and brazing.

(b) List the safety precautions that must be taken when using oxy-acetylene equipment.

2 With the aid of sketches show THREE faults that can occur when making riveted joints.

3 (a) Screw thread systems in 'inch' units are now obsolescent, but are still used for maintenance purposes on older plant. What screw thread systems do the following abbreviations stand for?
BSW, BSF, UNC, UNF, BA.

(b) A bolt is specified as M5 $\times$ 0.8. What does this mean?

4 State the precautions that must be taken during the design stage and during the manufacture of an adhesive bonded joint to ensure maximum strength.

5 (a) Describe the advantages of an active soldering flux over a passive soldering flux and explain why an active flux is unsuitable for building electronic circuits.

(b) Explain why silver soldering is unsuitable for securing components to a printed circuit board.

Heat treatment

The properties of many metals and alloys can be changed by heating them to specified temperatures and cooling them under controlled conditions at specified rates. These are called, respectively, critical temperatures and critical cooling rates. We are only going to consider the heat treatment of plain carbon steels.

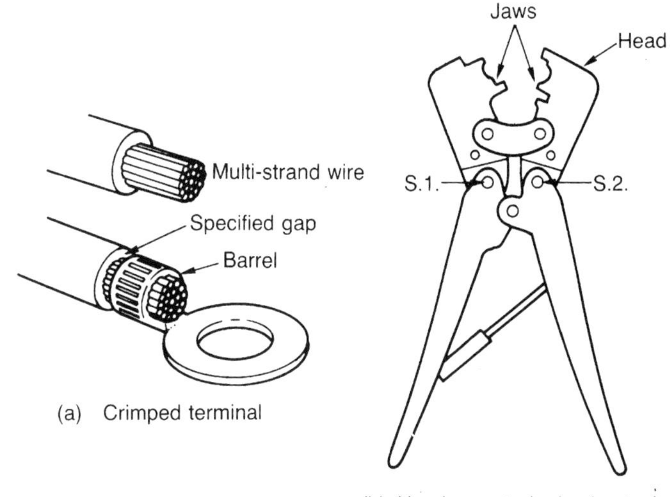

(a) Crimped terminal

(b) Hand operated crimping tool

Figure 1.110 *Crimping*

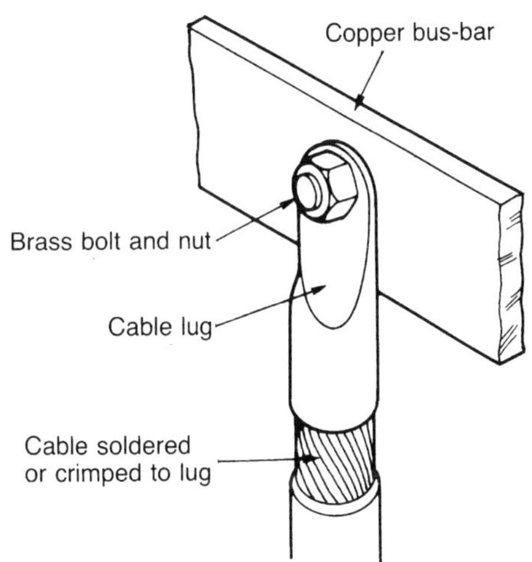

(a) Bolted connection

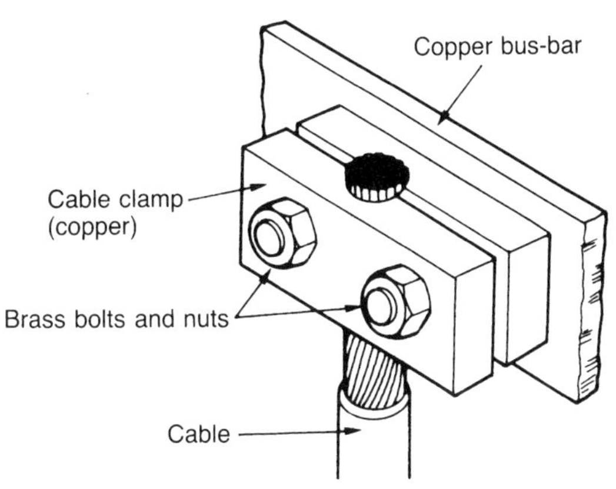

(b) Clamped connection

Figure 1.111 *Bolted and clamped connections*

Hardening

The degree of hardness that can be given to any plain carbon steel depends upon two factors: the amount of carbon present, and how quickly the steel is cooled from the hardening temperature. The hardening temperature for medium carbon steels containing up to 0.8% carbon is bright red heat. Above 0.8% carbon the hardening temperature is dull red (cherry red) heat. Table 1.12(a) relates hardness to carbon content. Table 1.12(b) relates hardness to rate of cooling.

If oil quenching is used, a number of safety rules must be observed because of its flammability.

- Never use lubricating oil
- Always use a good quality quenching oil
- Always use a metal quenching bath with an airtight metal lid to smother the flames should the oil ignite
- Always keep the bath covered when not in use to keep out foreign objects and to avoid the absorption of moisture from the atmosphere.

Hardening faults

Under heating
If the temperature of a steel does not reach its critical temperature, the steel won't harden.

Overheating
It is a common mistake to think that increasing the temperature from which the steel is quenched will increase its hardness. Once the correct temperature has been reached, the hardness will depend only upon the carbon content of the steel and its rate

Table 1.12

(a) Effect of carbon content

Type of steel	*Carbon content (%)*	*Effect of quench hardening*
Low carbon	Below 0.3	Negligible
Medium carbon	0.3–0.5	Becomes tougher
	0.5–0.9	Becomes hard
High carbon	0.9–1.3	Becomes very hard

(b) Rate of cooling

Carbon content (%)	*Quenching media*	*Required treatment*
0.3–0.5	Oil (see note (i))	Toughening
0.5–0.9	Oil	Toughening
0.5–0.9	Water	Hardening
0.9–1.2	Oil (see note (ii))	Hardening

of cooling. If the temperature of a steel exceeds its critical temperature, grain growth will occur and the steel will be weakened. Also overheating will slow the cooling rate and will actually reduce the hardness of the steel.

Cracking

There are many causes of hardening cracks. Some of the more important are: sharp corners, sudden changes of section, screw threads, holes too near the edge of a component. These should all be avoided at the design stage as should over rapid cooling for the type of steel being used.

Distortion

There are many causes of distortion. Some of the more important are as follows:

- lack of balance or symmetry in the shape of the component
- lack of uniform cooling. Long thin components should be dipped end-on into the quenching bath
- change in the grain structure of the steel causing shrinkage.

No matter how much care is taken when quench hardening, some distortion (movement) will occur. Also slight changes in the chemical composition may occur at the surface of the metal. Therefore, precision components should be finish ground after hardening. The components must be left slightly oversize before grinding to allow for this. That is, a grinding allowance must be left on such components before hardening.

Tempering

Quench-hardened plain carbon steels are very brittle and unsuitable for immediate use. Therefore, further heat treatment is required. This is called tempering. It greatly increases the toughness of the hardened steel at the expense of some loss of hardness.

Tempering consists of re-heating the hardened steel and again quenching it in oil or water. Typical tempering temperatures for various applications are summarized in Table 1.13. You can

Table 1.13 Tempering temperatures

Component	*Temper colour*	*Temperature (°C)*
Edge tools	Pale straw	220
Turning tools	Medium straw	230
Twist drills	Dark straw	240
Taps	Brown	250
Press tools	Brownish-purple	260
Cold chisels	Purple	280
Springs	Blue	300
Toughening (medium carbon steels)	–	450–600

Test your knowledge 1.27

Describe how you should harden and temper a small cold chisel made from 1.2% high carbon steel (silver steel).

judge the tempering temperature by the colour of the oxide film. First, the component must be polished after hardening and before tempering. Then heat the component gently and watch for the colour of the metal surface to change. When you see the appropriate colour appear, the component must be quenched immediately.

Chemical treatment

The chemical treatments that will be considered in this section are as follows.

- Chemical machining (etching) as used in the production of printed circuit boards
- The coating of metal components with decorative and/or corrosion resistant finishes.

Etching

Printed circuit boards have already been introduced in the section on assembly. First the circuit is drawn out by hand or designed using a computer aided design (CAD) package. A typical circuit master drawing is shown in Figure 1.112. The master drawing of the circuit is then photographed to produce a transparent copy called a negative. In a negative copy the light and dark areas are reversed.

The printed circuit board is made as follows:

- The copper coated laminated plastic (Tufnol) or fibre glass board is coated with a photoresist by dipping or spraying
- The negative of the circuit is placed in contact with the prepared circuit board. They are then exposed to ultraviolet light. The light passes through the transparent parts of the negative. The areas of the board exposed to the ultraviolet light will eventually become the circuit

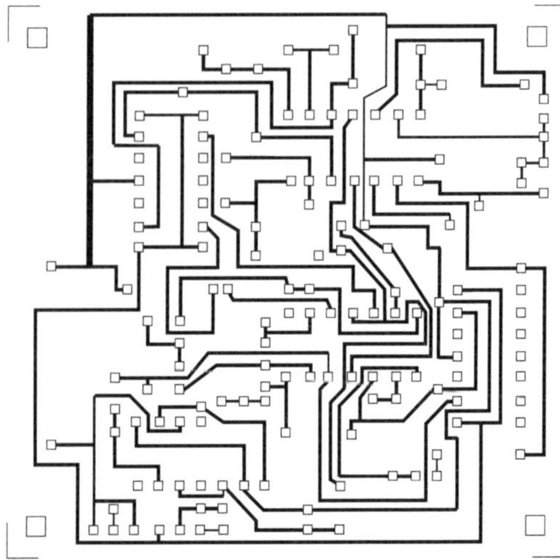

Figure 1.112 *A printed circuit layout*

Test your knowledge 1.28

I have just described how a p.c.b. is produced photographically. Now describe how you would produce a prototype p.c.b. by drawing directly onto the copper coated board.

- The exposed circuit board is then developed in a chemical solution that hardens the exposed areas
- The photoresist is stripped away from the unexposed areas of the circuit board
- The circuit board is then placed in a suitable etchant. Ferric chloride solution can be used as an etchant for copper. This eats away the copper where it is not protected by the hardened photoresist. The remaining copper is the required circuit
- The circuit boards are washed to stop the reaction. The remaining photoresist is removed so as not to interfere with the tinning of the circuit with soft solder and the soldering of the components into position.

SAFETY: This process is potentially dangerous. Ultraviolet light is very harmful to your skin and to your eyes. The ferric chloride solution is highly corrosive to your skin. The various processes also give off harmful fumes. Therefore you should only carry out this process under properly supervised, controlled and ventilated conditions. Appropriate protective clothing should be worn.

A similar process can be used for the chemical engraving of components with their identification numbers and other data.

Electroplating

The component to be plated is placed into a plating bath as shown in Figure 1.113. The component is connected to the negative terminal of a direct current supply. This operates at a low voltage but relatively heavy current. The anode is connected to the positive terminal of the supply. The anode is usually made from the same metal as that which is to be plated onto the component. The electrolyte is a solution of chemical salts. The composition depends upon the process being carried out.

When the current passes through the bath the component is coated with a thin layer of the protective and/or decorative metal. This metal comes from the chemicals in the electrolyte. The process is self-balancing. The anode dissolves into the electrolyte at the same rate as the metal taken from the electrolyte is being deposited on the component. This applies to most plating processes such as zinc, copper, tin and nickel plating.

An exception is chromium plating. A neutral anode is used that does not dissolve into the electrolyte. Additional salts have to be added to the electrolyte from time to time to maintain the solution strength. Chromium is not usually deposited directly onto the component. The component is usually nickel plated and polished and then a light film of chromium is plated over the nickel as a decorative and sealing coat.

The 'hard-chromium' plating used to build up worn gauges and tools is applied directly onto the steel. This is a specialized process outside the scope of this unit.

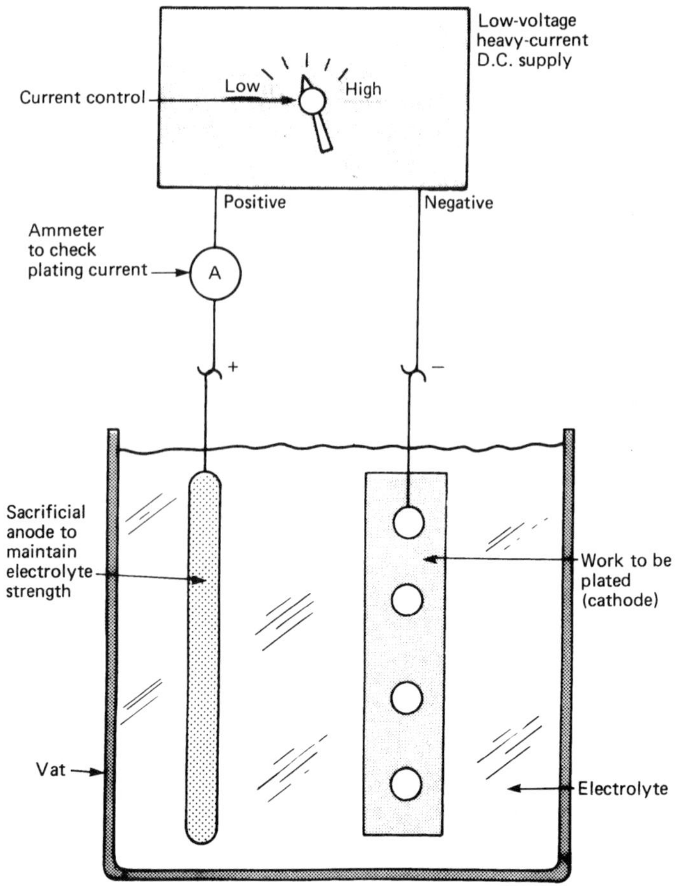

Figure 1.113 *Electroplating*

Electrolytic galvanizing

This is the coating of low carbon steels with a layer of zinc. It is an electroplating process as described earlier. The metal deposited is zinc and the process is usually limited to flat and corrugated sheet. It is quicker and cheaper than hot-dip galvanizing (see: coating), but the coating is thinner.

Any corrosive attack on galvanized products eats away the zinc in preference to the iron. The zinc is said to be sacrificial. To prolong the life of galvanized sheeting it should be painted to protect the zinc itself from the atmosphere.

Test your knowledge 1.29

Give two examples of where electroplating is used as a decorative and protective finish. Name the type of plating used. Do not repeat the examples in the text

Surface finishing and coating

Grinding

A grinding wheel consists of abrasive particles bonded together. It does not 'rub' the metal away, it cuts the metal like any other cutting tool. Each abrasive particle is a cutting tooth. Imagine an abrasive wheel to be a milling cutter with thousands of teeth.

Wheels are made in a variety of shapes and sizes. They are also available with a variety of abrasive particle materials and a variety of bonds. It is essential to choose the correct wheel for any given job.

Figure 1.114(a) shows two types of electrically powered portable grinding machines used for dressing welds and for fettling castings. Care must be taken in its use and the operator should wear a suitable grade of protective goggles and a filter type respirator.

Figure 1.114(b) shows a double-ended, off-hand tool grinder. This is used for sharpening drills and lathe tools and other small tools and marking out instruments. It is essential to check that the guard is in place and that the visor and tool rest are correctly adjusted before commencing to use the machine. Grinding wheels can only be changed by a trained and certificated person.

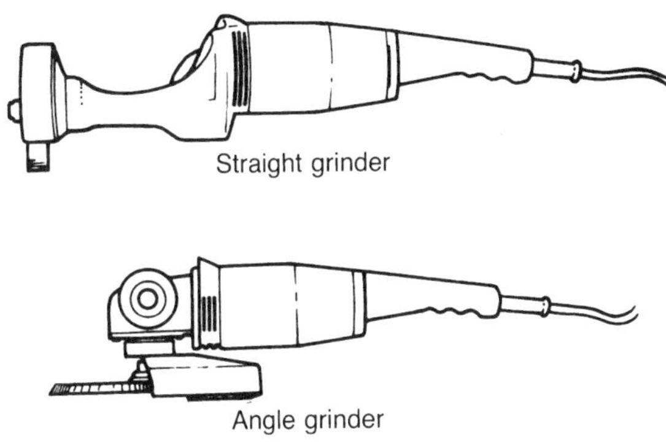

(a) Portable grinding machines

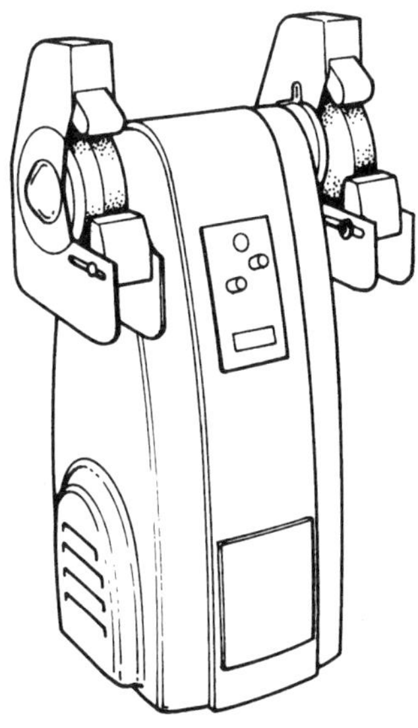

(b) Off-hand grinding machine

Figure 1.114 *Grinding machines*

Polishing

Polishing produces an even better finish than grinding but only removes the minutest amounts of metal. It only produces a smooth and shiny surface finish, the geometry of the surface is uncontrolled. Polishing is used to produce decorative finishes, to improve fluid flow through the manifolds of racing engines, and to remove machining marks from surfaces that cannot be precision ground. This is done to reduce the risk of fatigue failure in highly stressed components.

Figure 1.115 shows a typical polishing lathe. It consists of an electric motor with a double ended extended spindle. At each end of the spindle is a tapered thread onto which you screw the polishing mops. The mops may be made up from discs of leather (basils) or discs of cloth (calico mops). Polishing compound in the form of 'sticks' is pressed against the mops to impregnate them with the abrasive.

The components to be polished are held against the rapidly spinning mops by hand. Because the mops are soft and flexible they will follow the contours of complex shaped components. It is essential that dust extraction equipment is fitted to the polishing lathe and that the operator wears eye protection and a filter type dust mask. The process should only be carried out by a skilled polisher or under close supervision.

Coating

Electroplating has already been discussed and is the coating of metal components with another metal that is more decorative and/or corrosion resistant. Hot-dip galvanizing coats low carbon steels with zinc without using an electroplating process.

Hot-dip galvanizing

Hot-dip galvanizing is the original process used for zinc coating buckets, animal feeding troughs and other farming accessories. It is also used for galvanized sheeting. The work to be coated is chemically cleaned, fluxed and dipped into the molten zinc. This forms a coating on the work. A small percentage of aluminium is added to the zinc to give the traditional bright finish. The molten

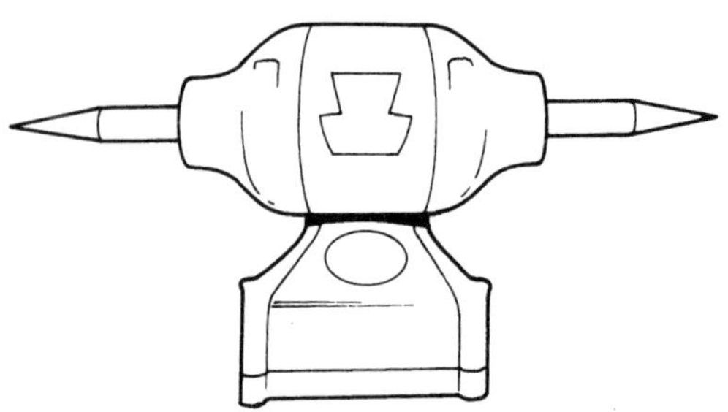

Figure 1.115 *A typical polishing lathe*

zinc also seals any cut edges and joints in the work and renders them fluid tight. Metal components may also be coated with non-metallic surfaces.

Oxidizing

Oil blueing

Steel components have a natural oxide film due to their reaction with atmospheric oxygen. This film can be thickened and enhanced by heating the steel component until it takes on a dark blue colour. Then immediately dip the component into oil to seal the oxide film. This process does not work if there is any residual mill scale on the metal surfaces.

Chemical blacking

Alternatively, an even more corrosion resistant oxide film can be applied to steel components by chemical blacking. The components are cleaned and degreased. They are then immersed in the oxidizing chemical solution until the required film thickness has been achieved. Finally the treated components are rinsed, dewatered and oiled. Again, the process only works on bright surfaces.

Plastic coating

Plastic coatings can be both functional, corrosion resistant and decorative. The wide range of plastic materials available in a wide variety of colours and finishes provides a designer with the means of achieving one or more of the following.

- abrasion resistance
- cushioning effects with coatings up to 6 mm thick
- electrical and thermal insulation
- flexibility over a wide range of temperatures
- non-stick properties (Teflon PTFE coatings)
- permanent protection against weathering and atmospheric pollution, resulting in reduced maintenance costs
- resistance to corrosion by a wide range of chemicals
- the covering of welds and the sealing of porous castings.

To ensure success, the surfaces of the work to be plasticized must be physically and chemically clean and free from oils and greases. The surfaces to be plasticized must not have been plated, galvanized or oxide treated.

Fluidized bed dipping

Finely powdered plastic particles are suspended in a current of air in a fluidizing bath as shown in Figure 1.116. The powder continually bubbles up and falls back and looks as though it is boiling. It offers no resistance to the work to be immersed in it. The work is preheated and immersed in the powder. A layer of powder melts onto the surface of the metal to form a homogeneous layer.

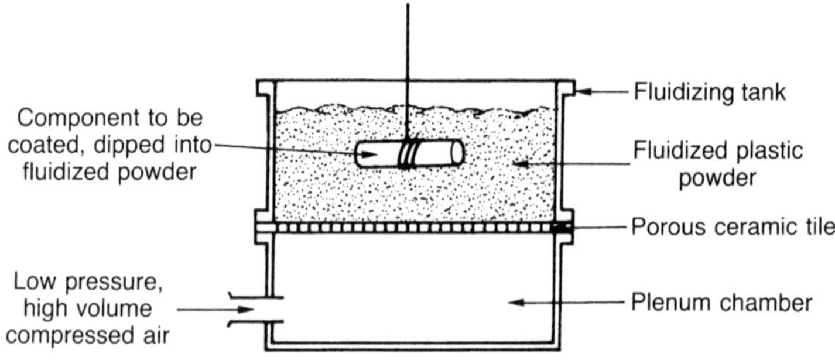

Figure 1.116 *Fluidized bed dipping*

Liquid plastisol dipping
This process is limited to PVC coating. A plastisol is a resin powder suspended in a plastisol and no dangerous solvents are present. The preheated work is suspended in the PVC plastisol until the required thickness of coating has adhered to the metal surface.

Painting

Painting is used to provide a decorative and corrosion resistant coating for metal surfaces. It is the easiest and cheapest means of coating that can be applied with any degree of permanence. A paint consists of three components:

- *Pigment* The finely powered pigment provides the paint with its opacity and colour.
- *Vehicle* This is a film-forming liquid or binder in a volatile solvent. This binder is a natural or synthetic resinous material. When dry (set) it must be flexible, adhere strongly to the surface being painted, corrosion resistant and durable.
- *Solvent (thinner)* This controls the consistency of the paint and its application. It forms no part of the final paint film as it totally evaporates. As it evaporates it increases the concentration of catalyst in the 'vehicle' causing it to change chemically and set.

A paint system consists of the following components:

- *Primer* This is designed to adhere strongly to the surface being painted and to provide a key for the subsequent coats. It also contains anti-corrosion compounds.
- *Putty or filler* This is used mainly on castings to fill up and repair blemishes. It provides a smooth surface for subsequent paint coats.
- *Undercoat* This builds up the thickness of the paint film. To produce a smooth finish, more than one undercoat is used with careful rubbing down between each coat. It also gives richness and opacity to the colour.
- *Top coat* This coat is decorative and abrasion resistant. It also seals the paint film with a waterproof membrane.

Modern top coats are usually based on acrylic resins or polyurethane rubbers.

There are four main groups of paint:

- *Group 1* The vehicle is polymerized (see thermosetting plastics) into a solid film by reaction with atmospheric oxygen. Paints that set naturally in this way include traditional linseed oil based paints, oleo-resinous paints, and modern general purpose air drying paints based on modified alkyd resins.
- *Group 2* This group of paints is based on amino-alkyd resins that do not set at room temperatures but they have to be stoved at 110–150°C. When set these paints are much tougher than group 1 air-drying paints.
- *Group 3* These are the 'two-pack' paints. Polymerization starts to occur as soon as the catalyst is mixed with the paint immediately before use. Modern 'one-pack' versions of these paints have the catalyst diluted with a volatile solvent as mentioned earlier. The solvent evaporates after the paint has been spread and, when the concentration of the catalyst reaches a critical level, polymerization takes place and the paint sets.
- *Group 4* These paints dry by evaporation of the solvent and no polymerization occurs. An example is the cellulose paint used widely at one time for spray painting motor car body panels. Lacquers also belong to this group but differ from all other paints in that dyes are used as the colorant instead of pigments.

Paints may be applied by brushing, spraying or dipping. What ever method is used, great care must be taken to ensure adequate ventilation. Not only can the solvents produce narcotic effects, but inhaling dried particles and liquid droplets of paint can cause serious respiratory diseases. The appropriate protective clothing, goggles and face masks must be used. Paints are also highly flammable and the local fire-prevention officer must be consulted over their storage and use. On no account can smoking be tolerated anywhere near the storage or use of paints.

Activity 1.9

1 State the essential differences between grinding and polishing.

2 Describe a suitable application process and a suitable coating medium for:

(a) painting refrigerator body panels;

(b) painting pressed steel angle brackets;

(c) plastic cladding metal tubing for bathroom towel rails.

Multiple-choice Questions

1 A suitable material for the pin contacts of standard 13A mains plugs is

A aluminium
B brass
C steel
D cast iron.

2 A metal that is to be drawn out into fine wire must have the property of

A ductility
B malleability
C elasticity
D compressibility.

3 The property that is tested by pressing a hard steel ball into a metal specimen by means of a standard force is

A hardness
B strength
C malleability
D toughness.

4 A resistor is marked with the following coloured bands; BROWN, RED, ORANGE, GOLD. The value of the resistor is

A 123 Ω 5%
B 123 Ω 10%
C 12 kΩ 5%
D 12 kΩ 10%.

5 A suitable machine for making cylindrical components is a

A lathe
B milling machine
C drilling machine
D portable grinding machine.

6 The correct sequence of operation for producing a large diameter hole is

A centre punch, mark out, main drill, pilot drill
B centre punch, mark out, pilot drill, main drill
C mark out, pilot drill, centre punch, main drill
D mark out, centre punch, pilot drill, main drill.

7 After heating and quenching a carbon steel, its final properties are given by

A tempering
B normalizing
C annealing
D case-hardening.

8 To produce a printed circuit from a copper clad laminated plastic board, the procedure is

A apply photoresist, develop, expose, etch
B develop, apply photoresist, expose, etch

C etch, expose, apply photoresist, develop
D apply photoresist, expose, develop, etch.

9 The correct sequence when painting a casting is

A degrease, undercoat, fill, prime, top-coat
B degrease, fill, undercoat, prime, top-coat
C prime, degrease, undercoat, fill, top-coat
D degrease, prime, fill, undercoat, top-coat.

10 Under the Health and Safety at Work Act

A only the employer is responsible for safe working practices
B only the employee is responsible for safe working practices
C the employer and the employee are equally responsible for safe working practices
D only the Health and Safety Excecutive is responsible for safe working practices.

Unit 2 Graphical communication in engineering

Summary

This unit intoduces you to the more widely used methods of graphical communication you may meet with in the engineering industry. It is based on the recommendations of the British Standards Institute (BSI) and uses current BSI conventions, symbols and abbreviations as appropriate. The full standards are very lengthy and expensive publications. However, low-cost, abridged editions are available for student use. These are: PP7307 Graphical Symbols for use in Schools and Colleges, and PP7308 Engineering Drawing Practice for Schools and Colleges. These are more than adequate for your requirements and you should either purchase copies or obtain access to copies to supplement this text. The general advantages and limitations of Computer Aided Drawing/Design (CAD) are also introduced in this text. However, no detailed instruction on CAD techniques is included, since this will depend upon the hardware and software available to you.

Communicating engineering information

My dictionary defines communication as 'the act or means of conveying information'. There are many ways of conveying information that are used in the engineering industry. We can group these broadly as:

- the spoken word (oral communication)
- the written word
- the use of graphical representations (drawings, sketches, graphs, etc.).

The spoken word

The spoken word is widely used in the following situations:

- informal discussions either on the telephone or face to face
- formal presentations to groups of persons who all require the same information.

Where a group of persons all require the same information, a formal presentation must be used. On no account should information be 'passed down the line' from person to person. Errors are bound to creep in. There is a story that during World War I the message 'send reinforcements, we are going to advance' arrived at headquarters, by word of mouth, as 'send three and four pence (old money), we are going to a dance'. I will let you decide on the truth behind this story, but I feel it makes the point.

It is important to remember that the spoken word is easily forgotten and oral communication should be reinforced by:

- notes taken at the time
- tape recording the conversation
- a written summary. For example the published 'proceedings' of formal lectures and presentations. Another example is the 'press release' to ensure factual accuracy of information intended for the public.

Oral communication must be presented in a manner appropriate to the audience. It must be brief and to the point. The key facts must be emphasized so that they can be easily remembered. The presentation must be interesting so that the attention of the audience does not wander.

When communicating by the spoken word, it is as equally important to be a good listener as it is to be a good speaker. This applies to conversations between two or three people as well as to formal presentations.

Written communication

This is a more reliable method of communication since it provides a permanent record of the key information. The same information is available for all the persons who require it.

Anyone who has ever marked an English comprehension test will know that the same written passage can mean very different things to different persons. Therefore, care must be taken in preparing written information. To avoid confusion, the normal conventions of grammar and punctuation must be used. Words must be correctly spelt. Use a dictionary if you are uncertain. If you are using a word processing package use the spell-checker. However, take care, many software packages originate in the USA and the spell-checker may reflect this.

Never use jargon terms and acronyms (go on, look it up) unless you are sure that the persons reading the message are as equally familiar with them as is the writer.

An engineer often has to write notes, memoranda, and reports. He/she often has to maintain logbooks and complete service sheets. An engineer may also have to communicate with other engineers, suppliers and customers by letter. Practice in the writing of clear and concise messages is of great importance.

Graphical communication

Engineers rely heavily upon graphical methods of communication. Drawings and charts produced to international standards using international symbols and conventions suffer no language barriers. They are not liable to be misinterpreted by translation errors. Graphical communication does not replace spoken and written communication. It is used to simplify, reinforce and complement other means of communication.

Design office

A designer will conceptualize or 'dream up' the design for a new product using free hand sketches and often simple cardboard models. These preliminary stages will include engineering the design of the product so that it will function correctly, and styling the product so that it is attractive to the customer.

This can be made easier by the use of advanced computer aided design (CAD) software. This advanced software enables three-dimensional representations to be prepared by 'wire-frame' or solid modelling. Such images can be rotated on the screen so that viewing is possible from all angles. Modifications to the design can be made at the touch of a key. Further, the function of the design can be checked by simulation techniques, before embarking on expensive prototype manufacture.

Once these preliminaries have been achieved, the design team will need to consult with representatives of the customer, with maintenance and field service engineers, and with the various bodies responsible for type approval. These may include insurance engineers and Local Authority environmental engineers.

Finally the design will need to be proved by the manufacture and testing of a full-scale prototype.

Drawing office

Once the design has been approved, technical drawings and specifications will be produced by draughtspersons. These drawings and specifications will vary depending upon for whom they are intended. The manufacturing engineer will want flow charts, and orthographic detail and general arrangement drawings. The service engineer will want circuit diagrams and exploded views. Industrial customers will want installation drawings and operating data. Drawings and specifications are also required for other purposes.

Marketing

Formal engineering drawings are not appropriate for most customers. Usually, sales material is produced by graphic artists who combine colour photographs with performance data, often in the form of graphs; for example the glossy brochures available in car showrooms. This type of material is closely geared to the potential customer. The pictorial material is usually enhanced by a graphic artist using airbrush techniques.

Estimating, costing and planning

These functions are closely linked and involve both engineers and accountants. Fully detailed drawings and specifications are required, together with block diagrams, flow diagrams and spreadsheets. The cost of every component and operation is estimated so that a provisional selling price can be arrived at. Cost targets (budgets) are set and actual production costs are constantly compared with these targets. If the targets are not achieved or improved upon, then either the selling price has to be increased or the manufacturing process modified. Planning involves the economical loading of the work onto the production shops so that the machines are kept fully employed and the work flows smoothly so that delivery dates are achieved and customer goodwill is maintained.

Manufacturing

The workshops will require general arrangement drawings and detail drawings. The former show the assembled unit and list all the parts that are to be manufactured together with the identification numbers of the detail drawings required. Also shown and listed will be all the standard 'bought in' components such as nuts and bolts. The detail drawings show all the information needed to manufacture the individual components; that is, the material to be used, the shape of the components, the dimensions of the component and any heat treatment or special surface finishing treatments. Figure 2.1 shows an example of a General Arrangement (GA) drawing and Figure 2.2 shows an example of a detail drawing.

In addition, assembly drawings will be required. These are often confused with general arrangment drawings that can fulfil this function for very simple assemblies. For more complex assemblies, 'exploded' drawings are often used together with flow charts showing the sequence in which the components should be fitted together.

Service/maintenance

Pictorial drawings (isometric or oblique) are often used to show the various lubrication points on a machine, the type of lubricant to be used and the frequency of lubrication. These pictorial drawings are often 'exploded' and list the spare parts numbers, as shown in Figure 2.3. Such drawings also show the relationships between the parts to aid dismantling and assembly.

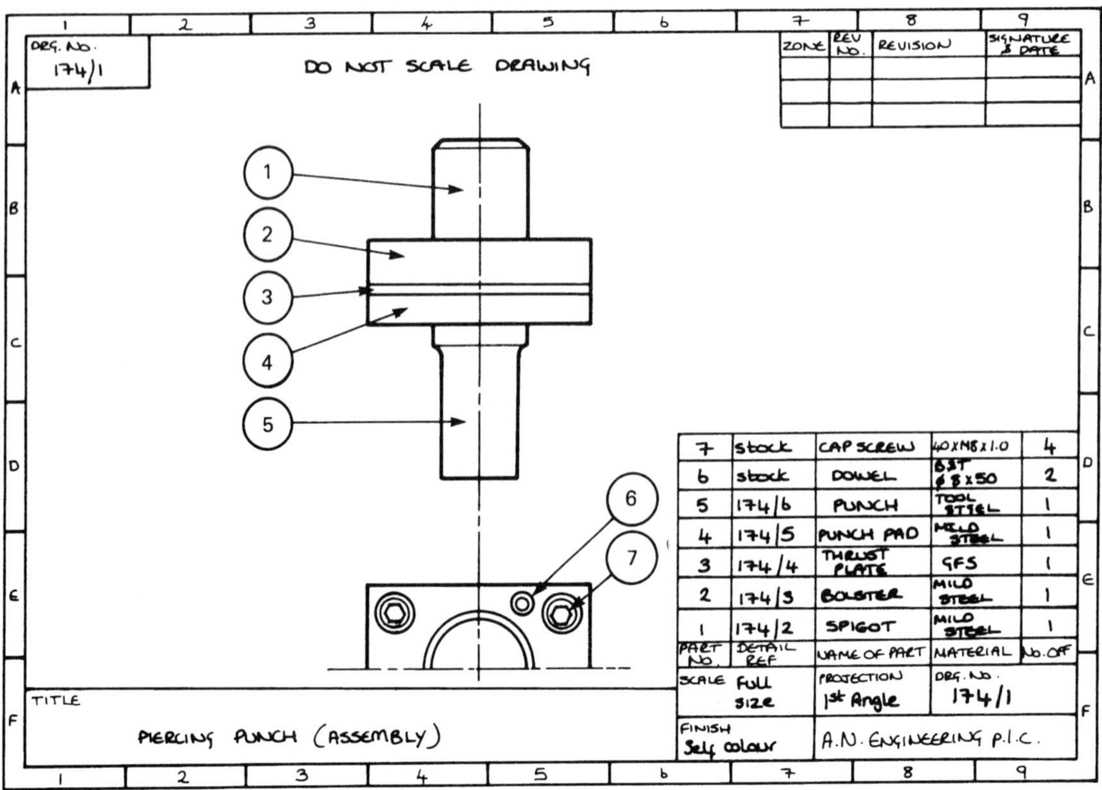

Figure 2.1 *General Arrangement (GA) drawing*

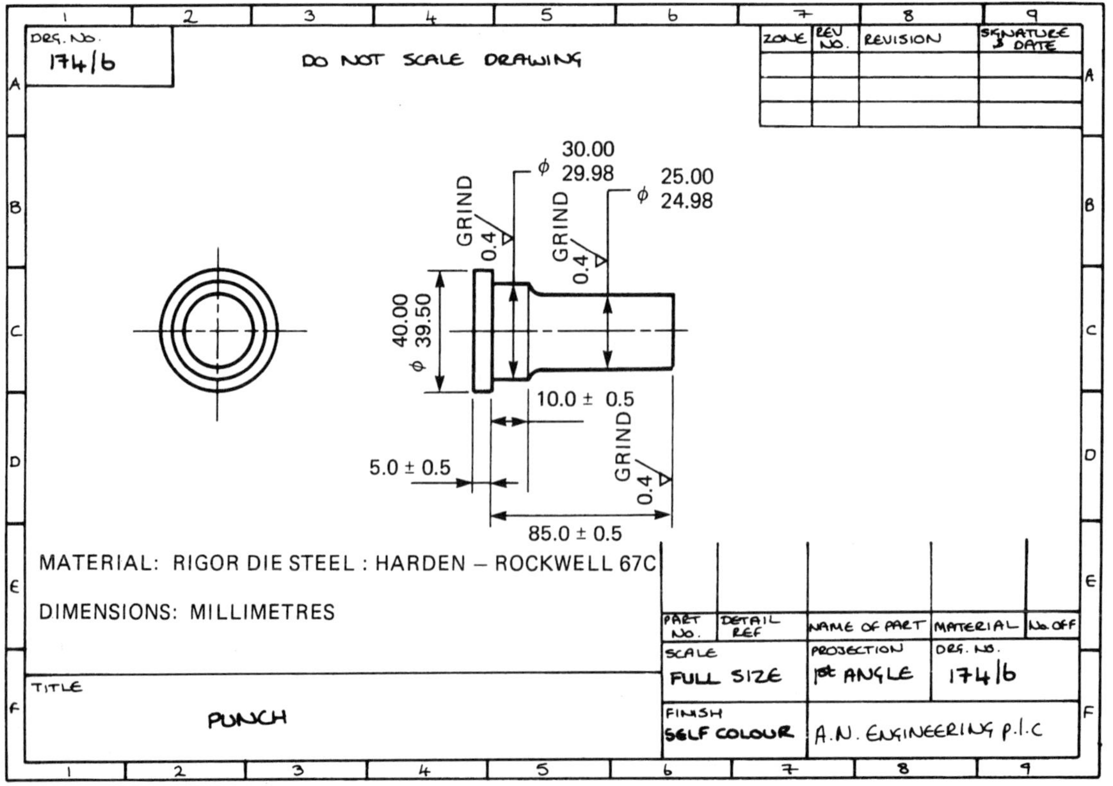

Figure 2.2 *Detail drawing*

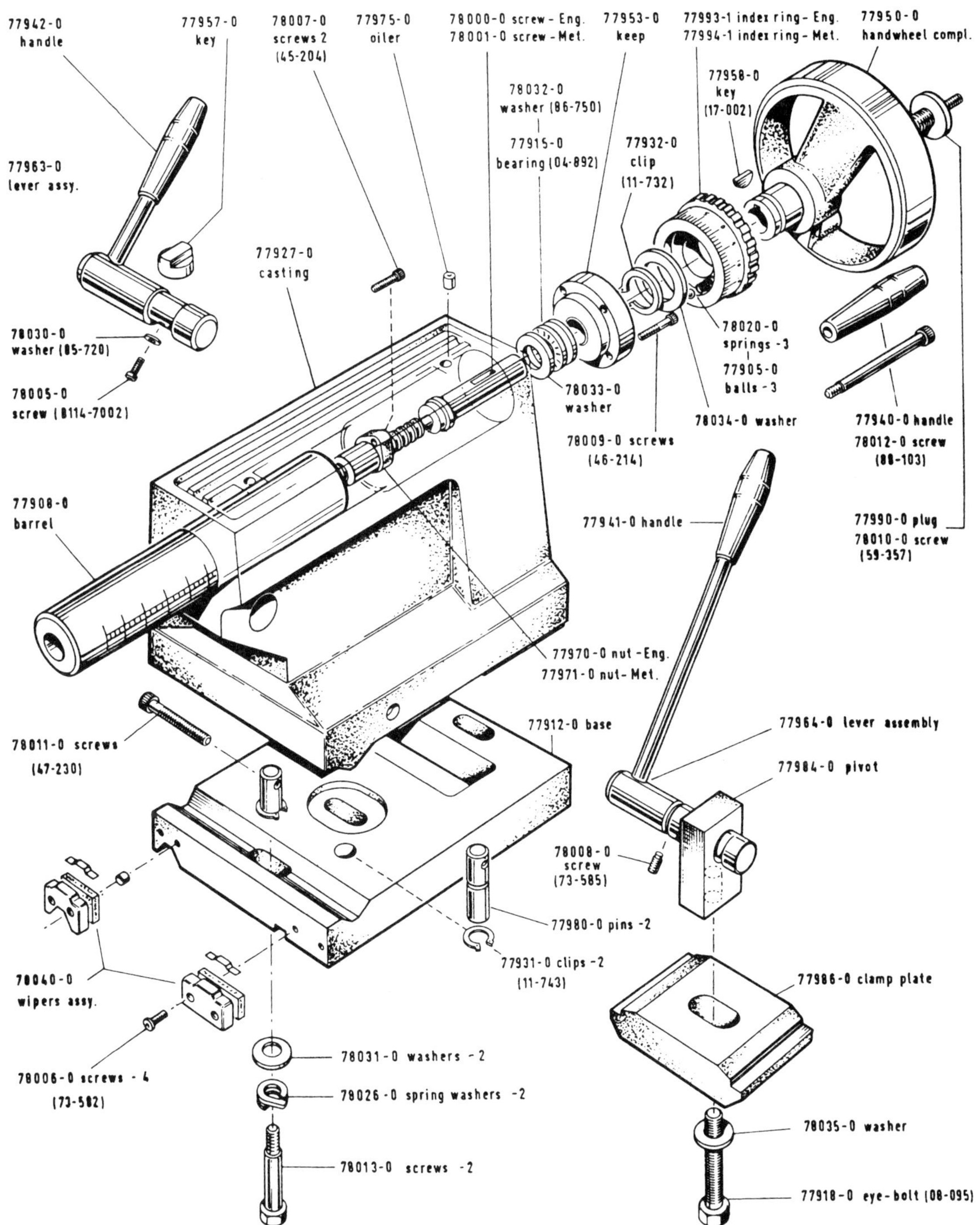

Figure 2.3 *Exploded view and parts list for tailstock assembly*

Activity 2.1

Compare and contrast the advantages and limitations of:

(a) oral communication

(b) written communication

(c) graphical communication.

Present your findings in the form of a three-minute talk recorded on a cassette tape to be broadcast by a local radio station.

Graphical methods for engineering information

Having now established the need for communicating engineering information, let's look at the various methods of graphical communication available. We can broadly divide engineering information into two categories: that which is mathematically based, and that which is technically based. Let's be logical and look at the former first.

Mathematical data

I expect that you will already have met most of the methods of expressing mathematical data by means of graphs. Nevertheless, let's revise the techniques available.

Line graphs

Just as engineering drawings are used as a clear and convenient way of describing complex components and assemblies, so can graphs be used to give a clear and convenient picture of the mathematical relationships between engineering and scientific quantities. Figure 2.4(a) shows a graph of the relationship between distance S and time t for the mathematical expression $S = (at^2)/2$ where the acceleration $a = 10\,\mathrm{m/s^2}$.

In this instance it is correct to use a continuous flowing curve to connect the points plotted. Not only do these points lie on the curve, but every corresponding value of s and t between the points plotted also lie on the curve.

However, this is not true for every type of line graph. Figure 2.4(b) shows a graph relating speed and distance for a journey. From A to B the vehicle is accelerating. From B to C the vehicle is travelling at a constant speed. From C to D the vehicle is decelerating (slowing down). In this example it is correct to join the points by straight lines This is because each stage of the journey is represented by a linear mathematical expression which is unrelated to the previous stage of the journey and unrelated to the following stage of the journey.

Test your knowledge 2.1

1 Given the formula $N = (1000S)/(yD)$ where:
N = spindle speed in rev/min
S = cutting speed = 33 m/min
D = drill diameter in mm
plot the graph connecting drill diameter and spindle speed for the following drill diameters: 2 mm, 4 mm, 6 mm, 8 mm, 10 mm.

2 From the graph determine the spindle speed of a drill 5 mm diameter.

3 Draw a graph connecting the following furnace temperatures taken at hourly intervals. 600°C, 700°C, 750°C, 700°C, 800°C, 600°C.

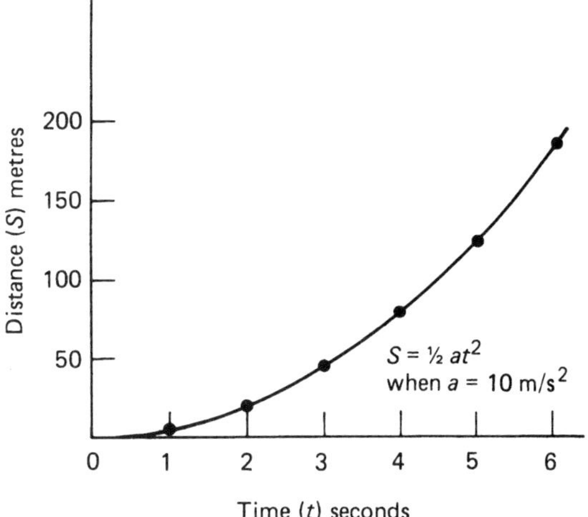

(a)

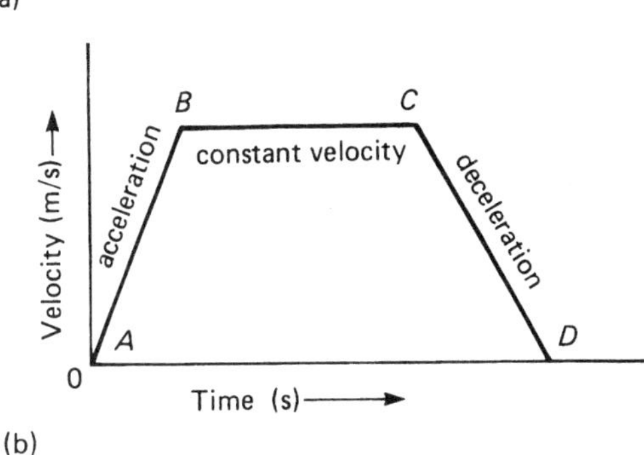

(b)

Figure 2.4 *Line graphs*

Histograms

Consider the student intake of a college over a number of years. The students enrol in September and leave in the following July. Between these times there is a negligible change in the number of students attending. Therefore to plot the enrolments for each September and connect these by a flowing curve between the points would be incorrect. Such graphs would imply that there is a continuous change in student numbers between one September and the next and that the change satisfies a mathematical equation.

It would also be incorrect to connect the plotted points by straight lines. This would imply that, although the points plotted are not following a mathematical expression, there is some continuous change in the number of students attending between one September and the next.

The correct way to show this sort of data is by means of a histogram as shown in Figure 2.5. This clearly compares the attendances for each college year. At the same time it shows that each year's attendance stands alone and is unrelated to the previous year and unrelated to the following year.

Test your knowledge 2.2

The total number of machines manufactured by a small firm each year is as follows.

1990	250 machines
1991	350 machines
1992	325 machines
1993	375 machines
1994	290 machines
1995	180 machines

Draw a histogram to compare these output quantities.

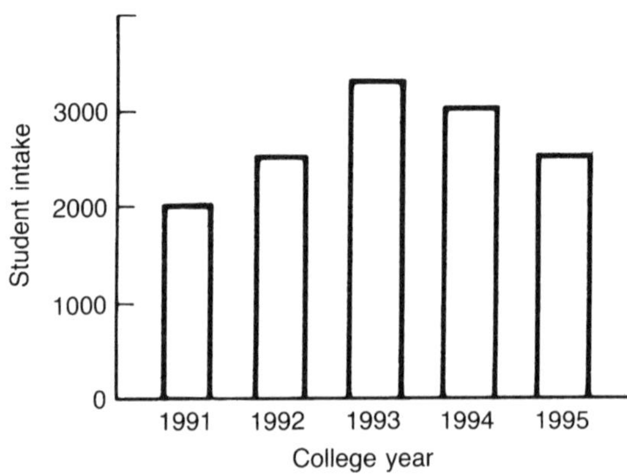

Figure 2.5 *A histogram*

Test your knowledge 2.3

Use an ideograph to show the following output statistics for an ice-cream manufacturer. Each symbol represents 10 000 cartons.

January	20 000 cartons
February	20 000 cartons
March	30 000 cartons
April	40 000 cartons
May	60 000 cartons
June	80 000 cartons

Bar charts

These are also used for displaying statistical data, but are usually plotted horizontally. They are widely used for recording work in progress, as shown in Figure 2.6.

Ideographs (pictograms)

These are frequently used to present statistical data to the general public. A typical example of the number of cars using a car park is shown in Figure 2.7. In this example each symbol represents 1000 cars. Therefore in 1994, 3000 cars used the car park each week (1000 cars multiplied by 3 symbols).

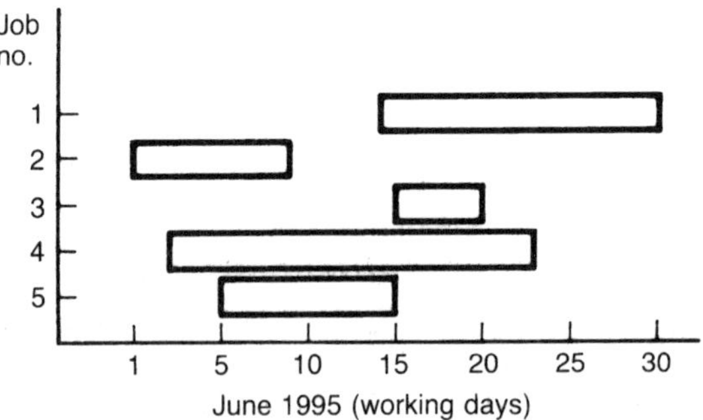

Figure 2.6 *A bar chart*

Figure 2.7 *An ideograph*

Test your knowledge 2.4

Draw a pie-chart of the following data which represents the breakdown of costs for a unit product of a company:

Labour costs	50%
Material costs	25%
Overhead costs	12.5%
Profit	12.5%

Pie-charts

These are used to show how a total quantity is divided up into its individual parts. Let's look at Figure 2.8(a). Since a complete circle is 360°, we can represent 25% of a complete circle as $360° \times 25/100 = 90°$. Figure 2.8(b) shows how the expenditure of a company can be represented by a pie-chart.

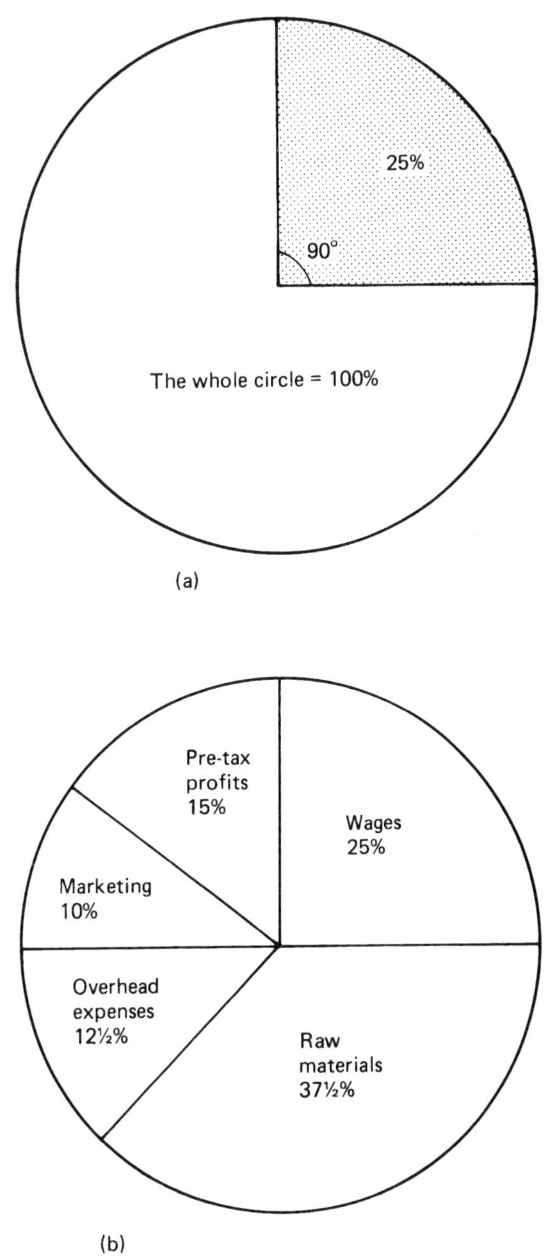

Figure 2.8 *Pie-charts*

Activity 2.2

(a) Figure 2.9 shows a graph relating electrical potential and current for a particular circuit.

(i) State the magnitude of the current in amperes when the potential is 30 V.

(ii) State the magnitude of the potential in volts when the current is 4 A.

(b) (i) Name the type of graph shown in Figure 2.10.

(ii) State how many students attended in: (i) 1990, (ii) 1991, (iii) 1994.

(iii) State how many symbols would represent 850 students.

(c) Draw a histogram to represent the attendances shown in Figure 2.10.

(d) The breakdown of costs for a product is: Materials = £500, Labour = £1000, Overheads = £250, Profit = £250

Represent these costs by means of a pie-chart.

Technical information and drawings.

Like the graphs that we have just considered, there are many different ways of representing and communicating technical information. To avoid confusion such information should make use of nationally and internationally recognized symbols, conventions and abbreviations. These are listed and their use explained in the appropriate British Standards. Such standards

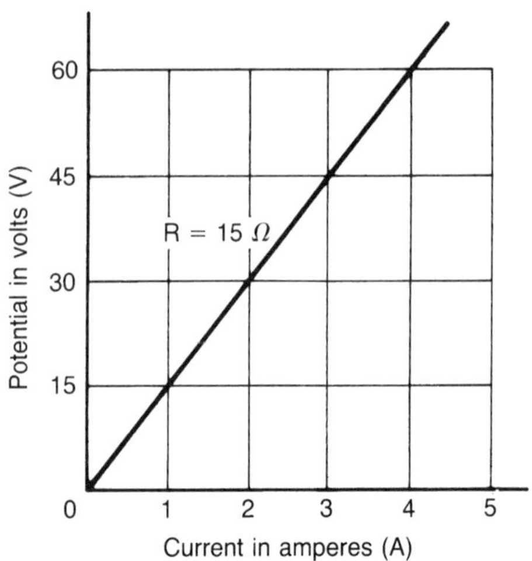

Figure 2.9 *See Activity 2.2*

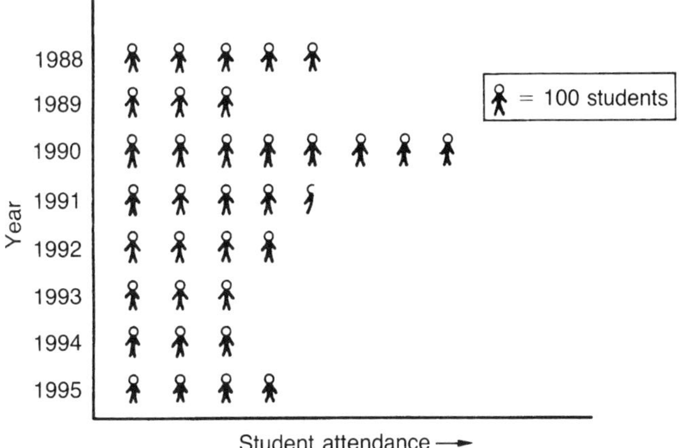

Figure 2.10 *See Activity 2.2*

are lengthy and costly. Low-cost summaries of these standards are available for students. These abridged editions are:

PP7307 Graphical Symbols for Use in Schools and Colleges.
PP7308 Engineering Drawing Practice for Schools and Colleges (abridged from BS 308).

> **Test your knowledge 2.5**
>
> Use a block diagram to show how the engine drives the road wheels of a car via such elements as the differential, clutch and gearbox.

Block diagrams

These show the relationship between the various elements of a system. Figure 2.11(a) shows a block diagram for the drive mechanism of a centre lathe and Figure 2.11(b) shows the block diagram for a simple radio receiver. These sorts of diagrams are used in the initial stages of conceptualizing a design.

> **Test your knowledge 2.6**
>
> Your bicycle tyre is flat. Draw up a flow chart for testing a bicycle tyre inner-tube for a puncture and repairing the puncture if there is one. Figure 2.14 shows you how to start the flow chart.

Flow diagrams

These are used to test the logic of a sequence of events. They are used for a variety of purposes by computer software engineers when planning a new program, and by production engineers in working out the best sequence of operations in which to manufacture a component. Figure 2.12 shows a flow chart for drilling a hole. The shape of the 'boxes' used in this flow chart have particular meanings, as shown in Figure 2.13. For the complete set of 'boxes', their meanings and uses see British Standard PP3707.

Circuit diagrams

These are used to show the functional relationships between the components in a circuit. The components are represented by symbols and their position in the circuit diagram does not represent their actual position in the final assembly. Circuit diagrams are also referred to as schematic diagrams or even schematic circuit diagrams.

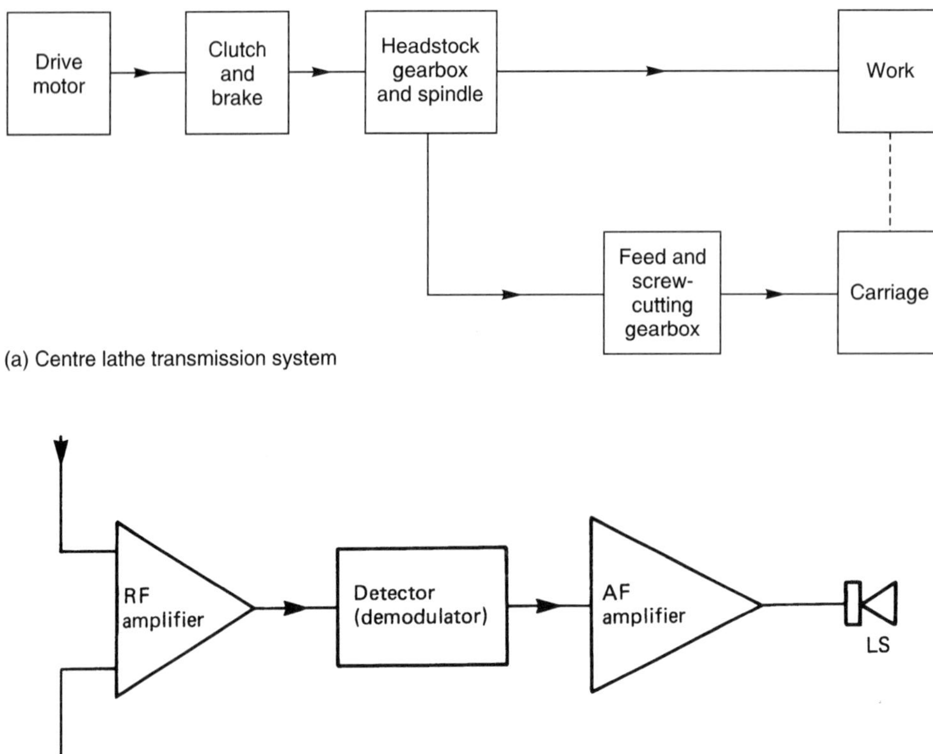

(a) Centre lathe transmission system

(b) Simple AM radio receiver

Figure 2.11 *Block diagrams*

Figure 2.15(a) shows the circuit for an electronic filter unit using standard component symbols. Figure 2.15(b) shows a layout diagram with the components correctly positioned. Figure 2.15(c) shows the printed circuit board. This is also called a track diagram or a wiring diagram.

However, it is more usual to use the term wiring diagram where the components are hard wired, as in the wiring up of a building or the manufacture of a control cubicle. Architects use circuit diagrams to show the electrical installation of buildings. They also provide installation drawings to show where the components are to be sited. They may also provide a wiring diagram to show how the cables are to be routed to and between the components. The symbols used in architectural installation drawings and wiring diagrams are not the same as those used in circuit diagrams. Examples of architectural and topographical electrical component symbols are shown in BS: PP7307.

Schematic circuit diagrams are also used to represent pneumatic (compressed air) circuits and hydraulic circuits. Pneumatic circuits and hydraulic circuits share the same symbols. You can tell which circuit is which because pneumatic circuits should have open arrow heads, whilst hydraulic circuits should have solid arrowheads. Also, pneumatic circuits exhaust to the atmosphere, whilst hydraulic circuits have to have a return path to the oil reservoir. Figure 2.16 shows a typical hydraulic circuit.

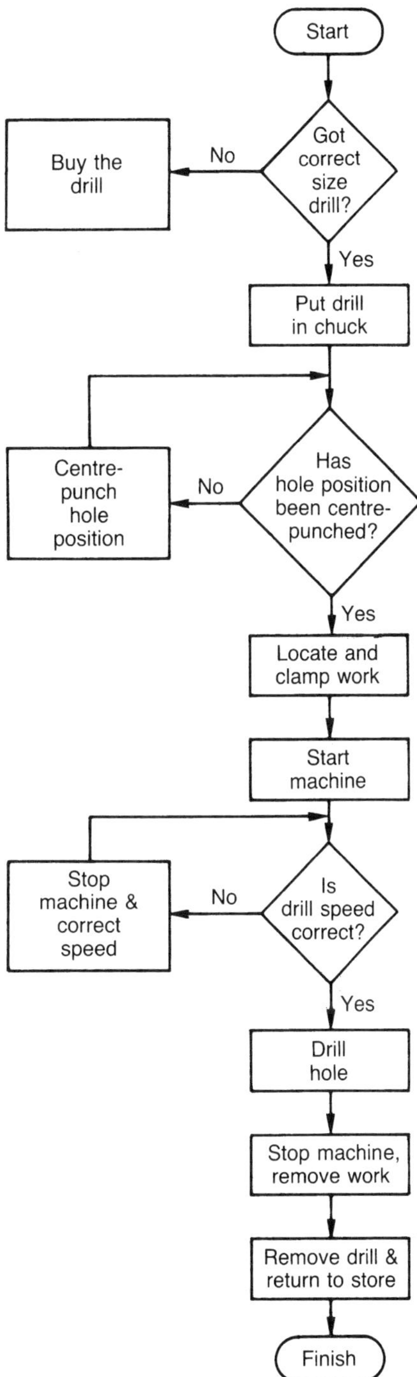

Figure 2.12 *Flow chart for drilling a hole*

Just as electrical circuit diagrams may have corresponding installation and wiring diagrams, so do hydraulic, pneumatic and plumbing circuits. Only this time the wiring diagram becomes a pipe-work diagram. A plumbing example is shown in Figure 2.17. As you may not be familiar with the symbols, I have named them for you. Normally this is not necessary and the symbols are recognized by their shapes.

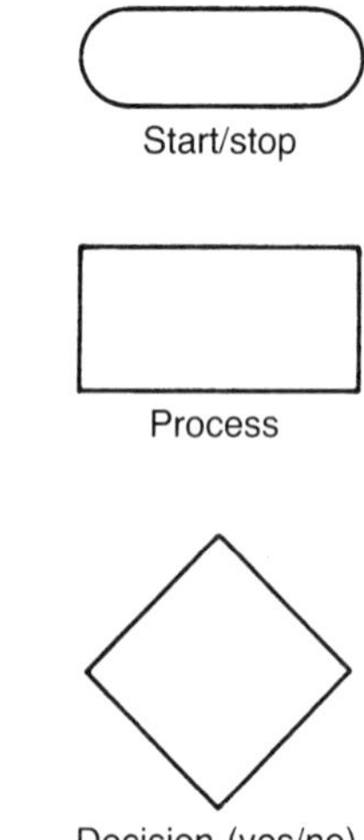

Figure 2.13 *Some common flow chart symbols*

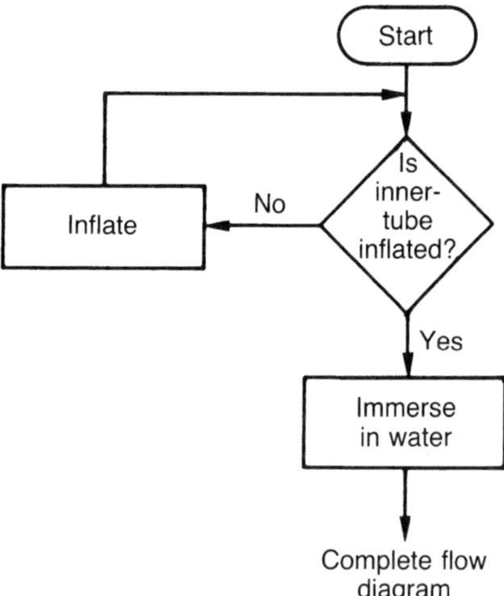

Figure 2.14 *See Test your knowledge 2.6*

General arrangement (GA) drawings

Figure 2.18 shows the layout of a typical drawing sheet. To save time these are printed to a standardized layout for a particular company, ready for the draughtsperson to add the drawing and complete the boxes and tables.

The basic information found on most drawing sheets consists of:

- the drawing number and name of the company
- the title and issue details
- scale
- method of projection (first or third angle)
- initials of persons responsible for: drawing, checking, approving, tracing, etc., together with the appropriate dates
- unit(s) of measurement (inches or millimetres) and general tolerances

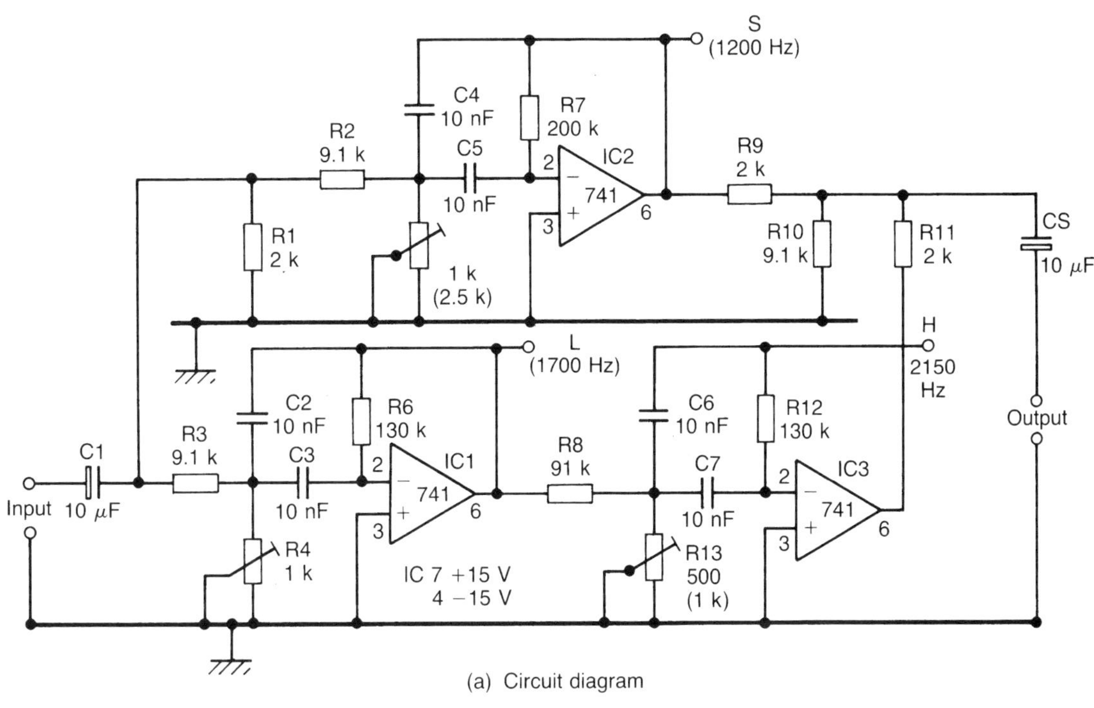

(a) Circuit diagram

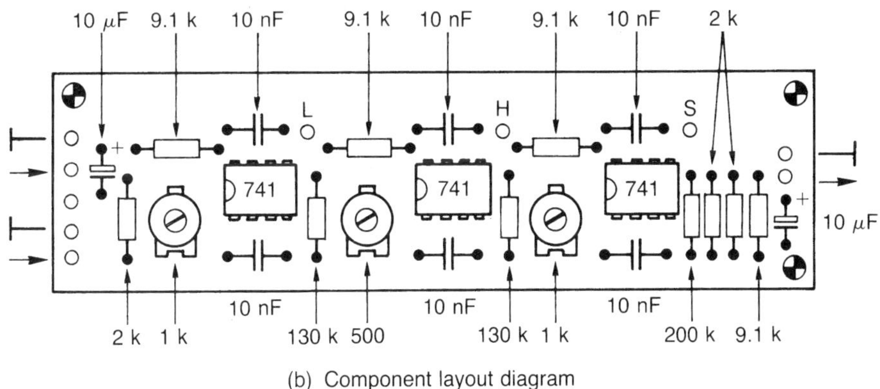

(b) Component layout diagram

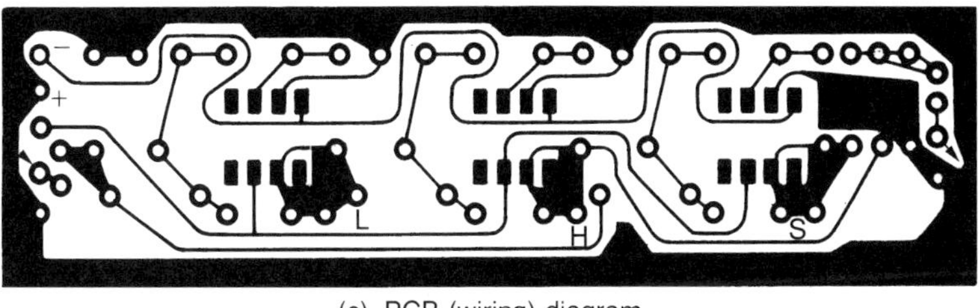

(c) PCB (wiring) diagram

Figure 2.15 *A typical electronic circuit diagram with corresponding component layout diagram and PCB (wiring) diagram*

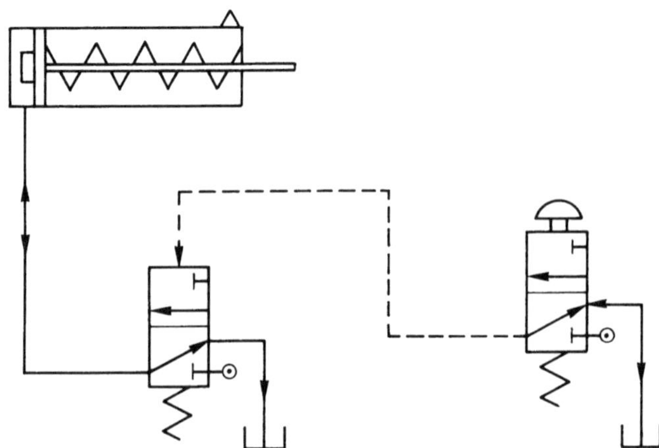

Figure 2.16 *A typical hydraulic circuit*

- material and finish
- copyright and standards reference
- guidance notes such as: 'do not scale'
- reference grids so that 'zones' on the drawing sheet can be quickly found
- modifications table for alterations which are reference related to the issue number on the drawing and identified by the means of the reference grid.

The following additional information may also be included:

- fold marks
- centre marks for camera alignment when microfilming
- line scale, so that the true size is not lost when enlarging or reducing copies
- trim marks
- orientation marks.

Figure 2.19 shows a typical General Arrangement (GA) drawing. This shows as many of the features listed above as are appropriate for this drawing. It shows all the components correctly assembled together. Dimensions are not usually given on GA drawings although, sometimes, overall dimensions will be given for reference when the GA drawing is of a large assembly drawn to a reduced scale.

The GA drawing shows all the parts. These are listed in a table together with the quantities required. Manufacturers' catalogue references are also given for bought-in components. The detail drawing numbers are also included for components that have to be manufactured as special items.

Detail drawings

As the name implies, detail drawings provide all the details required to make the component shown on the drawing. Referring to Figure 2.19, we see from the table that the detail drawing for the punch has the reference number 174/6. Figure

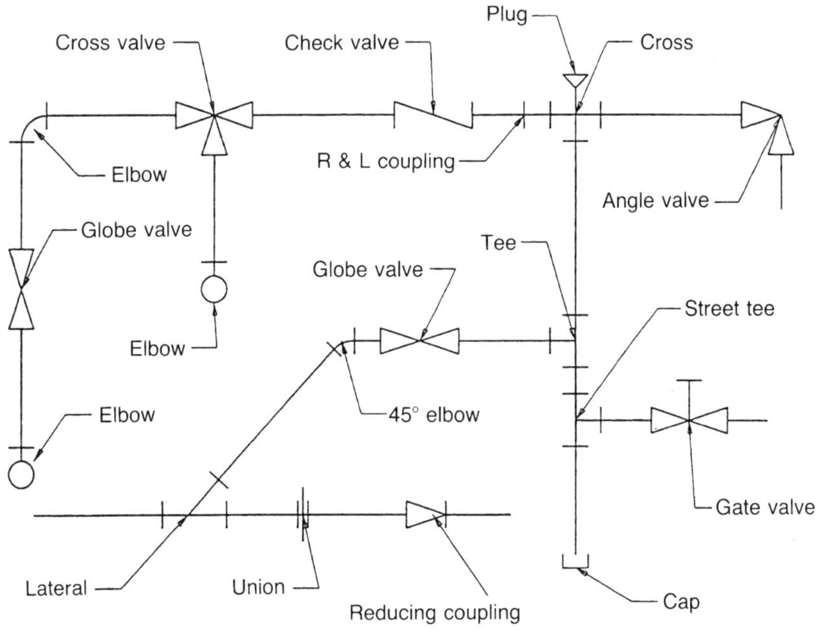

(a) Circuit diagram (schematic)

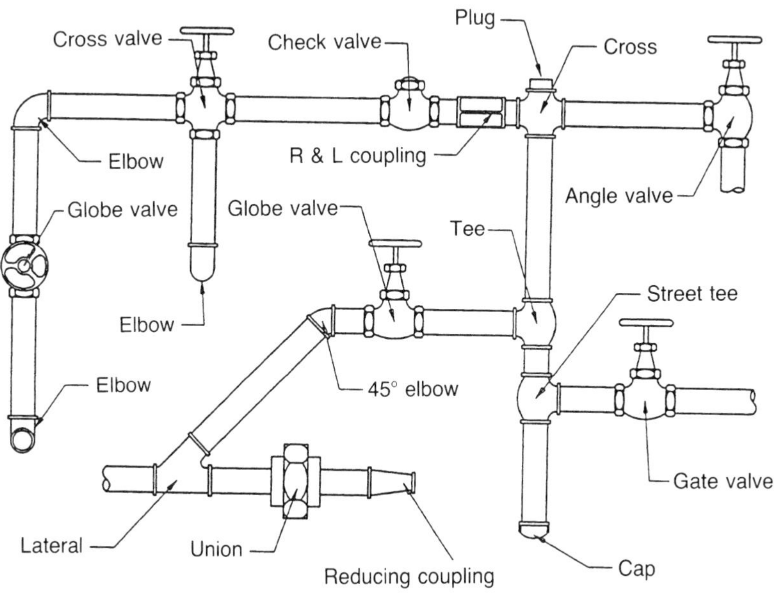

(b) Piping diagram

Figure 2.17 *A typical plumbing circuit with corresponding piping diagram*

2.20 shows this detail drawing. In this instance, the drawing provides the following information:

- the shape of the punch
- the dimensions of the punch and the manufacturing tolerances
- the material from which the punch is to be made and its subsequent heat treatment

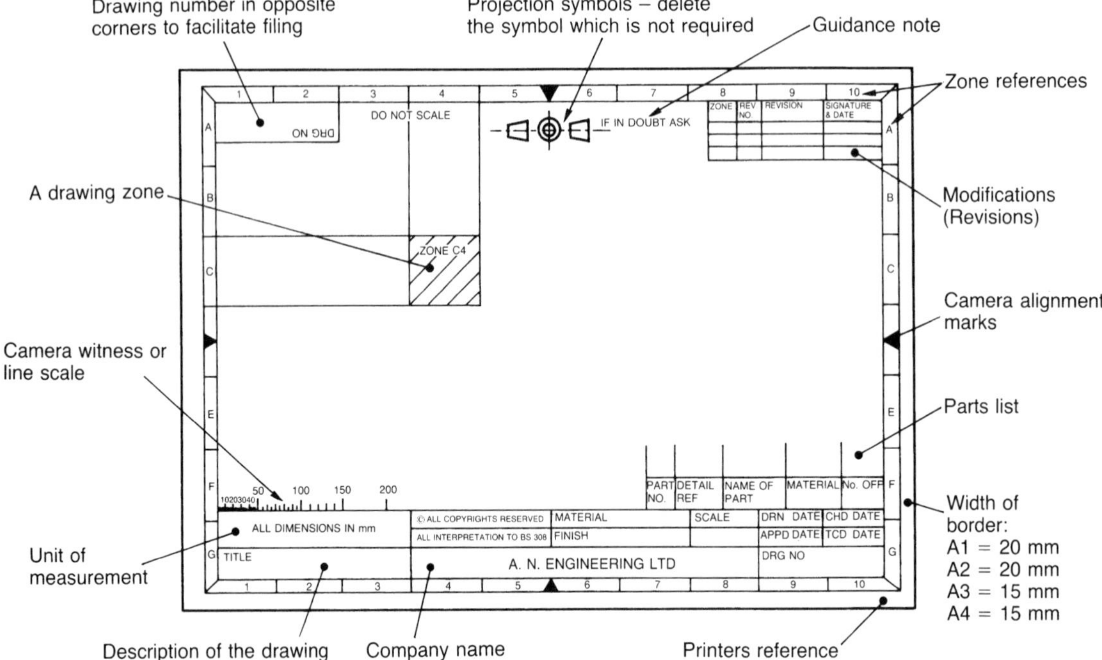

Figure 2.18 *Layout of a typical drawing sheet*

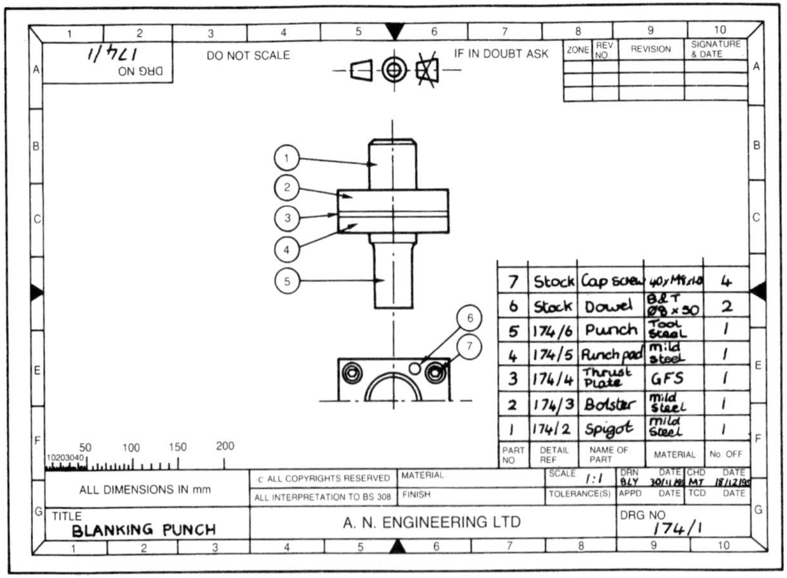

Figure 2.19 *A typical General Arrangement (GA) drawing*

- the unit of measurement (millimetre)
- the projection (first angle)
- the finish
- the guidance note 'do not scale drawing'
- the name of the company
- the name of the draughtsperson
- the name of the person checking the drawing.

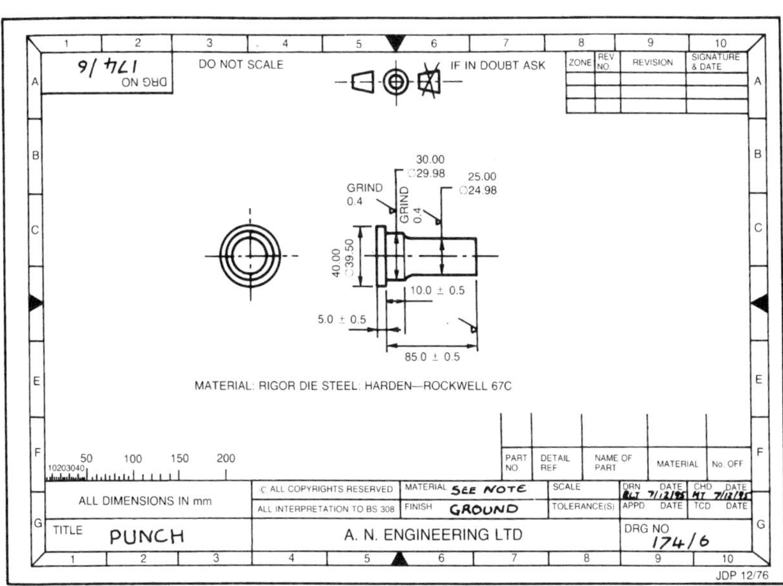

Figure 2.20 *A typical detail drawing*

The amount of information given will depend upon the job. Drawings for a critical aircraft component will be much more fully detailed than a drawing for a wheelbarrow component.

Data storage

There are many different ways of storing technical data.

- *Tracing linen* This was the traditional material for making technical drawings. It was strong and durable and stood up well to the effects of the ultra-violet arc lamps used for making 'blue-print' copies. They got their name from the fact that the print appeared as white lines on a blue background.
- *Tracing paper* This is widely used in conjunction with manual drawing techniques. It is cheap and readily available. It is also easy to draw on. Unfortunately the paper becomes brittle with age and requires careful handling. Therefore it is not suitable where print copies have to be made frequently.
- *Tracing film* This is tough plastic film that is shiny on one side and matt on the other. You draw on the matt surface. No special techniques are required in its use and it stands up to repeated handling without deterioration. It is more expensive than tracing paper.
- *Microfilm* The storage of full size 'negatives', as the tracings are called, when produced on linen, paper or film takes up a lot of room. These large drawings can be reduced photographically onto 16 mm or 35 mm film stock for storage. This saves considerable space. The microfilm

copies can be projection printed (enlarged) full-size when required for issue.

- *Microfiche* Libraries, offices and stores use microfiche systems. Data is stored photographically in a grid of frames on a large rectangle of film. A desktop viewer is used to select and enlarge a single frame. The frame is then projected onto a rear projection screen for easy reading. This system is more likely to be used for storing literal and numerical data than for drawings.
- *Electronic* Computer aided drawing and design (CAD) is rapidly taking over from manual drawing and tracing. The advantages of CAD will be discussed later in this Unit. CAD software is used in conjunction with a computer and the 'drawing' produced on the computer screen is saved in a computer file on disk. This is the most economical method of data storage. Very many complex drawings can be saved on a single 'floppy' disk. When required the file can be recalled for immediate viewing on a computer screen and hard copy can be printed out to any desired scale at the touch of a key.

Activity 2.3

(a) Compare the printed drawing sheets provided by your tutor with the model sheet shown in Figure 2.18. List any differences.

(b) Examine the general arrangement (GA) drawings provided by your tutor and compare them with the model GA shown in Figure 2.19. List any differences.

(c) Examine the detail drawings provided by your Tutor and compare them with the model detail drawing shown in Figure 2.20. List any differences.

Activity 2.4

Draw up a flow diagram for quench hardening a medium carbon steel. Don't forget to allow for decision making, e.g. is metal hot enough? Produce your flowchart in the form of an overhead projector transparency.

Activity 2.5

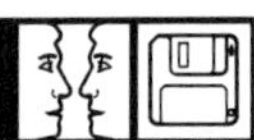

Draw a block diagram showing the transmission path for a milling machine. Produce your block diagram using a computer drawing or CAD package.

Activity 2.6 

List the basic equipment required for making a technical drawing manually and explain what each item is used for. Present your answer in the form of a word processed instruction card suitable for use by someone who is a complete newcomer to technical drawing.

Engineering drawing techniques

Engineering drawings can be produced either manually or on a computer using suitable CAD software. Drawings produced manually can range from freehand sketches to formally prepared drawings produced with the aid of a drawing board and conventional drawing instruments.

The choice of technique is dependent upon a number of factors such as:

- *Speed* How much time can be allowed for producing the drawing. How soon the drawing can be commenced.
- *Media* The choice will depend upon the equipment available (e.g. CAD or conventional drawing board and instruments) and the skill of the person producing the drawing.
- *Complexity* The amount of detail required and the anticipated amount and frequency of development modifications.
- *Cost* Engineering drawings are not works of art and have no intrinsic value. They are only a means to an end and should be produced as cheaply as possible. Both initial and on-going costs must be considered.
- *Presentation* This will depend upon who will see/use the drawings. Non-technical people can visualize pictorial representations better than orthographic drawings.

Nowadays technical drawings are increasingly produced using computer aided drawing techniques (CAD). Developments in software and personal computers have reduced the cost of CAD and made it more powerful. At the same time, it has become more 'user friendly'. Computer aided drawing does not require the high physical skill required for manual drawing and which takes years of practice to achieve. It also has a number of other advantages over manual drawing. Let's consider some of these advantages:

- *Accuracy* Dimensional control does not depend upon the draughtsperson's eyesight.
- *Radii* These can be made to blend with straight lines automatically.
- *Repetitive features* For example holes round a pitch circle do not have to be individually drawn but can be easily produced automatically by 'mirror imaging'. Again, some repeated, complex feature need only be drawn once and saved as a matrix. It can then be called up from the computer memory at each point in the drawing where it appears at the touch of a key.

- *Editing* Every time you erase and alter a manually produced drawing on tracing paper or plastic film the surface of the drawing is increasingly damaged. On a computer you can delete and redraw as often as you like with no ill effects.
- *Storage* No matter how large and complex the drawing, it can be stored digitally on floppy disk. Copies can be taken and transmitted between factories without errors or deterioration.
- *Prints* Hard copy can be produced accurately and easily on flat bed or drum plotters and to any scale. Colour prints can also be made.

Engineering drawings such as General Arrangement drawings and detail drawings are produced by a technique called orthographic drawing using the conventions set out in BS: 308. Since I will be asking you to make orthographic drawings from more easily recognized pictorial drawings, I will start by introducing you to the two pictorial techniques widely used by draughtspersons.

Oblique drawing

Figure 2.21 shows a simple oblique drawing. The front view (elevation) is drawn to true shape and size. Therefore this view should be chosen so as to include any circles or arcs so that these can be drawn with compasses. The lines forming the side views appear to travel away from you, so these are called 'receders'.

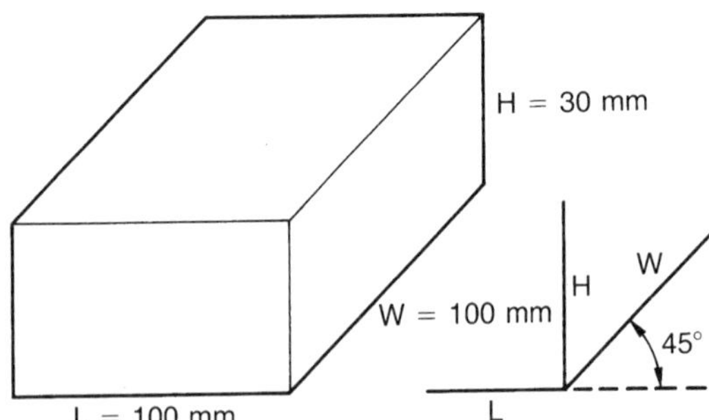

Cavalier oblique projection

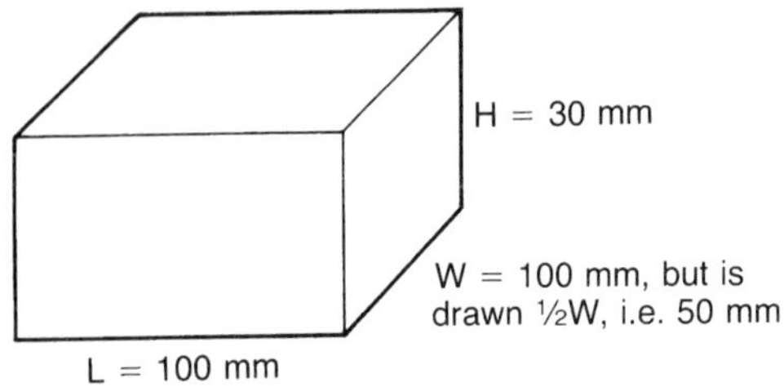

Cabinet oblique projection

Figure 2.21 *A simple oblique drawing*

Test your knowledge 2.7

(a) Obtain a sheet of quadrille ruled paper (maths paper with 5 mm squares) and draw the box shown in Figure 2.21 full size. Use cabinet oblique projection.

(b) Now use your compasses to draw a 50 mm diameter hole in the centre of the front (elevation) of the box.

(c) Can you think of a way to draw the same circle on the side (receding) face of the box? It will not be a true circle so you cannot use your compasses.

They are drawn at 45° to the horizontal using a 45° set-square. They may be drawn full length as in cavalier oblique drawing or they may be drawn half-length as in cabinet oblique drawing. This latter method gives a more realistic representation, and is the one you and I will be using.

Isometric drawing

Figure 2.22(a) shows an isometric drawing of our previous box. To be strictly accurate, the vertical lines should be drawn true length and the receders should be drawn to a special isometric scale. However this sort of accuracy is rarely required and, for all practical purposes, we draw all the lines full size. As you can see, the receders are drawn at 30° to the horizontal for both the elevation and the end view.

Although an isometric drawing is more pleasing to the eye, it has the disadvantage that all circles and arcs have to be constructed. They cannot be drawn with compasses. Figures 2.22(b), 2.22(c) and 2.22(d) show you how to construct an isometric curve. You could have used this technique in Test your

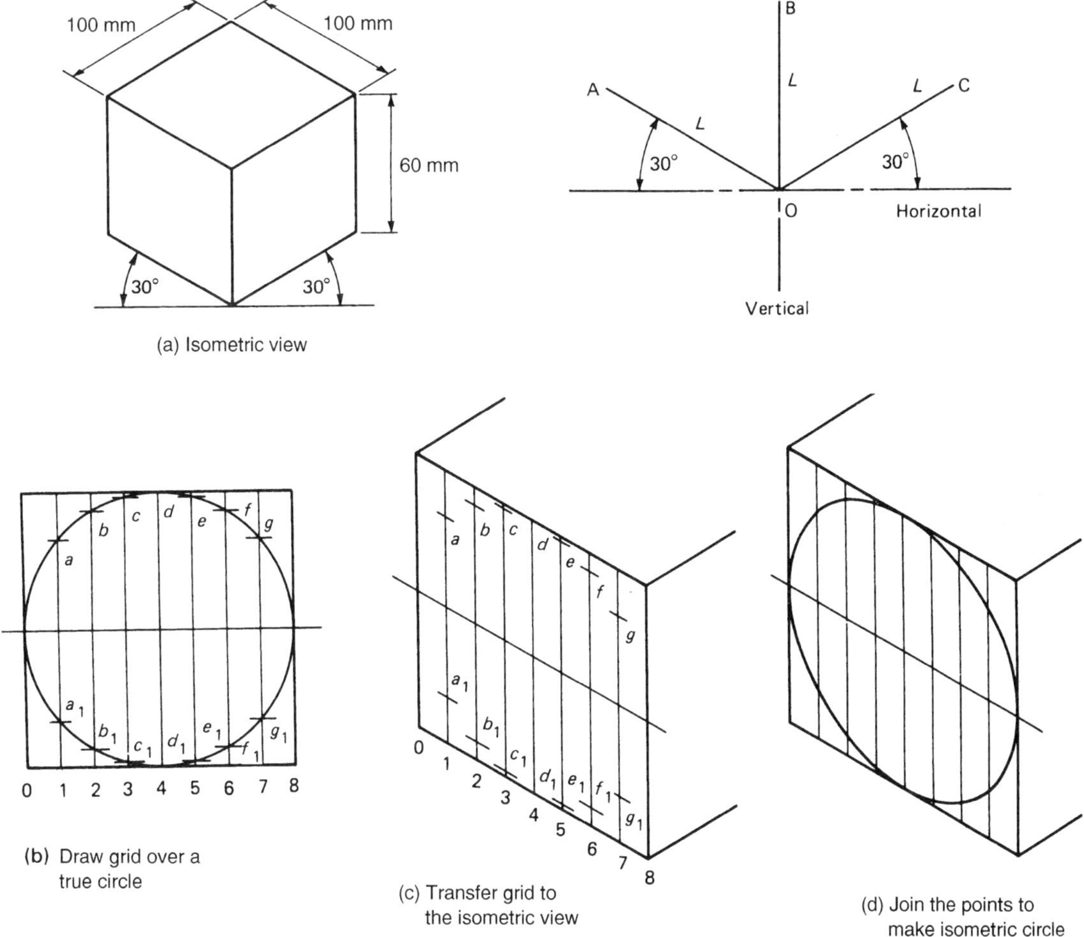

Figure 2.22 *Isometric drawing*

Test your knowledge 2.8

(a) Draw, full size, an isometric view of the box shown in Figure 2.22. Isometric ruled paper will be of great assistance if you can obtain some.

(b) Draw a 50 mm diameter isometric circle on the TOP face of the box. Remember that Figure 2.22 shows it on the side of the box.

knowledge 2.7 to draw the circle on the side of the box drawn in oblique projection.

First we draw the required circle. Then we draw a grid over it as shown in Figure 2.22(b). Next number or letter the points where the circle cuts the grid as shown. Now draw the grid on the side elevation of the box and step off the points where the circle cuts the grid with your compasses as shown in Figure 2.22(c). All that remains is to join up the dots and you have an isometric circle as shown in Figure 2.22(d).

Another way of drawing isometric circles and curves is the 'four-arcs' method. This does not produce true curves but they are near enough for all practical purposes and quicker and easier than the previous method for constructing true curves. The steps are shown in Figure 2.23.

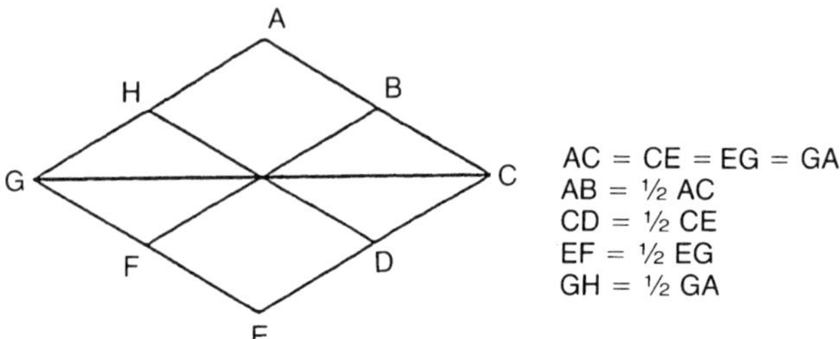

(a) Draw an isometric grid of appropriate size

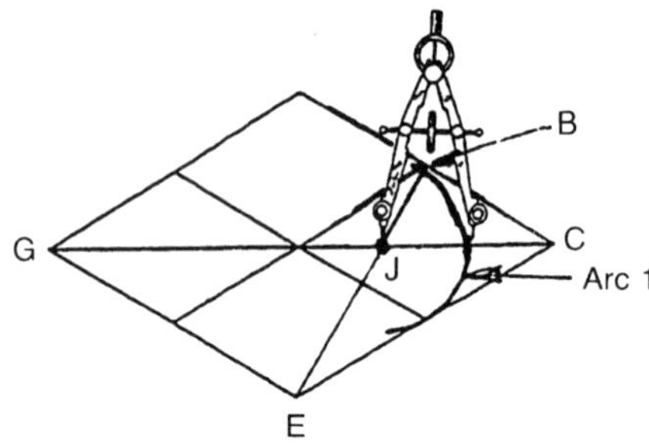

(b) Construct the 1st arc using a compass located as shown

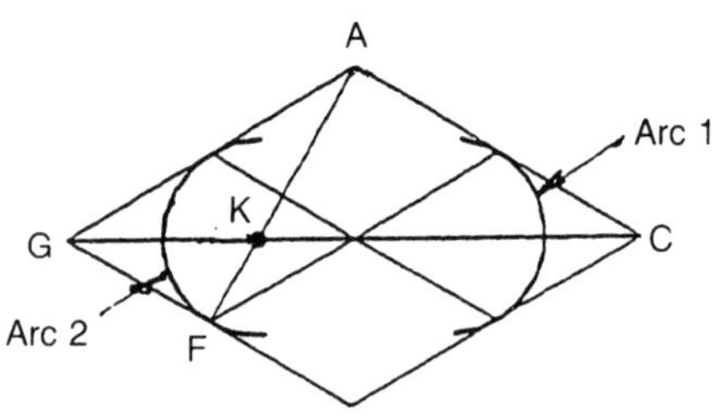

(c) Draw the 2nd arc using the construction process shown

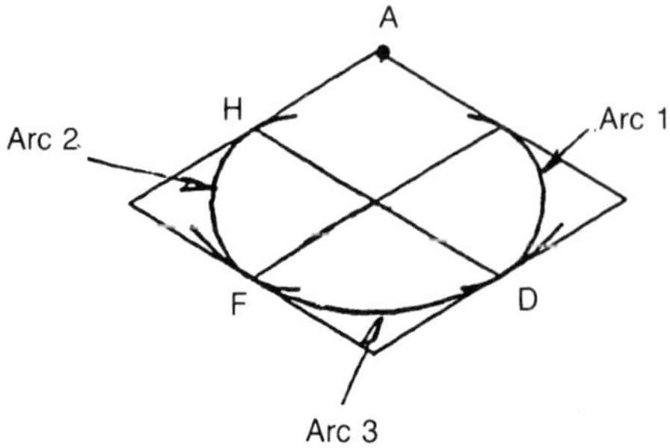

(d) Draw the 3rd arc through the points shown

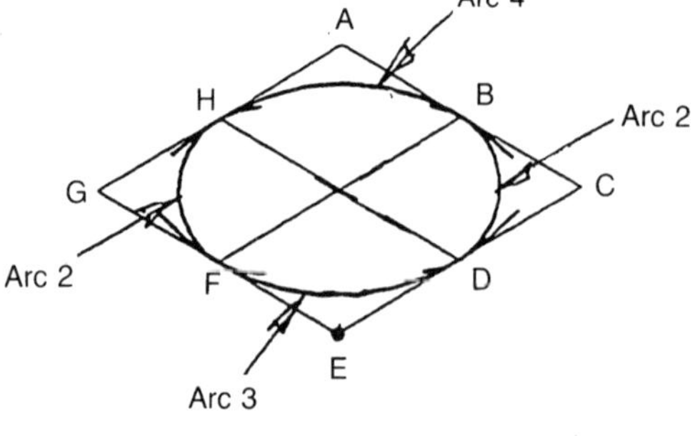

(e) Complete the process drawing the 4th arc from the opposite corner

Figure 2.23 *'Four-arcs' method*

Test your knowledge 2.9

Use the technique just decribed to draw a 40 mm diameter centre on the side face of our box. Start off by drawing a 40 mm isometric square in the middle of the side face.

Test your knowledge 2.10

(a) Figure 2.24(a) shows some further examples of isometric drawings. Redraw them as cabinet oblique drawings.

(b) Figure 2.24(b) shows some further examples of cabinet oblique drawings. Redraw them as isometric drawings. Any circles and arcs on the vertical surfaces should be drawn using the grid construction method. Any arcs and circles on the horizontal (plan) surfaces should be drawn using the 'four-arcs method'.

- Join points B and E, as shown in Figure 2.23(b). The line BE cuts the line GC at the point J. The point J is the centre of the first arc. With radius BJ set your compass to strike the first arc as shown.
- Join the points A and F, as shown in Figure 2.23(c). The line AF cuts the line GC at the point K. The point K is the centre of the second arc. With radius KF set your compasses to strike the second arc as shown. If your drawing is accurate both arcs should have the same radius.
- With centre A and radius AF or AD strike the third arc, as shown in Figure 2.23(d).
- With centre E and radius EH or EB strike the fourth and final arc, as shown in Figure 2.23(e). If your drawing is accurate, arcs 3 and 4 should have the same radius.

Here are some more examples for you to practise on. Then we will start on orthographic drawing.

Orthographic drawing

GA and detail drawings are produced by the use of a drawing technique called orthographic projection. This is used to represent three-dimensional solids on the two-dimensional surface of a sheet of drawing paper so that all the dimensions are true length and all the surfaces are true shape. To achieve this when surfaces are inclined to the vertical or the horizontal we have to use auxiliary views, but more about these later. Let's keep things simple for the moment.

First-angle projection

Figure 2.25(a) shows a simple component drawn in isometric projection. Figure 2.25(b) shows the same component as an orthographic drawing. This time we make no attempt to represent the component pictorially. Each view of each face is drawn out separately either full size or to the same scale. What is important is how we position the various views as this determines how we 'read' the drawing.

Engineers use two orthographic drawing techniques, either first-angle or third-angle projection. The former is called 'English' projection and the latter is called 'American' projection. The drawing in Figure 2.25 is in first-angle projection. The views are arranged as follows.

- *Elevation* This is the main view from which all the other views are positioned. You look directly at the side of the component and draw what you see.
- *Plan* To draw this, you look directly down on the top of the component and draw what you see below the elevation.
- *End view* This is sometimes called an 'end elevation'. To draw this you look directly at the end of the component and draw what you see at the opposite end of the elevation. There may be two end views, one at each end of the

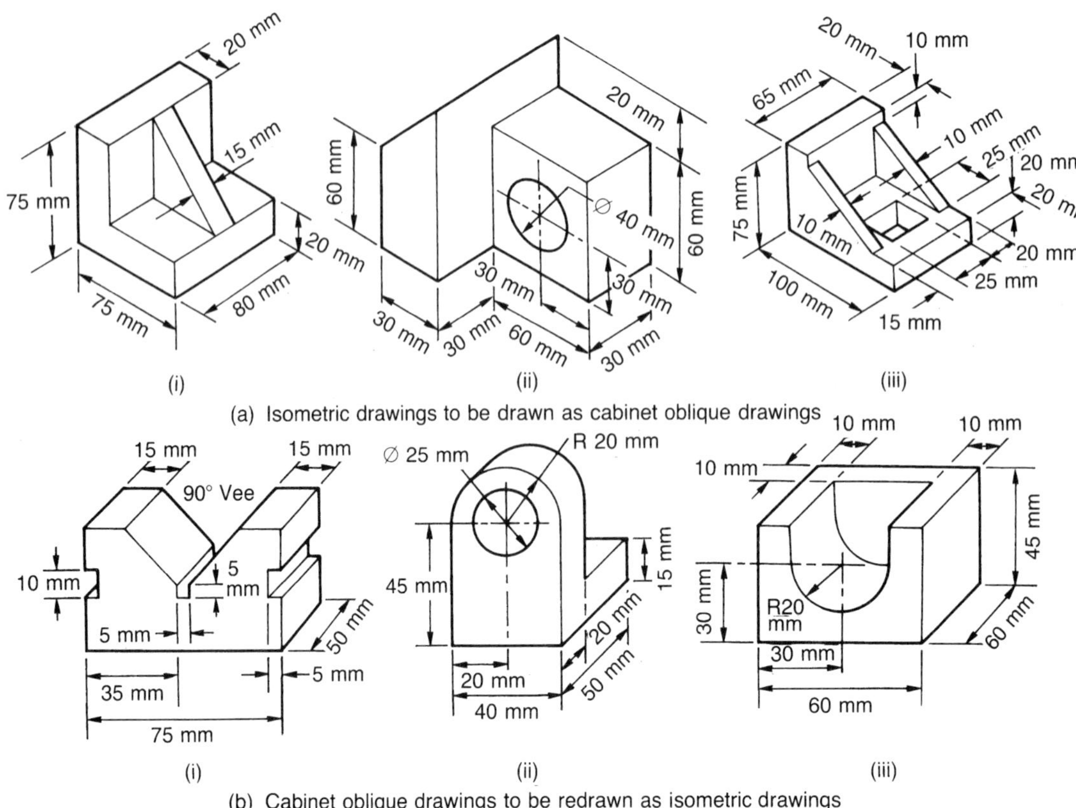

Figure 2.24 *See Test your knowledge 2.10*

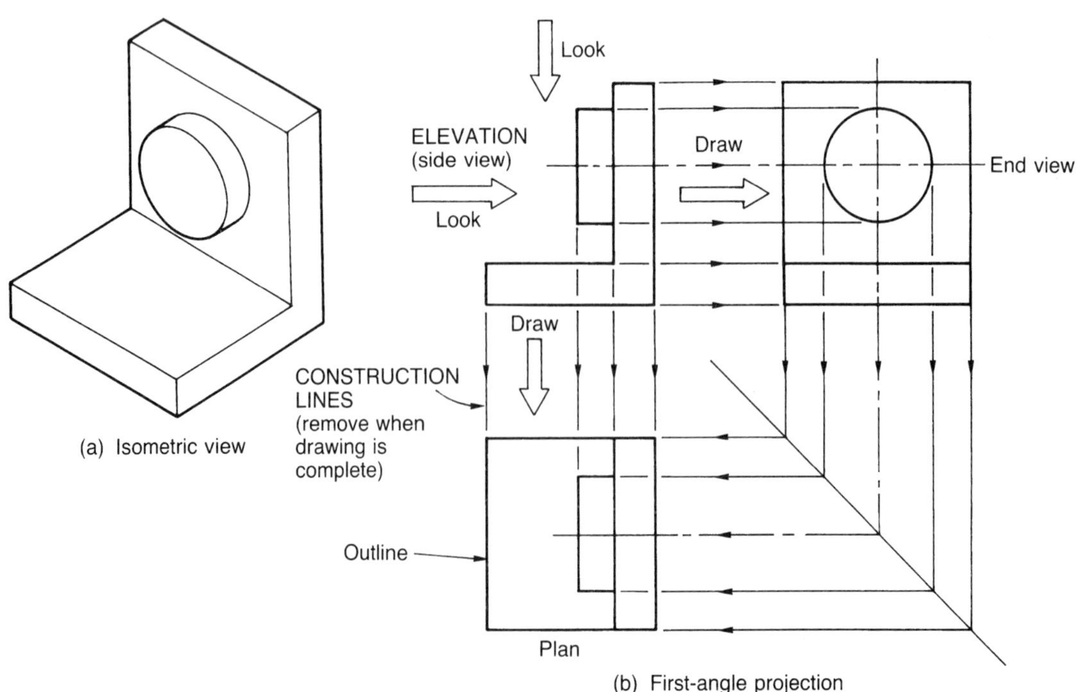

Figure 2.25 *An isometric view and its corresponding first-angle projection*

Test your knowledge 2.11

Figure 2.24 showed some components using pictorial projections. I now want you to redraw these components in first-angle orthographic projection. To start you off I have drawn the first one for you. This is shown in Figure 2.27. Note how I have positioned the end-view this time so that you can see the web.

elevation, or there may be only one end view if this is all that is required to completely depict the component. Figure 2.25 requires only one end view. When there is only one end view this can be placed at either end of the elevation depending upon which position gives the greater clarity and ease of interpretation. Whichever end is chosen, the rules for drawing this view must be obeyed.

Use feint construction lines to produce the drawing, as shown in Figure 2.25(b). When these are complete, 'line-out' the outline more heavily. Carefully remove the construction lines to leave the drawing uncluttered, thus improving the clarity. Sometimes, in examinations, you are asked to leave the construction lines so that the examiner can see how you have arrived at your answer. Figure 2.26 shows the finished drawing.

Third-angle projection

Figure 2.28 shows the same component, but this time I have drawn it in third-angle projection for you.

Elevation Again I have started with the elevation or side view of the component and, as you can see, there is no difference.

Plan Again we look down on top of the component to see what the plan view looks like. However, this time, we draw the plan view above the elevation. That is, in third-angle projection we draw all the views from where we look.

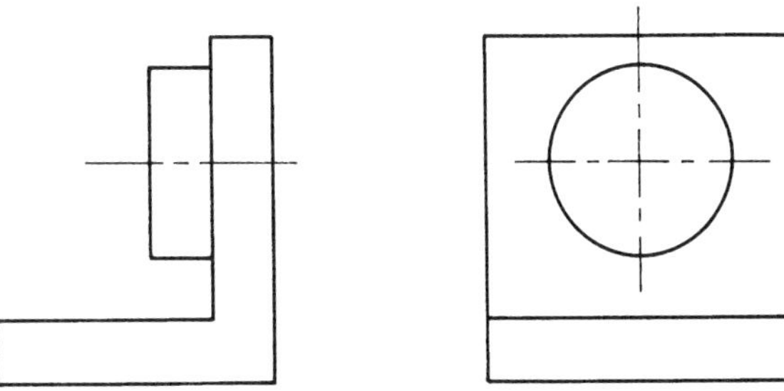

FIRST-ANGLE PROJECTION

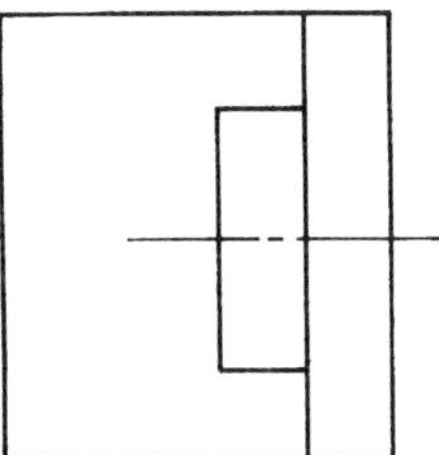

Figure 2.26 *Completed first-angle projection*

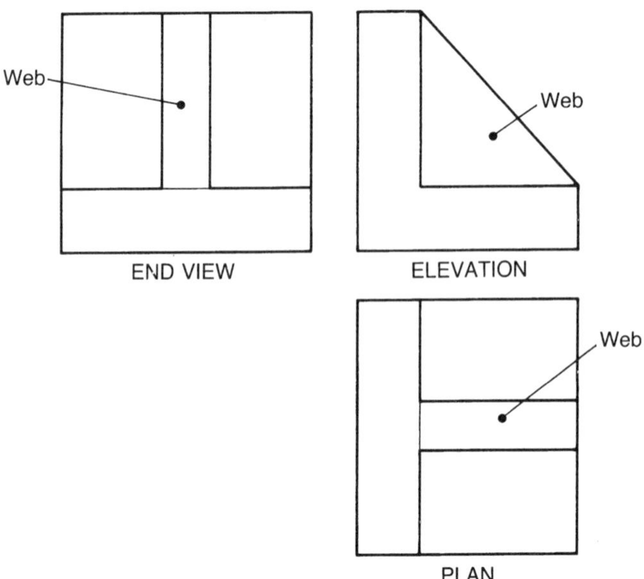

Figure 2.27 *See Test your knowledge 2.11*

Test your knowledge 2.12

Figure 2.24 showed some components using pictorial projections. I now want you to redraw these components in third-angle orthographic projection. To start you off, I have again the drawn the first one for you. This is shown in Figure 2.30. Note how I have posotioned the end-view so that you can see the web.

End view Note how the position of the end view is reversed compared with first-angle projection. This is because, like the plan view, we draw the end views at the same end from which we look at the component.

Again use feint construction lines to produce the drawing as shown in Figure 2.28. Then 'line-in' the outline more heavily and carefully remove the construction lines for clarity, unless you have been instructed otherwise. Figure 2.29 shows the finished drawing in third-angle projection.

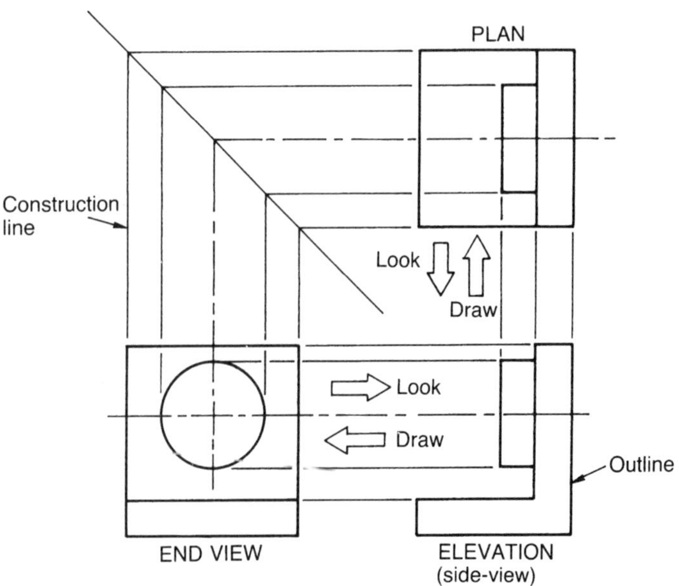

Figure 2.28 *Third-angle projection*

Test your knowledge 2.13

(a) Figure 2.31 shows some components drawn in first-angle projection and some in third-angle projection. I have not necessarily drawn all the views each time. I have only drawn as many of the views as are needed. State which is first-angle and which is third-angle.

(b) Two of the drawings are standard symbols for indicating whether the drawing is in first-angle or whether it is in third-angle. Which drawings do you think are these symbols?

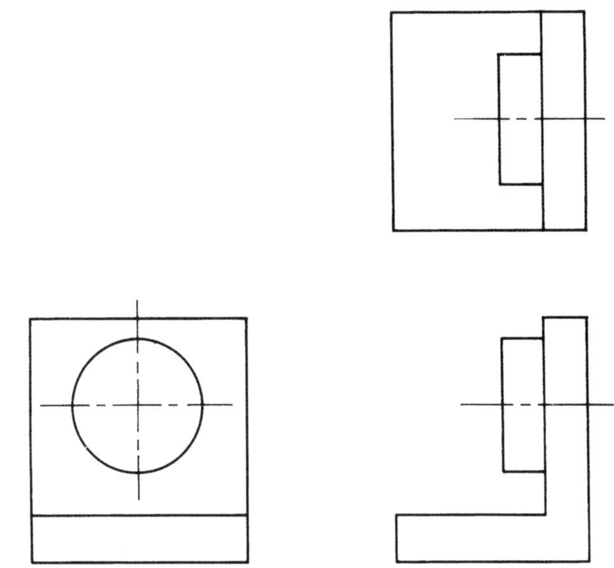

Figure 2.29 *Completed third-angle projection*

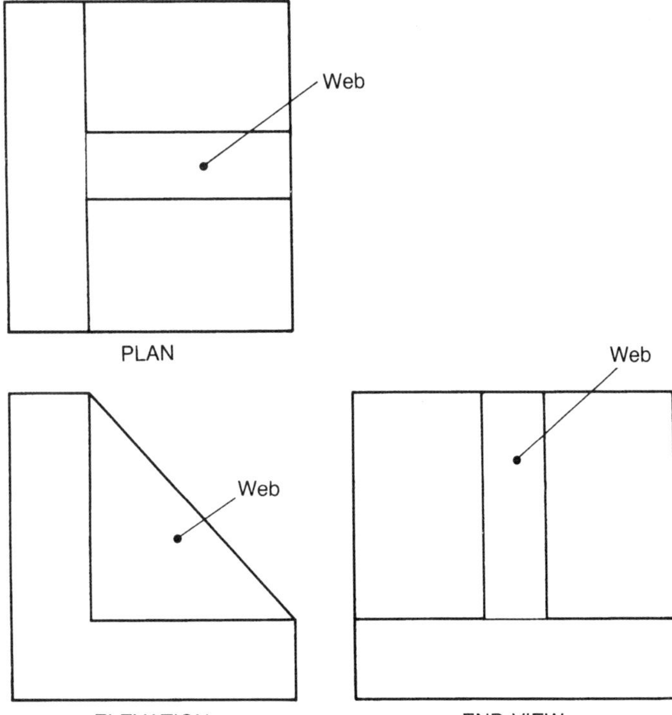

Figure 2.30 *See Test your knowledge 2.12*

Test your knowledge 2.14

Figure 2.32 shows pictorial views of some more solid objects.

(a) Draw these objects in first-angle orthographic projection and label the views.

(b) Draw these objects in third-angle orthographic projection and label the views.

Auxiliary views

In addition to the main views on which we have just been working, we sometimes have to use auxiliary views. We use auxiliary views when we cannot show the true outline of the component or a feature of the component in one of the main views; for

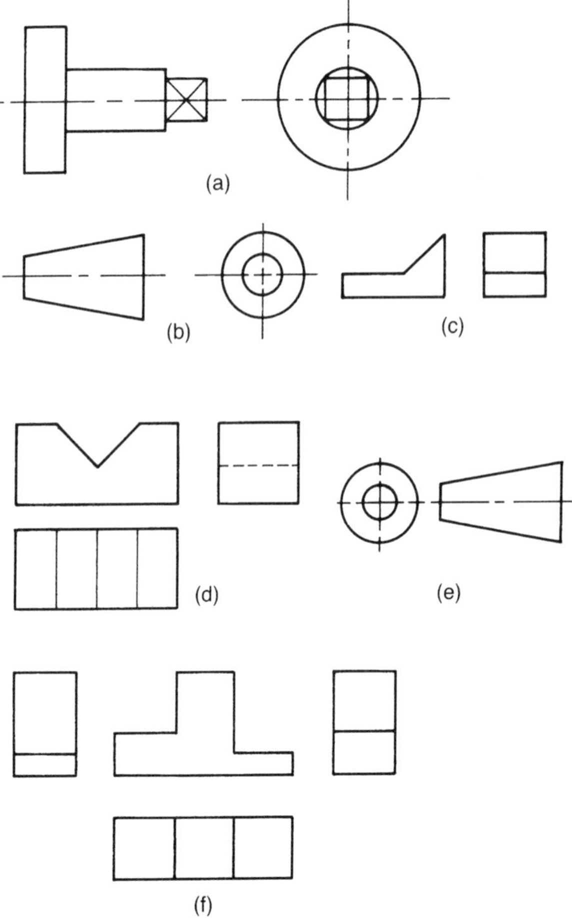

Figure 2.31 *See Test your knowledge 2.13*

example, when a surface of the component is inclined as shown in Figure 2.33.

Activity 2.7

Draw the component shown in Figure 2.34 in isometric projection. Each square has a side length of 10 mm.

Also draw the component in:

(a) first-angle orthographic projection (only two views are required)

(b) third-angle orthographic projection (only two views are required)

(c) cabinet oblique projection.

Present your work in the form of a portfolio of drawings. Clearly mark each drawing with the projection used.

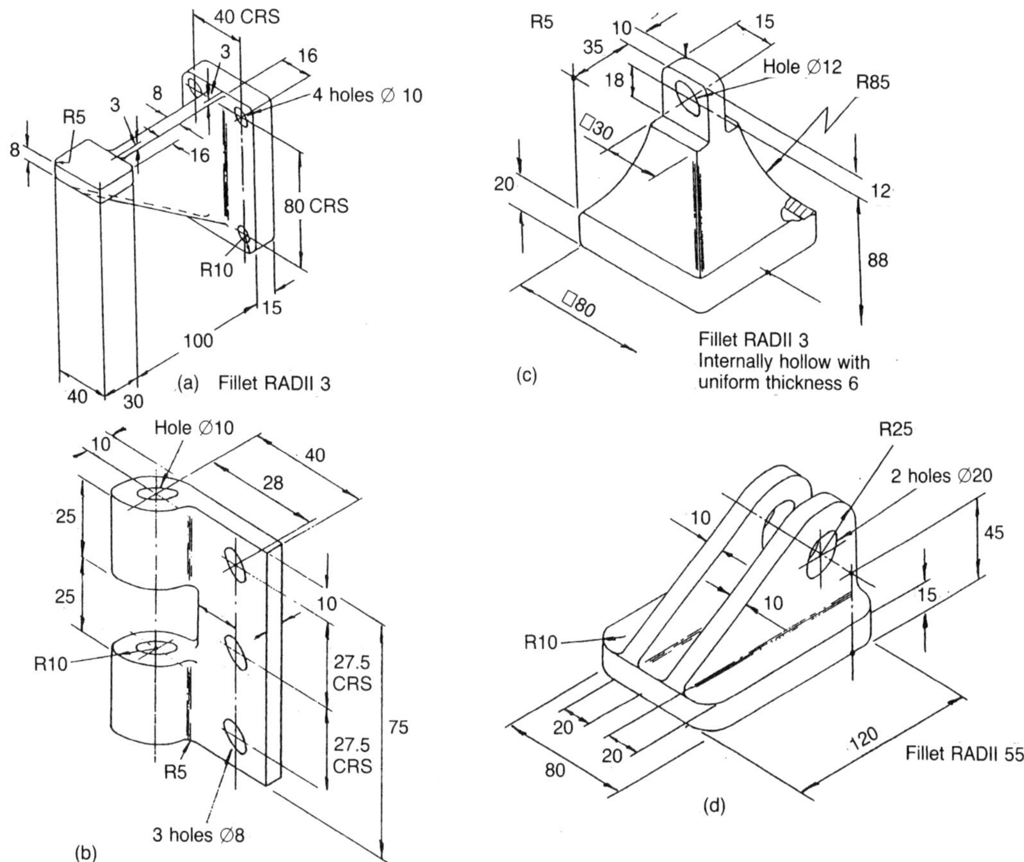

Figure 2.32 *See Test your knowledge 2.14*

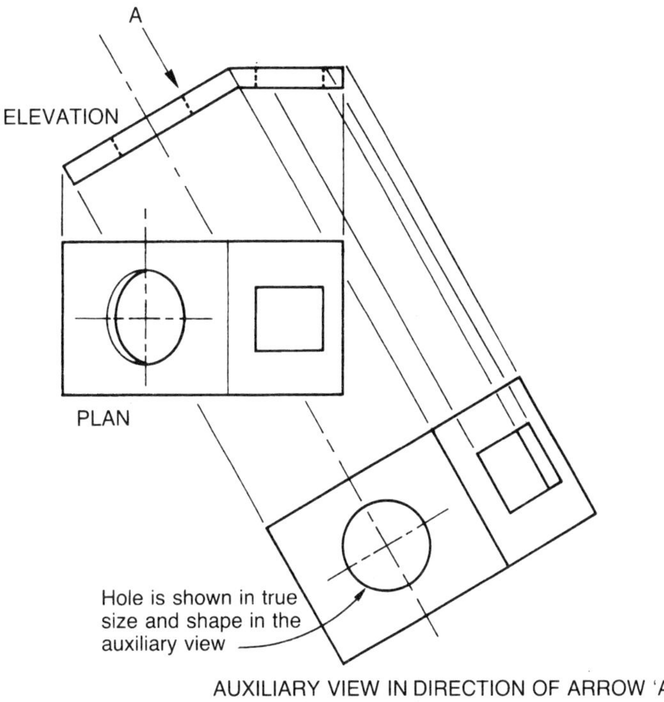

Figure 2.33 *An auxiliary view*

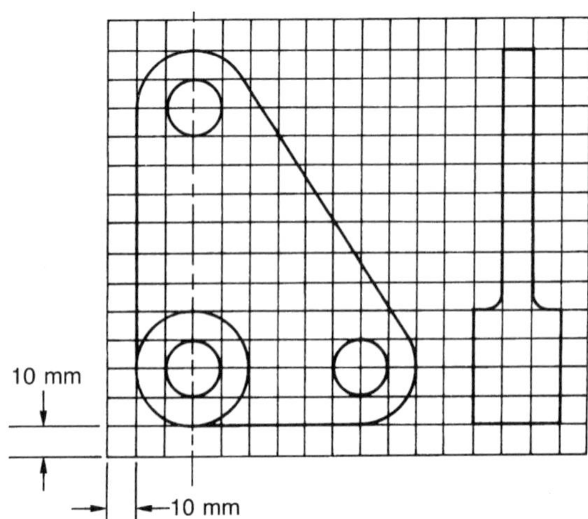

Figure 2.34 *See Activity 2.7*

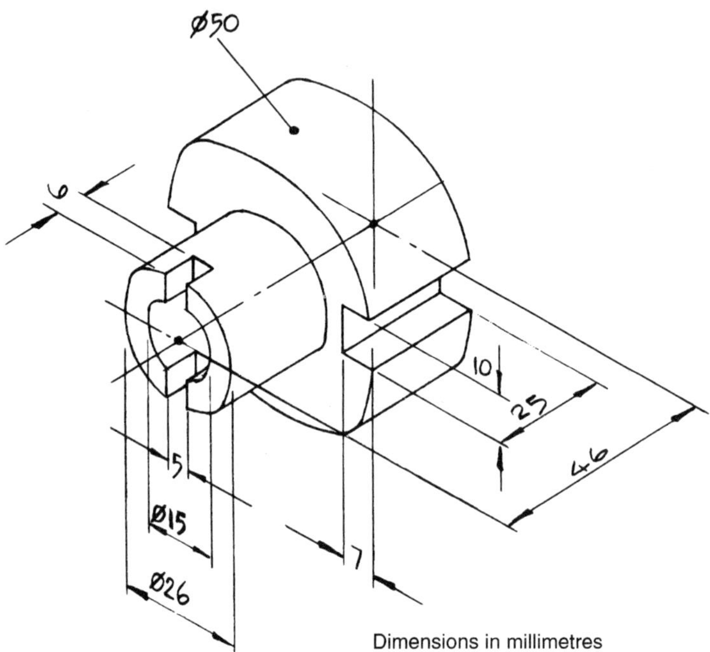

Figure 2.35 *Dimensioned drawing*

Production of engineering drawings

Standard conventions are used in order to avoid having to draw out, in detail, common features in frequent use. Figure 2.35 shows a typical dimensioned engineering drawing. Figure 2.36(a) shows a pictorial representation of a screw thread. Figure 2.36(b) shows the convention for a screw thread. The convention for the screw thread is much the quicker and easier to draw.

All engineering drawings should be produced using appropriate drawing standards and conventions for the following reasons.

- *Time* It speeds up the drawing process by making life easier for the draughtsperson as indicated above. This reduces costs and also reduces the 'lead-time' required to get a new product into production.

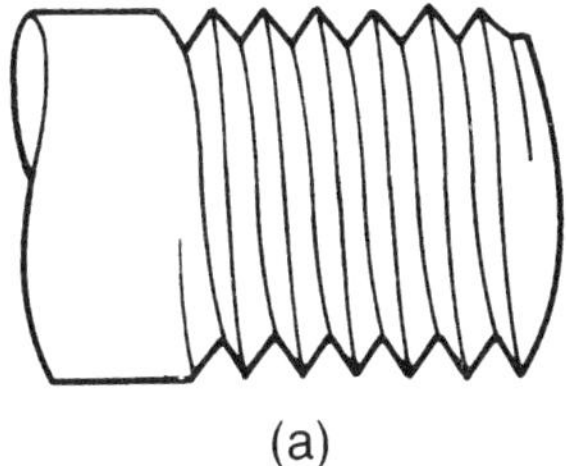
(a)

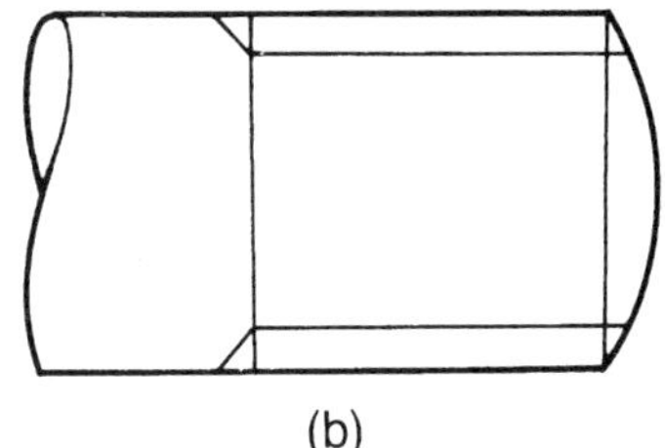
(b)

Figure 2.36 *Screw threads*

- *Appearance* It makes your drawings look more professional and improves the 'image' of yourself and your company. Badly presented drawings can send out the wrong messages and can call your competence into question.
- *Portability* Drawings produced to international standards and conventions can be read and correctly interpreted by any trained engineer, anywhere in this country or abroad. This avoids misunderstandings that could lead to expensive and complex components and asemblies being scrapped and dangerous situations arising. The only difficulties will arise from written notes that are language dependent.

Drawing conventions used by engineers in the UK are specified in BS 308. This is produced in three parts:

- Part 1 General principles.
- Part 2 Dimensioning and tolerancing of size.
- Part 3 Geometrical tolerancing.

These are all 'harmonized' with their appropriate ISO counterparts (ISO = International Standards Organization).

As has been stated earlier, you should get yourself a copy of the British Standards Institution's publication PP7308: Engineering Drawing Practice for Schools and Colleges. Also useful is PP7307: Graphical Symbols for Use in Schools and Colleges. This latter Standard contains symbols for use in electrical, electronic, pneumatic and hydraulic schematic circuit diagrams. It also contains many other useful symbols. Since they are intended for students, both PP7307 and PP7308 are sold at very low prices compared with the full standards.

Other British Standards of importance to engineering draughtspersons are:

- BS 4500: ISO Limits and Fits (these are used by mechanical and electronics engineers).
- BS 3939: Graphical symbols for electrical power, telecommunications and electronics diagrams.
- BS 2197: Specifications for graphical symbols used in diagrams for fluid power systems and components.

Planning the drawing

Before we start the actual drawing and lay pencil to paper we should plan what we are going to do. This saves having to alter the drawing or even start again later on. We have to decide whether the drawing is to be pictorial, orthographic or schematic. If orthographic we have to decide on the projection we are going to use. We also have to decide whether we need a formal drawing or whether a freehand sketch is all that is required. If a formal drawing is needed then we have to decide whether to use manual techniques or CAD.

Paper size

When you start to plan your drawing you have to decide on the paper size that you are going to use. Engineering drawings are usually produced on 'A' size paper. Paper size A0 is approximately one square metre in area and is the basis of the system. Size A1 is half the area of size A0, size A2 is half the area of size A1 and so on down to size A4. Smaller sizes are available but they are not used for drawing. All the 'A' size sheets have their sides in the ratio of 1 : 12. This gives the following paper sizes.

A0 = 841 mm × 1189 mm
A1 = 594 mm × 841 mm
A2 = 420 mm × 594 mm
A3 = 297 mm × 420 mm
A4 = 210 mm × 297 mm

These relationships are shown diagrammatically in Figure 2.37.

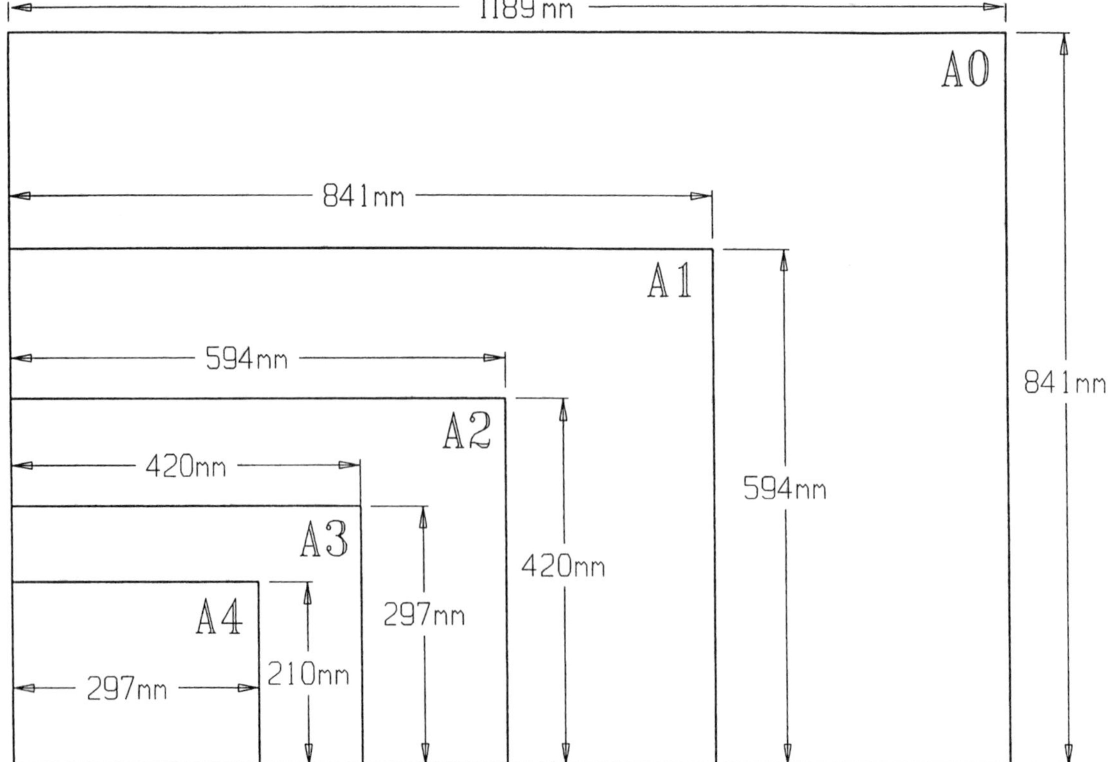

Figure 2.37 *Paper sizes*

The paper size you choose will depend upon the size of the drawing and the number of views required. Be generous, nothing looks worse than a cramped up drawing, and overcrowded dimensions. It is also false economy since overcrowding invariably leads to reading errors. As you will already have seen from some of the previous examples, the drawing should always have a border and a title block. This restricts the blank area available to draw on. Figures 2.38 and 2.39 show how the views should be positioned. These layouts are only a guide but they offer a good starting point until you become more experienced. If only one view is required then it is centred in the drawing space available.

Title block

A typical title block was shown in Figure 2.19. If you refer back to this figure you will see that it is expandable vertically and horizontally to accommodate any written information that is required. The title block should contain:

- the drawing number (which should be repeated in the top left-hand corner of the drawing)
- the drawing name (title)

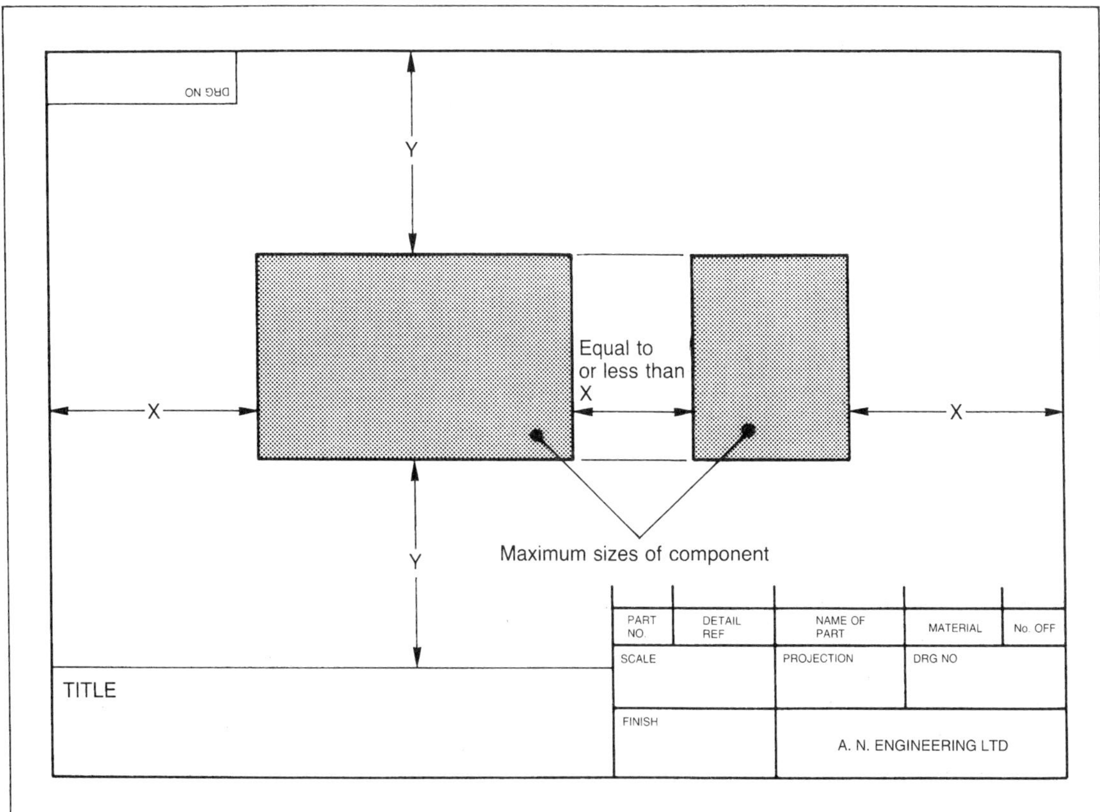

Figure 2.38

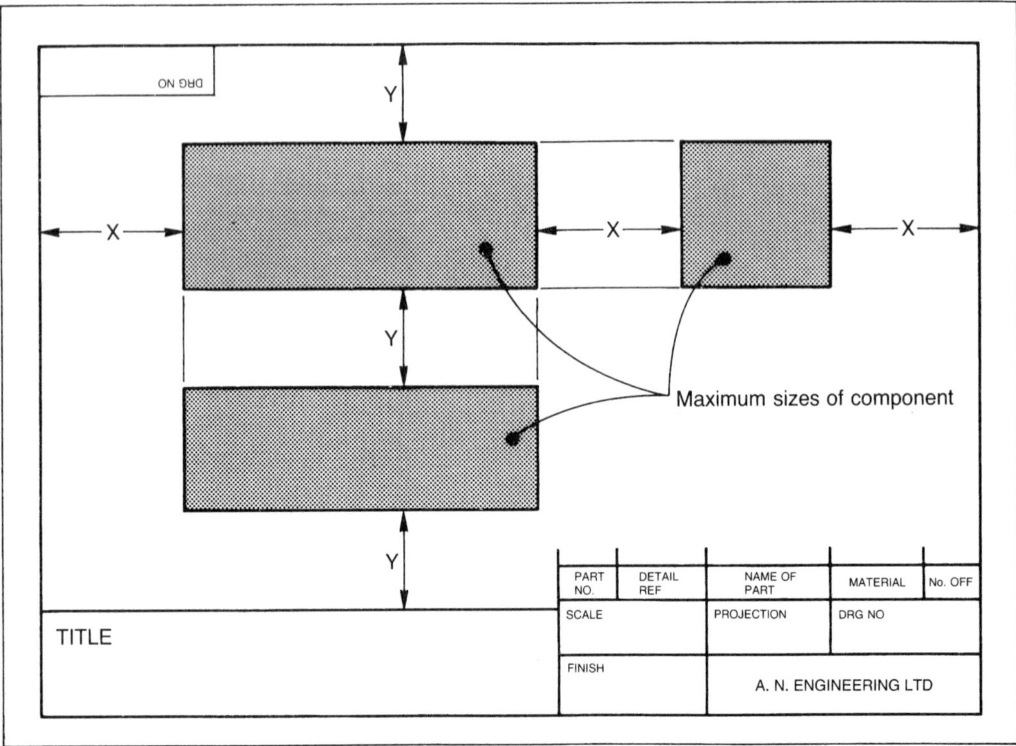

Figure 2.39

- the drawing scale
- the projection used (standard symbol)
- the name and signature of the draughtsperson together with the date on which the drawing was signed
- the name and signature of the person who checks and/or approves the drawing, together with the date of signing
- the issue number and its release date
- any other information as dictated by company policy.

Scale

The scale should be stated on the drawing as a ratio. The recommended scales are as follows:

- Full size = 1 : 1
- Reduced scales (smaller than full size) are:

1 : 2	1 : 5	1 : 10
1 : 20	1 : 50	1 : 100
1 : 200	1 : 500	1 : 1000

(NEVER use the words full-size, half-size, quarter-size, etc.).

- Enlarged scales (larger than full size) are:

2 : 1	5 : 1	10 : 1
20 : 1	50 : 1	

Production of the drawing

Lines and linework

The lines of a drawing should be uniformly black, dense and bold. On any one drawing they should all be in pencil or in black ink. Pencil is quicker to use but ink prints out more clearly. Lines should be thick or thin as recommended below. Thick lines should be twice as thick as thin lines. Figure 2.40 shows the types of lines recommended in BS 308 for use in engineering drawing and how the lines should be used. This is reinforced by Table 2.1.

Sometimes the lines overlap in different views. When this happens, as shown in Figure 2.41, the following order of priority should be observed.

- Visible outlines and edges (type A) take priority over all other lines.
- Next in importance are hidden outlines and edges (type E).
- Then cutting planes (type G).
- Next come centre lines (type F and B).
- Outlines and edges of adjacent parts, etc. (type H).
- Finally, projection lines and shading lines (type B).

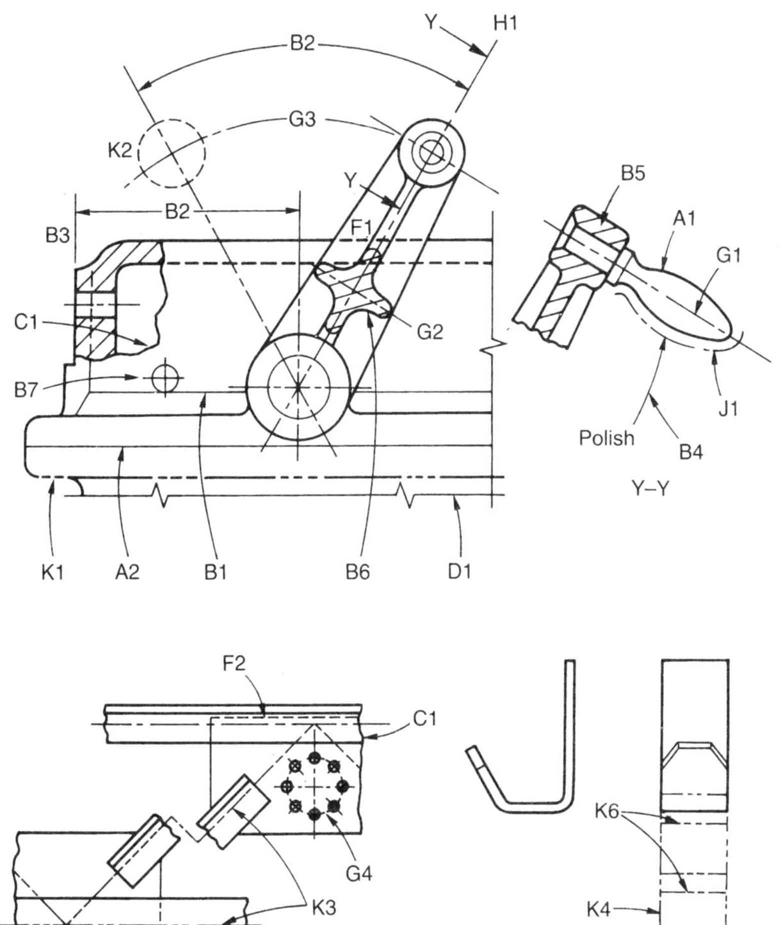

Figure 2.40 *Use of various line types*

Table 2.1 *Types of line*

Types of line		
Line	Description	Application
A	Continuous thick	A1 Visible outlines A2 Visible edges
B	Continuous thin	B1 Imaginary lines of intersection B2 Dimension lines B3 Projection lines B4 Leader lines B5 Hatching B6 Outlines of revolved sections B7 Short centre lines
C D	Continuous thin irregular Continous thin straight with zigzags	C1 Limits of partial or interrupted views and sections, if the limit is not an axis †D1 Limits of partial or interrupted views and sections, if the limit is not an axis
E F	Dashed thick Dashed thin‡	E1 Hideen outlines E2 Hidden edges F1 Hidden outlines F2 Hidden edges
G	Chain thin	G1 Centre lines G2 Lines of symmetry G3 Trajectories and loci G4 Pitch lines and pitch circles
H	Chain thin, thick at ends and changes of direction	H1 Cutting planes
J	Chain thick	J1 Indication of lines or surfaces to which a special requirement applies (drawn adjacent to surface)
K	Chain thin double dashed	K1 Outlines and edges of adjacent parts K2 Outlines and edges of alternative and extreme positions of movable parts K3 Centroidal lines K4 Initial outlines prior to forming § K5 Parts situated in front of a cutting plane K6 Bend lines on developed blanks or patterns

NOTE. The lengths of the long dashes shown for lines G, H, J and K are not necessarily typical due to the confines of the space available.

†This type of line is suited for production of drawings by machines.

‡ The thin F type line is more common in the UK, but on any one drawing or set of drawings only one type of dashed line should be used.

§Included in ISO 128-1982 and used mainly in the building industry.

Test your knowledge 2.15

Figure 2.42 shows a simple drawing using a variety of lines. I have numbered some of the lines. List the number and, against each number, write down whether the type of line chosen is correct or incorrect. If incorrect state what type of line should have been used.

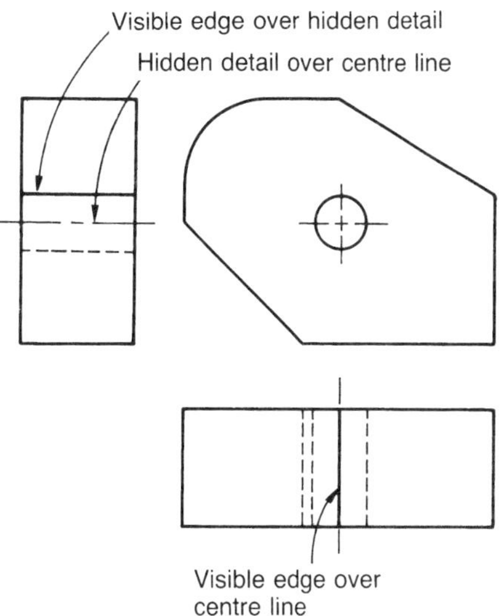

Figure 2.41 *Line priorities*

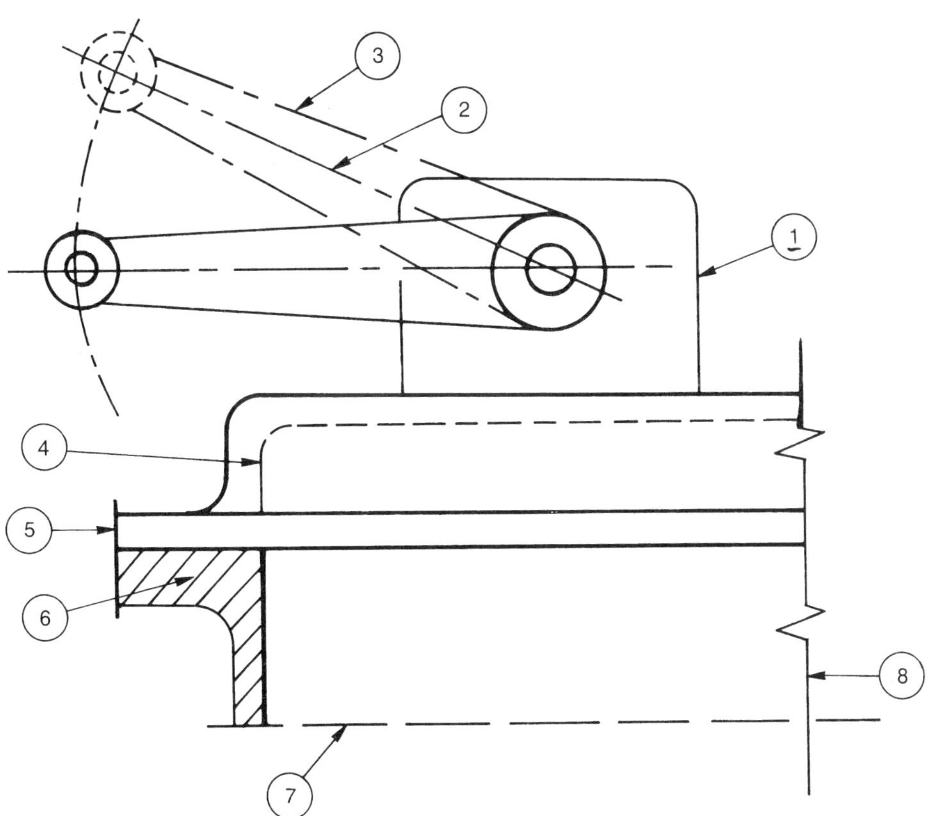

Figure 2.42 *See Test your knowledge 2.15*

Leader lines

Leader lines, as their name implies, lead written information or dimensions to the points where they apply. Leader lines are thin lines (type B) and they end in an arrowhead or in a dot, as shown in Figure 2.43(a). Arrowheads touch and stop on a line, whilst dots should always be used within an outline.

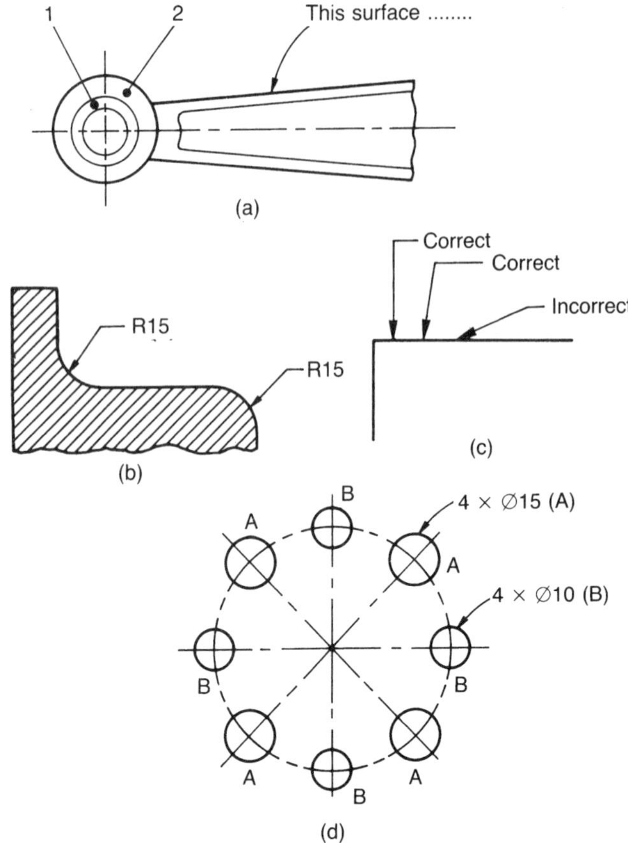

Figure 2.43 *Examples of the use of leader lines*

- When an arrowed leader line is applied to an arc it should be in line with the centre of the arc, as shown in Figure 2.43(b).
- When an arrowed leader line is applied to a flat surface, it should be nearly normal to the lines representing that surface, as shown in Figure 2.43(c).
- Long and intersecting leader lines should not be used, even if this means repeating dimensions and/or notes, as shown in Figure 2.43(d).
- Leader lines must not pass through the points where other lines intersect.
- Arrowheads should be triangular with their length some three times larger than the maximum width. They should be formed from straight lines and the arrowheads should be filled in. The arrowhead should be symmetrical about the leader line, dimension line or stem. It is recommended that arrowheads on dimension and leader lines should be some 3–5 mm long.
- Arrowheads showing direction of movement or direction of viewing should be some 7–10 mm long. The stem should be the same length as the arrowhead or slightly greater. It must never be shorter.

Test your knowledge 2.16

Figure 2.44 shows some applications of leader lines with arrowheads and leader lines with dots. List the numbers and state whether the application is correct or incorrect. If incorrect explain (with sketches if required) how the application should be corrected.

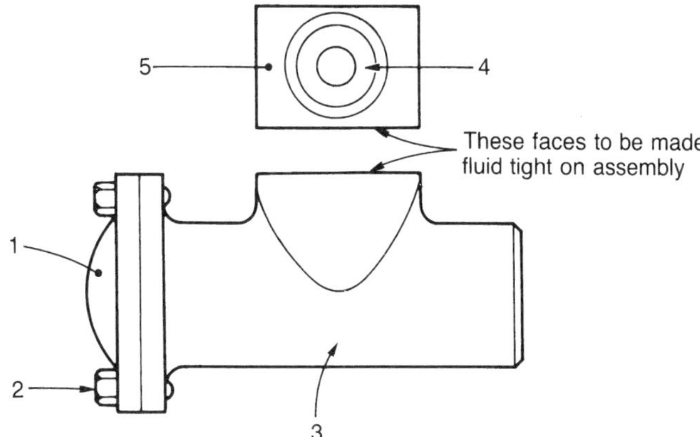

Figure 2.44 *See Test your knowledge 2.16*

Letters and numerals

Style

They style should be clear and free from embellishments. In general, capital letters should be used. A suitable style could be:

ABCDEFGHIJKLMNOPQRSTUVWXYZ
1234567890

Size

The characters used for dimensions and notes on drawings should be not less than 3 mm tall. Title and drawing numbers should be at least twice as big.

Direction of lettering

Notes and captions should be positioned so that they can be read in the same direction as the information in the title block. Dimensions have special rules and will be dealt with later.

Location of notes

General notes should all be grouped together and not scattered about the drawing. Notes relating to a specific feature should be placed adjacent to that feature.

Emphasis

Characters, words and/or notes should not be emphasized by underlining. Where emphasis is required the characters should be enlarged.

Symbols and abbreviations

Test your knowledge 2.17

With reference to BS 308 or PP7308, complete Table 2.2.

If all the information on a drawing was written out in full, the drawing would become very cluttered. Therefore symbols and abbreviations are used to shorten written notes. Those recommended for use on engineering drawings are listed in BS 308, and in the corresponding student version PP7308.

Table 2.2 *See Test your knowledge 2.17*

Abbreviation or symbol	Term
AF	
ASSY	
CRS	
℄	
	Countersink
	counterbore
ø	
DAG	
	Hexagon
	Internal
LH	
MATL	
	Maximum
	Minimum
	Number
PCD	
	Radius (in a note)
	Radius (preceding a dimension)
REQD	
RH	
SCR	
SH	
SK	
SPEC	
	Figure
	Diameter (in a note)
CYL	
CHAM	
CH HD	
	Equally spaced.

Test your knowledge 2.18

With reference to BS 308 or PP7308 complete Figure 2.45. You must take care to use the same types of line as shown in the standard or the conventions become meaningless. This applies particularly to line thickness.

Conventions

These are a form of 'shorthand' used to speed up the drawing of common features in regular use. The full range of conventions and examples of their use can be found in BS 308 or PP7308, so I will not waste space by listing them here. However by completing the next exercise you will use some of the more common conventions and this will help you to become familiar with them.

Dimensioning

When a component is being dimensioned, the dimension lines and the projection lines should be thin full lines (type B). Where possible dimensions should be placed outside the outline of the object, as shown in Figure 2.46(a). The rules are:

- Outline of object to be dimensioned in thick lines (type A).

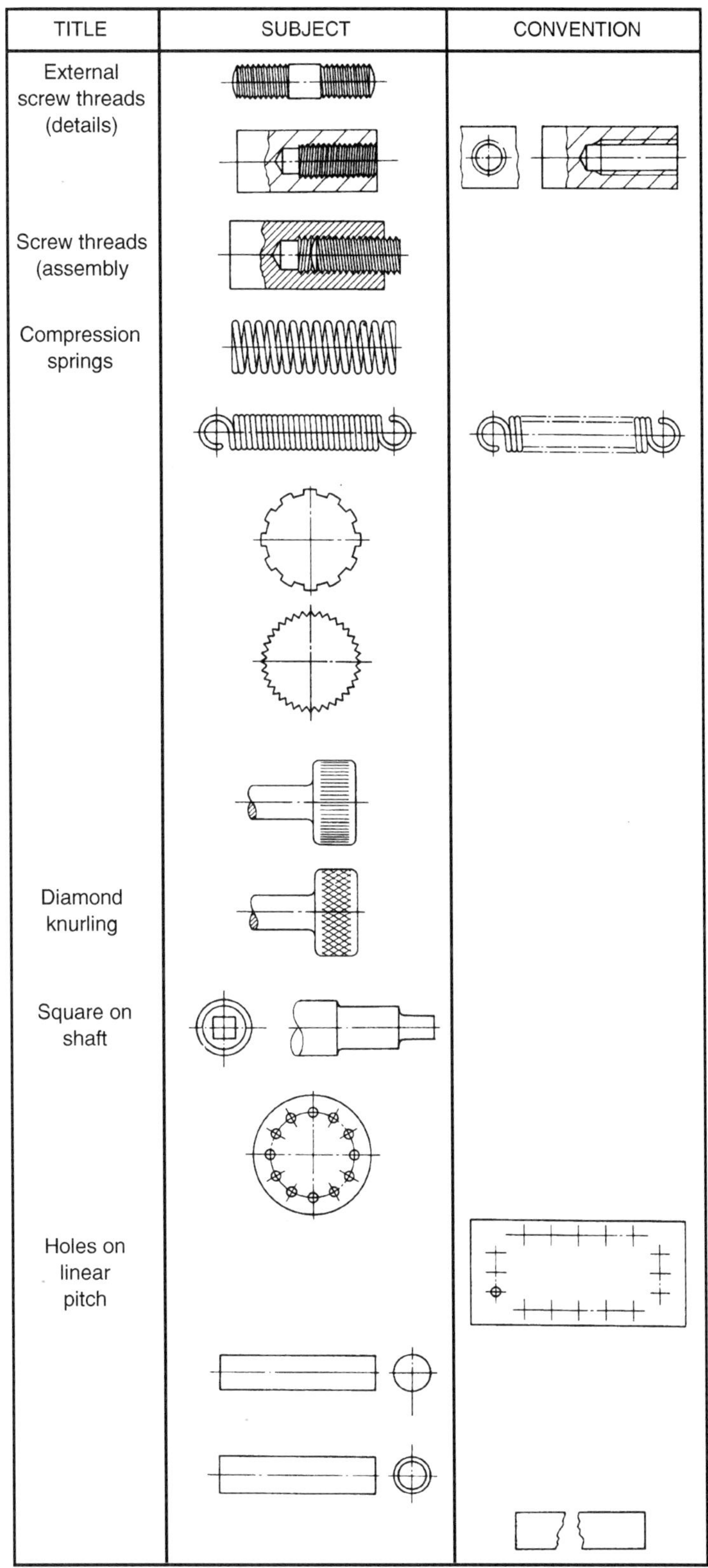

Figure 2.45 *See Test your knowledge 2.18*

- Dimension and projection lines should be half the thickness of the outline (type B).
- There should be a small gap between the projection line and the outline.
- The projection line should extend to just beyond the dimension line.
- Dimension lines end in an arrowhead that should touch the projection line to which it refers.
- All dimensions should be placed in such a way that they can be read from the bottom right-hand corner of the drawing.

The purpose of these rules is to allow the outline of the object to stand out prominently from all the other lines and to prevent confusion.

There are three ways in which a component can be dimensioned. These are:

(i) chain dimensioning, as shown in Figure 2.46(b)

(ii) absolute dimensioning (dimensioning from a datum) using parallel dimension lines, as shown in Figure 2.46(c)

(iii) absolute dimensioning (dimensioning from a datum using superimposed running dimensions, as shown in Figure 2.46(d)). Note the common origin (termination) symbol.

It is neither possible to manufacture an object to an exact size nor to measure an exact size. Therefore important dimensions have to be toleranced. That is, the dimension is given two sizes: an upper limit of size and a lower limit of size. Providing the component is made so that it lies between these limits it will function correction. Information on Limits and Fits can be found in BS 4500.

The method of dimensioning can also affect the accuracy of a component and produce some unexpected effects. Figure 2.46(b) shows the effect of chain dimensioning on a series of holes or other features. The designer specifies a common tolerance of ± 0.2 mm. However, since this tolerance is applied to each and every dimension, the cumulative tolerance becomes ± 0.6 mm by the time you reach the final, right-hand hole, which is not what was intended. Therefore, absolute dimensioning as shown in Figure 2.46(c) and (d) is to be preferred in this example. With absolute dimensioning, the position of each hole lies within a tolerance of ± 0.2 mm and there is no cumulative error. Further examples of dimensioning techniques are shown in Figure 2.47.

It is sometimes necessary to indicate machining processes and surface finish. The machining symbol, together with examples of process notes and the surface finishes in micrometres (μm), is shown in Figure 2.48.

Test your knowledge 2.19

Figure 2.49 shows a component draw in isometric projection. Redraw it in first-angle orthographic projection and add the dimensions, using the following techniques:

(a) Absolute dimensioning using parallel dimension lines.

(b) Absolute dimensioning using superimposed running dimensions.

Sectioning

Sectioning is used to show the hidden detail inside hollow objects more clearly than can be achieved using dashed thin (type E)

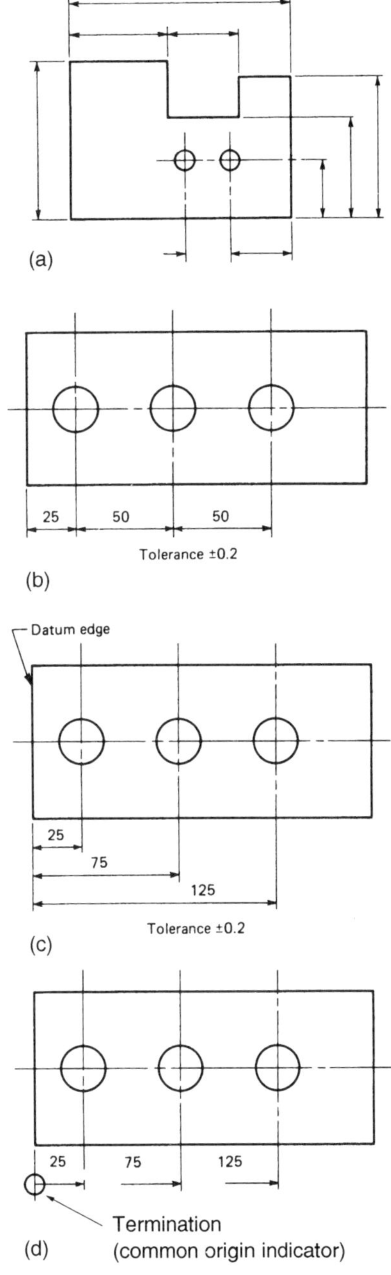

Figure 2.46 *Dimensioning*

lines. Figure 2.50(a) shows an example of a simple sectioned drawing. The cutting plane is the line A–A. In your imagination you remove everything to the left of the cutting plane, so that you only see what remains to the right of the cutting plane looking in the direction of the arrowheads. Another example is shown in Figure 2.50(b).

Figure 2.50(c) shows how to section an assembly. Note how solid shafts and the key are not sectioned. Also note that thin webs that lie on the section plane are not sectioned.

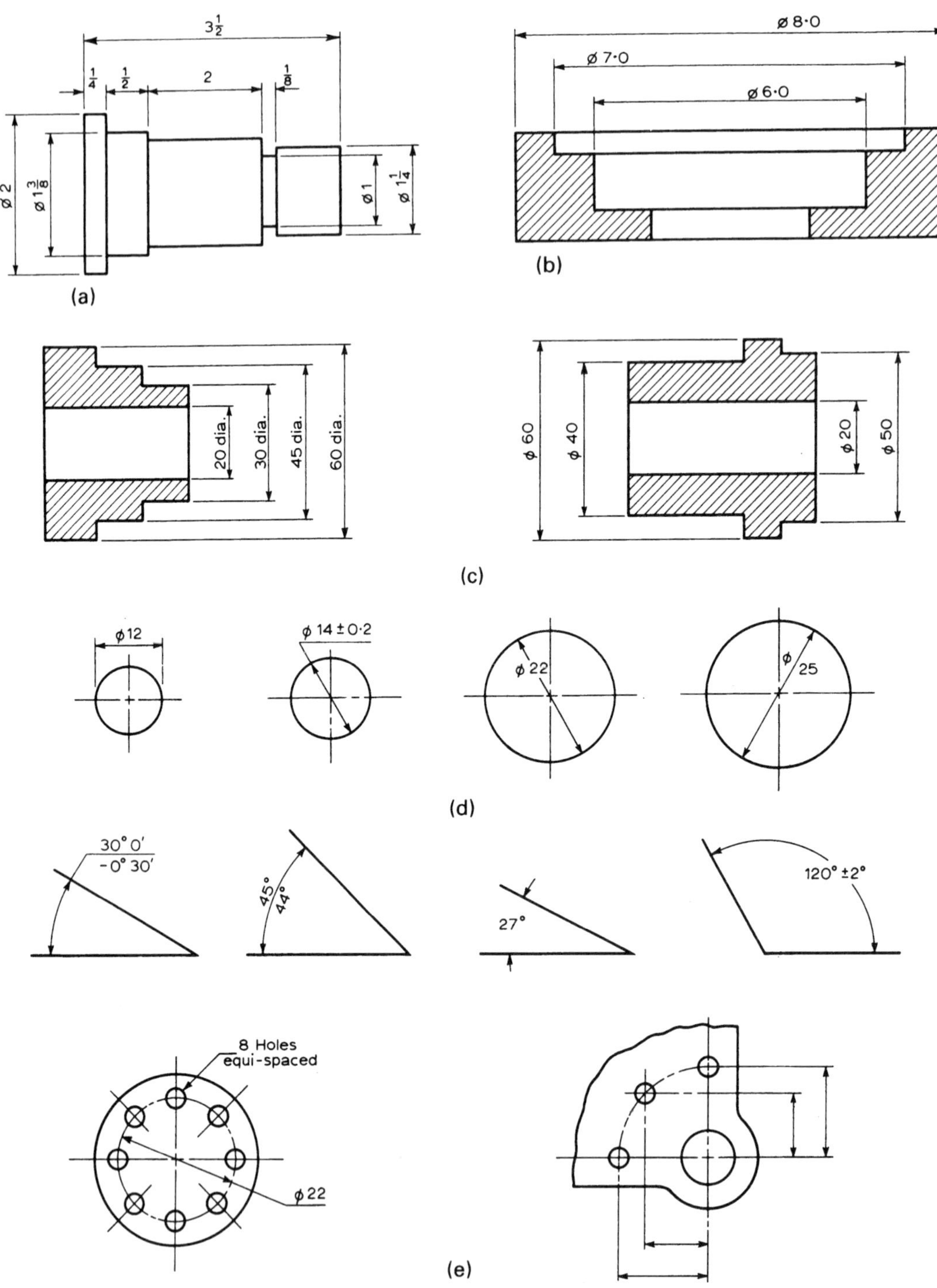

Figure 2.47 *More examples of dimensioning*

When interpreting sectioned drawings, some care is required. It is easy to confuse the terms Sectional view and Section.

Sectional view

In a sectional view you see the outline of the object at the cutting plane. You also see all the visible outlines seen beyond the cutting plane in the direction of viewing. Therefore, Figure 2.50(a) is a sectional view.

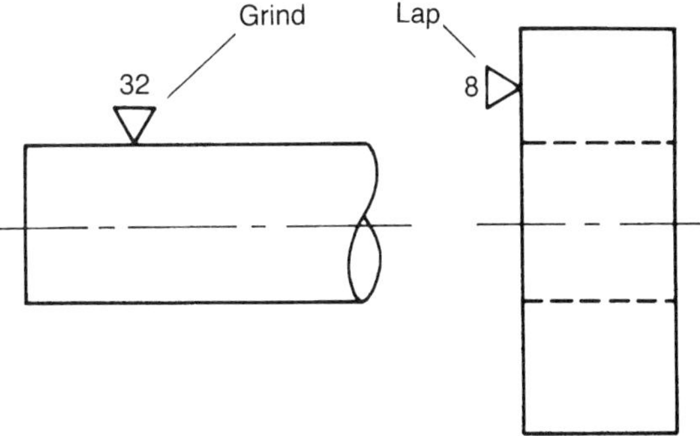

Figure 2.48 *Indicating surface finishes*

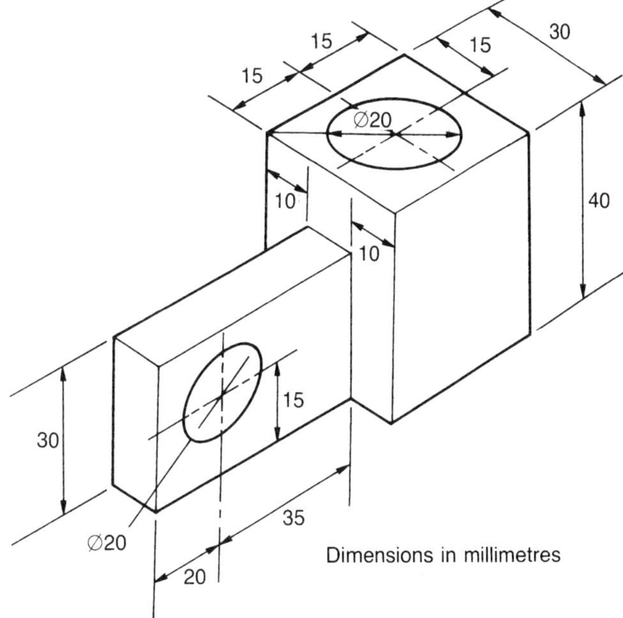

Figure 2.49 *See Test your knowledge 2.19*

Test your knowledge 2.20

(a) Redraw Figure 2.50(a) as a section (remember that my drawing is a sectional view).

(b) Explain why Figure 2.50(b) can be a section or a sectional view.

Section

A section only shows the outline of the object at the cutting plane. Visible outlines beyond the cutting plane in the direction of viewing are not shown. Therefore a section has no thickness.

Cutting planes

You have already been introduced to cutting planes in the previous examples. They consist of type G lines; that is, a thin chain line that is thick at the ends and at changes of direction. The direction of viewing is shown by arrows with large heads. The points of the arrowheads touch the thick portion of the cutting plane. The cutting plane is labelled by placing a capital letter close to the stems of the arrows. The same letters are used to identify the corresponding section or sectional view.

Hatching

You will have noticed that the shading of sections and sectional views consists of sloping, thin (type B) lines. This is called

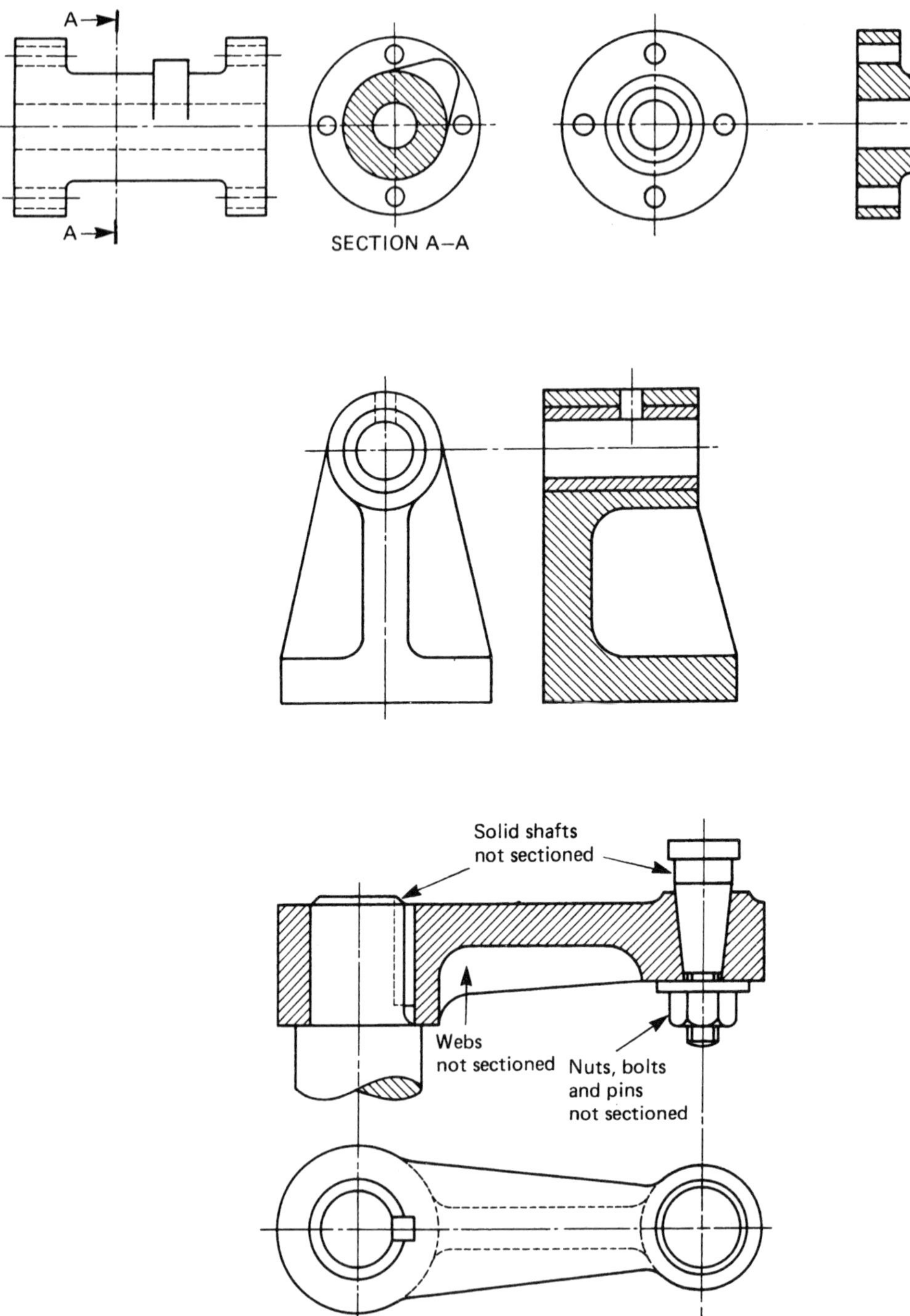

Figure 2.50 *See Test your knowledge 2.20*

hatching. The lines are equally spaced, slope at 45° and are not usually less than 4 mm apart. However when hatching very small areas the hatching can be reduced, but never less than 1 mm. The drawings in this book may look as though they do not obey these rules. Remember that they have been reduced from much bigger drawings to fit onto the pages.

Figure 2.51 shows the basic rules of hatching. The hatching of separated areas is shown in Figure 2.51(a). Separate sectioned

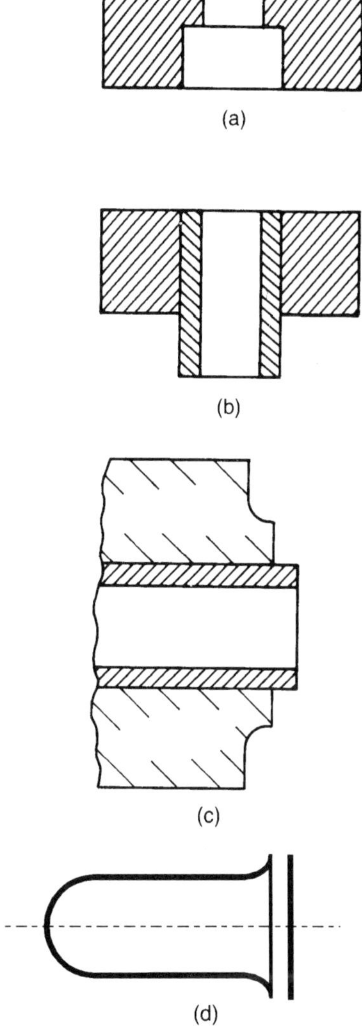

Figure 2.51 *Rules of hatching*

areas of the same component should be hatched in the same direction and with the same spacing.

Figure 2.51(b) shows how to hatch assembled parts. Where the different parts meet on assembly drawings, the direction of hatching should be reversed. The hatching lines should also be staggered. The spacing may also be changed.

Figure 2.51(c) shows how to hatch large areas. This saves time and avoids clutter. The hatching is limited to that part of the area that touches adjacent hatched parts or just to the outline of a large part.

Figure 2.51(d) shows how sections through thin materials can be blocked in solid rather than hatched. There should be a gap of not less than 1 mm between adjacent parts even when these are a tight fit in practice.

Finally I have included some further examples of sectioning in Figure 2.52. These include assemblies, half sections, part sections and revolved sections. Then it will be your turn to produce some engineering drawings including some or all of the features outlined in this section.

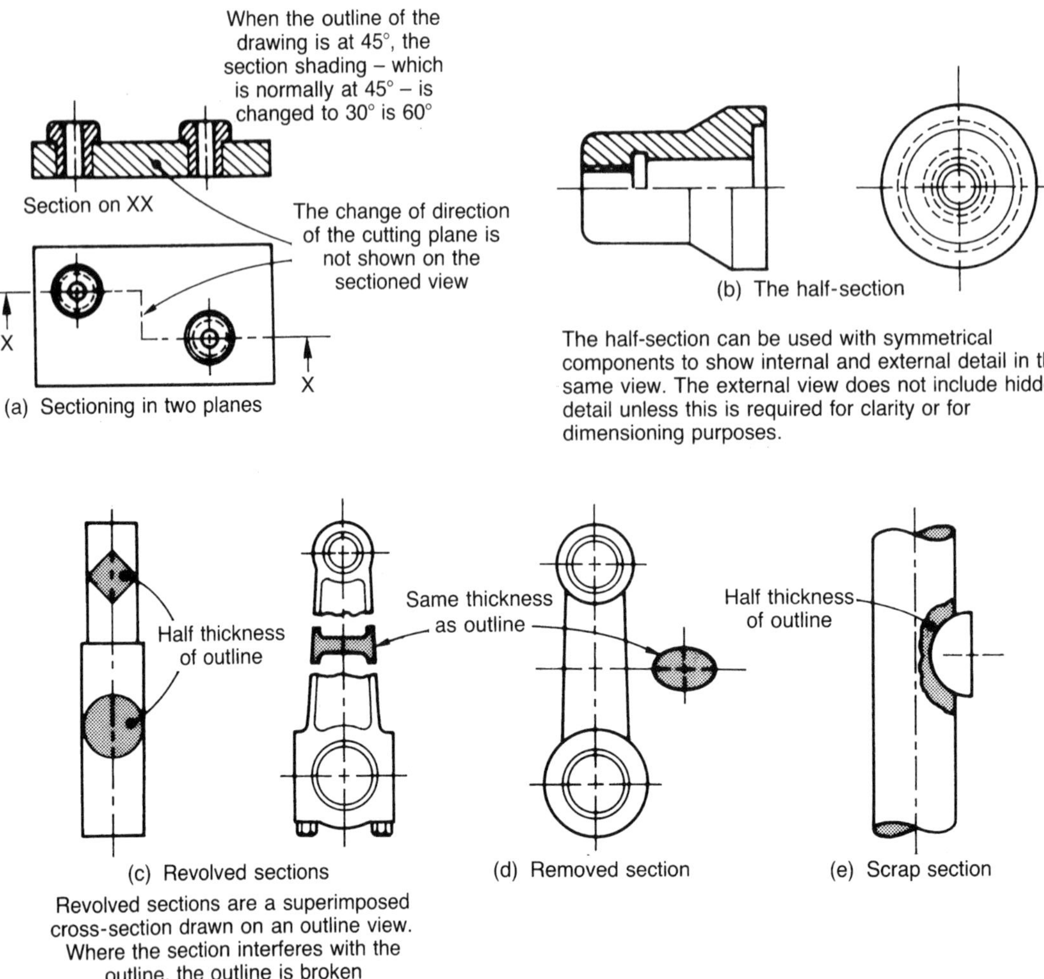

Figure 2.52 *Examples of sectioning*

Activity 2.8

(a) Redraw Figure 2.53 in THIRD-ANGLE projection. Include an end view looking in the direction of arrow 'A' , and section the elevation on the cutting plane XX.

(b) Figure 2.54 shows a cast iron pipe bend.

(i) Redraw, adding an end view looking in the direction of arrow A.

(ii) Section the elevation on the centre line.

(iii) Draw an auxiliary view of flange B.

Schematic circuit diagrams

I introduced you to examples of electrical, pneumatic and hydraulic schematic circuit diagrams earlier in this unit. We are now going to look at the drawing of such circuits in greater detail. In all these circuits, the circuit components are not drawn out in detail but are represented by symbols. To read the circuit

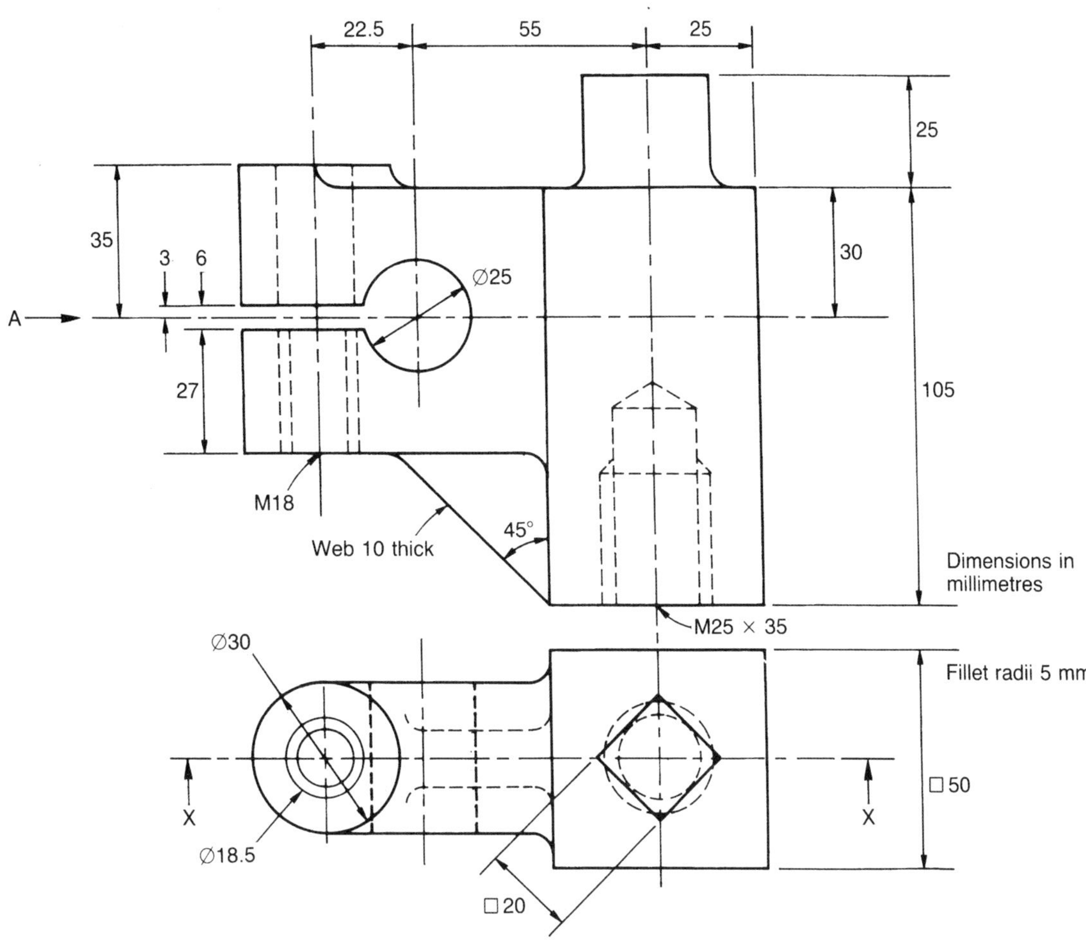

Figure 2.53 *See Activity 2.8*

diagram you must understand what the symbols stand for and how their function is interpreted.

You require access to BSI publication PP7307: Graphical Symbols for Use in Colleges and Schools. It summarizes the basic essentials of a number of full standard specifications relating to schematic circuit diagrams and their related wiring and piping diagrams.

Fluid power schematic circuit diagrams

These diagrams cover both pneumatic and hydraulic circuits. The symbols that we shall use do not illustrate the physical make-up, construction or shape of the components. Neither are the symbols to scale or orientated in any particular position. They are only intended to show the 'function' of the component they portray, the connections and the fluid flow path.

Complete symbols are made up from one or more basic symbols and from one or more functional symbols. Examples of some basic symbols are shown in Figure 2.55 and some functional symbols are shown in Figure 2.56.

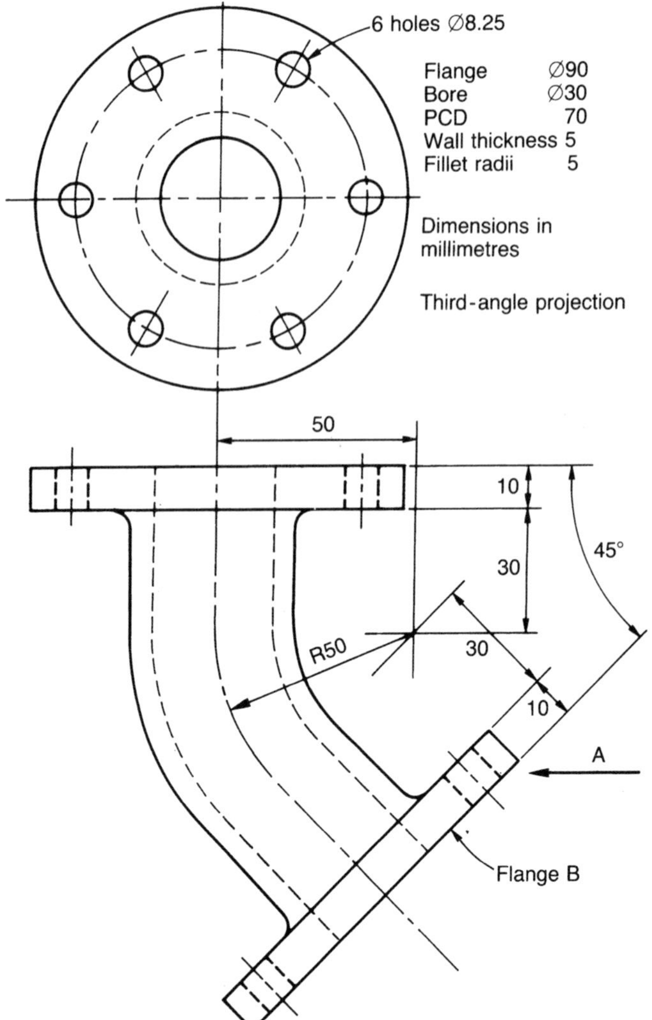

Figure 2.54 *See Activity 2.8*

Energy converters

Let's now see how we can combine some of these basic and functional symbols to produce a complete symbol representing a component. For example let's start with a motor. The complete symbol is shown in Figure 2.57.

The large circle indicates that we have an energy conversion unit such as a motor or pump. Notice that the fluid flow is into the device and that it is pneumatic. The direction of the arrowhead indicates the direction of flow. The fact that the arrowhead is clear (open) indicates that the fluid is air. Therefore the device must be a motor. If it was a pump the fluid flow would be out of the circle. The single line at the bottom of the circle is the outlet (exhaust) from the motor and the double line is the mechanical output from the motor.

Now let's analyse the symbol shown in Figure 2.58.

- The circle tells us that it is an energy conversion unit.

Description	Symbol
Flow lines Continuous: Working line return line feed line	———————
Long dashes: Pilot control lines	— — — —
Short dashes Drawn lines	- - - - - - -
Long chain enclosure line	—— - ——
Flow line connections	┴
Mechanical link, roller, etc.	○
Semi-rotary actuator	D
As a rule, control valves (valve) except for non- return valves	□
Conditioning apparatus (filter, separator, lubricator, heat exchanger)	◇

Description	Symbol
Spring	WV
Restriction: affected by viscosity unaffected by viscosity	)(V Λ
As a rule, energy conversion units (pump, com- pressor motor)	○
Measuring instruments	○
Non-return valve, rotary connection, etc.	○

Figure 2.55 *Basic symbols used in fluid power diagrams*

- The arrowheads show that the flow is from the unit so it must be a pump.
- The arrowheads are solid so it must be a hydraulic pump.
- The arrowheads point in opposite directions so the pump can deliver the hydraulic fluid in either direction depending upon its direction of rotation.
- The arrow slanting across the pump is the variability symbol, so the pump has variable displacement.
- The double lines indicate the mechanical input to the pump from some engine or motor.

Summing up, we have a variable displacement, hydraulic pump that is bi-directional.

Test your knowledge 2.21

(a) Draw the symbol for a uni-directional, fixed displacement pneumatic pump (compressor).

(b) Draw the symbol for a fixed capacity hydraulic motor.

Directional control valves

The function of a directional control valve is to open or close flow lines in a system. Control valve symbols are always drawn in square boxes or groups of square boxes to form a rectangle. This is how you recognize them. Each box indicates a discrete position for the control valve. Flow paths through a valve are known

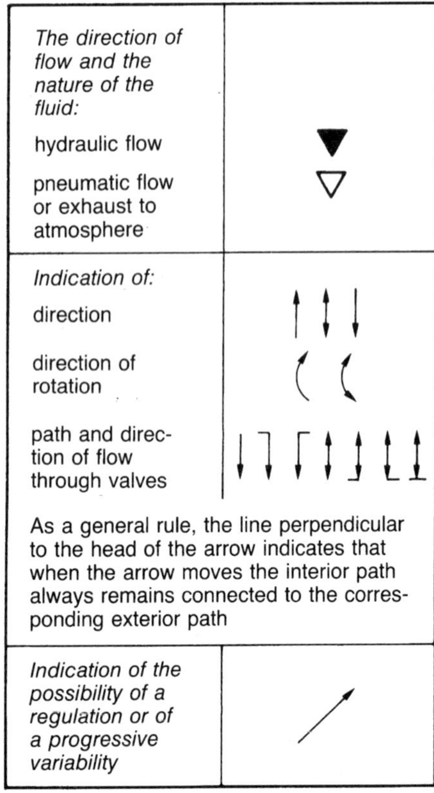

The direction of flow and the nature of the fluid:	
hydraulic flow	
pneumatic flow or exhaust to atmosphere	
Indication of:	
direction	
direction of rotation	
path and direction of flow through valves	
As a general rule, the line perpendicular to the head of the arrow indicates that when the arrow moves the interior path always remains connected to the corresponding exterior path	
Indication of the possibility of a regulation or of a progressive variability	

Figure 2.56 *Functional symbols used in fluid power diagrams*

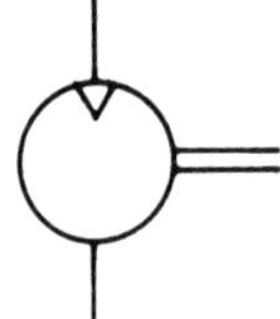

Figure 2.57 *Basic symbol for a motor*

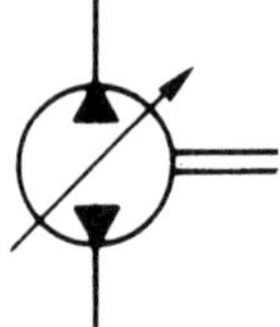

Figure 2.58 *Energy converter symbol (see text)*

as 'ways'. Thus a 4-way valve has four flow paths through the valve. This will be the same as the number of connections. We can, therefore, use a number code to describe the function of a valve. Figure 2.59 shows a 4/2 directional control valve. This valve has four flow paths, ports or connections and two positions. The two boxes indicate the two positions. The appropriate

Test your knowledge 2.22

(a) State the numerical code that describes the valve shown in Figure 2.60.

(b) (i) Describe the flow path drawn.
(ii) Sketch and describe the flow path when the valve is in its alternative position.

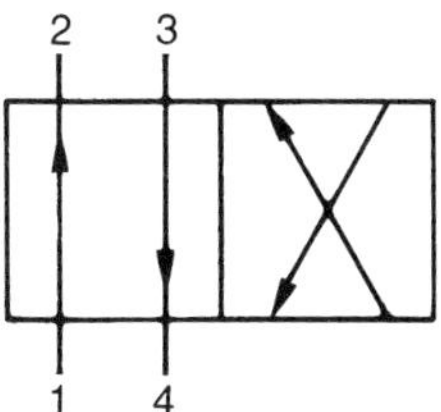

Figure 2.59 *4/2 directional control valve*

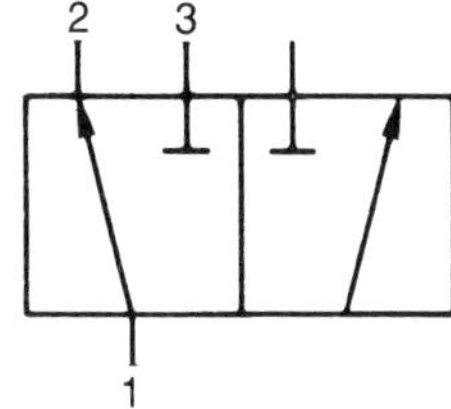

Figure 2.60 *See Test your knowledge 2.22*

box is shunted from side to side so that, in your imagination, the internal flow paths line up with the connections. Connections are shown by the lines that extend 'outside' the perimeters of the boxes.

As drawn, the fluid can flow into port 1 and out of port 2. Fluid can also flow into port 3 and out of port 4. In the second position, The fluid flows into port 3 and out of port 1. Fluid can also flow into port 4 and out of port 2.

Valve control methods

Before we look at other examples of directional control valves, let's see how we can control the positions of a valve. There are four basic methods of control, these are:

- manual control of the valve position
- mechanical control of the valve position
- electromagnetic control of the valve position
- pressure control of the valve positions (direct and indirect)
- combined control.

The methods of control are shown in Figure 2.61. With simple electrical or pressure control, it is only possible to move the valve to one, two or three discrete positions. The valve spool may be located in such positions by a spring loaded detent.

Combinations of the above control methods are possible. For example a single solenoid with spring return for a two position valve. Let's now look at some further directional control valves (DCVs).

- Figure 2.62(a) shows a 4/2 DCV controlled by a single solenoid with a spring return.
- Figure 2.62(b) shows a 4/3 DCV. That is, a directional control valve with 4 ports (connections) and 3 positions. It

Description	*Symbol*
Manual control:	
general symbol	
by push-button	
by lever	
by pedal	
Mechanical control:	
by plunger or tracer	
by spring	
by roller	
by roller, operating in one direction only	
Electrical control:	
by solenoid (one winding)	
(two windings operating in opposite directions)	
by electric motor	M

Description	*Symbol*
Control by application or release of pressure	
Direct acting control:	
by application of pressure	
by release of pressure	
by different control areas	
Indirect control, pilot actuated:	
by application of pressure	
by release of pressure	
Interior control paths (paths are inside the unit)	
Combined control:	
by solenoid and pilot directional valve (pilot directional valve is actuated by the solenoid)	
by solenoid or pilot direction valve (either may actuate the control independently)	

Figure 2.61 *Methods of control*

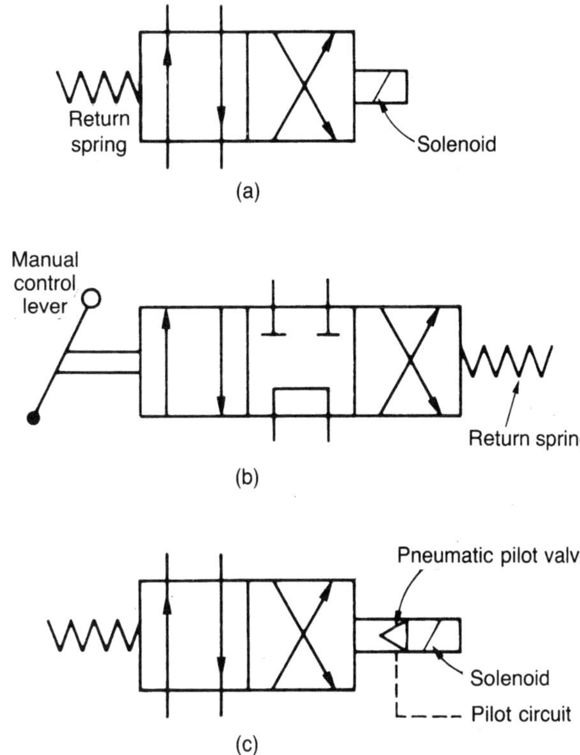

Figure 2.62 *Various types of DCV*

is operated manually by a lever with spring return to the centre. The service ports are isolated in the centre position. An application of this valve will be shown later.

- Figure 2.62(c) shows a 4/2 DCV controlled by pneumatic pressure by means of a pilot valve. The pilot valve is actuated by a single solenoid and a return spring.

Test your knowledge 2.23

Describe the DCV whose symbol is shown in Figure 2.63

Linear actuators

A linear actuator is a device for converting fluid pressure into a mechanical force capable of doing useful work and combining this force with limited linear movement. Put more simply, a piston in a cylinder. The symbols for linear actuators (also known as 'jacks' and 'rams') are simple to understand and some examples are shown in Figure 2.64.

- Figure 2.64(a) shows a single-ended, double-acting actuator. That is, the piston is connected by a piston rod to some

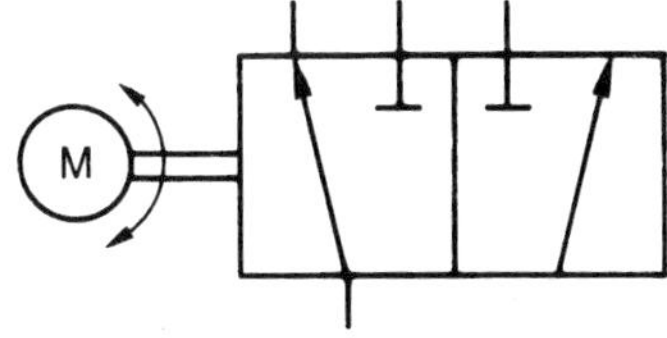

Figure 2.63 *See Test your knowledge 2.23*

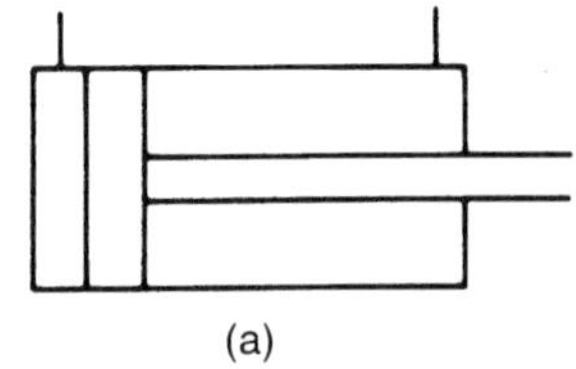

(a)

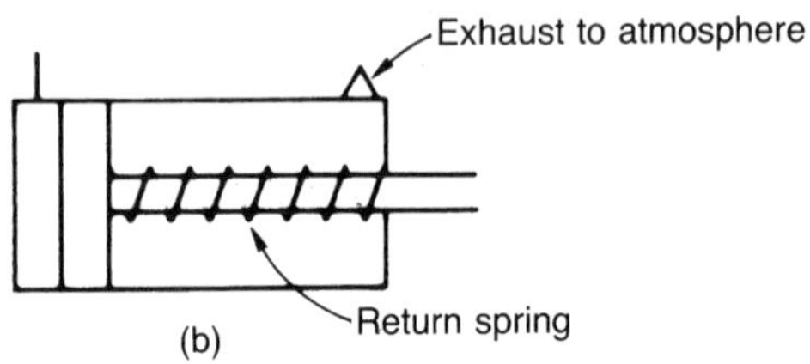

(b)

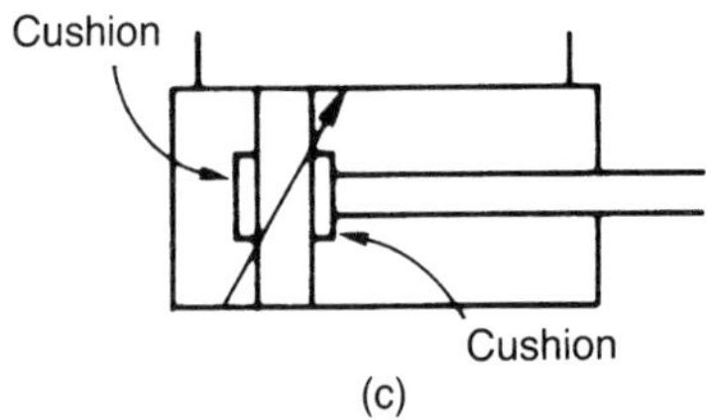

(c)

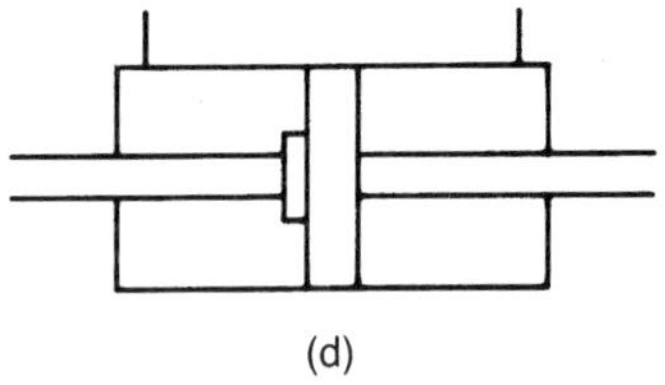

(d)

Figure 2.64 *Various types of linear actuator*

external mechanism through one end of the cylinder only. It is double acting because fluid pressure can be applied to either side of the piston.

- Figure 2.64(b) shows a single-ended, single-acting actuator with spring return. Here the fluid pressure is applied only to one side of the piston. Note the pneumatic exhaust to atmosphere so that the air behind the piston will not cause a fluid lock.
- Figure 2.64(c) shows a single-ended, single-acting actuator, with double variable cushion damping. The cushion damping prevents the piston impacting on the ends of the cylinder and causing damage.
- Figure 2.64(d) shows a double ended, double-acting actuator fitted with single, fixed cushion damping.

We are now in a position to use the previous component symbols to produce some simple fluid power circuits.

Figure 2.65 shows a single-ended, double-acting actuator controlled by a 4/3 tandem centre, manually operated DCV. Note

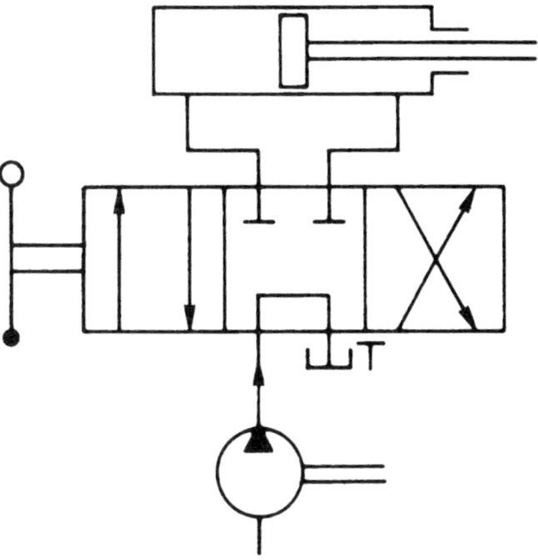

Figure 2.65 *Actuator controlled by a DCV*

that in the neutral position both sides of the actuator piston are blocked off, forming a hydraulic lock. In this position the pump flow is being returned directly to the tank. Note the tank symbol. This system is being supplied by a single direction fixed displacement hydraulic pump.

Figure 2.66 shows a simple pneumatic hoist capable of raising a load. The circuit uses two 2-port manually operated push-button valves connected to a single-ended, single-acting actuator. Supply pressure is indicated by the circular symbol with a black dot in its centre. Valve 'b' has a threaded exhaust port indicated by the extended arrow. When valve 'a' is operated, compressed air from the air line is admitted to the underside of the piston in the cylinder. This causes the piston to rise and to raise the load. Any air above the piston is exhausted to the atmosphere through the threaded exhaust port at the top of

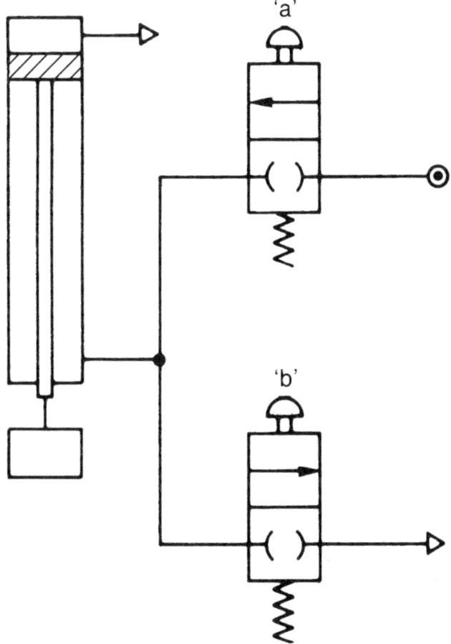

Figure 2.66 *A simple pneumatic hoist*

the cylinder. Again this is indicated by a long arrow. When valve 'b' is operated, it connects the cylinder to the exhaust and the actuator is vented to the atmosphere. The load is lowered by gravity.

Both these circuits are functional, but they do not have protection against over-pressurization, neither do they have any other safety devices fitted. Therefore, we need to increase our vocabulary of components before we can design a safe, practical circuit. We will now consider the function and use of pressure and flow control valves.

Pressure relief and sequence valves

Figure 2.67 shows an example of a pressure relief (safety) valve. In Figure 2.67(a) the valve is being used in a hydraulic circuit. Pressure is controlled by opening the exhaust port to the reservoir tank against an opposing force such as a spring. In Figure 2.67(b) the valve is being used in a pneumatic circuit so it exhausts to the atmosphere.

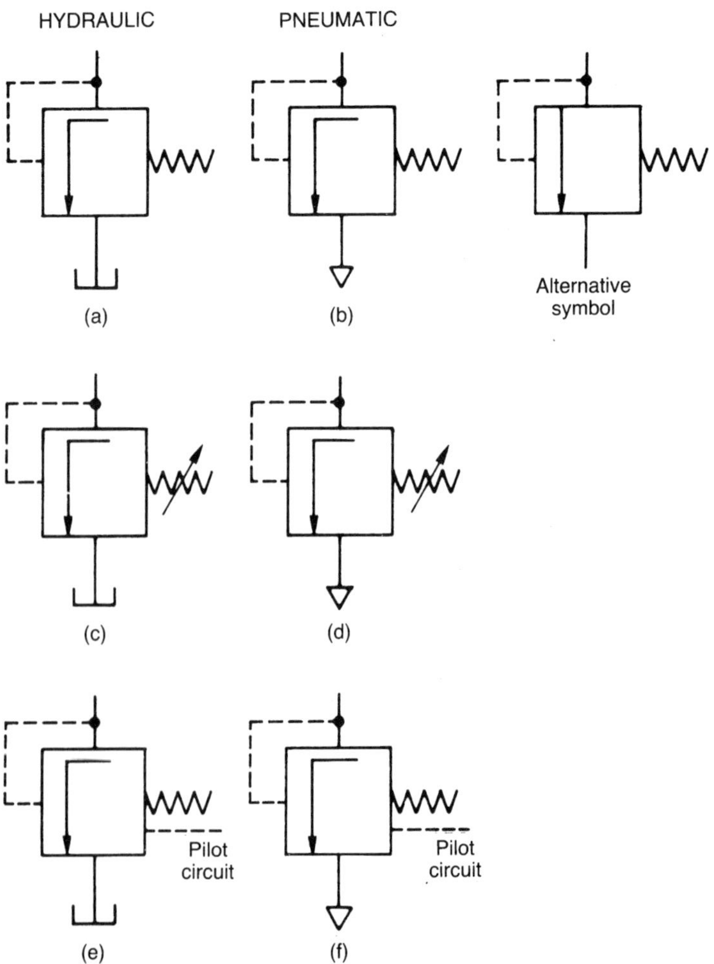

Figure 2.67 *Use of a pressure relief valve*

Figures 2.67(c) and 2.67(d) show the same valves except that this time the relief pressure is variable, as indicated by the arrow drawn across the spring. If the relief valve setting is used to control the normal system pressure as well as acting as an emergency safety valve, the adjustment mechanism for the valve must be designed so that the maximum safe working pressure for the circuit cannot be exceeded.

Figures 2.67(e) and 2.67(f) show the same valves with the addition of pilot control. This time the pressure at the inlet port is not only limited by the spring but also by the pressure of the pilot circuit superimposed on the spring. The spring offers a minimum pressure setting and this can be increased by increasing the pilot circuit pressure up to some pre-determined safe maximum. Sometimes the spring is omitted and only pilot pressure is used to control the valve.

Sequence valves are closely related to relief valves in both design and function and are represented by very similar symbols. They permit the hydraulic fluid to flow into a sub-circuit, instead of back to the reservoir, when the main circuit pressure reaches the setting of the sequence valve. You can see that Figure 2.68 is very similar to a pressure relief valve (PRV) except that, when it opens, the fluid is directed to the next circuit in the sequence instead of being exhausted to the reservoir tank or allowed to escape to the atmosphere.

Flow control valves

Flow control valves, as their name implies, are used in systems to control the rate of flow of fluid from one part of the system to another. The simplest valve is merely a fixed restrictor. For operational reasons this type of flow control valve is inefficient, so the restriction is made variable as shown in Figure 2.69(a). This is a throttling valve. The full symbol is shown in Figure 2.69(b). In this example the valve setting is being adjusted mechanically. The valve rod ends in a roller follower in contact with a cam plate.

Sometimes it is necessary to to ensure that the variation in inlet pressure to the valve does not affect the flow rate from

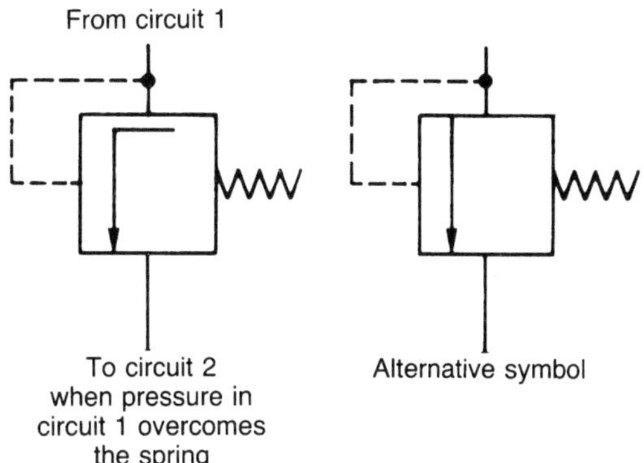

Figure 2.68 *Sequence valve*

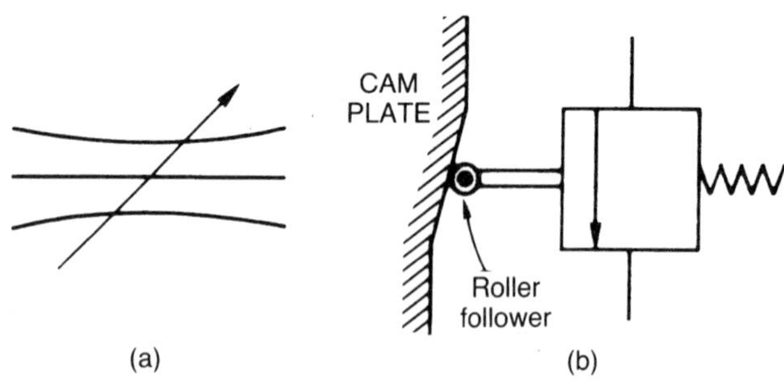

Figure 2.69 *Fluid control valves*

the valve. Under these circumstances we use a pressure compensated flow control valve (PCFCV). The symbol for this type of valve is shown in Figure 2.70. This symbol suggests that the valve is a combination of a variable restrictor and a pilot operated relief valve. The enclosing box is drawn using a long-chain line. This signifies that the components making up the valve are assembled as a single unit.

Non-return valves and shuttle valves

The non-return valve (NRV), or check valve as it is sometimes known, is a special type of directional control valve. It only allows the fluid to flow in one direction and it blocks the flow in the reverse direction. These valves may be operated directly or by a pilot circuit. Some examples are shown in Figure 2.71.

- Figure 2.71(a) shows a valve that opens (is free) when the inlet pressure is higher than the outlet pressure (back-pressure).
- Figure 2.71(b) shows a spring-loaded valve that only opens when the inlet pressure can overcome the combined effects of the outlet pressure and the force exerted by the spring.
- Figure 2.71(c) shows a pilot controlled NRV. It only opens if the inlet pressure is greater than the outlet pressure. However, these pressures can be augmented by the pilot circuit pressure.
  (i) The pilot pressure is applied to the inlet side of the NRV. We now have the combined pressures of the main

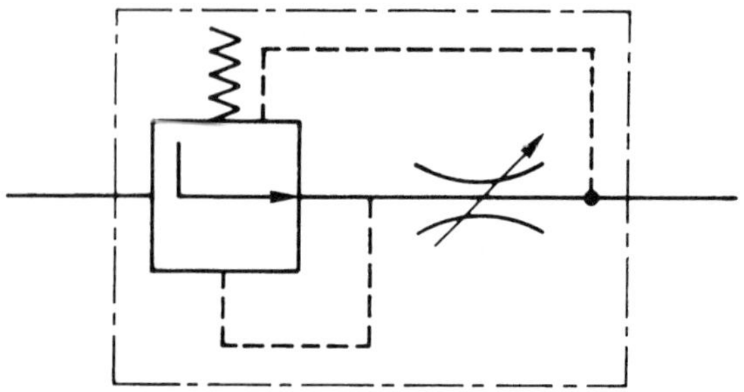

Figure 2.70 *Pressure compensated flow control valve*

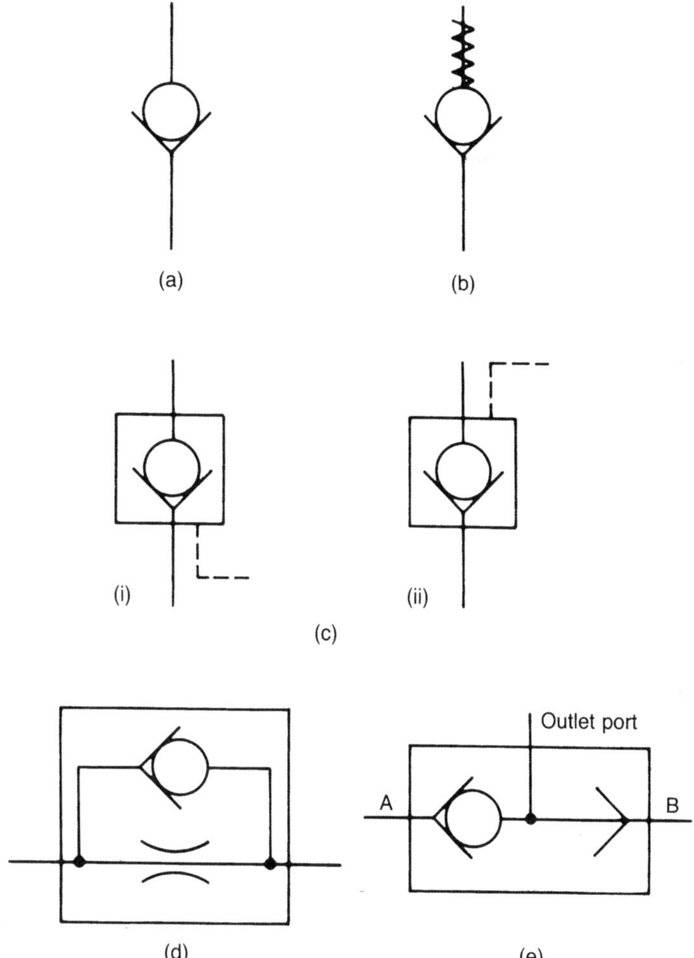

Figure 2.71 *Some examples of non-return valves*

(primary) circuit and the pilot circuit acting against the outlet pressure. This enables the valve to open at a lower main circuit pressure than would normally be possible.

(ii) The pilot pressure is applied to the outlet side of the NRV. This assists the outlet or back-pressure in holding the valve closed. Therefore it requires a greater main circuit pressure to open the valve. By adjusting the pilot pressure in these two examples we can control the circumstances under which the NRV opens.

- Figure 2.71(d) shows a valve that allows normal full flow in the forward direction but restricted flow in the reverse direction. The valves previously discussed did not allow any flow in the reverse direction.
- Figure 2.71(e) shows a simple shuttle valve. As its name implies, the valve is able to shuttle backwards and forwards. There are two inlet ports and one outlet port. Imagine that inlet port 'A' has the higher pressure. This pressure overcomes the inlet pressure at 'B' and moves the shuttle valve to the right. The valve closes inlet port 'B' and

connects inlet port 'A' to the outlet port. If the pressure at inlet port 'B' rises, or that at 'A' falls, the shuttle will move back to the left. This will close inlet port 'A' and connect inlet port 'B' to the outlet. Thus, the inlet port with the higher pressure is automatically connected to the outlet port.

Conditioning equipment

The working fluid, be it oil or air, has to operate in a variety of environments and it can become overheated and/or contaminated. As its name implies, conditioning equipment is used to maintain the fluid in its most efficient operating condition. A selection of conditioning equipment symbols are shown in Figure 2.72. Note that all conditioning device symbols are diamond shaped.

Filters and *strainers* have the same symbol. They are normally identified within the system by their position. The filter element (dotted line) is always positioned at 90° to the fluid path.

Filters, water traps, lubricators and miscellaneous apparatus

Description	*Symbol*
Filter or strainer	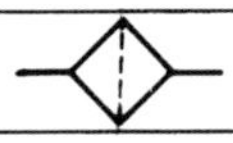
Water trap: with manual control	
automatically drained	
Filter with water trap: with manual control	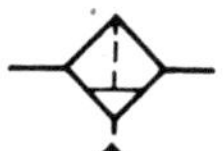
automatically drained	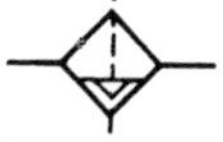
Air dryer	
Lubricator	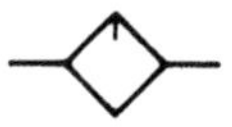
Conditioning unit detailed symbol	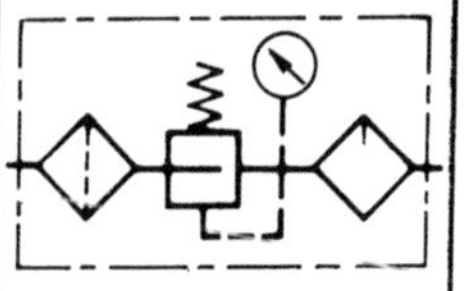
simplified symbol	

Heat exchangers

Description	*Symbol*
Temperature controller (arrows indicate that heat may be either introduced or dissipated)	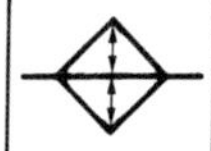
Cooler (arrows indicate the extraction of heat) without representation of the flow lines of the coolant	
with representation of the flow lines of the coolant	
Heater (arrows indicate the introduction of heat)	

Figure 2.72 *Symbols for conditioning devices*

Test your knowledge 2.24

Figure 2.73 shows a range of fluid circuit symbols. Name the symbol and explain the purpose of the device it represents.

Water traps are easily distinguished from filters since they have a drain connection and an indication of trapped water. Water traps are particularly important in pneumatic systems because of the humidity of the air being compressed.

Lubricators are particularly important in pneumatic systems. Hydraulic systems using oil are self-lubricating. Pneumatic systems use air, which has no lubricating properties so oil, in the form of a mist, has to be added to the compressed air line.

Heat exchangers can be either heaters or coolers. If the hydraulic oil becomes too cool it becomes thicker (more viscous)

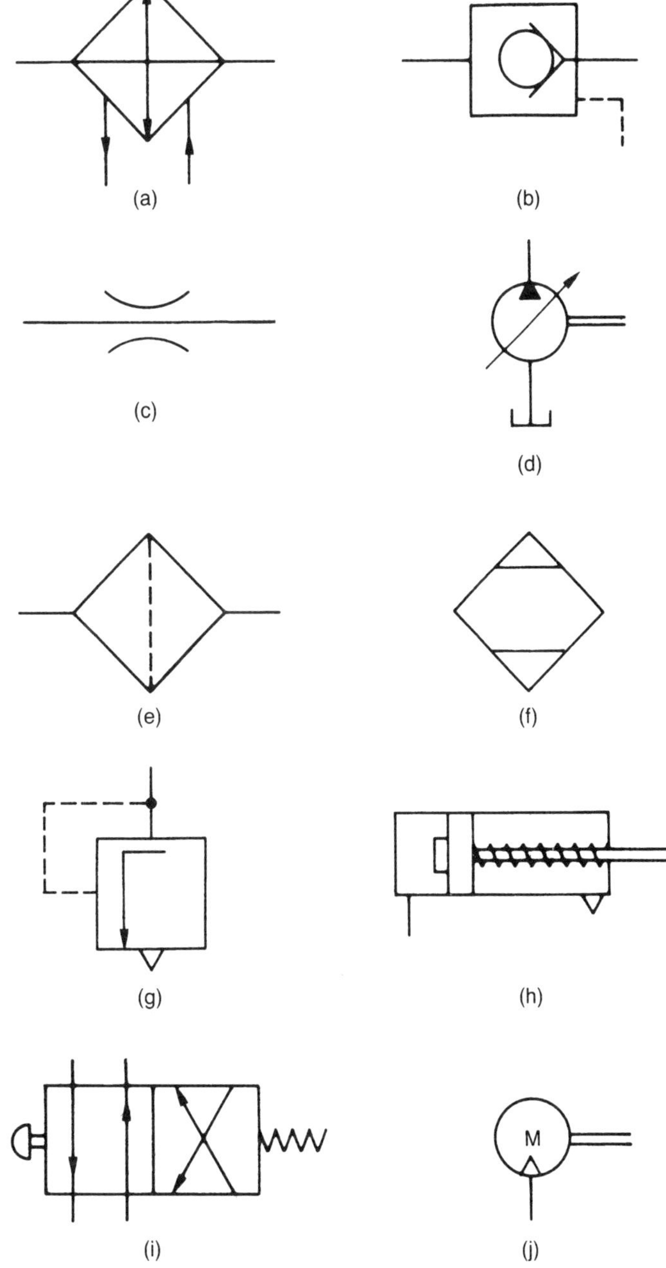

Figure 2.73 *See Test your knowledge 2.24*

and the system becomes sluggish. If the oil becomes too hot it will become too thin (less viscous) and not function properly. The direction of the arrows in the symbol indicates whether heat energy is taken from the fluid (cooler) or given to the fluid (heater). Notice that the cooler can show the flow lines of the coolant.

There is one final matter to be considered before you can try your hand at designing a circuit, and that is the pipework circuit to connect the various components together. The correct way of representing pipelines is shown in Figure 2.74.

- Figure 2.74(a) shows pipelines that are crossing each other but are not connected.
- Figure 2.74(b) shows three pipes connected at a junction. The junction (connection) is indicated by the solid circle (or large dot, if you prefer).
- Figure 2.74(c) shows four pipes connected at a junction. On no account can the connection be drawn as shown in Figure

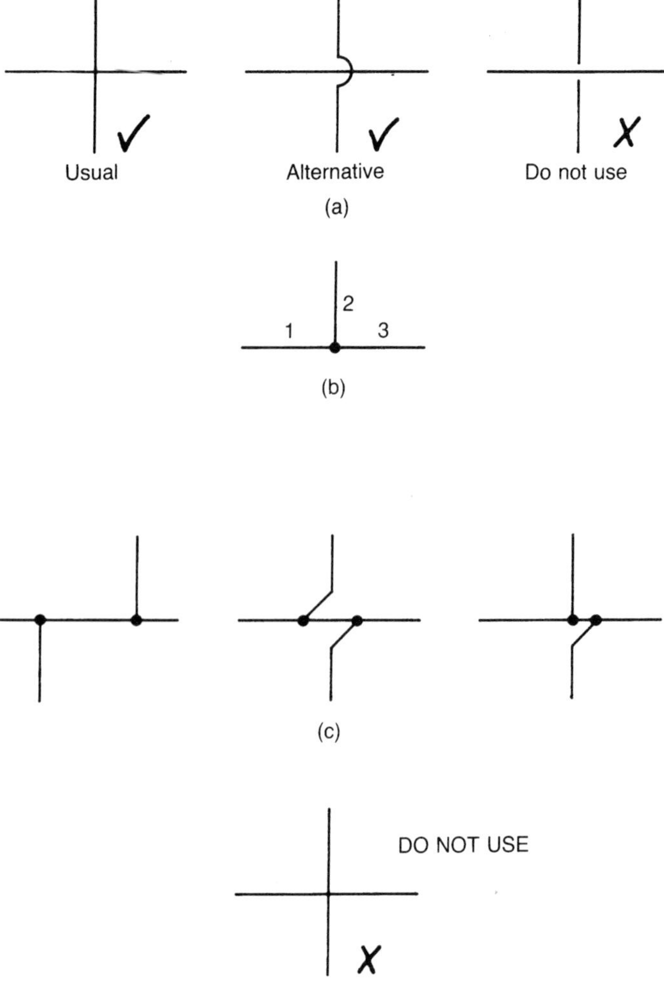

Figure 2.74 *Representing pipelines*

2.74(d). This is because there is always a chance of the ink running where lines cross on a drawing. The resulting 'blob' could then be misinterpreted as a connection symbol with disastrous results.

Activity 2.9 

Figure 2.75 shows the general principles for the hydraulic drive to the ram of a shaping machine. The ram is moved back and forth by a double-acting single-ended hydraulic actuator.

The drawing was made many years ago and uses outdated symbols. You are to draw a schematic hydraulic circuit diagram for this machine using current symbols and practices as set out in PP7307.BS.

Electrical and electronic circuits

Electrical and electronic circuits can also be drawn using schematic symbols to represent the various components. The full range of symbols and their useage can be found in BS 3939. This is a very extensive standard and well beyond the needs of this book. For our immediate requirements you should refer to PP7307 Graphical Symbols for Use in Schools and Colleges. Figure 2.76 shows a selection of symbols that will be used in the following examples.

- A *cell* is a source of direct current (d.c.) electrical energy. Primary cells have a nominal potential of 1.5 volt each. They cannot be recharged and are disposable. Secondary cells are rechargeable. Lead–acid cells have a nominal potential of 2 volts and NiFe cells have a nominal potential of 1.2 volt.

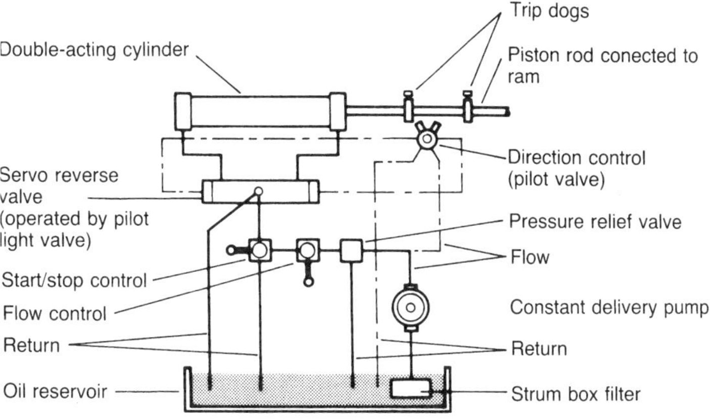

Figure 2.75 *See Activity 2.9*

Description	*Symbol*
Primary or secondary cell	
Battery of primary or secondary cells	
Alternative symbol	
Earth or ground	
Signal lamp, general symbol	
Electric bell	
Electric buzzer	
Fuse	
Resistor, general symbol	
Variable resistor	
Resistor with sliding contact	
Potentiometer with moving contact	
Capacitor, general symbol	
Polarized capacitor	
Voltage-dependent polarized	
Capacitor with pre-set adjustment	
Inductor, winding, coil, choke	
Inductor with magnetic core	

Description	*Symbol*
Transformer with magnetic core	
Ammeter	A
Voltmeter	V
Make contact, normally open. This symbol is also used as the general symbol for a switch	
Semiconductor diode, general symbol	
PNP transistor	
NPN transistor with collector connected to envelope	
Amplifier, simplified form	

Figure 2.76 *Electronic symbols*

- *Batteries* consist of a number of cells connected in series to increase the overall potential. A 12 volt car battery consists of 6 lead–acid secondary cells of 2 volts each.
- *Fuses* protect the circuit in which they are connected from excess current flow. This can result from a fault in the circuit, from a fault in an appliance connected to the circuit or from too many appliances being connected to the same circuit. The current flowing in the circuit tends to heat up the fuse wire. When the current reaches some pre-determined value the fuse wire melts and breaks the circuit so the current can no longer flow. Without a fuse the circuit wiring could overheat and cause a fire.
- *Resistors* are used to control the magnitude of the current flowing in a circuit. The resistance value of the resistor may be fixed or it may be variable. Variable resistors may be pre-set or they may be adjustable by the user. The electric current does work in flowing through the resistor and this heats up the resistor. The resistor must be chosen so that it can withstand this heating effect and sited so that it has adequate ventilation.
- *Capacitors*, like resistors, may be fixed in value or they may be preset or variable. Capacitors store electrical energy but, unlike secondary cells, they may be charged or discharged almost instantaneously. The stored charge is much smaller than the charge stored by a secondary cell. Large value capacitors are used to smooth the residual ripple from the rectifier in a power pack. Medium value capacitors are used for coupling and decoupling the stages of audio frequency amplifiers. Small value capacitors are used for coupling and decoupling radio frequency signals and they are also used in tuned (resonant) circuits.
- *Inductors* act like electrical 'flywheels'. They limit the build up of current in a circuit and try to keep the circuit running by putting energy back into it when the supply is turned off. They are used as current limiting devices in fluorescent lamp units. They are used as chokes in telecommunications equipment. They are also used together with capacitors to make up resonant (tuned) circuits in telecommunications equipment.
- *Transformers* are used to raise or lower the voltage of alternating currents. Inductors and transformers cannot be used in direct current circuits. You cannot get something for nothing, so if you increase the voltage you decrease the current accordingly so that (neglecting losses), $VA = C$ where C is a constant for the primary and secondary circuits of any given transformer.
- *Ammeters* measure the current flowing in a circuit. They are always wired in series with the circuit so that the current being measured can flow through the meter.
- *Voltmeters* measure the potential difference (voltage) between two points in a circuit. To do this they are always wired in parallel across that part of the circuit where the potential is to be measured.

- *Switches* are used to control the flow of current in a circuit. They can only open or close the circuit. So the current either flows or it doesn't.
- *Diodes* are like the non-return valves in hydraulic circuits. They allow the current to flow in one direction only as indicated by the arrowhead of the symbol. They are used to rectify alternating current (a.c.) and convert it into d.c.
- *Transistors* are used in high-speed switching circuits and to magnify radio and audio frequency signals.
- *Integrated circuits* consist of all the components necessary to produce amplifiers, oscillators, central processor units, computer memories and a host of other devices fabricated onto a single slice of silicon; each chip being housed in a single compact package.

Let's look at some examples of schematic circuit diagrams using these symbols. All electric circuits consist of:

- a source of electrical energy (e.g. a battery or a generator)
- a means of controlling the flow of electric current (e.g. a switch or a variable resistor)
- an appliance to convert the electrical energy into useful work (e.g. a heater, a lamp, or a motor)
- except for low-power battery operated circuits, an over-current protection device (fuse or circuit breaker)
- conductors (wires) to connect these various circuit elements together. Note that the rules for drawing conductors that are connected and conductors that are crossing but not connected are the same as for drawing pipework as previously described in Figure 2.74.

Figure 2.77 shows a very simple circuit that satisfies the above requirements. In Figure 2.77(a) the switch is 'closed' therefore the circuit as a whole is also a closed loop. This enables the electrons that make up the electric current to flow from the source of electrical energy through the appliance (lamp) and back to the source of energy ready to circulate again. Rather like the fluid in our earlier hydraulic circuits. In Figure 2.77(b) the switch is 'open' and the circuit is no longer a closed loop. The circuit is broken. The electrons can no longer circulate. The circuit ceases to function. We normally draw our circuits with the switches in the 'open' position so that the circuit is not functioning and is 'safe'.

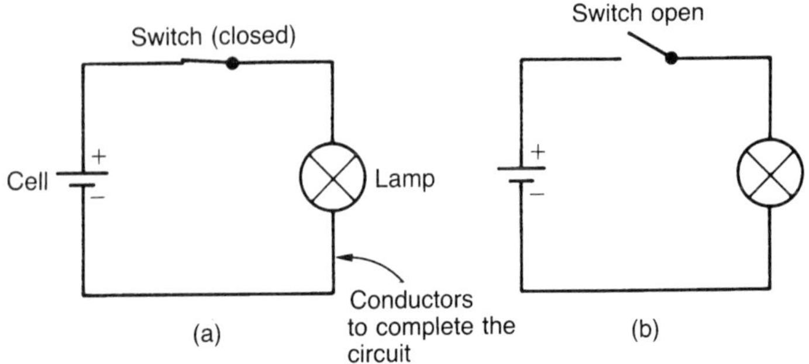

Figure 2.77 *A simple electric circuit*

Figure 2.78 shows a simple battery operated circuit for determining the resistance of a fixed value resistor. The resistance value is obtained by substituting the values of current and potential into the formula $R = V/I$. The current in amperes is read from the ammeter and the potential in volts is read from the voltmeter. Note that the ammeter is wired in series with the resistor so that the current can flow through it. The voltmeter is wired in parallel with the resistor so that the potential can be read across it. This is always the way these instruments are connected.

Figure 2.79 shows a circuit for operating the light over the stairs in a house. The light can be operated either by the switch at the bottom of the stairs or by the switch at the top of the stairs. Can you work out how this is achieved? The switches are of a type called 'two-way, single-pole'. The circuit is connected to the mains supply. It is protected by a fuse in the the 'consumer unit'. This unit contains the main switch and all the fuses for the house and is situated adjacent to the supply company's meter and main fuse.

Test your knowledge 2.25

Figure 2.82 shows some typical electrical and electronic component symbols. Name them and briefly explain what they do.

Figure 2.80 shows a two-stage transistorized amplifier. It also shows a suitable power supply. Table 2.3 lists and names the components.

Figure 2.81 shows a similar amplifier using a single chip. Such an amplifier would have the same performance but fewer components are required. Therefore it is cheaper and quicker to make.

Activity 2.10 

Draw a simple circuit for powering four electric lamps from a battery. It must be possible to turn each lamp on or off independently. A master switch must also be provided so that the whole circuit can be turned on or off. A fuse must also be provided to prevent the battery being overloaded.

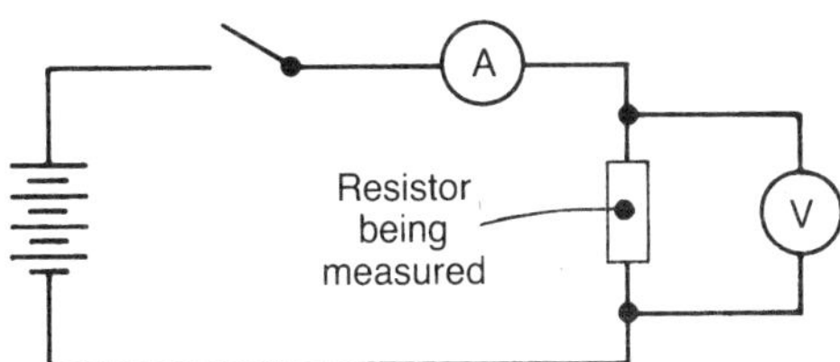

Figure 2.78 *Circuit for determining resistance*

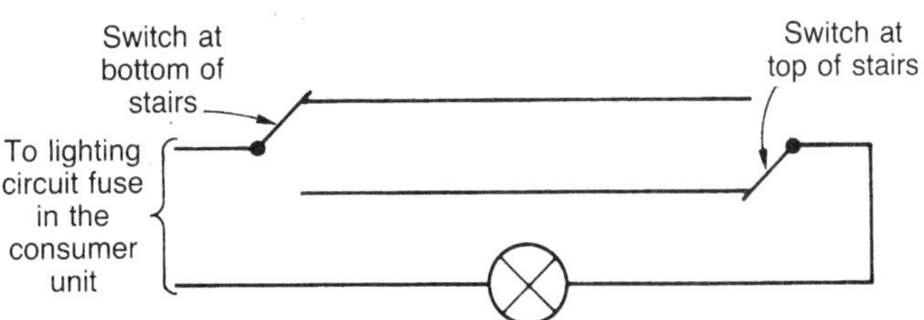

Figure 2.79 *Two-way light switch*

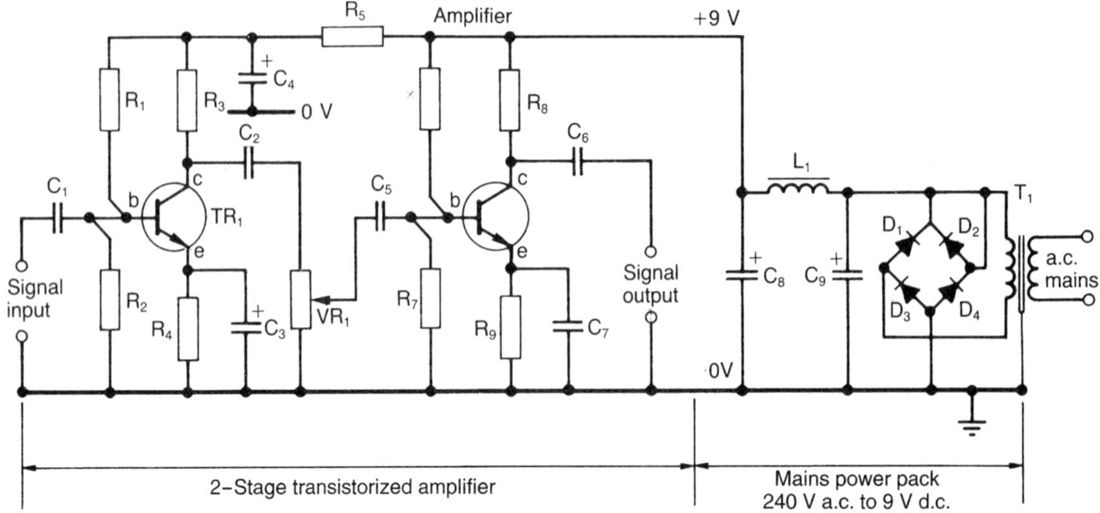

Figure 2.80 *A two-stage transistor amplifier*

Table 2.3 *Components list for the two-stage transistor amplifier*

Component	*Name*
R_1 R_2 R_3 R_4 R_5 R_6 R_7 R_8 R_9	Fixed resistors
VR_1 (volume control)	Variable resistor
C_1 C_2 C_3 C_4 C_5 C_6 C_7 C_8 C_9	Capacitors
D_1 D_2 D_3 D_4	Diodes
TR_1 TR_2	Transistors
T_1	Mains transformer
L_1	Choke (inductor)

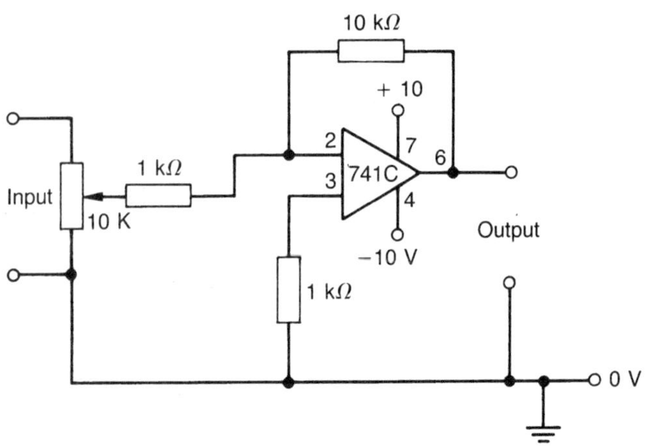

Figure 2.81 *A single-chip amplifier circuit*

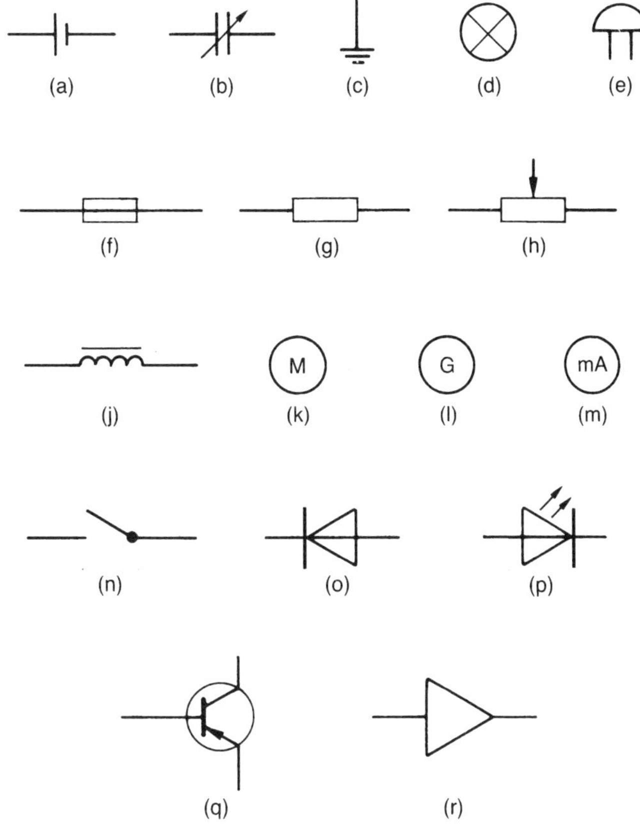

Figure 2.82 *See Test your knowledge 2.25*

Activity 2.11

Draw a schematic circuit diagram for a battery charger having the following features:

- The primary circuit of the transformer to have an on/off switch, a fuse and an indicator lamp.
- The secondary circuit to have a bridge rectifier, a variable resistor to control the charging current, a fuse and an ammeter to indicate the charging current.

Activity 2.12

Figure 2.83 shows an electronic circuit.

(a) List and name the components.

(b) Suggest what the circuit is used for.

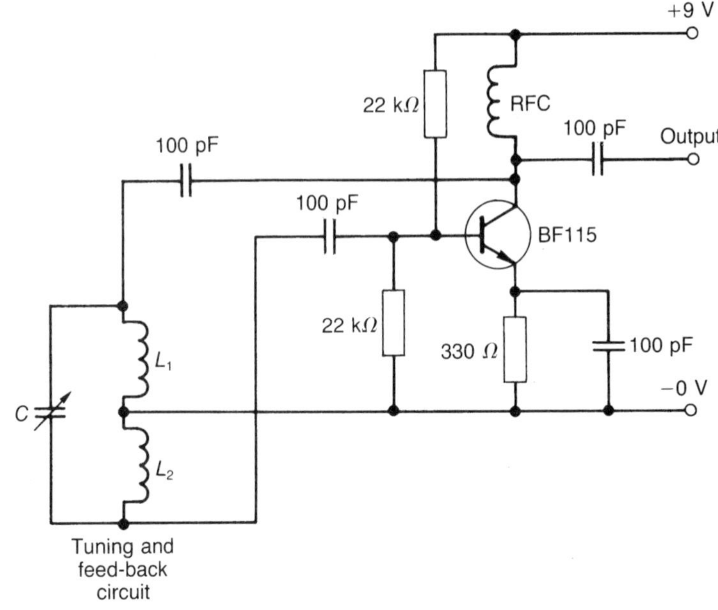

Figure 2.83 *See Activity 2.12*

Multiple-choice questions

1 Three dimensional solid objects are usually represented on a detail drawing using

A oblique projection
B orthographic projection
C isometric projection
D axonometric projection.

2 Drawings for service engineers must show the relationship between the assembled parts clearly and in such a way that the parts can be quickly identified by using

A first-angle projection
B third-angle projection
C perspective drawing
D exploded views.

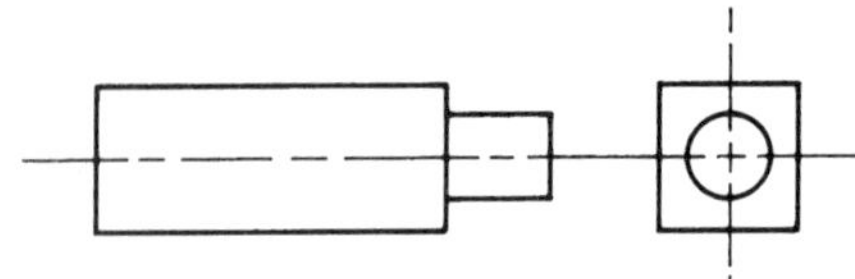

Figure 2.84

3 Figure 2.84 has been drawn in

A first-angle orthographic projection
B third-angle orthographic projection
C oblique projection
D isometric projection.

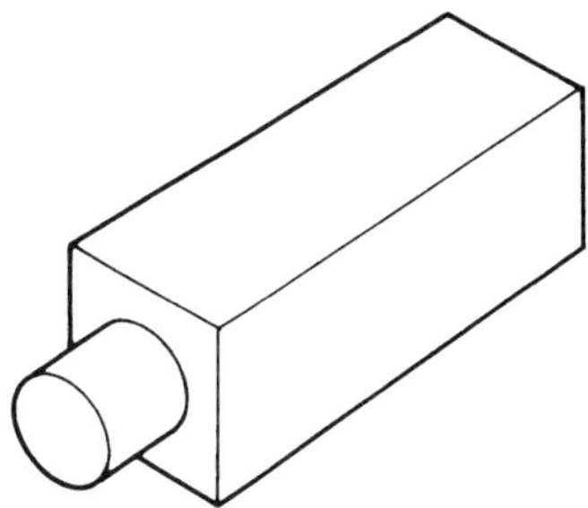

Figure 2.85

4 Figure 2.85 shows one method of presenting statistical data. It is called a

A bar chart
B histogram
C pie chart
D pictogram (ideograph).

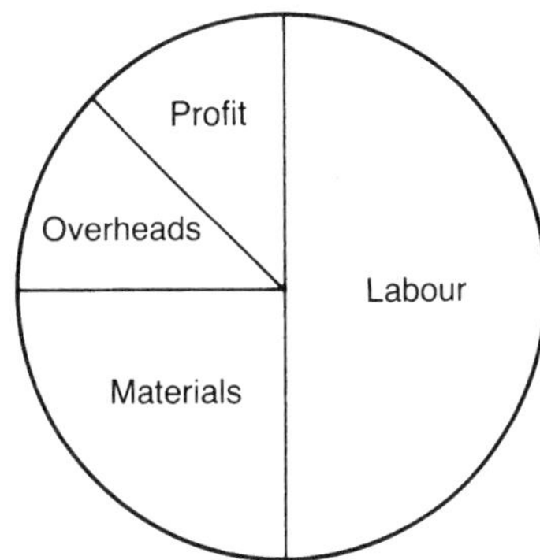

Figure 2.86

5 With reference to Figure 2.86, the proportion of the total cost represented by the materials is

A 10%
B 15%
C 25%
D 50%.

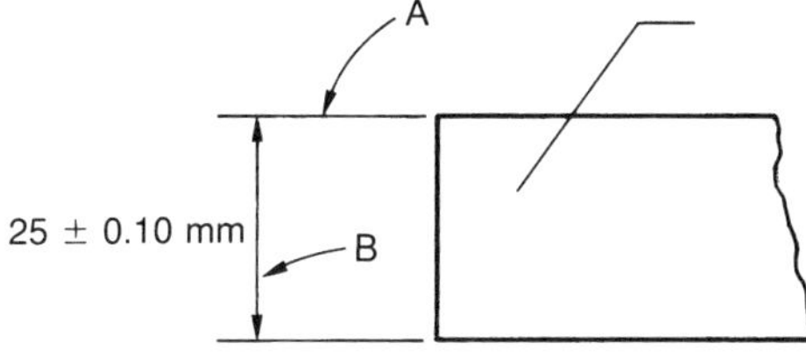

Figure 2.87

6 In Figure 2.87 the line A is referred to as
A a dimension line
B a leader line
C an outline
D a projection line.

7 In Figure 2.87 the line B is referred to as

A a dimension line
B a leader line
C an outline
D a construction line.

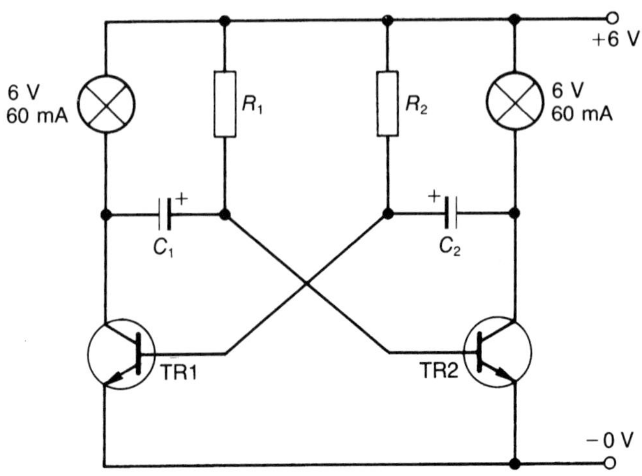

Figure 2.88

8 Figure 2.88 represents an electronic device. This type of drawing is called

A a wiring diagram
B an installation diagram
C a layout diagram
D a schematic circuit diagram.

9 The device shown in Figure 2.88 is

A a two-stage amplifier
B a single-chip amplifier
C a power supply
D a pulse generator.

10 In Figure 2.88 the components are represented by

A abbreviations
B conventions
C symbols
D standards.

Unit 3 Science and mathematics for engineering (intermediate)

Summary

This unit aims to provide knowledge and understanding of basic mathematical and scientific tools underpinning engineering activities. More specifically, the aims are to use mathematical techniques to investigate the behaviour of linear systems in engineering, to describe engineering systems in terms of scientific laws and principles and to describe how some physical quantities used in engineering are measured. It is assumed that the reader has little previous knowledge of each topic that is considered.

The mathematical techniques investigated include basic arithmetic operations, fractions, ratio and proportion, decimals, percentages, indices, standard form, use of calculator, the solution of simple equations, evaluation and transposition of formulae and straight line graph plotting.

The scientific laws and principles described include linear motion in a straight line, static equilibrium laws, force–extension for a spring, force–acceleration for a given mass, pressure, work done, power, energy, efficiency, linear expansion, basic change of state and simple relationships in d.c. and a.c. electrical circuits.

Typical variables and constants encountered in engineering are identified and include mass, length, time, acceleration, velocity, displacement, force, energy, stress, strain, modulus of elasticity, temperature, pressure, voltage, current, resistance, frequency, power, charge, magnetic flux and magnetic flux density.

Measuring devices described include multimeter, oscilloscope, micrometer, dead-weight, spring-balance, thermometers and thermocouples.

Mathematics for engineering

Arithmetic operations

Whole numbers are called **integers**. $+3$, $+5$, $+72$ are called **positive integers**; -13, -6, -51 are called **negative integers**. Between positive and negative integers is the number 0 which is neither positive nor negative.

The four basic arithmetic operators are: add ($+$), subtract ($-$), multiply ($\times$) and divide ($\div$).

For addition and subtraction, when **unlike signs** are together in a calculation, the overall sign is **negative**. Thus, adding minus 4 to 3 is $3 + -4$ and becomes $3 - 4 = -1$. **Like signs** together give an overall **positive sign**. Thus subtracting minus 4 from 3 is $3 - -4$ and becomes $3 + 4 = 7$.

For multiplication and division, when the numbers have **unlike signs**, the answer is **negative**, but when the numbers have **like signs** the answer is **positive**. Thus $3 \times -4 = -12$, whereas $-3 \times -4 = +12$. Similarly

$$\frac{4}{-3} = -\frac{4}{3} \quad \text{and} \quad \frac{-4}{-3} = +\frac{4}{3}$$

Problem 3.1 Add 27, −74, 81 and −19.

This problem is written as $27 - 74 + 81 - 19$

Adding the positive integers: 27, 81

Sum of positive integers is: 108

Adding the negative integers: 74, 19

Sum of negative integers is: 93

Taking the sum of the negative integers from the sum of the positive integers gives:

$$\begin{array}{r} 108 \\ -93 \\ \hline 15 \end{array}$$

Thus **27 − 74 + 81 − 19 = 15**

Problem 3.2 Subtract 89 from 123

This is written mathematically as $123 - 89$

$$\begin{array}{r} 123 \\ -89 \\ \hline 34 \end{array}$$

Thus **123 − 89 = 34**

Problem 3.3 Subtract −74 from 377

This problem is written as $377 - -74$. Like signs together give an overall positive sign, hence

$377 - -74 = 377 + 74$

$$\begin{array}{r} 377 \\ +74 \\ \hline 451 \end{array}$$

Thus **377 − −74 = 451**

Problem 3.4 Subtract 243 from 126

The problem is $126 - 243$. When the second number is larger than the first, take the small number from the larger and make the result negative. Thus $126 - 243 = -(243 - 126)$

$$\begin{array}{r} 243 \\ -126 \\ \hline 117 \end{array}$$

Thus **126 − 243 = −117**

Problem 3.5 Subtract 318 from −269

−269 − 318. The sum of the negative integers is

$$\begin{array}{r} 269 \\ +318 \\ \hline 587 \\ \hline \end{array}$$

Thus **−269** − **317** = **−587**

Problem 3.6 Multiply 74 by 13

This is written as 74 × 13

$$\begin{array}{rl} & 74 \\ & \underline{13} \\ & 222 \quad \leftarrow 74 \times 3 \\ & \underline{740} \quad \leftarrow 74 \times 10 \\ \text{Adding:} & \underline{962} \end{array}$$

Problem 3.7 Multiply 178 by −46

When the numbers have different signs, the result will be negative. (With this in mind, the problem can now be solved by multiplying 178 by 46.)

$$\begin{array}{r} 178 \\ \underline{46} \\ 1068 \\ \underline{7120} \\ \underline{8188} \end{array}$$

Thus 178 × 46 = 8188 and **178** × (−**46**) = −**8188**

Problem 3.8 Divide 1043 by 7

When dividing by the numbers 1 to 12, it is usual to use a method called **short division**.

$$\begin{array}{r} 1\ \ 4\ \ 9 \\ 7\overline{)10\ {}^{3}4\ {}^{6}3} \end{array}$$

Step 1: 7 into 10 goes 1, remainder 3. Put 1 above the 0 of 1043 and carry the 3 remainder to the next digit on the right, making it 34

Step 2: 7 into 34 goes 4, remainder 6. Put 4 above the 4 of 1043 and carry the 6 remainder to the next digit on the right, making it 63

Step 3: 7 into 63 goes 9, remainder 0. Put 9 above the 3 of 1043

Thus **1043** ÷ **7** = **149**

Problem 3.9 Divide 378 by 14

When dividing by numbers which are larger than 12, it is usual to use a method called **long division**.

$$\begin{array}{r} 27 \\ 14\overline{)378} \end{array}$$

(1) 14 into 37 goes twice. Put 2 above the 7 of 378

(2) 2 × 14 — $\underline{28}$

98 — (3) Subtract. Bring down the 8. 14 into 98 goes 7 times. Put 7 above the 8 of 378

(4) 7 × 14 — $\underline{98}$

$\underline{\cdot\cdot}$ — (5) Subtract

Thus **378** ÷ **14** = **27**

Test your knowledge 3.1

Evaluate the following:

1 −2461 + 4177 + 29 −1101

2 374 × 7

3 747 ÷ 9

Order of precedence and brackets

When a particular arithmetic operation is to be performed first, the numbers and the operator(s) are placed in brackets. Thus 3 times the result of 6 minus 2 is written as $3 \times (6-2)$.

In arithmetic operations, the order in which operations are performed are:

(i) to determine the values of operations contained in brackets

(ii) multiplication and division (the word 'of' also means multiply)

(iii) addition and subtraction.

This **order of precedence** can be remembered by the word **BODMAS**, standing for **B**rackets, **O**f, **D**ivision, **M**ultiplication, **A**ddition and **S**ubtraction, taken in that order.

The basic laws governing the use of brackets and operators are shown by the following examples:

(i) $2+3=3+2$, i.e. the order of numbers when adding does not matter

(ii) $2 \times 3 = 3 \times 2$, i.e. the order of numbers when multiplying does not matter

(iii) $2+(3+4)=(2+3)+4$, i.e. the use of brackets when adding does not affect the result

(iv) $2 \times (3 \times 4) = (2 \times 3) \times 4$, i.e. the use of brackets when multiplying does not affect the result

(v) $2 \times (3+4) = 2(3+4) = (3+4)2 = 2 \times 3 + 2 \times 4$, i.e. a number placed outside of a bracket indicates that the whole contents of the bracket must be multiplied by that number

(vi) $(2+3)(4+5)=(5)(9)=45$, i.e. adjacent brackets indicate multiplication

(vii) $2[3+(4 \times 5)] = 2[3+20] = 2 \times 23 = 46$, i.e. when an expression contains inner and outer brackets, the inner brackets are removed first.

Problem 3.10 Find the value of $6+4 \div (5-3)$

The order of precedence of operations is remembered by the word BODMAS

$$
\begin{aligned}
\text{Thus } 6+4 \div (5-3) &= 6+4 \div 2 && \text{(Brackets)} \\
&= 6+2 && \text{(Division)} \\
&= 8 && \text{(Addition)}
\end{aligned}
$$

Problem 3.11 Determine the value of $13-2 \times 3+14 \div (2+5)$

$$
\begin{aligned}
&13-2 \times 3+14 \div (2+5) \\
&= 13-2 \times 3+14 \div 7 \qquad \text{(B)}
\end{aligned}
$$

$= 13 - 2 \times 3 + 2$ (D)
$= 13 - 6 + 2$ (M)
$= 15 - 6$ (A)
$= \mathbf{9}$ (S)

Problem 3.12 Evaluate $16 \div (2 + 6) + 18[3 + (4 \times 6) - 21]$

$16 \div (2 + 6) + 18[3 + (4 \times 6) - 21]$
$= 16 \div (2 + 6) + 18[3 + 24 - 21]$
$= 16 \div 8 + 18 \times 6$ (B)
$= 2 + 18 \times 6$ (D)
$= 2 + 108$ (M)
$= \mathbf{110}$ (A)

Test your knowledge 3.2

Evaluate the following:

1 $54 - 21(16 + 2) + 111$

2 $23 - 4(2 \times 7) + \dfrac{144 \div 4}{14 - 8}$

Fractions

When 2 is divided by 3, it may be written as $\frac{2}{3}$ or 2/3. $\frac{2}{3}$ is called a fraction. The number above the line, i.e. 2, is called the **numerator** and the number below the line, i.e. 3, is called the **denominator**.

When the value of the numerator is less than the value of the denominator, the fraction is called a **proper fraction**; thus $\frac{2}{3}$ is a proper fraction. When the value of the numerator is greater than the denominator, the fraction is called an **improper fraction**. Thus $\frac{7}{3}$ is an improper fraction and can also be expressed as a **mixed number**, that is, an integer and a proper fraction. Thus the improper $\frac{7}{3}$ fraction is equal to a mixed number $2\frac{1}{3}$.

When a fraction is simplified by dividing the numerator and denominator by the same number, the process is called **cancelling**. Cancelling by 0 is not permissible.

Problem 3.13 Simplify $\frac{1}{3} + \frac{2}{7}$

The LCM (i.e. the lowest common multiple) of the two denominators is 3×7, i.e. 21

Expressing each fraction so that their denominators are 21, gives:

$$\frac{1}{3} + \frac{2}{7} = \frac{1}{3} \times \frac{7}{7} + \frac{2}{7} \times \frac{3}{3} = \frac{7}{21} + \frac{6}{21}$$
$$= \frac{7 + 6}{21} = \mathbf{\frac{13}{21}}$$

Alternatively,

Step (2) Step (3)
↓ ↓

$$\frac{1}{3} + \frac{2}{7} = \frac{(7 \times 1) + (3 \times 2)}{21}$$

↑
Step (1)

Step 1: the LCM of the two denominators
Step 2: for the fraction $\frac{1}{3}$, 3 into 21 goes 7 times, $7 \times$ the numerator is 7×1

Step 3: for the fraction $\frac{2}{7}$, 7 into 21 goes 3 times, $3 \times$ the numerator is 3×2

Thus $\dfrac{1}{3} + \dfrac{2}{7} = \dfrac{7+6}{21} = \mathbf{\dfrac{13}{21}}$ as obtained previously.

Problem 3.14 Find the value of $3\frac{2}{3} - 2\frac{1}{6}$

One method is to split the mixed numbers into integers and their fractional parts. Then

$$\begin{aligned} 3\tfrac{2}{3} - 2\tfrac{1}{6} &= (3 + \tfrac{2}{3}) - (2 + \tfrac{1}{6}) \\ &= 3 + \tfrac{2}{3} - 2 - \tfrac{1}{6} = 1 + \tfrac{2}{3} - \tfrac{1}{6} = 1 + \tfrac{4}{6} - \tfrac{1}{6} \\ &= 1\tfrac{3}{6} = \mathbf{1\tfrac{1}{2}} \end{aligned}$$

Another method is to express the mixed numbers as improper fractions.

$$\begin{aligned} &\text{Since } 3 = \tfrac{9}{3}, \text{ then } 3\tfrac{2}{3} = \tfrac{9}{3} + \tfrac{2}{3} &&= \tfrac{11}{3} \\ &\text{Similarly, } 2\tfrac{1}{6} = \tfrac{12}{6} + \tfrac{1}{6} &&= \tfrac{13}{6} \\ &\text{Thus} \quad 3\tfrac{2}{3} - 2\tfrac{1}{6} &&= \tfrac{11}{3} - \tfrac{13}{6} \\ & &&= \tfrac{22}{6} - \tfrac{13}{6} = \tfrac{9}{6} = \mathbf{1\tfrac{1}{2}} \end{aligned}$$

as obtained previously.

Problem 3.15 Determine the value of $4\frac{5}{8} - 3\frac{1}{4} + 1\frac{2}{5}$

$$\begin{aligned} 4\tfrac{5}{8} - 3\tfrac{1}{4} + 1\tfrac{2}{5} &= (4 - 3 + 1) + (\tfrac{5}{8} - \tfrac{1}{4} + \tfrac{2}{5}) \\ &= 2 + \frac{5 \times 5 - 10 \times 1 + 8 \times 2}{40} \\ &= 2 + \frac{25 - 10 + 16}{40} = 2 + \frac{31}{40} = \mathbf{2\frac{31}{40}} \end{aligned}$$

Problem 3.16 Find the value of $\frac{3}{7} \times \frac{14}{15}$

Dividing numerator and denominator by 3 gives:

$$\frac{{}^{1}\not{3}}{7} \times \frac{14}{\not{15}_{5}} = \frac{1 \times 14}{7 \times 5}$$

Dividing numerator and denominator by 7 gives:

$$\frac{1 \times \not{14}^{2}}{{}_{1}\not{7} \times 5} = \mathbf{\frac{2}{5}}$$

This process of dividing both the numerator and denominator of a fraction by the same factor(s) is called **cancelling**.

Problem 3.17 Evaluate $1\frac{3}{5} \times 2\frac{1}{3} \times 3\frac{3}{7}$

Mixed numbers must be expressed as improper fractions before multiplication can be performed. Thus,

$$\begin{aligned} 1\frac{3}{5} \times 2\frac{1}{3} \times 3\frac{3}{7} &= \left(\frac{5}{5} + \frac{3}{5}\right) \times \left(\frac{6}{3} \times \frac{1}{3}\right) \times \left(\frac{21}{7} + \frac{3}{7}\right) \\ &= \frac{8}{5} \times \frac{{}^{1}\not{7}}{{}_{1}\not{3}} \times \frac{\not{24}^{8}}{\not{7}^{1}} = \frac{8 \times 8}{5} \\ &= \frac{64}{5} = \mathbf{12\frac{4}{5}} \end{aligned}$$

Problem 3.18 Simplify $\frac{3}{7} \div \frac{12}{21}$

$$\frac{3}{7} \div \frac{12}{21} = \frac{\frac{3}{7}}{\frac{12}{21}}$$

Multiplying both numerator and denominator by the reciprocal of the denominator gives:

$$\frac{\frac{3}{7}}{\frac{12}{21}} = \frac{\frac{{}^{1}\cancel{3}}{{}_{1}\cancel{7}} \times \frac{\cancel{21}^{3}}{\cancel{12}_{4}}}{\frac{{}^{1}\cancel{12}}{{}_{1}\cancel{21}} \times \frac{\cancel{21}^{1}}{\cancel{12}_{1}}} = \frac{\frac{3}{4}}{1} = \mathbf{\frac{3}{4}}$$

This method can be remembered by the rule: invert the second fraction and change the operation from division to multiplication. Thus:

$$\frac{3}{7} \div \frac{12}{21} = \frac{{}^{1}\cancel{3}}{{}_{1}\cancel{7}} \times \frac{\cancel{21}^{3}}{\cancel{12}_{4}} = \mathbf{\frac{3}{4}} \text{ as obtained previously.}$$

Problem 3.19 Find the value of $5\frac{3}{5} \div 7\frac{1}{3}$

The mixed numbers must be expressed as improper fractions. Thus,

$$5\frac{3}{5} \div 7\frac{1}{3} = \frac{28}{5} \div \frac{22}{3}$$
$$= \frac{{}^{14}\cancel{28}}{5} \times \frac{3}{\cancel{22}_{11}} = \mathbf{\frac{42}{55}}$$

Problem 3.20 Simply $\frac{1}{3} - (\frac{2}{5} + \frac{1}{4}) \div (\frac{3}{8} \times \frac{1}{3})$

The order of precedence of operations for problems containing fractions is the same as that for integers, i.e. remembered by BODMAS (Brackets, Of, Division, Multiplication, Addition and Subtraction). Thus,

$$\frac{1}{3} - \left(\frac{2}{5} + \frac{1}{4}\right) \div \left(\frac{3}{8} \times \frac{1}{3}\right)$$
$$= \frac{1}{3} - \frac{4 \times 2 + 5 \times 1}{20} \div \frac{\cancel{3}^{1}}{\cancel{24}_{8}} \qquad \text{(B)}$$
$$= \frac{1}{3} - \frac{13}{\cancel{20}_{5}} \times \frac{\cancel{8}^{2}}{1} \qquad \text{(D)}$$
$$= \frac{1}{3} - \frac{26}{5} \qquad \text{(M)}$$
$$= \frac{(5 \times 1) - (3 \times 26)}{15} \qquad \text{(S)}$$
$$= \frac{-73}{75} = \mathbf{-4\frac{13}{15}}$$

Test your knowledge 3.3

Evaluate the following:

1 $\frac{3}{4} + \frac{1}{8} - \frac{7}{16}$

2 $3\frac{2}{5} + 1\frac{1}{10} - 2\frac{4}{15}$

3 $\frac{1}{4}$ of $(15 \times \frac{3}{5}) + (\frac{1}{2} \div \frac{5}{8})$

Ratio and proportion

The **ratio** of one quantity to another is a fraction, and is the number of times one quantity is contained in another quantity **of the same kind**.

If one quantity is **directly proportional** to another, then as one quantity doubles, the other quantity also doubles. When a quantity is **inversely proportional** to another, then as one quantity doubles, the other quantity is halved.

Problem 3.21 Divide 126 in the ratio of 5 to 13

Because the ratio is to be 5 parts to 13 parts, then the total number of parts is $5 + 13$, that is 18. Then,

18 parts correspond to 126

Hence 1 part corresponds to $\frac{126}{18} = 7$, 5 parts correspond to $5 \times 7 = 35$ and 13 parts correspond to $13 \times 7 = \mathbf{91}$
(Check: the parts must add up to the total $35 + 91 = \mathbf{126} =$ the total.)

Problem 3.22 A piece of timber 273 cm long is cut into three pieces in the ratio of 3 to 7 to 11. Determine the lengths of the three pieces.

The total number of parts is $3 + 7 + 11$, that is, 21.
Hence 21 parts correspond to 273 cm.

1 part corresponds to $\frac{273}{21} = 13$ cm
3 parts correspond to $3 \times 13 = 39$ cm
7 parts correspond to $7 \times 13 = 91$ cm
11 parts correspond to $11 \times 13 = 143$ cm

i.e. the lengths of the three pieces are 39 cm, 91 cm and 143 cm
(Check: $39 + 91 + 143 = 273$.)

Problem 3.23 If 3 people can complete a task in 4 h, find how long it will take 5 people to complete the task, assuming the rate of work remains constant.

The greater the number of people, the more quickly the task is done, hence inverse proportion exists.

3 people complete the task in 4 h
1 person takes three times as long, i.e. $4 \times 3 = 12$ h
5 people can do it in one fifth of the time that one person takes, that is

$$\frac{12}{5}\text{ h or } \mathbf{2\,h\,24\,min}$$

Test your knowledge 3.4

1 Express 25p as a ratio of £4.25

2 When mixing a quantity of paints, dyes of four different colours are used in the ratio of 7 : 3 : 19 : 5. If the mass of the first dye used is $3\frac{1}{2}$ g, determine the total mass of the dyes used.

Decimals

The decimal system of numbers is based on the **digits** 0 to 9. A number such as 53.17 is called a **decimal fraction**, a **decimal point** separating the integer part, i.e. 53, from the fractional part, i.e. 0.17

A number which can be expressed exactly as a decimal fraction is called a **terminating decimal** and those which cannot be expressed exactly as a decimal fraction are called **non-terminating decimals**. Thus, $\frac{3}{2} = 1.5$ is a **terminating decimal**, but $\frac{4}{3} = 1.33333\ldots$ is a **non-terminating decimal**. 1.33333... can be written as $1.\dot{3}$, called 'one point-three recurring'.

The answer to a non-terminating decimal may be expressed in two ways, depending on the accuracy required:

(i) correct to a number of **significant figures**, that is, figures which signify something, and

(ii) correct to a number of **decimal places**, that is, the number of figures after the decimal point.

The last digit in the answer is unaltered if the next digit on the right is in the group of numbers 0, 1, 2, 3 or 4, but is increased by 1 if the next digit on the right is in the group of numbers 5, 6, 7, 8 or 9. Thus the non-terminating decimal 7.6183... becomes 7.62, correct to 3 significant figures, since the next digit on the right is 8, which is in the group of numbers 5, 6, 7, 8 or 9. Also 7.6183... becomes 7.618, correct to 3 decimal places, since the next digit on the right is 3, which is in the group of numbers 0, 1, 2, 3 or 4.

Problem 3.24 Evaluate $42.7 + 3.04 + 8.7 + 0.06$

The numbers are written so that the decimal points are under each other. Each column is added, starting from the right.

$$\begin{array}{r} 42.7 \\ 3.04 \\ 8.7 \\ \underline{0.06} \\ \underline{54.50} \end{array}$$

Thus $\mathbf{42.7 + 3.04 + 8.7 + 0.06 = 54.50}$

Problem 3.25 Take 81.7 from 87.23

The numbers are written with the decimal points under each other.

$$\begin{array}{r} 87.23 \\ \underline{-81.70} \\ \underline{5.53} \end{array}$$

Thus $\mathbf{87.23 - 81.7 = 5.53}$

Problem 3.26 Determine the value of 74.3×3.8

When muliplying decimal fractions: (i) the numbers are multiplied as if they are integers, and (ii) the position of the decimal point in the answer is such that there are as many digits to the right of it as the sum of the digits to the right of the decimal point of the two numbers being multiplied together. Thus

$$\text{(i)} \quad \begin{array}{r} 743 \\ \underline{38} \\ 5944 \\ \underline{22290} \\ \underline{28234} \end{array}$$

(ii) As there are $(1 + 1) = 2$ digits to the right of the decimal points of the two numbers being multiplied together $(74.\underline{3} \times 3.\underline{8})$, then **74.3 × 3.8 = 282.34**

Problem 3.27 Convert (a) 0.4375 to a proper fraction and (b) 4.285 to a mixed number.

(a) 0.4375 can be written as $\frac{0.4375 \times 10\,000}{10\,000}$ without changing its value, i.e. $0.4375 = \frac{4375}{10\,000}$

By cancelling $\frac{4375}{10\,000} = \frac{875}{2000} = \frac{175}{400} = \frac{35}{80} = \frac{7}{16}$

i.e. $0.4375 = \mathbf{\frac{7}{16}}$

(b) Similarly, $4.285 = 4\frac{285}{1000} = \mathbf{4\frac{57}{200}}$

Problem 3.28 Express as decimal fractions (a) $\frac{9}{16}$ and (b) $5\frac{7}{8}$

(a) To convert a proper fraction to a decimal fraction, the numerator is divided by the denominator. Division by 16 can be done by the long division method, or, more simply, by dividing by 2 and then 8:

$$2\overline{)9.{}^{1}00}\;=4.5 \qquad 8\overline{)4.5\;{}^{5}0{}^{2}0{}^{4}0}\;=0.5625 \qquad \text{Thus, } \mathbf{\frac{9}{16} = 0.5625}$$

(b) For mixed numbers, it is only necessary to convert the proper fraction part of the mixed number to a decimal fraction. Thus, dealing with the $\frac{7}{8}$ part gives:

$$8\overline{)7.0{}^{6}0{}^{4}0}\;=0.875 \quad \text{i.e. } \frac{7}{8} = 0.875 \quad \text{Thus } \mathbf{5\frac{7}{8} = 5.875}$$

Test your knowledge 3.5

1 Evaluate:
(a) 41.72 + 14.179 − 25.16
(b) 21.46 × 7.32

2 Evaluate, correct to 4 significant figures:
(a) 329.8 ÷ 14
(b) 29.142 − 11.68 × 1.23

Percentages

Percentages are used to give a common standard and are fractions having the number 100 as their denominators. For example, 25 per cent means $\frac{25}{100}$, i.e. $\frac{1}{4}$, and is written 25%.

Problem 3.29 Express as percentages (a) 1.875 and (b) 0.0125

A decimal fraction is converted to a percentage by multiplying by 100. Thus,

(a) 1.875 corresponds to 1.875 × 100%, i.e. **187.5%**

(b) 0.0125 corresponds to 0.0125 × 100%, i.e. **1.25%**

Problem 3.30 Express as percentages:

(a) $\frac{5}{16}$ and (b) $1\frac{2}{5}$

To convert fractions to percentages, they are (i) converted to decimal fractions and (ii) multiplied by 100.

(a) By division, $\frac{5}{16} = 0.3125$, hence $\frac{5}{16}$ corresponds to $0.3125 \times 100\%$, i.e. **31.25%**

(b) Similarly, $1\frac{2}{5} = 1.4$ when expressed as a decimal fraction. Hence $1\frac{2}{5} = 1.4 \times 100\% =$ **140%**

Problem 3.31 Find $12\frac{1}{2}\%$ of £378

$12\frac{1}{2}\%$ of £378 means $\dfrac{12\frac{1}{2}}{100} \times 378$, since per cent means 'per hundred'.

Hence $12\frac{1}{2}\%$ of £378 $= \dfrac{^{1}\cancel{12\frac{1}{2}}}{_{8}\cancel{100}} \times 378 = \dfrac{378}{8} =$ **£47.25**

Problem 3.32 Express 25 min as a percentage of 2 h, correct to the nearest 1%

Working in minute units, 2 h = 120 min
Hence 25 minutes is $\frac{25}{120}$ths of 2 h
By cancelling, $\frac{25}{120} = \frac{5}{24}$
Expressing $\frac{5}{24}$ as a decimal fraction gives 0.2083
Multiplying by 100 to convert the decimal fraction to a percentage gives:

$$0.2083 \times 100 = 20.83\%$$

thus **25 min is 21% of 2 h**, correct to the nearest 1%.

Problem 3.33 A German silver alloy consists of 60% copper, 25% zinc and 15% nickel. Determine the masses of the copper, zinc and nickel in a 3.74 kg block of the alloy.

By direct proportion:

100% corresponds to 3.74 kg
1% corresponds to $\dfrac{3.74}{100} = 0.0374$ kg
60% corresponds to $60 \times 0.0374 = 2.244$ kg
25% corresponds to $25 \times 0.0374 = 0.935$ kg
15% corresponds to $15 \times 0.0374 = 0.561$ kg

Thus, the masses of the copper, zinc and nickel are **2.444 kg**, **0.935 kg** and **0.561 kg**, respectively.
(Check: $2.244 + 0.935 + 0.561 = 3.74$.)

Test your knowledge 3.6

1 Express as percentages:
(a) 0.462
(b) $2\frac{3}{4}$

2 Determine, correct to 2 decimal places
(a) 16% of 15.62 g
(b) 27% of 28.31 t

Indices

The lowest factors of 2000 are $2 \times 2 \times 2 \times 2 \times 5 \times 5 \times 5$. These factors are written as $2^4 \times 5^3$, where 2 and 5 are called **bases** and the numbers 4 and 3 are called **indices**.

When an index is an integer it is called a **power**. Thus, 2^4 is called 'two to the power of four', and has a base of 2 and an index of 4. Similarly, 5^3 is called 'five to the power of 3' and has a base of 5 and an index of 3. Special names may be used when the indices are 2 and 3, these being called 'squared' and 'cubed', respectively. Thus 7^2 is called 'seven squared' and 9^3 is called 'nine cubed'. When no index is shown, the power is 1, i.e. 2^1 means 2.

Reciprocal

The **reciprocal** of a number is when the index is -1 and its value is given by 1 divided by the base. Thus the reciprocal of 2 is 2^{-1} and its value is $\frac{1}{2}$ or 0.5. Similarly, the reciprocal of 5 is 5^{-1} which means $\frac{1}{5}$ or 0.2.

Square root

The **square root** of a number is when the index is $\frac{1}{2}$, and the square root of 2 is written as $2^{(1/2)}$ or $\sqrt{2}$. The value of a square root is the value of the base which when multipled by itself gives the number. Since $3 \times 3 = 9$, then $\sqrt{9} = 3$. However, $(-3) \times (-3) = 9$, so $\sqrt{9} = -3$. There are always two answers when finding the square root of a number and this is shown by putting both a + and a − sign in front of the answer to a square root problem. Thus $\sqrt{9} = \pm 3$ and $4^{(1/2)} = \sqrt{4} = \pm 2$, and so on.

Laws of indices

When simplifying calculations involving indices, certain basic rules or laws can be applied, called the **laws of indices**. These are given below.

(i) When multiplying two or more numbers having the same base, the indices are added. Thus $3^2 \times 3^4 = 3^{2+4} = 3^6$

(ii) When a number is divided by a number having the same base, the indices are subtracted. Thus

$$\frac{3^5}{3^2} = 3^{5-2} = 3^3$$

(iii) When a number which is raised to a power is raised to a further power, the indices are multiplied. Thus $(3^5)^2 = 3^{5\times 2} = 3^{10}$

(iv) When a number has an index of 0, its value is 1. Thus $3^0 = 1$

(v) A number raised to a negative power is the reciprocal of that number raised to a positive power. Thus

$$3^{-4} = \frac{1}{3^4}. \text{ Similarly, } \frac{1}{2^{-3}} = 2^3$$

(vi) When a number is raised to a fractional power the denominator of the fraction is the root of the number and the numerator is the power. Thus
$8^{(2/3)} = \sqrt[3]{8}^2 = (2)^2 = 4$
and
$25^{(1/2)} = \sqrt{25^1} = \pm 5.$

Problem 3.34 Evaluate: (a) $5^2 \times 5^3$, (b) $3^2 \times 3^4 \times 3$ and (c) $2 \times 2^2 \times 2^5$

From law (i):

(a) $5^2 \times 5^3 = 5^{(2+3)} = 5^5 = 5 \times 5 \times 5 \times 5 \times 5 = \mathbf{3125}$

(b) $3^2 \times 3^4 \times 3 = 3^{(2+4+1)} = 3^7 = 3 \times 3 \times 3 \times \ldots$ to 7 terms $= \mathbf{2187}$

(c) $2 \times 2^2 \times 2^5 = 2^{(1+2+5)} = 2^8 = \mathbf{256}$

Problem 3.35 Find the value of (a) $\frac{7^5}{7^3}$ and (b) $\frac{5^7}{5^4}$

From law (ii):

(a) $\frac{7^5}{7^3} = 7^{(5-3)} = 7^2 = \mathbf{49}$

(b) $\frac{5^7}{5^4} = 5^{(7-4)} = 5^3 = \mathbf{125}$

Problem 3.36 Evaluate (a) $5^2 \times 5^3 + 5^4$ and (b) $(3 \times 3^5) + (3^2 \times 3^3)$

From laws (i) and (ii):

$$\text{(a)}\quad 5^2 \times 5^3 + 5^4 = \frac{5^2 \times 5^3}{5^4} = \frac{5^{(2+3)}}{5^4} = \frac{5^5}{5^4} = 5^{(5-4)} = 5^1 = \mathbf{5}$$

$$\text{(b)}\quad (3 \times 3^5) + (3^2 \times 3^3) = \frac{3 \times 3^5}{3^2 \times 3^3} = \frac{3^{(1+5)}}{3^{(2+3)}} = \frac{3^6}{3^5} = 3^{6-5} = 3^1 = \mathbf{3}$$

Problem 3.37 Simplify (a) $(2^3)^4$ and (b) $(3^2)^5$, expressing the answers in index form.

From law (iii):

(a) $(2^3)^4 = 2^{3\times 4} = \mathbf{2^{12}}$

(b) $(3^2)^5 = 3^{2\times 5} = \mathbf{3^{10}}$

Problem 3.38 Evaluate: $\frac{(10^2)^3}{10^4 \times 10^2}$

From the laws of indices:

$$\frac{(10^2)^3}{10^4 \times 10^2} = \frac{10^{(2\times 3)}}{10^{(4+2)}} = \frac{10^6}{10^6} = 10^{6-6} = 10^0 = \mathbf{1}$$

Problem 3.39 Find the value of

$$\text{(a)}\ \frac{2^3 \times 2^4}{2^7 \times 2^5} \text{ and (b) } \frac{(3^2)^3}{3 \times 3^9}$$

$$\text{(a)}\ \frac{2^3 \times 2^4}{2^7 \times 2^5} = \frac{2^{(3+4)}}{2^{(7+5)}} = \frac{2^7}{2^{12}}$$

$$= 2^{7-12} = 2^{-5} = \frac{1}{2^5} = \mathbf{\frac{1}{32}}$$

$$\text{(b)}\ \frac{(3^2)^3}{3 \times 3^9} = \frac{3^{2\times 3}}{3^{1+9}} = \frac{3^6}{3^{10}}$$

$$= 3^{6-10} = 3^{-4} = \frac{1}{3^4} = \mathbf{\frac{1}{81}}$$

Problem 3.40 Evaluate (a) $4^{1/2}$ (b) $16^{3/4}$ (c) $27^{2/3}$ (d) $9^{-1/2}$

(a) $4^{1/2} = \sqrt{4} = \mathbf{\pm 2}$ (b) $16^{3/4} = \sqrt[4]{16^3} = (2)^3 = \mathbf{8}$

(Note that it does not matter whether the 4th root of 16 is found first or whether 16 cubed is found first – the same answer will result.)

(c) $27^{2/3} = \sqrt[3]{27^2} = (3)^2 = \mathbf{9}$

(d) $9^{-1/2} = \frac{1}{9^{1/2}} = \frac{1}{\sqrt{9}} = \frac{1}{\pm 3} = \pm\mathbf{\frac{1}{3}}$

Test your knowledge 3.7

Evaluate the following:

1 $\frac{(5^2)^3}{5^4}$

2 $\frac{3^3 \times 3^2}{3}$

3 $4^{3/2} \times \left(\frac{1}{2}\right)^{-1}$

Standard form

A number written with one digit to the left of the decimal point and multiplied by 10 raised to some power is said to be written in **standard form**. Thus: 5837 is written as 5.837×10^3 in standard form, and 0.0415 is written as 4.15×10^{-2} in standard form.

The laws of indices are used when multiplying or dividing numbers given in standard form. For example,

$$(2.5 \times 10^3) \times (5 \times 10^2) = (2.5 \times 5) \times (10^{3+2})$$
$$= 12.5 \times 10^5 \text{ or } 1.25 \times 10^6$$

Similarly,

$$\frac{6 \times 10^4}{1.5 \times 10^2} = \frac{6}{1.5} \times (10^{4-2}) = 4 \times 10^2$$

Problem 3.41 Express in standard form (a) 38.71 (b) 3746 (c) 0.0124

For a number to be in standard form, it is expressed with only one digit to the left of the decimal point. Thus:

(a) 38.71 must be divided by 10 to achieve one digit to the left of the decimal point and it must also be multiplied by 10 to maintain the equality, i.e.

$$38.71 = \frac{38.71}{10} \times 10 = \mathbf{3.871 \times 10} \text{ in standard form}$$

(b) $3746 = \frac{3746}{1000} \times 1000 = \mathbf{3.746 \times 10^3}$ in standard form

(c) $0.0124 = 0.0124 \times \frac{100}{100} = \frac{1.24}{100} = \mathbf{1.24 \times 10^{-2}}$ in standard form.

Problem 3.42 Express the following numbers, which are in standard form, as decimal numbers:

(a) 1.725×10^{-2} (b) 5.491×10^4 (c) 9.84×10^0

(a) $1.725 \times 10^{-2} = \frac{1.725}{100} = \mathbf{0.01725}$

(b) $5.491 \times 10^4 = 5.491 \times 10\,000 = \mathbf{54\,910}$

(c) $9.84 \times 10^0 = 9.84 \times 1 = \mathbf{9.84}$ (since $10^0 = 1$)

Problem 3.43 Express in standard form, correct to 3 significant figures:

(a) $\frac{3}{8}$ (b) $19\frac{2}{3}$ (c) $741\frac{9}{16}$

(a) $\frac{3}{8} = 0.375$, and expressing it in standard form gives: $0.375 = \mathbf{3.75 \times 10^{-1}}$

(b) $19\frac{2}{3} = 19.6 = \mathbf{1.97 \times 10}$ in standard form, correct to 3 significant figures

(c) $741\frac{9}{16} = 741.5625 = \mathbf{7.42 \times 10^2}$ in standard form, correct to 3 significant figures.

Problem 3.44 Express the following numbers, given in standard form, as fractions or mixed numbers:

(a) 2.5×10^{-1} (b) 6.25×10^{-2} (c) 1.354×10^2

(a) $2.5 \times 10^{-1} = \frac{2.5}{10} = \frac{25}{100} = \mathbf{\frac{1}{4}}$

(b) $6.25 \times 10^{-2} = \frac{6.25}{100} = \frac{625}{10\,000} = \mathbf{\frac{1}{16}}$

(c) $1.354 \times 10^2 = 135.4 = 135\frac{4}{10} = \mathbf{135\frac{2}{5}}$.

Problem 3.45 Evaluate (a) $(3.75 \times 10^3)(6 \times 10^4)$ and

(b) $\frac{3.5 \times 10^5}{7 \times 10^2}$ expressing the answers in standard form.

(a) $(3.75 \times 10^3)(6 \times 10^4) = (3.75 \times 6)(10^{3+4}) = 22.50 \times 10^7$ $= \mathbf{2.25 \times 10^8}$

(b) $\frac{3.5 \times 10^5}{7 \times 10^2} = \frac{3.5}{7} \times 10^{5-2} = 0.5 \times 10^3 = \mathbf{5 \times 10^2}$

Test your knowledge 3.8

1 Express in standard form:
(a) 47.6 (b) 0.0032 (c) 512 60

2 Express in standard form, correct to 3 significant figures:
(a) $15\frac{7}{16}$ (b) $\frac{4}{7}$

3 Evaluate
$$\frac{(5.6 \times 10^3)(2 \times 10^2)}{4 \times 10^{-2}}$$

Use of calculator

The most modern aid to calculations is the pocket-sized electronic calculator. With one of these, calculations can be quickly and accurately performed, correct to about 9 significant figures. The scientific type of calculator has made the use of tables and logarithms largely redundant.

To help you to become competent at using your calculator check that you agree with the answers to the following problems.

Problem 3.46 Evaluate the following, correct to 4 significant figures:

(a) $4.7826 + 0.027\,13$

(b) $17.6941 - 11.8762$

(c) $21.93 \times 0.012\,981$

(a) $4.7826 + 0.027\,13 = 4.809\,73 = \mathbf{4.810}$, correct to 4 significant figures

(b) $17.6941 - 11.8762 = 5.8179 = \mathbf{5.818}$, correct to 4 significant figures

(c) $21.93 \times 0.0129\,81 = 0.0284\,673\,3\ldots = \mathbf{0.2847}$, correct to 4 significant figures.

Problem 3.47 Evaluate the following, correct to 4 decimal places:

(a) $46.32 \times 97.17 \times 0.012\,58$

(b) $\dfrac{4.621}{23.76}$ (c) $\frac{1}{2}(62.49 \times 0.0172)$

(a) $46.32 \times 97.17 \times 0.012\,58 = 56.621\,503\,1\ldots = \mathbf{56.6215}$, correct to 4 decimal places

(b) $\dfrac{4.621}{23.76} = 0.194\,486\,53\ldots = \mathbf{0.1945}$, correct to 4 decimal places

(c) $\frac{1}{2}(62.49 \times 0.0172) = 0.537\,414 = \mathbf{0.5374}$, correct to 4 decimal places.

Problem 3.48 Evaluate the following, correct to 3 decimal places:

(a) $\dfrac{1}{52.73}$ (b) $\dfrac{1}{0.0275}$ (c) $\dfrac{1}{4.92} + \dfrac{1}{1.97}$

(a) $\dfrac{1}{52.73} = 0.018\,964\,53\ldots = \mathbf{0.019}$, correct to 3 decimal places

(b) $\dfrac{1}{0.0275} = 36.363\,636\,3\ldots = \mathbf{36.364}$, correct to 3 decimal places

(c) $\dfrac{1}{4.92} + \dfrac{1}{1.97} = 0.710\,866\,24\ldots = \mathbf{0.711}$, correct to 3 decimal places.

Problem 3.49 Evaluate the following, expressing the answers in standard form, correct to 4 significant figures.

(a) $(0.004\,51)^2$ (b) $541.7 - (6.21 + 2.95)^2$
(c) $46.27^2 - 31.79^2$

(a) $(0.00451)^2 = 2.034\,01 \times 10^{-5} = \mathbf{2.034 \times 10^{-5}}$, correct to 4 significant figures

(b) $541.7 - (6.21 + 2.95)^2 = 547.7944 = 5.477\,944 \times 10^2 =$ $\mathbf{5.478 \times 10^2}$, correct to 4 significant figures

(c) $46.27^2 - 31.79^2 = 1130.3088 = \mathbf{1.130 \times 10^3}$, correct to 4 significant figures.

Problem 3.50 Evaluate the following, correct to 3 decimal places:

(a) $\dfrac{(2.37)^2}{0.0526}$ (b) $\left(\dfrac{3.60}{1.92}\right)^2 + \left(\dfrac{5.40}{2.45}\right)^2$ (c) $\dfrac{15}{7.6^2 - 4.8^2}$

(a) $\dfrac{(2.37)^2}{0.0526} = 106.785\,171\ldots = \mathbf{106.785}$, correct to 3 decimal places

(b) $\left(\dfrac{3.60}{1.92}\right)^2 + \left(\dfrac{5.40}{2.45}\right)^2 = 8.373\,600\,84\ldots = \mathbf{8.374}$, correct to 3 decimal places

(c) $\dfrac{15}{7.6^2 - 4.8^2} = 0.432\,027\,64\ldots = \mathbf{0.432}$, correct to 3 decimal places.

Problem 3.51 Evaluate the following, correct to 4 significant figures:

(a) $\sqrt{5.462}$ (b) $\sqrt{54.62}$ (c) $\sqrt{546.2}$

(a) $\sqrt{5.462} = 2.337\,092\,2\ldots = \mathbf{2.337}$, correct to 4 significant figures

(b) $\sqrt{54.62} = 7.390\,534\,48\ldots = \mathbf{7.391}$, correct to 4 significant figures

(c) $\sqrt{546.2} = 23.370\,922\ldots = \mathbf{23.37}$, correct to 4 significant figures.

Problem 3.52 Evaluate the following, correct to 3 decimal places:

(a) $\sqrt{0.007\,328}$ (b) $\sqrt{52.91} - \sqrt{31.76}$ (c) $\sqrt{(1.6291 \times 10^4)}$

(a) $\sqrt{0.007\,328} = 0.0856\,037\,3 = \mathbf{0.086}$, correct to 3 decimal places

(b) $\sqrt{52.91} = \sqrt{31.76} = 1.638\,324\,91\ldots = \mathbf{1.638}$, correct to 3 decimal places

(c) $\sqrt{(1.6291 \times 10^4)} = \sqrt{(16\,291)} = 127.636\,201\ldots = \mathbf{127.636}$, correct to 3 decimal places.

Problem 3.53 Evaluate the following, correct to 4 significant figures:

(a) 4.72^3 (b) $(0.8316)^4$ (c) $\sqrt{(76.21^2 - 29.10^2)}$

(a) $4.72^3 = 105.154\,04\ldots = \mathbf{105.2}$, correct to 4 significant figures

(b) $(0.8316)^4 = 0.478\,253\,24\ldots = \mathbf{0.4783}$, correct to 4 significant figures

(c) $\sqrt{(76.21^2 - 29.10^2)} = 70.435\,460\,5\ldots = \mathbf{70.44}$, correct to 4 significant figures.

Problem 3.54 Evaluate the following, correct to 3 significant figures:

(a) $\sqrt{\left(\dfrac{(6.09)^2}{25.2 \times \sqrt{7}}\right)}$ (b) $\sqrt[3]{(47.291)}$

(c) $\sqrt{(7.213^2 + 6.418^3 + 3.291^4)}$

(a) $\sqrt{\left(\dfrac{(6.09)^2}{25.2 \times \sqrt{7}}\right)} = 0.745\,834\,57\ldots = \mathbf{0.746}$, correct to 3 significant figures

(b) $\sqrt[3]{(47.291)} = 3.616\,258\,76\ldots = \mathbf{3.62}$, correct to 3 significant figures

(c) $\sqrt{(7.213^2 + 6.418^3 + 3.291^4)} = 20.825\,299\,1\ldots = \mathbf{20.8}$, correct to 3 significant figures.

Problem 3.55 Evaluate the following, expressing the answers in standard form, correct to 4 decimal places:

(a) $(5.176 \times 10^{-3})^2$ (b) $\left(\dfrac{1.974 \times 10^1 \times 8.61 \times 10^{-2}}{3.462}\right)^4$

(c) $\sqrt{(1.792 \times 10^{-4})}$

(a) $(5.176 \times 10^{-3})^2 = 2.678\,097\ldots \times 10^{-5} = \mathbf{2.6781 \times 10^{-5}}$, correct to 4 decimal places

(b) $\left(\dfrac{1.974 \times 10^1 \times 8.61 \times 10^{-2}}{3.462}\right)^4 = 0.058\,088\,87\ldots$

$= \mathbf{5.8089 \times 10^{-2}}$, correct to 4 decimal places

(c) $\sqrt{(1.792 \times 10^{-4})} = 0.013\,386\,5\ldots = \mathbf{1.3387 \times 10^{-2}}$, correct to 4 decimal places.

Test your knowledge 3.9

Evaluate, correct to 3 significant figures:

1 $\dfrac{15.2}{\sqrt{27.21 - (1.778)^2}}$

2 $\dfrac{\sqrt{(0.004\,76)}}{(0.016)^3 + \frac{1}{2.46}}$

3 Below is a table of some metric to imperial conversions.
Length 2.54 cm = 1 in
1.61 km = 1 mile
Weight 1 kg = 2.2 lb (1 lb = 16 ounces)
Capacity 1 litre = 1.76 pints (8 pints = 1 gallon)

Use the table to determine (a) the number of millimetres in 15 in, (b) a speed of 35 mph in km/h, (c) the number of kilometres in 235 miles, (d) the number of pounds and ounces in 24 kg (correct to the nearest ounce, (e) the number of kilograms in 15 lb, (f) the number of litres in 12 gallons, and (g) the number of gallons in 25 litres.

Simple equations

An equation is simply a statement that two quantities are equal. For example, 1 m = 1000 mm or $y = mx + c$.

To '**solve an equation**' means 'to find the value of the unknown quantity'. For example, if $2x = 6$, then $x = 3$ is the solution of the equation.

Problem 3.56 Solve the equation $4x = 20$

Dividing each side of the equation by 4 gives:

$$\frac{4x}{4} = \frac{20}{4}$$

(Note that the same operation has been applied to both the left hand side (LHS) and the right hand side (RHS) of the equation so the equality has been maintained.)

Cancelling gives $\boldsymbol{x = 5}$, which is the solution to the equation. Solutions to simple equations should always be checked and this is accomplished by substituting the solution into the original equation.

In this case, LHS = 4(5) = 20 = RHS.

Problem 3.57 Solve $\frac{2x}{5} = 6$

The LHS is a fraction and this can be removed by multiplying both sides of the equation by 5.

$$\text{Hence } 5\left(\frac{2x}{5}\right) = 5(6)$$

Cancelling gives: $2x = 30$

Dividing both sides of the equation by 2 gives:

$$\frac{2x}{2} = \frac{30}{2}$$

$$\text{i.e. } \boldsymbol{x = 15}$$

Problem 3.58 Solve $a - 5 = 8$

Adding 5 to both sides of the equation gives:

$$a - 5 + 5 = 8 + 5$$

$$\text{i.e. } \boldsymbol{a = 13}$$

The result of the above procedure is to move the '-5' from the LHS of the original equation, across the equals sign, to the RHS, but the sign is changed to $+$.

Problem 3.59 Solve $x + 3 = 7$

Subtracting 3 from both sides of the equation gives:

$$x + 3 - 3 = 7 - 3$$

$$\text{i.e. } \boldsymbol{x = 4}$$

The result of the above procedure is to move the '$+3$' from the LHS of the original equation, across the equals sign, to the RHS, but the sign is changed to $-$. Thus a term can be moved from one side of an equation to the other as long as a change in sign is made.

Problem 3.60 Solve $6x + 1 = 2x + 9$

In such equations the terms containing x are grouped on one side of the equation and the remaining terms grouped on the other side of the equation. As in problems 3.3 and 3.4, changing from one side of an equation to the other must be accompanied by a change of sign.

Thus since $6x + 1 = 2x + 9$

then $6x - 2x = 9 - 1$

$$4x = 8$$

$$\frac{4x}{4} = \frac{8}{4}$$

i.e. $\boldsymbol{x = 2}$

Check: LHS of original equation $= 6(2) + 1 = 13$
RHS of original equation $= 2(2) + 9 = 13$
Hence the solution $x = 2$ is correct.

Problem 3.61 Solve $3(x - 2) = 9$

Removing the bracket gives: $3x - 6 = 9$

Rearranging gives: $3x = 9 + 6$

$$3x = 15$$

$$\frac{3x}{3} = \frac{15}{3}$$

i.e. $\boldsymbol{x = 5}$

Check: LHS $= 3(5 - 2) = 3(3) = 9 =$ RHS
Hence the solution $x = 5$ is correct.

Problem 3.62 Solve $\frac{3}{x} = \frac{4}{5}$

The lowest common multiple (LCM) of the denominators, i.e. the lowest algebraic expression that both x and 5 will divide into, is $5x$.

Multiplying both sides by $5x$ gives: $5x\left(\frac{3}{x}\right) = 5x\left(\frac{4}{5}\right)$

Cancelling gives: $15 = 4x$ (1)

$$\frac{15}{4} = \frac{4x}{4}$$

i.e. $\boldsymbol{x = 3\frac{3}{4}}$

Check: LHS $= \frac{3}{3\frac{3}{4}} = \frac{3}{15/4} = 3\left(\frac{4}{15}\right) = \frac{12}{15} = \frac{4}{5} =$ RHS

(Note that when there is only one fraction on each side of an equation, 'cross-multiplication' can be applied.
In this example if $\frac{3}{x} = \frac{4}{5}$ then $(3)(5) = 4x$, which is a quicker way of arriving at equation (1) above.)

Problem 3.63 Solve $\sqrt{x} = 2$

Whenever square root signs are involved with the unknown quantity, both sides of the equation must be squared.

Hence $(\sqrt{x})^2 = (2)^2$

i.e. $\boldsymbol{x = 4}$

Problem 3.64 Solve $2\sqrt{d} = 8$

To avoid possible errors it is usually best to arrange the term containing the square root on its own.

$$\text{Thus } \frac{2\sqrt{d}}{2} = \frac{8}{2}$$

$$\text{i.e. } \sqrt{d} = 4$$

Squaring both sides gives $\boldsymbol{d} = 16$, which may be checked in the orignal equation.

Problem 3.65 Solve $x^2 = 25$

This problem involves a square term and thus is not a simple equation (it is, in fact, a quadratic equation). However, the solutions of such equations are often required and are therefore included for completeness. Whenever a square of the unknown is involved, the square root of both sides of the equation is taken. Hence $\sqrt{x^2} = \sqrt{25}$

$$\text{i.e. } \boldsymbol{x = 5}$$

However, $x = -5$ is also a solution of the equation because $(-5) \times (-5) = +25$. Therefore, whenever the square root of a number is required there are always two answers, one positive, the other negative.
The solution of $x^2 = 25$ is thus written as $\boldsymbol{x = \pm 5}$.

Test your knowledge 3.10

Solve the following equations:

1 $9 - 2t = 3$

2 $5x - 3 - 6x = 4x - 8$

3 $\frac{3}{c} = \frac{2}{5}$

4 $2\sqrt{x} = 4$

5 $4 = \frac{a^2}{9}$

Practical problems involving simple equations

Problem 3.66 A copper wire has a length l of 1.5 km, a resistance R of $5\,\Omega$ and a resistivity ρ of 17.2×10^{-6} mm. Find the cross-sectional area, a, of the wire, given that $R = \rho l/a$

$$\text{Since } R = \frac{\rho l}{a} \text{ then } 5\,\Omega = \frac{(17.2 \times 10^{-6}\,\Omega\text{mm})(1500 \times 10^3\,\text{mm})}{a}$$

From the units given, a is measured in mm^2. Thus

$$5a = 17.2 \times 10^{-6} \times 1500 \times 10^3$$

$$a = \frac{17.2 \times 10^{-6} \times 1500 \times 10^3}{5}$$

$$= \frac{17.2 \times 1500 \times 10^3}{10^6 \times 5} = \frac{17.2 \times \cancel{15}^3}{10 \times \cancel{5}_1} = 5.16$$

Hence the cross-sectional area of the wire is 5.16 mm^2.

Problem 3.67 The temperature coefficient of resistance α may be calculated from the formula $R_t = R_0(1 + \alpha t)$. Find α given $R_t = 0.928$, $R_0 = 0.8$ and $t = 40$

Since $R_t = R_0(1 + \alpha t)$ then $0.928 = 0.8[1 + \alpha(40)]$

$$0.928 = 0.8 + (0.8)(\alpha)(40)$$

$$0.928 - 0.8 = 32\alpha$$

$$0.128 = 32\alpha$$

Hence $$\alpha = \frac{0.128}{32} = \mathbf{0.004}.$$

Problem 3.68 The distance s metres travelled in time t seconds is given by the formula $s = ut + \frac{1}{2}at^2$, where u is the initial velocity in m/s and a is the acceleration in m/s^2. Find the acceleration of a body if it travels 168 m in 6 s, with an initial velocity of 10 m/s.

$s = ut + \frac{1}{2}at^2$, and $s = 168$, $u = 10$ and $t = 6$

Hence $$168 = (10)(6) + \frac{1}{2}a(6)^2$$

$$168 = 60 + 18a$$

$$168 - 60 = 18a$$

$$108 = 18a$$

$$a = \frac{108}{18} = 6$$

Hence the acceleration of the body is 6 m/s^2.

Problem 3.69 When three resistors in an electrical circuit are connected in parallel the total resistance R_T is given by:

$$\frac{1}{R_T} = \frac{1}{R_1} + \frac{1}{R_2} + \frac{1}{R_3}$$

Find the total resistance when $R_1 = 5\,\Omega$, $R_2 = 10\,\Omega$ and $R_3 = 30\,\Omega$

$$\frac{1}{R_T} = \frac{1}{5} + \frac{1}{10} + \frac{1}{30} = \frac{6+3+1}{30} = \frac{10}{30} = \frac{1}{3}$$

Taking the reciprocal of both sides gives: $\boldsymbol{R_T = 3\,\Omega}$

Alternatively, if $\frac{1}{R_T} = \frac{1}{5} + \frac{1}{10} + \frac{1}{30}$ the LCM of the denominators is $30R_T$

Hence $$30R_T\left(\frac{1}{R_T}\right) = 30R_T\left(\frac{1}{5}\right) + 30R_T\left(\frac{1}{10}\right) + 30R_T\left(\frac{1}{30}\right)$$

Cancelling gives: $30 = 6R_T + 3R_T + R_T$

i.e. $30 = 10R_T$

$$\boldsymbol{R_T} = \frac{30}{10} = \mathbf{3\,\Omega}, \text{ as above.}$$

Problem 3.70 The extension x m of an aluminium tie bar of length l m and cross-sectional area $A\,\text{m}^2$ when carrying a load of F newtons is given by the modulus of elasticity $E = Fl/Ax$. Find the extension of the tie bar (in mm) if $E = 70 \times 10^9\,\text{N/m}^2$, $F = 20 \times 10^6\,\text{N}$, $A = 0.1\,\text{m}^2$ and $l = 1.4\,\text{m}$

$$E = \frac{Fl}{Ax}. \quad \text{Hence } 70 \times 10^9\,\frac{\text{N}}{\text{m}^2} = \frac{(20 \times 10^6\,\text{N})(1.4\,\text{m})}{(0.1\,\text{m}^2)(x)}$$

(the unit of x is thus metres)

$$70 \times 10^9 \times 0.1 \times x = 20 \times 10^6 \times 1.4$$

$$x = \frac{20 \times 10^6 \times 1.4}{70 \times 10^9 \times 0.1}$$

$$\text{Cancelling gives: } x = \frac{2 \times 1.4}{7 \times 100}\,\text{m} = \frac{2 \times 1.4}{7 \times 100} \times 1000\,\text{mm}$$

Hence the extension of the tie bar, x = 4 mm.

Problem 3.71 Power in a d.c. circuit is given by $P = V^2/R$, where V is the supply voltage and R is the circuit resistance. Find the supply voltage if the circuit resistance is $1.25\,\Omega$ and the power measured is 320 W.

$$\text{Since } P = \frac{V^2}{R} \text{ then } 320 = \frac{V^2}{1.25}$$

$$(320)(1.25) = V^2$$

$$\text{i.e.} \quad V^2 = 400$$

$$\text{Supply voltage,} \quad V = \sqrt{400} = \pm\mathbf{20\,V}$$

Test your knowledge 3.11

1 If $v^2 = u^2 + 2as$, evaluate u given $v = 20$, $a = -10$ and $s = 6.45$

2 A formula relating initial and final states of pressures, P_1 and P_2, volumes V_1 and V_2, and absolute temperatures T_1 and T_2, of an ideal gas is

$$\frac{P_1V_1}{T_1} = \frac{P_2V_2}{T_2}$$

Find the value of P_2 given $P_1 = 100 \times 10^3$, $V_1 = 1.0$, $V_2 = 0.266$, $T_1 = 423$ and $T_2 = 293$

Evaluation of formulae

The statement $v = u + at$ is said to be a **formula** for v in terms of u, a and t. v, u, a and t are called **symbols**. The single term on the left hand side of the equation, v, is called the **subject of the formula**.

Provided values are given for all the symbols in a formula except one, the remaining symbol can be made the subject of the formula and may be evaluated using a calculator.

Problem 3.72 In an electrical circuit the voltage V is given by Ohm's law, i.e. $V = IR$. Find, correct to 4 significant figures, the voltage when $I = 5.36\,\text{A}$ and $R = 14.76\,\Omega$

$$V = IR = (5.36)(14.76)$$

Hence **voltage V = 79.11 V**, correct to 4 significant figures.

Problem 3.73 The surface area A of a hollow cone is given by $A = \pi rl$. Determine the surface area when $r = 3.0\,\text{cm}$, $l = 8.5\,\text{cm}$ and $\pi = 3.14$.

$$A = \pi rl = (3.14)(3.0)(8.5)\,\text{cm}^2$$

Hence **surface area A = 80.07 cm²**.

Problem 3.74 Velocity v is given by $v = u + at$. If $u = 9.86\,\text{m/s}$, $a = 4.25\,\text{m/s}^2$ and $t = 6.84\,\text{s}$, find v, correct to 3 significant figures

$$\begin{aligned} v = u + at &= 9.86 + (4.25)(6.84) \\ &= 9.86 + 29.07 \\ &= 38.93 \end{aligned}$$

Hence **velocity** $\mathbf{v = 38.9\,m/s}$, correct to 3 significant figures.

Problem 3.75 The area, A, of a circle is given by $A = \pi r^2$. Determine the area correct to 2 decimal places, given $\pi = 3.142$ and $r = 5.23\,\text{m}$

$$\begin{aligned} A = \pi r^2 &= (3.142)(5.23)^2 \\ &= (3.142)(27.35) \end{aligned}$$

Hence **area** $\mathbf{A = 85.94\,m^2}$, correct to 2 decimal places.

Problem 3.76 The power P watts dissipated in an electrical circuit may be expressed by the formula $P = V^2/R$. Evaluate the power, correct to 3 significant figures, given that $V = 17.48\,\text{V}$ and $R = 36.12\,\Omega$

$$P = \frac{V^2}{R} = \frac{(17.48)^2}{36.12} = \frac{305.6}{36.12}$$

Hence **power,** $\mathbf{P = 8.46\,W}$, correct to 3 significant figures.

Problem 3.77 The volume $V\,\text{cm}^3$ of a right circular cone is given by $V = \frac{1}{3}\pi r^2 h$. Given that $r = 4.321\,\text{cm}$, $h = 18.35\,\text{cm}$ and $\pi = 3.142$, find the volume correct to 4 significant figures

$$\begin{aligned} V = \frac{1}{3}\pi r^2 h &= \frac{1}{3}(3.142)(4.321)^2(18.35) \\ &= \frac{1}{3}(3.142)(18.67)(18.35) \end{aligned}$$

Hence **volume,** $\mathbf{V = 358.8\,cm^3}$, correct to 4 significant figures.

Problem 3.78 Force F newtons is given by the formula $F = (Gm_1 m_2)/d^2$, where m_1 and m_2 are masses, d their distance apart and G is a constant. Find the value of the force given that $G = 6.67 \times 10^{-11}$, $m_1 = 7.36$, $m_2 = 15.5$ and $d = 22.6$. Express the answer in standard form, correct to 3 significant figures.

$$\begin{aligned} F = \frac{Gm_1 m_2}{d^2} = \frac{(6.67 \times 10^{-11})(7.36)(15.5)}{(22.6)^2} &= \frac{(6.67)(7.36)(15.5)}{(10^{11})(510.8)} \\ &= \frac{1.490}{10^{11}} \end{aligned}$$

Hence **force** $\mathbf{F = 1.49 \times 10^{-11}}$ **newtons**, correct to 3 significant figures.

Test your knowledge 3.12

1 The time of swing, t s, of a simple pendulum is given by $t = 2\pi\sqrt{(l/g)}$. Determine the time, correct to 3 decimal places, given that $\pi = 3.142$, $l = 12.0$ and $g = 9.81$

2 Resistance, $R\ \Omega$, varies with temperature according to the formula $R + R_0(1 + \alpha t)$. Evaluate R, correct to 3 significant figures, given $R_0 = 14.59$, $\alpha = 0.0043$ and $t = 80$

3 Find the distance, s, given that $s = \frac{1}{2}gt^2$, when $t = 0.032$ s and acceleration due to gravity $g = 9.81$ m/s^2

4 The energy stored in a capacitor is given by $E = \frac{1}{2}CV^2$ J. Determine the energy when capacitance $C = 5 \times 10^{-6}$ F and voltage $V = 240$ V.

Transposition of formulae

When a symbol other than the subject is required to be calculated it is usual to rearrange the formula to make a new subject. This rearranging process is called **transposing the formula** or **transposition**.

The rules used for transposition of formulae are the same as those used for the solution of simple equations (see earlier) – basically, that the equality of an equation must be maintained.

Problem 3.79 Transpose $p = q + r + s$ to make r the subject.

The aim is to obtain r on its own on the left hand side (LHS) of the equation. Changing the equation around so that r is on the LHS gives:

$$q + r + s = p \qquad (1)$$

Subtracting $(q + s)$ from both sides of the equation gives:

$$q + r + s - (q + s) = p - (q + s)$$

Thus $q + r + s - q - s = p - q - s$

$$\text{i.e. } \boldsymbol{r = p - q - s} \qquad (2)$$

It is shown with simple equations, that a quantity can be moved from one side of an equation to the other with an appropriate change of sign. Thus equation (2) follows immediately from equation (1) above.

Problem 3.80 If $a + b = w - x + y$, express x as the subject

Rearranging gives: $w - x + y = a + b$ and $-x = a + b - w - y$
Multiplying both sides by -1 gives:

$$(-1)(-x) = (-1)(a + b - w - y) \text{ i.e. } x = -a - b + w + y$$

The result of multiplying each side of the equation by -1 is to change all the signs in the equation.

It is conventional to express answers with positive quantities first. Hence rather than $x = -a - b + w + y$, $\boldsymbol{x = w + y - a - b}$, since the order of terms connected by + and – signs is immaterial.

Problem 3.81 Transpose $v = f\lambda$ to make λ the subject

Rearranging gives: $f\lambda = v$

Dividing both sides by f gives : $\frac{f\lambda}{f} = \frac{v}{f}$ i.e. $\boldsymbol{\lambda = \frac{v}{f}}$

Problem 3.82 When a body falls freely through a height h, the velocity v is given by $v^2 = 2gh$. Express this formula with h as the subject.

Rearranging gives: $2gh = v^2$

Dividing both sides by $2g$ gives: $\frac{2gh}{2g} = \frac{v^2}{2g}$ i.e. $\boldsymbol{h = \frac{v^2}{2g}}$

Problem 3.83 If $R = V/I$, rearrange to make V the subject.

Rearranging gives: $\frac{V}{I} = R$

Multiplying both sides by I gives: $I\left(\frac{V}{I}\right) = I(R)$

Hence $\boldsymbol{V = IR}$

Problem 3.84 Transpose $a = \frac{F}{m}$ for m

Rearranging gives: $\frac{F}{m} = a$

Multiplying both sides by m gives: $m\left(\frac{F}{m}\right) = m(a)$ i.e. $F = ma$

Rearranging gives: $ma = F$

Dividing both sides by a gives: $\frac{ma}{a} = \frac{F}{a}$ i.e. $\boldsymbol{m = \frac{F}{a}}$

Problem 3.85 Rearrange the formula $R = (\rho l)/a$ to make (i) a the subject, and (ii) l the subject

(i) Rearranging gives: $\frac{\rho l}{a} = R$

Multiplying both sides by a gives: $a\left(\frac{\rho l}{a}\right) = a(R)$

i.e. $\rho l = aR$

Rearranging gives: $aR = \rho l$

Dividing both sides by R gives: $\frac{aR}{R} = \frac{\rho l}{R}$

i.e. $\boldsymbol{a = \frac{\rho l}{R}}$

(ii) $\rho l/a = R$

Multiplying both sides by a gives: $\rho l = aR$

Dividing both sides by ρ gives: $\frac{\rho l}{\rho} = \frac{aR}{\rho}$

i.e. $\boldsymbol{l = \frac{aR}{\rho}}$

Problem 3.86 Transpose the formua $v = u + (ft)/m$, to make f the subject

Rearranging gives: $u + \frac{ft}{m} = v$

and $\frac{ft}{m} = v - u$

Mutiplying each side by m gives: $m\left(\frac{ft}{m}\right) = m(v - u)$

i.e. $ft = m(v - u)$

Dividing both sides by t gives: $\frac{ft}{t} = \frac{m}{t}(v - u)$

i.e. $\boldsymbol{f = \frac{m}{t}(v - u)}$

Problem 3.87 The final length, l_2, of a piece of wire heated through $\theta\,°C$ is given by the formula $l_2 = l_1(1 + \alpha\theta)$. Make the coefficient of expansion, α, the subject

Rearranging gives: $l_1(1 + \alpha\theta) = l_2$

Removing the bracket gives: $l_1 + l_2\alpha\theta = l_2$

Rearranging gives: $l_1\alpha\theta = l_2 - l_1$

Dividing both sides by $l_1\theta$ gives: $\frac{l_1\alpha\theta}{l_1\theta} = \frac{l_2 - l_1}{l_1\theta}$

i.e. $\boldsymbol{\alpha = \frac{l_2 - l_1}{l_1\theta}}$

Problem 3.88 A formula for kinetic energy is $k = \frac{1}{2}mv^2$. Transpose the formula to make v the subject

Rearranging gives: $\frac{1}{2}mv^2 = k$

Whenever the prospective new subject is a squared term, that term is isolated on the LHS, and then the square root of both sides of the equation is taken.

Multiplying both sides by 2 gives: $mv^2 = 2k$

Dividing both sides by m gives: $\frac{mv^2}{m} = \frac{2k}{m}$

i.e. $v^2 = \frac{2k}{m}$

Taking the square root of both sides gives: $\sqrt{v^2} = \sqrt{\left(\frac{2k}{m}\right)}$

i.e. $\boldsymbol{v = \sqrt{\left(\frac{2k}{m}\right)}}$

Test your knowledge 3.13

1. Rearrange $I = \frac{E}{R}$ to make R the subject
2. Transpose $F = \frac{9}{5}C + 32$ for C
3. A formula for the distance moved by a body is given by $s = \frac{1}{2}(v + u)t$. Rearrange the formula to make u the subject.

Problem 3.89 In a right angled triangle having sides x, y and hypotenuse z, Pythagoras' theorem states $z^2 = x^2 + y^2$. Transpose the formula to find x

Rearranging gives: $x^2 + y^2 = z^2$

and $x^2 = z^2 - y^2$

Taking the square root of both sides gives: $\boldsymbol{x = \sqrt{(z^2 - y^2)}}$

Introduction to straight line graphs

A graph is a pictorial representation of information showing how one quantity varies with another related quantity.

The most common method of showing a relationship between two **sets of data** is to use **cartesian** or **rectangular axes** as shown in Figure 3.1.

The points on a graph are called **co-ordinates**. Point A in Figure 3.1 has the co-ordinates (3, 2), i.e. 3 units in the x direction and 2 units in the y direction. Similarly, point B has co-ordinates (−4, 3) and C has co-ordinates (−3, −2). The origin has co-ordinates (0, 0).

Typical examples where the plotting of straight line graphs is used are with distance/time graphs, speed/time graphs and load/extension graphs.

The straight line graph

Let a relationship between two variables x and y be $y = 3x + 2$. When $x = 0$, $y = 3(0) + 2 = 2$, when $x = 1$, $y = 3(1) + 2 = 5$, when $x = 2$, $y = 3(2) + 2 = 8$, and so on. Thus co-ordinates (0, 2), (1, 5) and (2, 8) have been produced from the equation by selecting arbitrary values of x, and are shown plotted in Figure 3.2. When the points are joined together a **straight-line graph** results.

The **gradient** or **slope** of a straight line is the ratio of the change in th value of y to the change in the value of x between any two points on the line. If, as x increases (→), y also increases (↑), then the gradient is positive.

In Figure 3.3(a) the gradient of AC $= \dfrac{\text{change in } y}{\text{change in } x} = \dfrac{CB}{BA}$

$$= \frac{7-3}{3-1} = \frac{4}{2} = 2$$

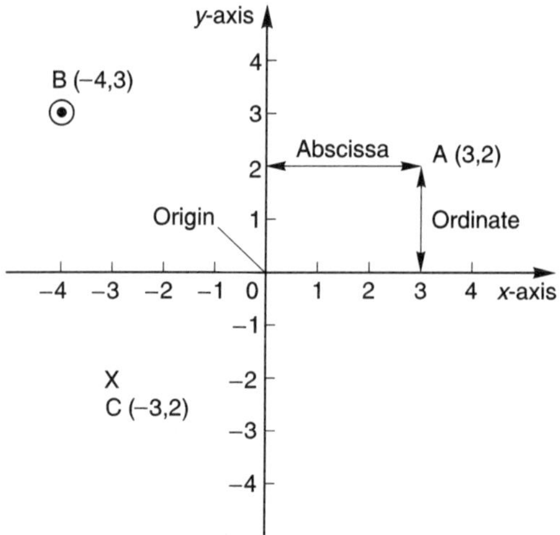

Figure 3.1

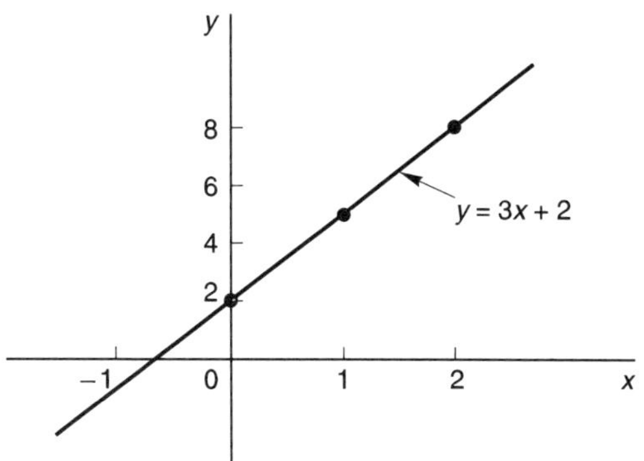

Figure 3.2

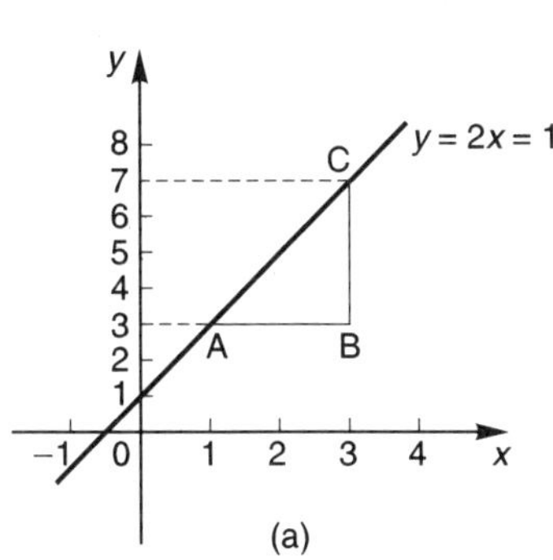

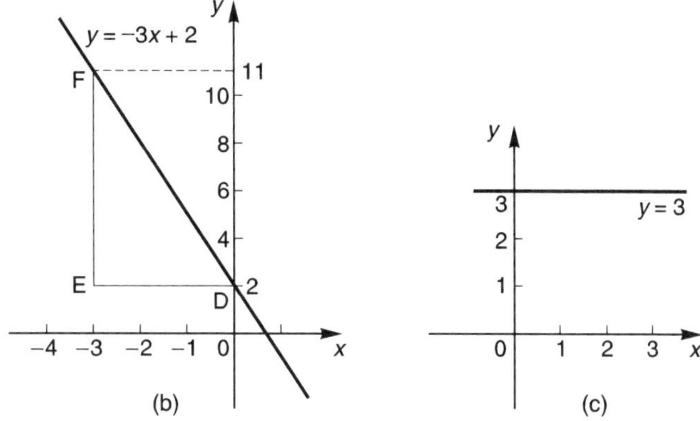

Figure 3.3

If as x increases ($\rightarrow$), y decreases ($\downarrow$), then the gradient is negative.

$$\text{In Figure 3.3(b), the gradient of DF} = \frac{\text{change in } y}{\text{change in } x} = \frac{\text{FE}}{\text{ED}} = \frac{11-2}{-3-0} = \frac{9}{-3} = -3$$

Figure 3.3(c) shows a straight line graph $y = 3$. Since the straight line is horizontal the gradient is zero.

The value of y when $x = 0$ is called the ***y*-axis intercept**.

In Figure 3.3(a) the y-axis intercept is 1 and in Figure 3.3(b) it is 2. If the equation of a graph is of the form $y = mx + c$, where m and c are constants, the graph will always be a straight line, ***m* representing the gradient** and ***c* the *y*-axis intercept**. Thus $y = 5x + 2$ represents a straight line of gradient 5 and y-axis intercept 2. Similary, $y = -3x - 4$ represents a straight line of gradient -3 and y-axis intercept -4.

Summary of general rules to be applied when drawing graphs

(i) Give the graph a title clearly explaining what is being illustrated.

(ii) Choose scales such that the graph occupies as much space as possible on the graph paper being used.

(iii) Choose scales so that interpolation is made as easy as possible. Usually scales such as 1 cm = 1 unit, or 1 cm = 2 units, or 1 cm = 10 units are used. Awkward scales such as 1 cm = 3 units or 1 cm = 7 units should not be used.

(iv) The scales need not start at zero, particularly when starting at zero produces an accumulation of points within a small area of the graph paper.

(v) The co-ordinates, or points, should be clearly marked. This may be done either by a cross, or by a dot and circle, or just by a dot (see Figure 3.1).

(vi) A statement should be made next to each axis explaining the numbers represented with their appropriate units.

(vii) Sufficient numbers should be written next to each axis without cramping.

Problem 3.90 Plot the graph $y = 4x + 3$ in the range $x = -3$ to $x = +4$. From the graph, find (a) the value of y when $x = 2.2$ and (b) the value of x when $y = -3$

Whenever an equation is given and a graph is required, a table giving corresponding values of the variable is necessary.
The table is achieved as follows:

When $x = -3$, $y = 4x + 3 = 4(-3) + 3 = -12 + 3 = -9$
When $x = -2$, $y = 4(-2) + 3 = -8 + 3 = -5$, and so on

Such a table is shown below:

x	-3	-2	-1	0	1	2	3	4
y	-9	-5	-1	3	7	11	15	19

The co-ordinates $(-3, -9)$, $(-2, -5)$, $(-1, -1)$, and so on, are plotted and joined together to produce the straight line graph shown in Figure 3.4. (Note that the scales used on the x- and y-axes do not have to be the same.) From the graph:

(a) when $x = 2.2$, $\mathbf{y = 11.8}$, and

(b) when $y = -3$, $\mathbf{x = -1.5}$

Problem 3.91 Plot the following graphs on the same axes between the values $x = -3$ to $x = +3$ and determine the gradient and y-axis intercept of each.

(a) $y = 3x$ (b) $y = 3x + 7$ (c) $y = -4x + 4$ (d) $y = -4x - 5$

A table of co-ordinates is drawn up for each equation.

(a) $y = 3x$

x	-3	-2	-1	0	1	2	3
y	-9	-6	-3	0	3	6	9

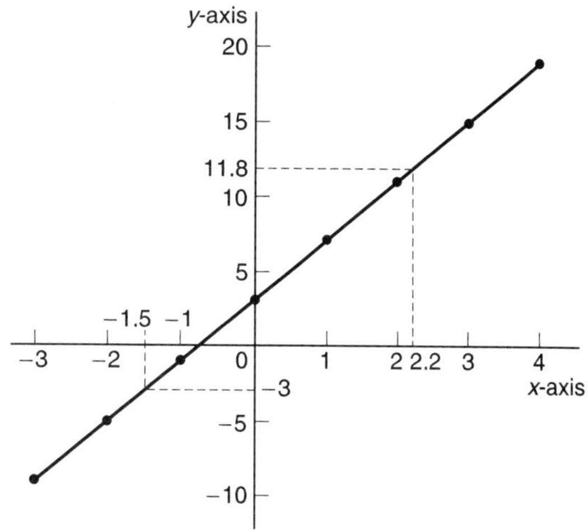

Figure 3.4

(b) $y = 3x + 7$

x	−3	−2	−1	0	1	2	3
y	−2	1	4	7	10	13	16

(c) $y = -4x + 4$

x	−3	−2	−1	0	1	2	3
y	16	12	8	4	0	−4	−8

(d) $y = -4x - 5$

x	−3	−2	−1	0	1	2	3
y	7	3	−1	−5	−9	−13	−17

Each of the graphs is plotted as shown in Figure 3.5, and all are straight lines. $y = 3x$ and $y = 3x + 7$ are parallel to each other and thus have the same gradient.

$$\text{Gradient of AC} = \frac{\text{BC}}{\text{AB}} = \frac{16 - 7}{3 - 0} = \frac{9}{3} = 3$$

Hence the gradient of both $y = 3x$ and $y = 3x + 7$ is 3
$y = -4x + 4$ and $y = -4x - 5$ are parallel to each other and thus have the same gradient. The gradient of DF is given by

$$\text{DF} = \frac{\text{EF}}{\text{ED}} = \frac{-5 - (-17)}{0 - 3} = \frac{12}{-3} = -4$$

Hence the gradient of both $y = -4x + 4$ and $y = -4x - 5$ is −4
The y-axis intercept means the value of y where the straight line cuts the y-axis.
From Figure 3.5, $y = 3x$ cuts the y-axis at $\boldsymbol{y = 0}$
$y = 3x + 7$ cuts the y-axis at $\boldsymbol{y = +7}$
$y = -4x + 4$ cuts the **y-axis at $y = +4$**
and $y = -4x - 5$ cuts the **y-axis at $y = -5$**

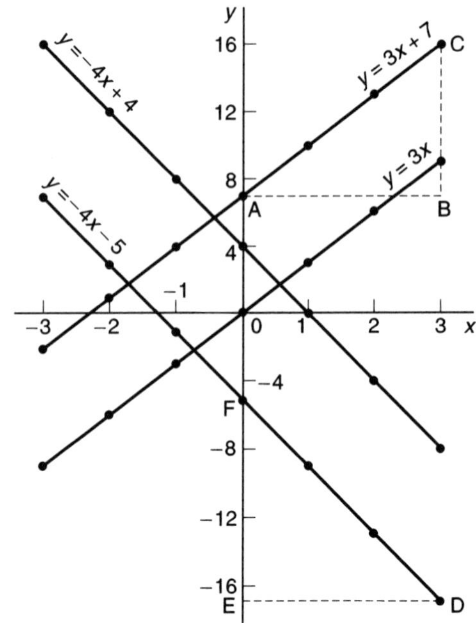

Figure 3.5

Some general conclusions can be drawn from the graphs shown in Figures 3.4 and 3.5.
When an equation is of the form $y = mx + c$, where m and c are constants, then

(i) a graph of y against x produces a straight line,

(ii) m represents the slope or gradient of the line, and

(iii) c represents the y-axis intercept.

Thus, given an equation such as $y = 3x + 7$, it may be deduced 'on sight' that its gradient is $+3$ and its y-axis intercept is $+7$, as shown in Figure 3.5. Similarly, if $y = -4x - 5$, then the gradient is -4 and the y-axis intercept is -5, as shown in Figure 3.5.

When plotting a graph of the form $y = mx + c$, only two co-ordinates need be determined. When the co-ordinates are plotted a straight line is drawn between the two points. Normally, three co-ordinates are determined, the third one acting as a check.

Problem 3.92 The following equations represent straight lines. Determine, without plotting graphs, the gradient and y-axis intercept for each.

(a) $y = 3$ (b) $y = 2x$ (c) $y = 5x - 1$ (d) $2x + 3y = 3$

(a) $y = 3$ (which is of the form $y = 0x + 3$) represents a horizontal straight line intercepting the **y-axis at 3**. Since the line is horizontal **its gradient is zero**.

(b) $y = 2x$ is of the form $y = mx + c$, where c is zero. Hence **gradient = 2** and **y-axis intercept = 0** (i.e. the origin)

(c) $y = 5x - 1$ is of the form $y = mx + c$. Hence **gradient = 5** and **y-axis intercept = −1**

(d) $2x + 3y = 3$ is not in the form $y = mx + c$ as it stands. Transposing to make y the subject gives

$$3y = -3 - 2x$$

$$\text{i.e. } y = \frac{3 - 2x}{3} = \frac{3}{3} - \frac{2x}{3}$$

$$\text{i.e. } y = -\frac{2x}{3} + 1,$$

which is of the form $y = mx + c$

Hence **gradient** $= -\frac{\mathbf{2}}{\mathbf{3}}$ **and y-axis intercept** $= +\mathbf{1}$

Problem 3.93 Determine the gradient of the straight line graph passing through the co-ordinates (a) $(-2, 5)$ and $(3, 4)$, (b) $(-2, -3)$ and $(-1, 3)$

A straight line graph passing through co-ordinates (x_1, y_1) and (x_2, y_2) has a gradient given by

$$m = \frac{y_2 - y_1}{x_2 - x_1} \text{ (see Figure 3.6)}$$

(a) A straight line passes through $(-2, 5)$ and $(3, 4)$, hence $x_1 = -2$, $y_1 = 5$, $x_2 = 3$ and $y_2 = 4$

Hence **gradient** $\boldsymbol{m} = \frac{y_2 - y_1}{x_2 - x_1} = \frac{4 - 5}{3 - (-2)} = -\frac{\mathbf{1}}{\mathbf{5}}$

(b) A straight line passes through $(-2, -3)$ and $(-1, 3)$, hence $x_1 = -2$, $y_1 = -3$, $x_2 = -1$ and $y_2 = 3$

Hence **gradient** $\boldsymbol{m} = \frac{y_2 - y_1}{x_2 - x_1} = \frac{3 - (-3)}{-1 - (-2)} = \frac{3 + 3}{-1 + 2}$

$$= \frac{6}{1} = \mathbf{6}$$

Test your knowledge 3.14

1 Corresponding values obtained experimentally for two quantities are

x	−2.0	−0.5	0	1.0	2.5	3.0	5.0
y	−13.0	−5.5	−3.0	2.0	9.5	12.0	22.0

Use a horizontal scale for x of 1 cm $= \frac{1}{2}$ unit and a vertical scale for y of 1 cm $=$ 2 units and draw a graph of x against y. Label the graph and each of its axes. By interpolation, find from the graph the value of y when x is 3.5.

2 Determine the gradient and intercept on the y-axis for each of the following equations:
(a) $y = 4x - 2$ (b) $y = -x$ (c) $y = -3x - 4$ (d) $y = 4$

3 Plot the graphs of $3x + y + 1 = 0$ and $2y - 5 = x$ on the same axes and find their point of intersection.

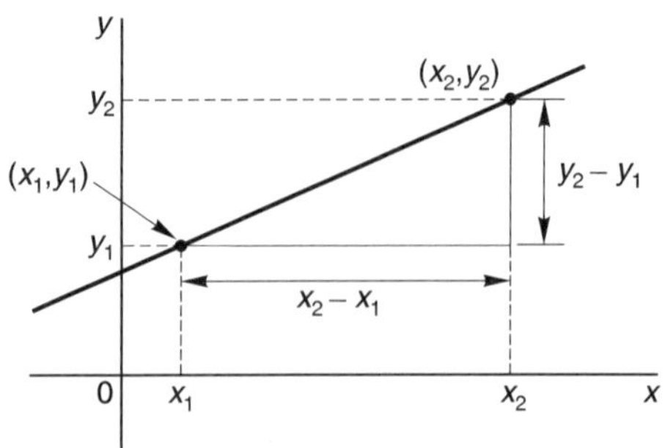

Figure 3.6

Practical problems involving straight line graphs

When a set of co-ordinate values are given or are obtained experimentally and it is believed that they follow a law of the form $y = mx + c$, then if a straight line can be drawn reasonably close to most of the co-ordinate values when plotted, this verifies tha a law of the form $y = mx + c$ exists. From the graph, constants m (i.e. gradient) and c (i.e. y-axis intercept) can be determined.

Problem 3.94 The temperature in degrees Celsius and the corresponding values in degrees Fahrenheit are shown in the table below. Construct rectangular axes, choose a suitable scale and plot a graph of degrees Celsius (on the horizontal axis) against degrees Fahrenheit (on the vertical axis).

°C	10	20	40	60	80	100
°F	50	68	104	140	176	212

From the graph find (a) the temperature in degrees Fahrenheit at 55°C, (b) the temperature in degrees Celsius at 167°F, (c) the Fahrenheit temperature at 0°C, and (d) the Celsius temperature at 230°F.

The co-ordinates (10, 50), (20, 68), (40, 104), and so on are plotted as shown in Figure 3.7. When the co-ordinates are joined, a straight line is produced. Since a straight line results there is a linear relationship between degrees Celsius and degrees Fahrenheit.

(a) To find the Fahrenheit temperature at 55°C a vertial line AB is constructed from the horizontal axis to meet the straight line at B. The point where the horizontal line BD meets the vertical axis indicates the equivalent Fahrenheit temperature. **Hence 55°C is equivalent to 131°F**.
This process of finding an equivalent value in between the given information in the above table is called interpolation.

(b) To find the Celsius temperature at 167°F, a horizontal line EF is constructed as shown in Figure 3.7. The point where

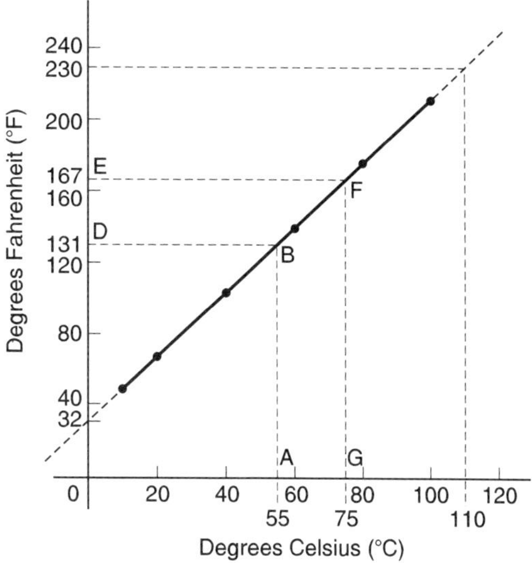

Figure 3.7

the vertical line FG cuts the horizontal axis indicates the equivalent Celsius temperature. **Hence 167°F is equivalent to 75°C.**

(c) If the graph is assumed to be linear even outside of the given data, then the graph may be extended at both ends (shown by broken lines in Figure 3.7). **From Figure 3.7, 0°C corresponds to 32°F.**

(d) **230°F is seen to correspond to 110°C**. The process of finding equivalent values outside of the given range is called extrapolation.

Problem 3.95 In an experiment on Charles' law, the value of the volume of gas, $V\,\text{m}^3$, was measured for various temperatures T°C. Results are shown below.

$V\,\text{m}^3$	25.0	25.8	26.6	27.4	28.2	29.0
T°C	60	65	70	75	80	85

Plot a graph of volume (vertical) against temperature (horizontal) and from it find (a) the temperature when the volume is $28.6\,\text{m}^3$ and (b) the volume when the temperature is 67°C.

If a graph is plotted with both the scales starting at zero then the result is as shown in Figure 3.8. All of the points lie in the top right-hand corner of the graph, making interpolation difficult. A more accurate graph is obtained if the temperature axis starts at 55°C and the volume axis starts at $24.5\,\text{m}^3$. The axes corresponding to these values are shown by the broken lines in Figure 3.8 and are called **false axes**, since the origin is not now at zero. A magnified version of this relevant part of the graph is shown in Figure 3.9. From the graph:

(a) when the volume is $28.6\,\text{m}^3$, the equivalent temperature is 82.5°C, and

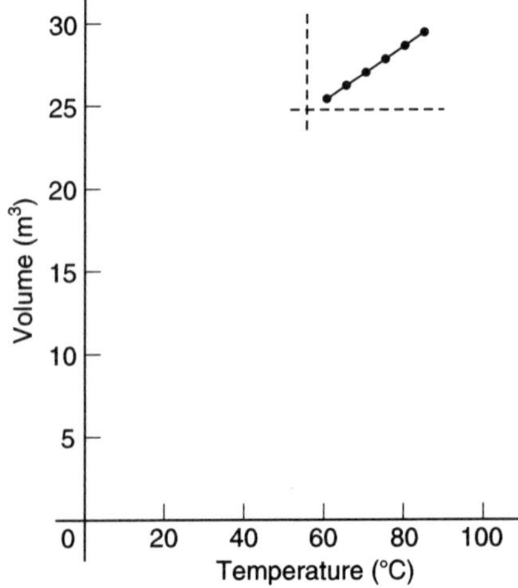

Figure 3.8

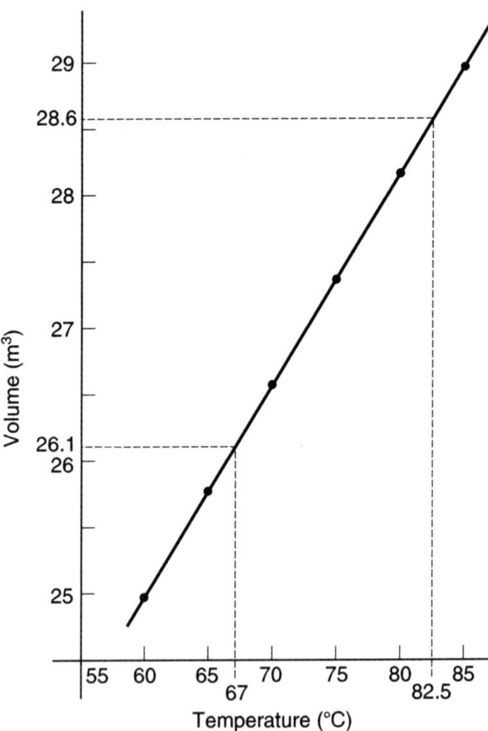

Figure 3.9

(b) when the temperature is 67°C, the equivalent volume is $26.1\,m^3$.

Problem 3.96 In an experiment demonstrating Hooke's law, the strain in an aluminium wire was measured for various stresses. The results were:

Stress N/mm^2	4.9	8.7	15.0	18.4	24.2	27.3
Strain	0.000 07	0.000 13	0.000 21	0.000 27	0.000 34	0.000 39

Plot a graph of stress (vertically) against strain horizontally. Find

(a) Young's modulus of elasticity for aluminium, which is given by the gradient of the graph,

(b) the value of the strain at a stress of $20\,\text{N/mm}^2$, and

(c) the value of the stress when the strain is 0.000 20.

The co-ordinates (0.000 07, 4.9), (0.000 13, 8.7), and so on, are plotted as shown in Figure 3.10. The graph produced is the best straight line which can be drawn corresponding to these points. (With experimental results it is unlikely that all the points will lie exactly on a straight line.) The graph, and each of its axes, are labelled. Since the straight line passes through the origin, then stress is directly proportional to strain for the given range of values.

(a) The gradient of the straight line,

$$AC = \frac{AB}{BC} = \frac{28 - 7}{0.000\,40 - 0.000\,10} = \frac{21}{0.000\,30}$$

$$= \frac{21}{3 \times 10^{-4}} = \frac{7}{10^{-4}} = 7 \times 10^4$$

$$= 70\,000\ \text{N/mm}^2$$

Thus, Young's modulus of elasticity for aluminium is $70\,000\ \text{N/mm}^2$

Since $1\,\text{m}^2 = 10^6\,\text{mm}^2$, $70\,000\,\text{N/mm}^2$ is equivalent to $70\,000 \times 10^6\,\text{N/m}^2$, i.e. $70 \times 10^9\,\text{N/m}^2$ (or Pascals).

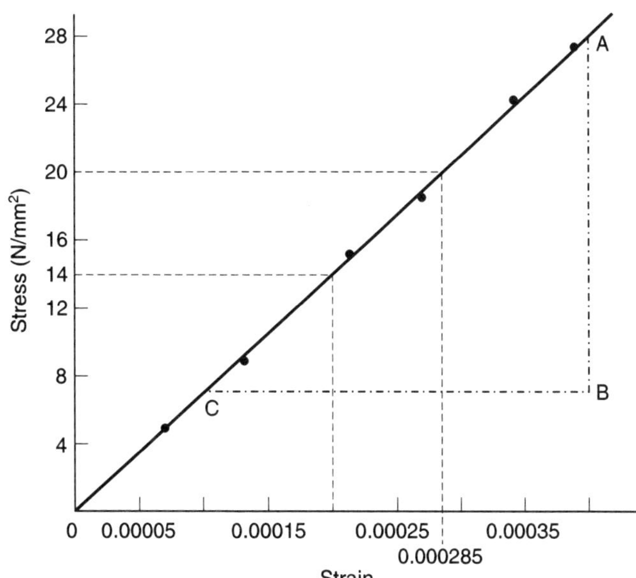

Figure 3.10

From Figure 3.10:

(a) the value of the strain at a stress of $20\,\text{N/mm}^2$ is **0.000 285**, and

(b) the value of the stress when the strain is 0.000 20 is **$14\,\text{N/mm}^2$**.

Problem 3.97 The following values of resistance $R\,\Omega$ and corresponding voltage V volts are obtained from a test on a filament lamp.

$R\,\Omega$	30	48.5	73	107	128
V volts	16	29	52	76	94

Choose suitable scales and plot a graph with R representing the vertical axis and V the horizontal axis. Determine (a) the slope of the graph, (b) the R axis intercept value, (c) the equation of the graph, (d) the value of resistance when the voltage is 60 V, and (e) the value of the voltage when the resistance is $40\,\Omega$. (f) If the graph were to continue in the same manner, what value of resistance would be obtained at 110 V?

The co-ordinates (16, 30), (29, 48.5), and so on, are shown plotted in Figure 3.11 where the best straight line is drawn through the points.

(a) The slope or gradient of the straight line,

$$AC = \frac{AB}{BC} = \frac{135 - 10}{100 - 0} = \frac{125}{100} = \mathbf{1.25}$$

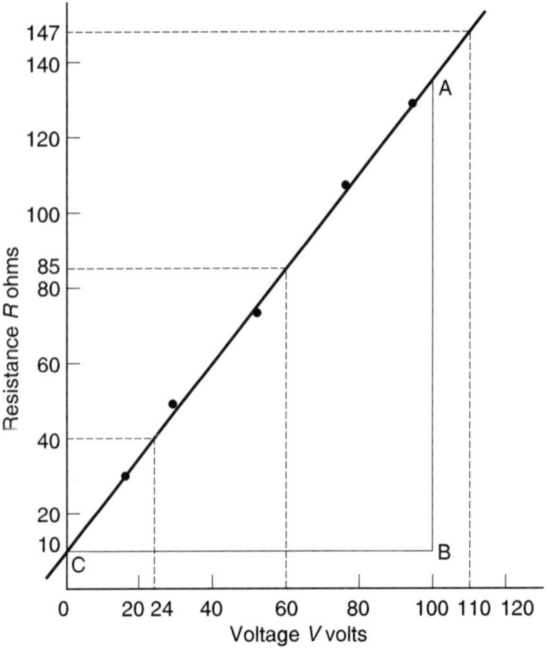

Figure 3.11

(Note that the vertical line AB and the horizontal line BC may be constructed anywhere along the length of the straight line. However, calculations are made easier if the horizontal line BC is carefully chosen, in this case, 100)

(b) The R axis intercept is at **R** $=$ **10 Ω** (by extrapolation)

(c) The equation of a straight line is $y = mx + c$; when y is plotted on the vertical axis and x on the horizontal axis, m represents the gradient and c the y-axis intercept. In this case, R corresponds to y, V corresponds to x, $m = 1.25$ and $c = 10$. Hence the equation of the graph is $\boldsymbol{R = (1.25V + 10)}$

From Figure 3.11:

(d) when the voltage is 60 V, the resistance is **85 Ω**

(e) when the resistance is 40 ohms, the voltage is **24 V**, and

(f) by extrapolation, when the voltage is 110 V, the resistance is **147 Ω**.

Problem 3.98 Experimental tests to determine the breaking stress σ of rolled copper at various temepratures t gave the following results:

Stress σ N/cm^2	8.42	8.02	7.75	7.35	7.06	6.63
Temperature t°C	70	200	280	410	500	640

Show that the values obey the law $\sigma = at + b$, where a and b are constants and determine approximate values for a and b. Use the law to determine the stress at 250°C and the temperature when the stress is 7.54 N/cm^2.

The co-ordinates (70, 8.46), (200, 8.04), and so on are plotted as shown in Figure 3.12. Since the graph is a straight line then the values obey the law $\sigma = at + b$.

$$\text{Gradient of straight line, } a = \frac{\text{AB}}{\text{BC}} = \frac{8.36 - 6.76}{100 - 600} = \frac{1.60}{-500} = -0.0032$$

Vertical axis intercept, $b = 8.68$

Hence the law of the graph is $\sigma = \mathbf{-0.0032}\boldsymbol{t} + \mathbf{8.68}$

When the temperature is 250°C,

stress $\sigma = -0.0032(250) + 8.68 =$ **7.88 N/cm^2**

Rearranging $\sigma = -0.0032t + 8.68$ gives

$$0.0032t = 8.68 - \sigma, \text{ i.e. } t = \frac{8.68 - \sigma}{0.0032}$$

Hence when the stress $\sigma = 7.54$ N/cm^2, temperature t

$$= \frac{8.68 - 7.54}{0.0032} = \mathbf{356.3°C}$$

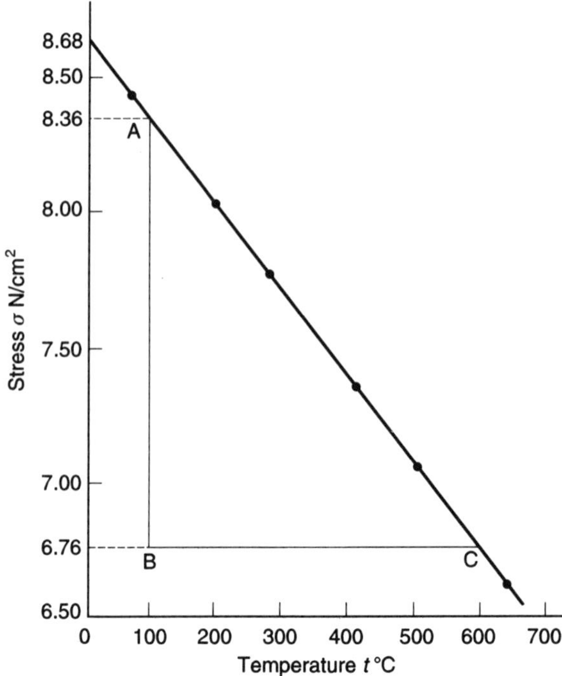

Figure 3.12

Test your knowledge 3.15

1 The resistance R ohms of a copper winding is measured at various temperatures t°C and the results are as follows:

R ohms	112	120	126	131	136
t° C	20	36	48	58	64

Plot a graph of R (vertically) against t (horizontally) and find from it (a) the temperature when the resistance is 122 Ω and (b) the resistance when the temperature is 52°C.

2 The following table gives the force F newtons which, when applied to a lifting machine overcomes a corresponding load of L newtons.

Force F newtons	25	47	64	120	149	187
Load L newtons	50	140	210	430	550	700

Choose suitable scales and plot a graph of F (vertically) against L (horizontally). Draw the best straight line through the points. Determine from the graph (a) the gradient, (b) the F-axis intercept, (c) the equation of the graph, (d) the force applied when the load is 310 N, and (e) the load that a force of 160 N will overcome. (f) If the graph were to continue in the same manner, what value of force will be needed to overcome a 800 N load?

Science for engineering

SI units

The system of units used in engineering and science is the **Système Internationale d'Unités** (International system of units), usually abbreviated to SI units, and is based on the metric system. This was introduced in 1960 and is now adopted by the majority of countries as the official system of measurement. The basic units in the SI system are listed below with their symbols:

Quantity	*Unit and symbol*
length	metre, m
mass	kilogram, kg
time	second, s
electric current	ampere, A
thermodynamic temperature	kelvin, K
luminous intensity	candela, cd
amount of substance	mole, mol

SI units may be made larger or smaller by using prefixes which denote multiplication or division by a particular amount. The eight most common multiples, with their meaning, are listed below:

Prefix	*Name*	*Meaning*
T	tera	multiply by 1 000 000 000 000 (i.e. $\times 10^{12}$)
G	giga	multiply by 1 000 000 000 (i.e. $\times 10^{9}$)
M	mega	multiply by 1 000 000 (i.e. $\times 10^{6}$)
k	kilo	multiply by 1 000 (i.e. $\times 10^{3}$)
m	milli	divide by 1 000 (i.e. $\times 10^{-3}$)
μ	micro	divide by 1 000 000 (i.e. $\times 10^{-6}$)
n	nano	divide by 1 000 000 000 (i.e. $\times 10^{-9}$)
p	pico	divide by 1 000 000 000 000 (i.e. $\times 10^{-12}$)

Length is the distance between two points. The standard unit of length is the **metre**, although the **centimetre, cm, millimetre, mm and kilometre, km**, are often used.

1 cm = 10 mm; 1 m = 100 cm = 1000 mm;
1 km = 1000 m

In an engineering workshop an instrument called a **micrometer** is used for making precise measurements of dimensions such as diameters, thickness and lengths of solid bodies.

Area is a measure of the size or extent of a plane surface and is measured by multiplying a length by a length. If the lengths are in metres then the unit of area is the **square metre, m^2.**

$$1\,m^2 = 1\,m \times 1\,m = 100\,cm \times 100\,cm = 10\,000\,cm^2 \text{ or } 10^4\,cm^2$$
$$= 1000\,mm \times 1000\,mm$$
$$= 1\,000\,000\,mm^2 \text{ or } 10^6\,mm^2$$

Conversely, $1\,cm^2 = 10^{-4}\,m^2$ and $1\,mm^2 = 10^{-6}\,m^2$.

Volume is a measure of the space occupied by a solid and is measured by multiplying a length by a length by a length. If the lengths are in metres then the unit of volume is in **cubic metres, m³.**

$$1\,m^3 = 1\,m \times 1\,m \times 1\,m$$
$$= 100\,cm \times 100\,cm \times 100\,cm = 10^6\,cm^3$$
$$= 1000\,mm \times 1000\,mm \times 1000\,mm$$
$$= 10^9\,mm^3$$

Conversely, $1\,cm^3 = 10^{-6}\,m^3$ and $1\,mm^3 = 10^{-9}\,m^3$.

Another unit used to measure volume, particularly with liquids, is the litre, l, where $1\,l = 1000\,cm^3$.

Mass is the amount of matter in a body and is measured in **kilograms, kg**.

1 kg = 1000 g (or conversely, $1\,g = 10^{-3}\,kg$)
and 1 tonne (t) = 1000 kg

Problem 3.99 Express (a) a length of 36 mm in metres, (b) 32 400 mm² in square metres, and (c) 8 540 000 mm³ in cubic metres

(a) $1\,m = 10^3\,mm$ or $1\,mm = 10^{-3}\,m$.

$$\text{Hence } 36\,mm = 36 \times 10^{-3}\,m = \frac{36}{10^3}\,m = \frac{36}{1000}\,m = \mathbf{0.036\,m}$$

(b) $1\,m^2 = 10^6\,mm^2$ or $1\,mm^2 = 10^{-6}\,m^2$.

$$\text{Hence } 32\,400\,mm^2 = 32\,400 \times 10^{-6}\,m^2$$
$$= \frac{32\,400}{10^6} = \mathbf{0.0324\,m^2}$$

(c) $1\,m^3 = 10^9\,mm^3$ or $1\,mm^3 = 10^{-9}\,m^3$.

$$\text{Hence } 8\,540\,000\,mm^3 = 8\,540\,000 \times 10^{-9}\,m^3$$
$$= \frac{8\,540\,000\,m}{10^9}$$
$$= \mathbf{8.54 \times 10^{-3}\,m^3 \text{ or } 0.008\,54\,m^3}$$

Problem 3.100 A cube has sides each of length 50 mm. Determine the volume of the cube in cubic metres.

$$\text{Volume of cube} = 50\,mm \times 50\,mm \times 50\,mm = 125\,000\,mm^3$$
$$1\,mm^3 = 10^{-9}m, \text{ thus volume} = 125\,000 \times 10^{-9}\,m^3$$
$$= \mathbf{0.125 \times 10^{-3}\,m^3}$$

Problem 3.101 A container has a capacity of 2.5 litres. Calculate its volume in (a) m³, (b) mm³.

Since 1 litre = 1000 cm³, 2.5 litres = 2.5 × 1000 cm³ = 2500 cm³

Test your knowledge 3.16

1 Determine the area of a room 15 m long by 8 m wide in (a) m^2, (b) cm^2 and (c) mm^2.

2 A bottle contains 4 litres of liquid. Determine the volume in (a) cm^3, (b) m^3 and (c) mm^3.

(a) $2500\,cm^3 = 2500 \times 10^{-6}\,m^3 = \mathbf{2.5 \times 10^{-3}\,m^3}$ or $\mathbf{0.0025\,m^3}$

(b) $2500\,cm^3 = 2500 \times 10^3\,mm^3 = \mathbf{2\,500\,000\,mm^3}$ or $\mathbf{2.5 \times 10^6\,mm^3}$

Density

Density is the mass per unit volume of a substance. The symbol used for density is ρ (Greek letter rho) and its units are kg/m^3.

$$\text{Density} = \frac{\text{mass}}{\text{volume}}, \text{ i.e. } \rho = \frac{m}{V} \text{ or } m = \rho V \text{ or } V = \frac{m}{\rho}$$

where m is the mass in kg, V is the volume in m^3 and ρ is the density in kg/m^3. Some typical values of densities include:

Aluminium	$2700\,kg/m^3$	Steel	$7800\,kg/m^3$
Cast iron	$7000\,kg/m^3$	Petrol	$700\,kg/m^3$
Cork	$2500\,kg/m^3$	Lead	$11\,400\,kg/m^3$
Copper	$8900\,kg/m^3$	Water	$1000\,kg/m^3$

The relative density of a substance is the ratio of the density of the substance to the density of water, i.e.

$$\text{relative density} = \frac{\text{density of substance}}{\text{density of water}}$$

Relative density has no units, since it is the ratio of two similar quantities. Typical values of relative densities can be determined from the above (since water has a density of $1000\,kg/m^3$), and include:

Aluminium	2.7	Steel	7.8
Cast iron	7.0	Petrol	0.7
Cork	0.25	Lead	11.4
Copper	8.9		

The relative density of a liquid may be measured using a hydrometer.

Problem 3.102 Determine the density of $50\,cm^3$ of copper if its mass is 445 g.
Volume $= 50\,cm^3 = 50 \times 10^{-6}\,m^3$;
mass $= 445\,g = 445 \times 10^{-3}\,kg$

$$\text{Density} = \frac{\text{mass}}{\text{volume}} = \frac{445 \times 10^3\,kg}{50 \times 10^{-6}\,m^3}$$
$$= \frac{445}{50} \times 10^3$$
$$= \mathbf{8.9 \times 10^3\,kg/m^3 \text{ or } 8900\,kg/m^3}$$

Problem 3.103 The density of aluminium is $2700\,kg/m^3$. Calculate the mass of a block of aluminium which has a volume of $100\,cm^3$.

Density $= 2700\,kg/m^3$; volume $V = 100\,cm^3 = 100 \times 10^{-6}\,m^3$
Since density = mass/volume, then mass = density × volume.

$$\text{Hence } m = \rho V = 2700\,\text{kg/m}^3 \times 100 \times 10^{-6}\,\text{m}^3$$
$$= \frac{2700 \times 100}{10^6}\,\text{kg} = \mathbf{0.270\,kg\ or\ 270\,g}$$

Problem 3.104 Determine the volume, in litres, of 20 kg of paraffin oil of density $800\,\text{kg/m}^3$.

Density = mass/volume hence volume = mass/density. Thus

$$\text{volume} = \frac{m}{\rho} = \frac{20\,\text{kg}}{800\,\text{kg/m}^3} = \frac{1}{40}\,\text{m}^3$$
$$= \frac{1}{40} \times 10^6\,\text{cm}^3$$
$$= 25\,000\,\text{cm}^3$$

$$1\text{ litre} = 1000\,\text{cm}^3 \text{ hence } 25\,000\,\text{cm}^3 = \frac{\mathbf{250\,000}}{\mathbf{1000}} = \mathbf{25\ litres}$$

Test your knowledge 3.17

1. Determine the density of $80\,\text{cm}^3$ of cast iron if its mass is 560 g.
2. Determine the volume, in litres, of 35 kg of petrol of density $700\,\text{kg/m}^3$.
3. A piece of metal 200 mm long, 150 mm wide and 10 mm thick has a mass of 2700 g. What is the density of the metal?

Scalar and vector quantities

Quantities used in engineering and science can be divided into two groups:

(a) **Scalar quantities** have a size (or magnitude) only and need no other information to specify them. Thus, 10 centimetres, 50 seconds, 7 litres and 3 kilograms are all examples of scalar quantities.

(b) **Vector quantities** have both a size or magnitude and a direction, called the line of action of the quantity. Thus, a velocity of 50 kilometres per hour due east, an acceleration of 9.81 metres per second squared vertically downwards and a force of 15 newtons at an angle of 30 degrees are all examples of vector quantities.

Force

When forces are all acting in the same plane, they are called **coplanar**. When forces act at the same time and at the same point, they are called **concurrent forces**.

Force is a **vector quantity** and thus has both a magnitude and a direction. A vector can be represented graphically by a line drawn to scale in the direction of the line of action of the force. Vector quantities may be shown by using bold, lower case letters, thus **ab** in Figure 3.13 represents a force of 5 newtons acting in a direction due east.

The resultant of two coplanar forces

For two forces acting at a point, there are three possibilities.

(a) For forces acting in the same direction and having the samel line of action, the single force having the same effect

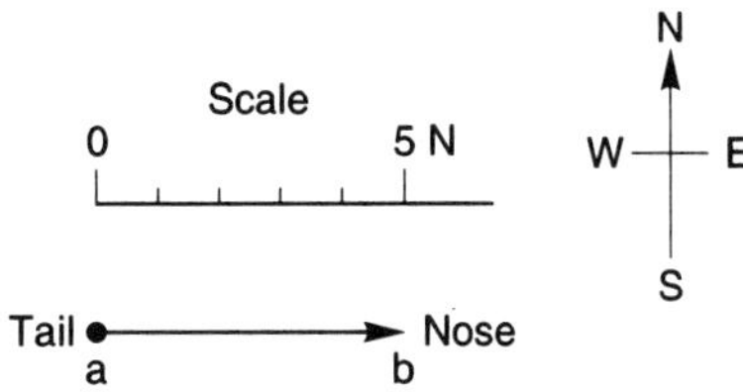

Figure 3.13

as both of the forces, called the **resultant force** or just the **resultant**, is the arithmetic sum of the separate forces. Forces of $\mathbf{F}_1$ and $\mathbf{F}_2$ acting at point P, as shown in Figure 3.14(a), have exactly the same effect on point P as force **F** shown in Figure 3.14(b), where $\mathbf{F} = \mathbf{F}_1 + \mathbf{F}_2$.

(b) For forces acting in opposite directions along the same line of action, the resultant force is the arithmetic difference between the two forces. Forces of $\mathbf{F}_1$ and $\mathbf{F}_2$ acting at point P as shown in Figure 3.15(a) have exactly the same effect on point P as force **F** shown in Figure 3.15(b), where $\mathbf{F} + \mathbf{F}_2 - \mathbf{F}_1$ and acts in the direction of $\mathbf{F}_2$, since $\mathbf{F}_2$ is greater than $\mathbf{F}_1$. Thus **F** is the resultant of $\mathbf{F}_1$ and $\mathbf{F}_2$.

Problem 3.105 Determine the resultant force of two forces of 5 kN and 8 kN,

(a) acting in the same direction and having the same line of action,

(b) acting in opposite directions but having the same line of action.

(a) The vector diagram of the two forces acting in the same direction is shown in Figure 3.16(a) which assumes that

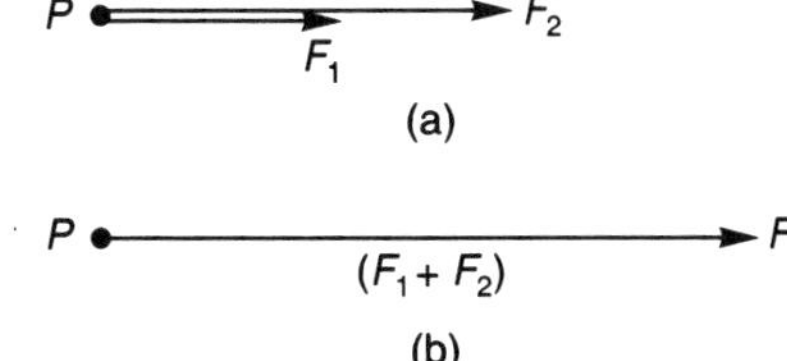

Figure 3.14

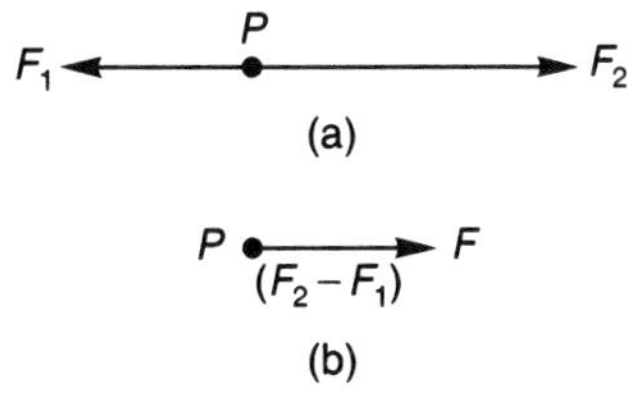

Figure 3.15

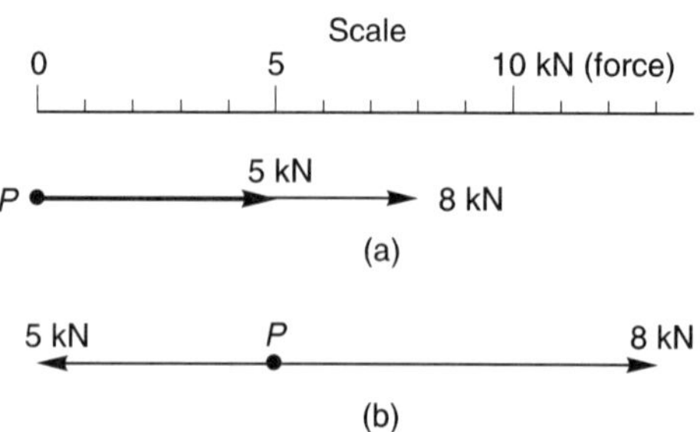

Figure 3.16

the line of action is horizontal, although since it is not specified, could be in any direction. From above, the resultant force **F** is given by:

$\mathbf{F} = \mathbf{F_1} + \mathbf{F_2}$, i.e. $\mathbf{F} = (\mathbf{5}+\mathbf{8})$ kN = **13 kN** in the direction of the original forces.

(b) The vector diagram of the two forces acting in opposite directions is shown in Figure 3.16(b), again assuming that the line of action is in a horizontal direction. From above, the resultant force **F** is given by:

$\mathbf{F} = \mathbf{F_2} - \mathbf{F_1}$, i.e. $\mathbf{F} = (\mathbf{8}-\mathbf{5})$ kN = **3 kN** in the direction of the 8 kN force.

(c) When two forces do not have the same line of action, the magnitude and direction of the resultant force may be found by a procedure called vector addition of forces. There are two graphical methods of performing **vector addition**, known as the **triangle of forces** method and the **parallelogram of forces** method.

Triangle of forces method

A simple procedure for the triangle of forces method of vector addition is as follows:

(i) Draw a vector representing one of the forces, using an appropriate scale and in the direction of its line of action.

(ii) From the nose of this vector and using the same scale, draw a vector representing the second force in the direction of its line of action.

(iii) The resultant vector is represented in both magnitude and direction by the vector drawn from the tail of the first vector to the nose of the second vector.

Problem 3.106 Determine the magnitude and direction of the resultant of a force of 15 N acting horizontally to the right and a force of 20 N, inclined at an angle of 60° to the 15 N force. Use the triangle of forces method.

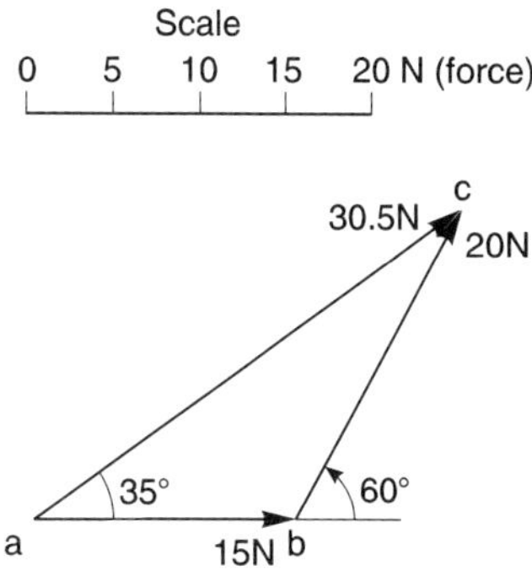

Figure 3.17

Using the procedure given above and with reference to Figure 3.17

(i) **ab** is drawn 15 units long horizontally

(ii) From *b*, **bc** is drawn 20 units long, inclined at an angle of 60° to **ab**. (Note, in angular measure, an angle of 60° from **ab** means 60° is in an anticlockwise direction.)

(iii) By measurement, the resultant **ac** is 30.5 units long inclined at an angle of 35° to **ab**. That is, the resultant force is **30.5 N**, inclined at an angle of 35° to the 15 N force.

The parallelogram of forces method

A simple procedure for the parallelogram of forces method of vector addition is as follows:

(i) Draw a vector representing one of the forces, using an appropriate scale and in the direction of its line of action.

(ii) From the tail of this vector and using the same scale draw a vector representing the second force in the direction of its line of action.

(iii) Complete the parallelogram using the two vectors drawn in (i) and (ii) as two sides of the parallelogram.

(iv) The resultant force is represented in both magnitude and direction by the vector corresponding to the diagonal of the parallelogram drawn from the tail of the vectors in (i) and (ii).

Problem 3.107 Use the parallelogram of forces method to find the magnitude and direction of the resultant of a force of 250 N acting at an angle of 135° and a force of 400 N acting at an angle of −120°.

From the procedure given above and with reference to Figure 3.18:

(i) **ab** is drawn at an angle of 135° and 250 units in length

Test your knowledge 3.18

1 Using the triangle of forces method, find the magnitude and direction of the two forces given.
First force: 1.5 kN acting at an angle of 30°
Second force: 3.7 kN acting at an angle of −45°

2 Determine the magnitude and direction of the resultant of forces of 23.8 N at −50° and 14.4 N at 215° using the parallelogram of forces method.

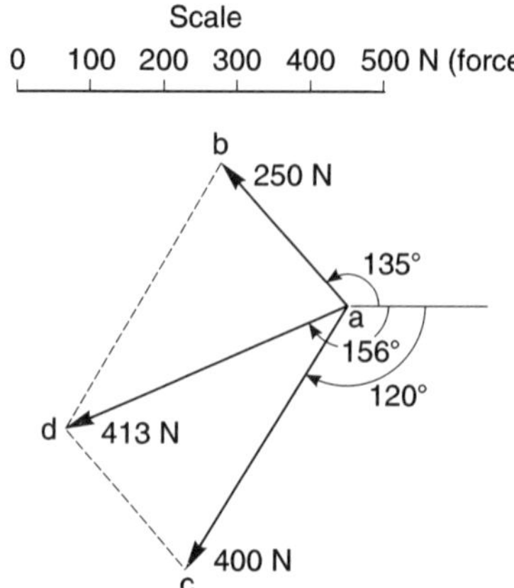

Figure 3.18

(ii) **ac** is drawn at an angle of −120° and 400 units in length

(iii) **bd** and **cd** are drawn to complete the parallelogram

(iv) **ad** is drawn. By measurement, **ad** is 413 units long at an angle of −156°. That is, the resultant force is **413 N** at an angle of **−156°**.

Resultant of more than two coplanar forces

For the three coplanar forces $\mathbf{F_1}$, $\mathbf{F_2}$ and $\mathbf{F_3}$ acting at a point as shown in Figure 3.19, the vector diagram is drawn using the nose-to-tail method. The procedure is:

(i) Draw **oa** to scale to represent force $\mathbf{F_1}$ in both magnitude and direction (see Figure 3.20).

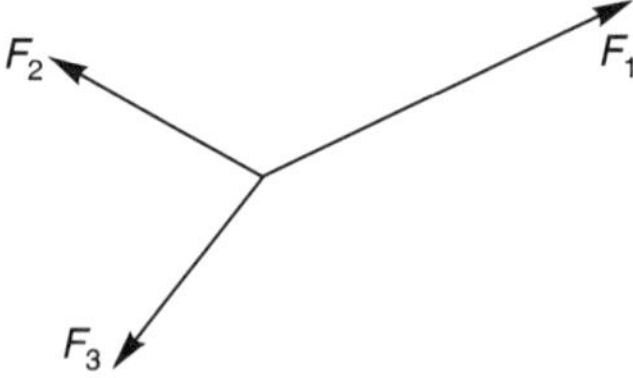

Figure 3.19

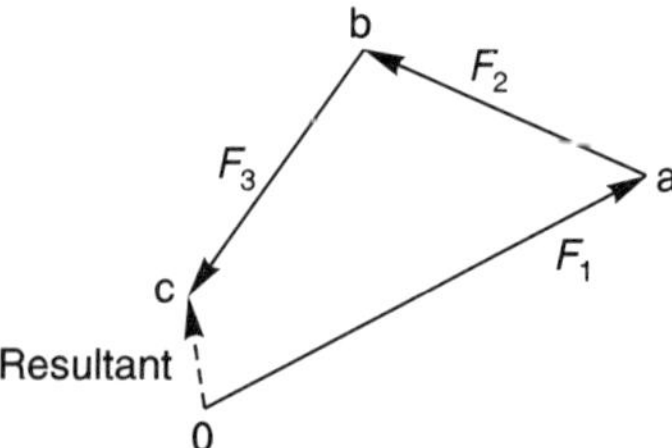

Figure 3.20

(ii) From the nose of **oa**, draw **ab** to represent force $\mathbf{F_2}$.

(iii) From the nose of **ab**, draw **bc** to represent force $\mathbf{F_3}$.

(iv) The resultant vector is given by length **oc** in Figure 3.20. The direction of resultant **oc** is from where we started, i.e. point **o**, to where we finished, i.e. point **c**. When acting by itself, the resultant force, given by **oc**, has the same effect on the point as force $\mathbf{F_1}$, $\mathbf{F_2}$ and $\mathbf{F_3}$ have when acting together. The resulting vector diagram of Figure 3.20 is called the **polygon of forces**.

Problem 3.108 Determine graphically the magnitude and direction of the resultant of these three coplanar forces, which may be considered as acting at a point. Force **A**, 12 N acting horizontally to the right; force **B**, 7 N inclined at 60° to force **A**; force **C**, 15 N inclined at 150° to force **A**.

The space diagram is shown in Figure 3.21. The vector diagram (Figure 3.22) is produced as follows:

(i) **oa** represents the 12 N force in magnitude and direction

(ii) from the nose of **oa**, **ab** is drawn inclined at 60° to **oa** and 7 units long

(iii) from the nose of **ab**, **bc** is drawn 15 units long inclined at 150° to **oa** (i.e. 150° to the horizontal)

(iv) **oc** represents the resultant. By measurement, the resultant is 13.8 N inclined at 80° to the horizontal.

Thus the resultant of the three forces, $\mathbf{F_A}$, $\mathbf{F_B}$ and $\mathbf{F_C}$ is a force of 13.8 N at 80° to the horizontal.

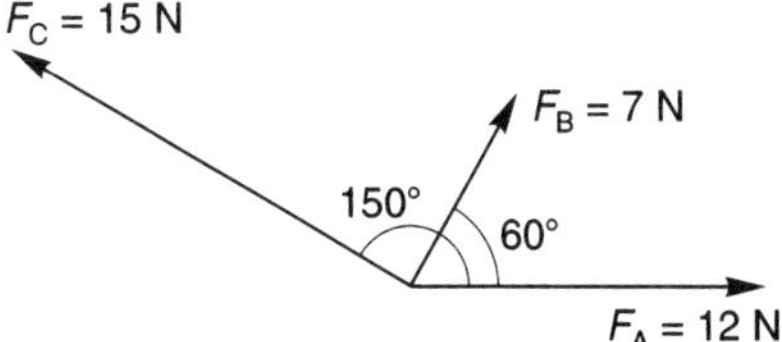

Figure 3.21

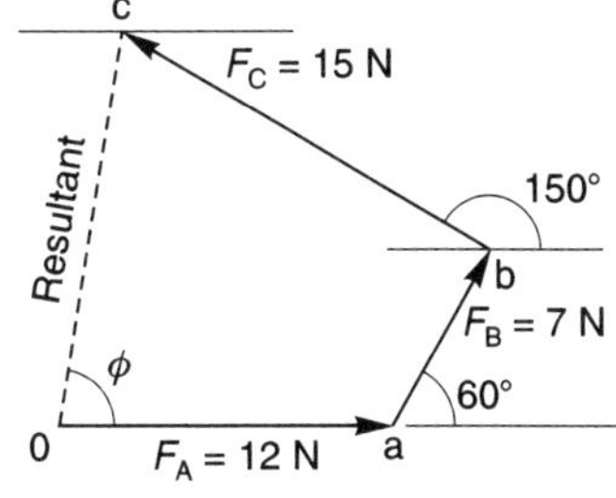

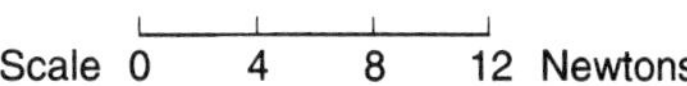

Figure 3.22

Coplanar forces in equilibrium

When three or more coplanar forces are acting at a point and the vector diagram closes, there is no resultant. The forces at the point are in **equilibrium**.

Problem 3.109 A load of 200 N is lifted by two ropes connected to the same point on the load, making angles of 40° and 35° with the vertical. Determine graphically the tensions in each rope when the system is in **equilibrium**.

The space diagram is shown in Figure 3.23. Since the system is in equilibrium, the vector diagram must close. The vector diagram (Figure 3.24) is drawn as follows:

(i) The load of 200 N is drawn vertically as shown by **oa**.

(ii) The direction only of force $\mathbf{F_1}$ is known, so from point **a**, **ad** is drawn 40° to the vertical.

(iii) The direction only of force $\mathbf{F_2}$ is known, so from point **o**, **oc** is drawn at 35° to the vertical.

Test your knowledge 3.19

1 Calculate the magnitude and direction of the resultant of the two forces 1.7 N at 45° and 2.4 N at −60°.

2 Determine the magnitude and direction of the resultant of the following coplanar forces:

force 1, 23 kN acting at 80° to the horizontal
force 2, 30 kN acting at 37° to force 1,
force 3, 15 kN acting at 70° to force 2.

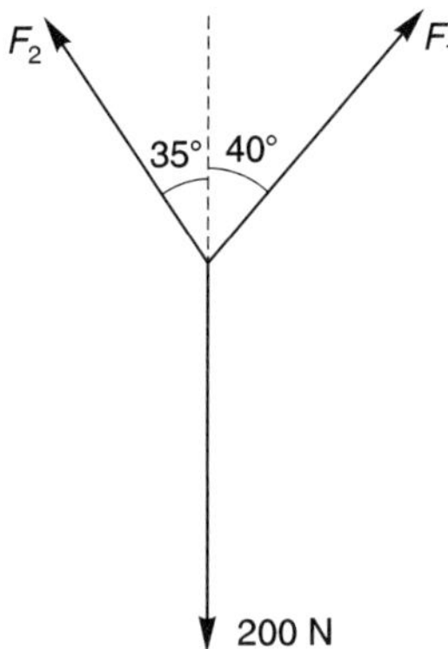

Figure 3.23

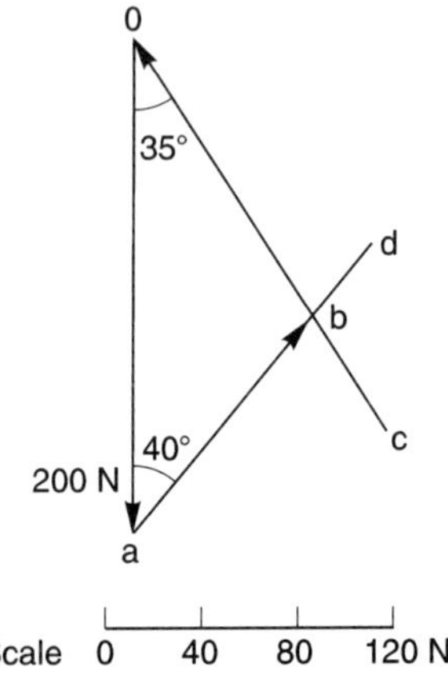

Figure 3.24

(iv) Lines **ad** and **oc** cross at point b. Hence the vector diagram is given by triangle oab. By measurement, **ab** is 119 N and **ob** is 133 N.

Thus the tensions in the ropes are $F_1 = 119$ N and $F_2 = 133$ N.

Speed

Speed is the rate of covering distance and is given by:

$$\text{speed} = \frac{\text{distance travelled}}{\text{time taken}}$$

The usual units for speed are metres per second (m/s or m s^{-1}), or kilometres per hour (km/h or km h^{-1}). Thus if a person walks 5 km in 1 h, the speed of the person is 5/1, that is 5 km/h. The symbol for the SI unit of speed (and velocity) is written as m s^{-1}, called the 'index notation'. However, engineers usually use the symbol m/s, called the 'oblique notation'.

Problem 3.110 A man walks 600 m in 5 min. Determine his speed in

(a) metres per second and (b) kilometres per hour.

$$\text{(a) Speed} = \frac{\text{distance travelled}}{\text{time taken}} = \frac{600\text{ m}}{5\text{ min}} = \frac{600\text{ m}}{5\text{ min}} \times \frac{1\text{ min}}{60\text{ s}} = \mathbf{2\ m/s}$$

$$\text{(b)} \quad \frac{2\text{ m}}{1\text{ s}} = \frac{2\text{ m}}{1\text{ s}} \times \frac{1\text{ km}}{1000\text{ m}} \times \frac{3600\text{ s}}{1\text{ h}} = \mathbf{7.2\ km/h}$$

(Note: to change from m/s to km/h, multiply by 3.6.)

Problem 3.111 A car travels at 50 km/h for 24 min. Find the distance travelled in this time.

$$\text{Since speed} = \frac{\text{distance travelled}}{\text{time taken}}$$

then, distance travelled = speed × time taken
Time = 24 min = (24/60) h, hence

$$\text{distance travelled} = 50\frac{\text{km}}{\text{h}} \times \frac{24}{60}\text{ h} = \mathbf{20\ km}$$

Test your knowledge 3.20

1 A train is travelling at a constant speed of 25 metres per second for 16 kilometres. Find the time taken to cover this distance.

2 A train covers a distance of 96 km in $1\frac{1}{3}$ h. Determine the average speed of the train (a) in km/h and (b) in m/s.

Distance/time graph

One way of giving data on the motion of an object is graphically. A graph of distance travelled (the scale on the vertical axis of the graph) against time (the scale on the horizontal axis of the graph) is called **a distance/time graph**. Thus if an aeroplane travels 500 kilometres in its first hour of flight and 750 km in its second hour of flight, then after 2 h, the total distance travelled is (500 + 750) km, that is, 1250 km. The distance/time graph for this flight is shown in Figure 3.25.

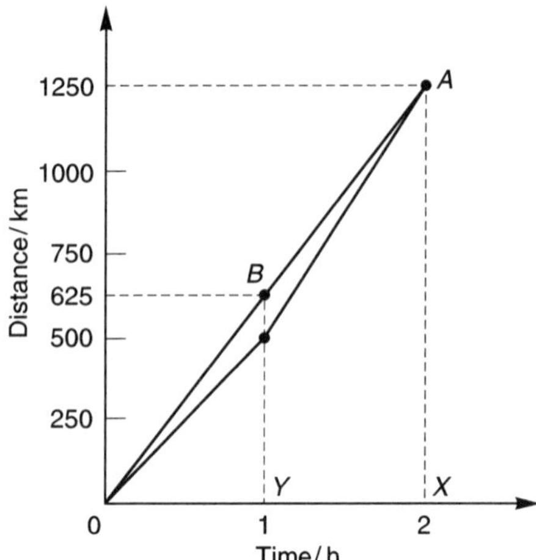

Figure 3.25

The average speed is given by

$$\frac{\text{total distance travelled}}{\text{total time taken}}$$

Thus, the average speed of the aeroplane is:

$$\frac{(500 + 750)\ \text{km}}{(1 + 1)\ \text{h}}, \text{ i.e. } \frac{1250}{2} \text{ or } 625\ \text{km/h}$$

If points O and A are joined in Figure 3.25, the slope of line OA is defined as

$$\frac{\text{change in distance (vertical)}}{\text{change in time (horizontal)}}$$

for any two points on line OA. For point A, the change in distance is AX, that is, 1250 km, and the change in time is OX, that is, 2 h. Hence the average speed is 1250/2, i.e. 625 km/h.

Alternatively, for point B on line OA, the change in distance is BY, that is, 625 km and the change in time is OY, that is 1 h, hence the average speed is 625/1, i.e. 625 km/h.

In general, the average speed of an object travelling between points M and N is given by the slope of line MN on the distance/time graph.

Problem 3.112 A person travels from point O to A, then from A to B and finally from B to C. The distances of A, B and C from O and the times, measured from the start to reach A, B and C are as shown:

	A	B	C
Distance (m)	100	200	250
Time (s)	40	60	100

Plot the distance/time graph and determine the speed of travel for each of the three parts of the journey.

The vertical scale of the graph is distance travelled and the scale is selected to span 0–250 m, the total distance travelled from the start. The horizontal scale is time and spans 0–100 s, the total time taken to cover the whole journey. Co-ordinates

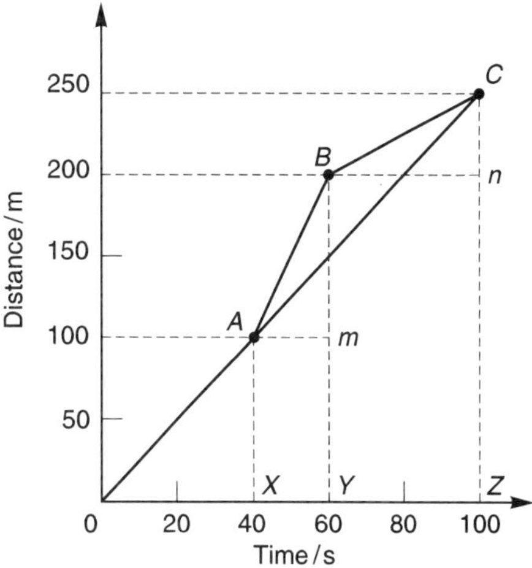

Figure 3.26

corresponding to A, B and C are plotted and OA, AB and BC are joined by straight lines. The resulting distance/time graph is shown in Figure 3.26.

The speed is given by the slope of the distance/time graph. Speed for part OA of the journey = slope of OA = AX/OX

$$= \frac{100\,\text{m}}{40\,\text{s}} = \mathbf{2\frac{1}{2}\,m/s}$$

Speed for part AB of the journey = slope of $AB = Bm/Am$

$$= \frac{(200 - 100)\,\text{m}}{(60 - 40)\,\text{s}} = \frac{100\,\text{m}}{20\,\text{s}} = \mathbf{5\,m/s}$$

Speed for part BC of the journey = slope of $BC = Cn/Bn$

$$= \frac{(250 - 200)\,\text{m}}{(100 - 60)\,\text{s}} = \frac{50\,\text{m}}{40\,\text{s}} = \mathbf{1\frac{1}{4}\,m/s}$$

Problem 3.113 Determine the average speed (both in m/s and km/h) for the whole journey for the information given in Problem 3.112.

Average speed = (total distance travelled)/(total time taken)
= slope of line OC.

From Figure 3.26,

$$\text{slope of line } OC = \frac{Cz}{Oz} = \frac{250\,\text{m}}{100\,\text{s}} = \mathbf{2.5\,m/s}$$

$$2.5\,\text{m/s} = \frac{2.5\,\text{m}}{1\,\text{s}} \times \frac{1\,\text{km}}{1000\,\text{m}} \times \frac{3600\,\text{s}}{1\,\text{h}}$$
$$= 2.5 \times 3.6\ \text{km/h} = \mathbf{9\,km/h}$$

Speed/time graph

If a graph is plotted of speed against time, **the area under the graph gives the distance travelled**. This is demonstrated in Problem 3.114.

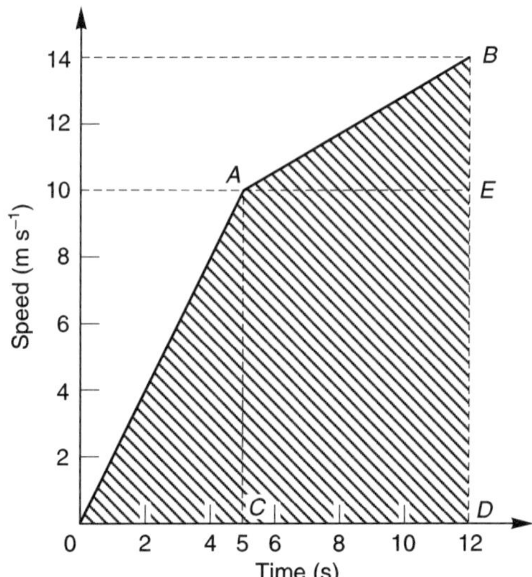

Figure 3.27

Problem 3.114 The motion of an object is described by the speed/time graph given in Figure 3.27. Determine the distance covered by the object when moving from O to B.

The distance travelled is given by the area beneath the speed/time graph, shown shaded in Figure 3.27.

Area of triangle OAC

$$= \frac{1}{2} \times \text{base} \times \text{perpendicular height}$$

$$= \frac{1}{2} \times 5\,\text{s} \times 10\,\frac{\text{m}}{\text{s}} = 25\,\text{m}$$

Area of rectangle AEDC = base × height

$$= (12 - 5)\,\text{s} \times (10 - 0)\,\frac{\text{m}}{\text{s}} = 70\,\text{m}$$

Area of triangle ABE

$$= \frac{1}{2} \times \text{base} \times \text{perpendicular height}$$

$$= \frac{1}{2} \times (12 - 5)\,\text{s} \times (14 - 10)\,\frac{\text{m}}{\text{s}}$$

$$= \frac{1}{2} \times 7\,\text{s} \times 4\,\frac{\text{m}}{\text{s}} = 14\,\text{m}$$

Hence the distance covered by the object moving from O to B is $(25 + 70 + 14)$ m, i.e. **109 m**.

Velocity

The velocity of an object is the speed of the object **in a specified direction**. Thus if a plane is flying due south at 500 km/h, its speed is 500 km/h, but its velocity is 500 km/h **due south**. It follows that if the plane had flown in a circular path for one hour at a speed of 500 km/h, so that 1 h after taking off it is again over the airport, its average velocity in the first hour of flight is zero.

The average velocity is given by:

$$\frac{\text{distance travelled in a specific direction}}{\text{time taken}}$$

If a plane flies from place O to place A, a distance of 300 km in 1 h, A being due north of O, then OA in Figure 3.28 represents the first hour of flight. It then flies from A to B, a distance of 400 km during the second hour of flight, B being due east of A, thus AB in Figure 3.28 represents its second hour of flight. Its average velocity for the 2 h flight is

$$\frac{\text{distance OB}}{\text{2 hours}}, \text{ i.e. } \frac{500\,\text{km}}{2\,\text{h}}$$

or 250 km/h in direction OB.

A graph of velocity (scale on the vertical axis) against time (scale on the horizontal axis) is called a **velocity/time graph**. The graph shown in Figure 3.29 represents a plane flying for 3 h at a constant speed of 600 km/h in a specified direction. The shaded area represents velocity (vertically) multiplied by time (horizontally), and has units of

$$\frac{\text{kilometres}}{\text{hours}} \times \text{hours}$$

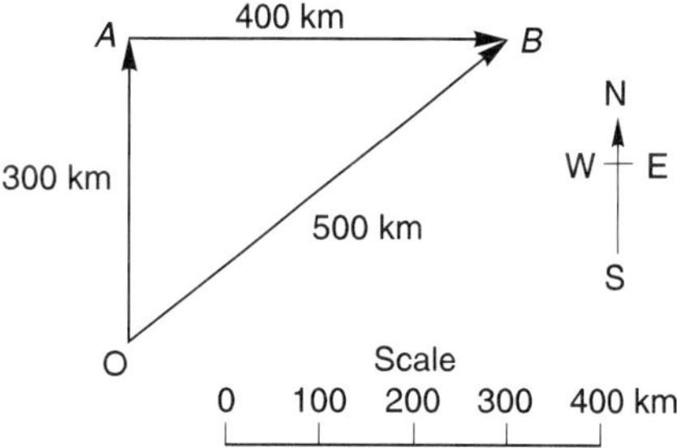

Figure 3.28

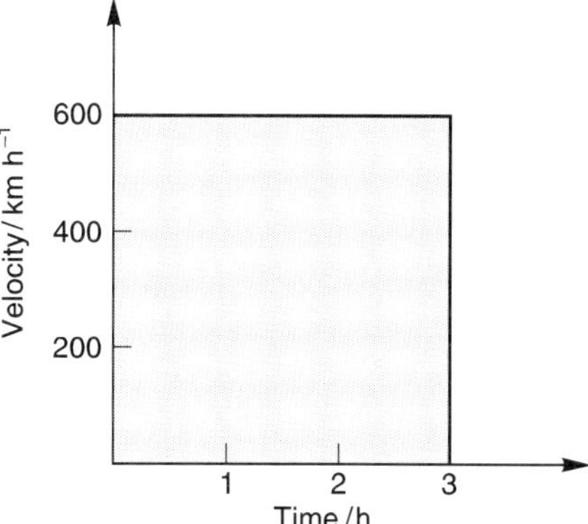

Figure 3.29

Test your knowledge 3.21

A coach travels from town A to town B, a distance of 40 km at an average speed of 55 km/h. It then travels from town B to town C, a distance of 25 km in 35 min. Finally, it travels from town C to town D at an average speed of 60 km/h in 45 min. Determine:

(a) the time taken to travel from A to B,

(b) the average speed of the coach from B to C,

(c) the distance from C to D, and

(d) the average speed of the whole journey from A to D.

i.e. kilometres, and represents the distance travelled in a specified direction. In this case,

$$\text{distance} = 600\,\frac{\text{km}}{\text{h}} \times (3\text{h}) = 1800\,\text{km}$$

Another method of determining the distance travelled is from:

$$\textbf{distance travelled} = \textbf{average velocity} \times \textbf{time}$$

Thus if a plane travels due south at 600 kilometres per hour for 20 minutes, the distance covered is

$$\frac{600\,\text{km}}{1\,\text{h}} \times \frac{20}{60}\,\text{h}, \text{ i.e. } 200\,\text{km}$$

Introduction to acceleration

Acceleration is the rate of change of velocity with time. The average acceleration, a, is given by:

$$a = \frac{\text{change in velocity}}{\text{time taken}}$$

The usual units arc metres per second squared (m/s^2 or m s^{-2}). If u is the initial velocity of an object in metres per second, v is the final velocity in metres per second and t is the time in seconds elapsing between the velocities of u and v, then

$$\textbf{average acceleration, } \boldsymbol{a} = \frac{\boldsymbol{v - u}}{\boldsymbol{t}}\ \text{m/s}^2$$

Velocity/time graph

A graph of velocity (scale on the vertical axis) against time (scale on the horizontal axis) is called a **velocity/time graph**, as introduced above. From the velocity/time graph shown in Figure 3.30, the slope of line OA is given by AX/OX. AX is the change in velocity from an initial velocity u of zero to a final velocity, v, of 4 m/s. OX is the time taken for this change in velocity, thus

$$\frac{AX}{OX} = \frac{\text{change in velocity}}{\text{time taken}}$$
$$= \text{the acceleration in the first 2 s}$$

From the graph:

$$\frac{AX}{OX} = \frac{4\,\text{m/s}}{2\,\text{s}} = 2\,\text{m/s}^2$$

i.e. the acceleration is 2 m/s^2. Similarly, the slope of linc AB in Figure 3.30 is given by BY/AY, i.e. the acceleration between 2 s and 5 s is

$$\frac{8-4}{5-2} = \frac{4}{3} = 1\frac{1}{3}\,\text{m/s}^2$$

In general, the slope of a line on a velocity/time graph gives the acceleration. The words ‘velocity’ and ‘speed’ are commonly interchanged in everyday language. Acceleration is a vector

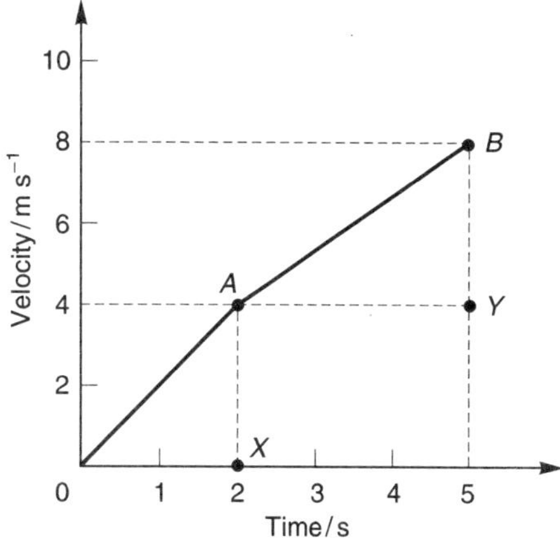

Figure 3.30

quantity and is correctly defined as the rate of change of velocity with respect to time. However, acceleration is also the rate of change of speed with respect to time in a certain specified direction.

Problem 3.115 The speed of a car travelling along a straight road changes uniformly from zero to 50 km/h in 20 s. It then maintains this speed for 30 s and finally reduces speed uniformly to rest in 10 s. Draw the speed/time graph for this journey.

The vertical scale of the speed/time graph is speed (km h^{-1}) and the horizontal scale is time (s). Since the car is initially at rest, then at time 0 s, the speed is 0 km/h. After 20 s, the speed is 50 km/h, which corresponds to point A on the speed/time graph shown in Figure 3.31. Since the change in speed is uniform, a straight line is drawn joining points O and A. The speed is constant at 50 km/h for the next 30 s, hence, horizontal line AB is drawn in Figure 3.31 for the time period 20 s to 50 s. Finally, the speed falls from 50 km/h at 50 s to zero in 10 s, hence point C on the speed/time graph in Figure 3.31 corresponds to a speed of

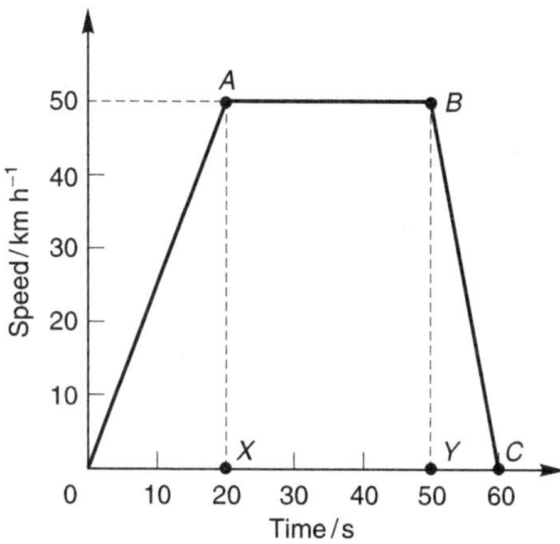

Figure 3.31

zero and a time of 60 s. Since the reduction in speed is uniform, a straight line is drawn joining BC. Thus, the speed/time graph for the journey is as shown in Figure 3.31.

Problem 3.116 For the speed/time graph shown in Figure 3.31, find the acceleration for each of the three stages of the journey.

From above, the slope of line OA gives the uniform acceleration for the first 20 s of the journey

$$\text{Slope of OA} = \frac{AX}{OX} = \frac{(50-0)\ \text{km/h}}{(20-0)\ \text{s}}$$
$$= 50\ \frac{\text{km}}{\text{h}} \times \frac{1\ \text{h}}{20\ \text{s}}$$

Expressing 50 km/h in metre-second units gives:

$$50\ \frac{\text{km}}{\text{h}} = \frac{50\ \text{km}}{1\ \text{h}} \times \frac{1000\ \text{m}}{1\ \text{km}} \times \frac{1\ \text{h}}{3600\ \text{s}}$$
$$= \frac{50}{3.6}\ \text{m/s}$$

(Note: to change from km/h to m/s, divide by 3.6.)
Thus,

$$50\ \text{km/h} \times \frac{1}{20\ \text{s}} = \frac{50}{3.6}\ \text{m/s} \times \frac{1}{20\ \text{s}}$$
$$= 0.694\ \text{m/s}^2$$

i.e. the acceleration during the first 20 s is **0.694 m/s²**.
Acceleration is defined as

$$\frac{\text{change of velocity}}{\text{time taken}} \text{ or } \frac{\text{change of speed}}{\text{time taken}}$$

since the car is travelling along a straight road. Since there is no change in speed for the next 30 s (line AB in Figure 3.31 is horizontal), **then the acceleration for this period is zero**.

From the above, the slope of line BC gives the uniform deceleration for the final 10 s of the journey.

$$\text{Slope of BC} = \frac{BY}{YC} = \frac{50\ \text{km/h}}{10\ \text{s}}$$
$$= \frac{50\ \text{m}}{3.6\ \text{s}} \times \frac{1}{10\ \text{s}}$$
$$= 1.39\ \text{m/s}^2$$

i.e. the deceleration during the final 10 s is 1.39 m/s².
Alternatively, the **the acceleration is −1.39 m/s².**

Free-fall and equation of motion

If a dense object such as a stone is dropped from a height, called **free-fall**, it has a constant acceleration of approximately 9.81 m/s². In a vacuum, all objects have this same constant acceleration; vertically downwards, that is, a feather has the same acceleration as a stone. However, if free-fall takes place in air, dense objects have the constant acceleration of 9.81 m/s² over short distances, but objects which have a low density, such as feathers, have little or no acceleration.

For bodies moving with a constant acceleration, the average acceleration is the constant value of the acceleration, and since from earlier:

$$a = \frac{v - u}{t}$$

then $a \times t = v - u$ or $v = u + at$

where u is the initial velocity in m/s,
v is the final velocity in m/s,
a is the constant acceleration in m/s^2,
t is the time in s.

When symbol 'a' has a negative value, it is called **deceleration** or **retardation**. The equation $v = u + at$ is called an **equation of motion**.

Problem 3.117 A stone is dropped from an aeroplane. Determine (a) its velocity after 2 s and (b) the increase in velocity during the third second, in the absence of all forces except that due to gravity.

The stone is free-falling and thus has an acceleration, a, of approximately 9.81 m/s^2 (taking downward motion as positive). From above:

$$\text{final velocity, } v = u + at$$

(a) The initial downward velocity of the stone, u, is zero. The acceleration, a, is 9.81 m/s^2 downwards and the time during which the stone is accelerating is 2 s. Hence, final velocity, $v = 0 + 9.81 \times 2 = 19.62$ m/s, i.e. **the velocity of the stone after 2 s is approximately 19.62 m/s**.

(b) From part (a), the velocity after 2 s, u, is 19.62 m/s. The velocity after 3 s, applying $v = u + at$, is

$$v = 19.62 + 9.81 \times 3 = 49.05 \text{ m/s.}$$

Thus, **the change in velocity during the third second is**

$$(49.05 - 19.62) \text{ m/s, that is } \mathbf{29.43\ m/s}$$

Problem 3.118 A train travelling at 30 km/h accelerates uniformly to 50 km/h in 2 min. Determine the acceleration.

$$30 \text{ km/h} = \frac{30}{3.6} \text{ m/s (see Problem 3.116)}$$

$$50 \text{ km/h} = \frac{50}{3.6} \text{ m/s}$$

$$2 \text{ min} = 2 \times 60 = 120 \text{ s}$$

From above, $v = u + at$, i.e.

$$\frac{50}{3.6} = \frac{50}{3.6} + a \times 120$$

Test your knowledge 3.22

1. A ship changes velocity from 15 km/h to 20 km/h in 25 min. Determine the average acceleration in m/s^2 of the ship during this time.
2. Determine how long it takes an object, which is free-falling, to change its speed from 100 km/h to 150 km/h, assuming all other forces, except that due to gravity, are neglected.
3. A car travelling at 50 km/h applies its brakes for 6 s and decelerates uniformly at 0.5 m/s. Determine its velocity in km/h after the 6 s braking period.

Transposing, gives

$$120 \times a = \frac{50 - 30}{3.6}$$

$$a = \frac{20}{3.6 \times 120} = 0.0463\,\mathrm{m/s^2}$$

i.e. **the uniform acceleration of the train is 0.0463 m/s^2.**

Introduction to force, mass and acceleration

When an object is pushed or pulled, a force is applied to the object. **This force is measured in newtons (N)**. The effects of pushing or pulling an object are:

(i) to cause a change in the motion of the object, and

(ii) to cause a change in the shape of the object.

If there is a change in the motion of the object, that is, its velocity changes from u to v, then the object accelerates. Thus, it follows that acceleration results from a force being applied to an object. If a force is applied to an object and it does not move, then the object changes shape, that is, deformation of the object takes place. Usually the change in shape is so small that it cannot be detected by just watching the object. However, when very sensitive measuring instruments are used, very small changes in dimensions can be detected.

A force of attraction exists between all objects. The factors governing the size of this force **F** are the masses of the objects and the distances between their centres.

Thus, if a person is taken as one object and the earth as a second object, a force of attraction exists between the person and the earth. This force is called the **gravitational force** and is the force which gives a person a certain weight when standing on the earth's surface. It is also this force which gives freely falling objects a constant acceleration in the absence of other forces.

Newton's laws of motion

To make a stationary object move or to change the direction in which the object is moving requires a force to be applied externally to the object. This concept is known as **Newton's first law of motion** and may be stated as:

An object remains in a state of rest, or continues in a state of uniform motion in a straight line, unless it is acted on by an externally applied force.

Since a force is necessary to produce a change in motion, an object must have some resistance to a change in its motion. The force necessary to give a stationary pram a given acceleration is far less than the force necessary to give a stationary car the same acceleration. The resistance to a change in motion is called the **inertia** of an object and the amount of inertia depends on the mass of the object. Since a car has a much larger mass than a pram, the inertia of a car is much larger than that of a pram.

Newton's second law of motion may be stated as:

The acceleration of an object acted upon by an external force is proportional to the force and is in the same direction as the force.

Thus, force $\propto$ acceleration, or force = a constant $\times$ acceleration, this constant of proportionality being the mass of the object,

i.e. **force = mass × acceleration**

The unit of force is the newton (N) and is defined in terms of mass and acceleration. One newton is the force required to give a mass of 1 kilogram an acceleration of 1 metre per second squared. Thus

$$\mathbf{F} = \boldsymbol{ma}$$

where **F** is the force in newtons (N), m is the mass in kilograms (kg) and a is the acceleration in metres per second squared (m/s^2), i.e.

$$1\,\text{N} = \frac{1\,\text{kg}\,\text{m}}{\text{s}^2}$$

It follows that $1\,\text{m/s}^2 = 1\,\text{N/kg}$. Hence a gravitational acceleration of 9.81 m/s^2 is the same as a gravitational field of 9.81 N/kg.

Newton's third law of motion may be stated as:

For every force, there is an equal and opposite reacting force.

Thus, an object on, say, a table, exerts a downward force on the table and the table exerts an equal upward force on the object, known as a **reaction force** or just a **reaction**.

Problem 3.119 Calculate the force needed to accelerate a boat of mass 20 t uniformly from rest to a speed of 21.6 km/h in 10 min.

The mass of the boat, m, is 20 t, that is 20 000 kg. The law of motion, $v = u + at$ can be used to determine the acceleration a. The initial velocity, u, is zero. The final velocity, v is 21.6 km/h, that is, 21.6/3.6 or 6 m/s. The time, t, is 10 min, that is, 600 s.

$$\text{Thus } 6 = 0 + a \times 600 \text{ or } a = \frac{6}{600} = \mathbf{0.01\,m/s^2}$$

From Newton's second law, $\mathbf{F} = \boldsymbol{ma}$, i.e.

$$\text{Force} = 20\,000 \times 0.01\,\text{N}$$
$$= \mathbf{200\,N}$$

Problem 3.120 The moving head of a machine tool requires a force of 1.2 N to bring it to rest in 0.8 s from a cutting speed of 30 m/min. Find the mass of the moving head.

From Newton's second law, $\mathbf{F} = ma$, thus $m = \mathbf{F}/a$, where the forc eis given as 1.2 N. The law of motion $v = u + at$ can be used to find acceleration a, where $v = 0$, $u = 30\,\text{m/min}$, that is 30/60 or $0.5\,\text{m/s}$, and $t = 0.8\,\text{s}$. Thus, $0 = 0.5 + a \times 0.8$,

$$\text{i.e. } a = -\frac{0.5}{0.8} = -0.625\,\text{m/s}^2$$

or a retardation of $0.625\,\text{m/s}^2$
Thus the mass, $m = 1.2/0.625 = \mathbf{1.92\,kg}$.

Problem 3.121 Find the weight of an object of mass 1.6 kg at a point on the earth's surface, where the gravitational field is 9.81 N/kg.

The weight of an object is the force acting vertically downwards due to the force of gravity acting on the object. Thus:

$$\begin{aligned}\text{Weight} &= \text{force acting vertically downwards}\\ &= \text{mass} \times \text{gravitational field}\\ &= 1.6 \times 9.81 = \mathbf{15.696\,N}\end{aligned}$$

Problem 3.122 A bucket of cement of mass 40 kg is tied to the end of a rope connected to a hoist. Calculate the tension in the rope when the bucket is suspended but stationary. Take the gravitational field, g, as 9.81 N/kg.

The tension in the rope is the same as the force acting in the rope. The force acting vertically downwards due to the weight of the bucket must be equal to the force acting upwards in the rope, i.e. the tension.

Weight of bucket of cement, $\mathbf{F} = mg = 40 \times 9.81 = 392.4\,\text{N}$. Thus, the tension in the rope is also **392.4 N**.

Problem 3.123 The bucket of cement in Problem 3.122 is now hoisted vertically upwards with a uniform acceleration of $0.4\,\text{m/s}^2$. Calculate the tension in the rope during the period of acceleration.

With reference to Figure 3.32(a), the forces acting on the bucket are:

(i) a tension (or force) of T acting in the rope;

(ii) a force of mg acting vertically downwards, i.e. the weight of the bucket and cement

The resultant force $\mathbf{F} = T - mg$. Hence $ma = T - mg$.

$$40 \times 0.4 = T - 40 \times 9.81 \text{ giving } \boldsymbol{T} = \mathbf{408.4\,N}$$

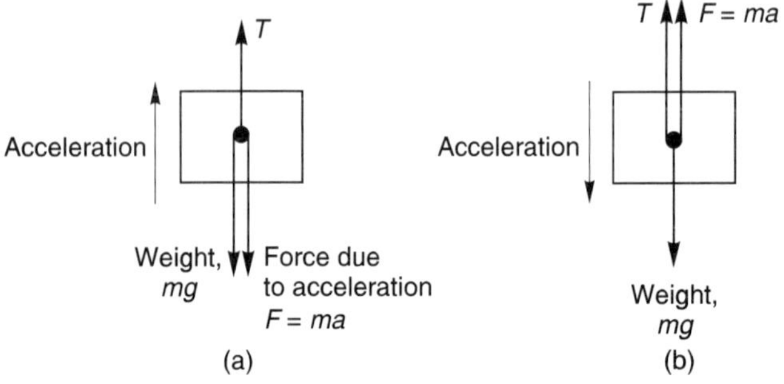

Figure 3.32

Test your knowledge 3.23

1. A lorry of mass 1350 kg accelerates uniformly from 9 km/h to reach a velocity of 45 km/h in 18 s. Determine (a) the acceleration of the lorry, (b) the uniform force needed to accelerate the lorry.
2. The tension in a rope lifting a crate vertically upwards is 2.8 kN. Determine its acceleration if the mass of the crate is 270 kg.

By comparing this result with that of Problem 3.122, it can be seen that there is an increase in the tension in the rope when an object is accelerating upwards.

The moment of a force

When using a spanner to tighten a nut, a force tends to turn the nut in a clockwise direction. This turning effect of a force is called the **moment of a force** or more briefly, just a **moment**. The size of the moment acting on the nut depends on two factors:

(a) the size of the force acting at right angles to the shank of the spanner, and

(b) the perpendicular distance between the point of application of the force and the centre of the nut.

In general, with reference to Figure 3.33, the moment M of a force acting about a point P is force × perpendicular distance between the line of action of the force and P, i.e.

$$M = \mathbf{F} \times d$$

The unit of a moment is the newton metre (N m). Thus, if force **F** in Figure 3.33 is 7 N and distance d is 3 m, then the moment M is 7 N × 3 m, i.e. 21 N m.

Problem 3.124 A force of 15 N is applied to a spanner at an effective length of 140 mm from the centre of a nut. Calculate (a) the moment of the force applied to the nut, (b) the magnitude of the force required to produce the same moment if the effective length is reduced to 100 mm.

From the above, $M = \mathbf{F} \times d$, where M is the turning moment, **F** is the force applied at right angles to the spanner and d is the effective length between the force and the centre of the nut. Thus, with reference to Figure 3.34(a):

(a) Turning moment,

$$M = 15\,\text{N} \times 140\,\text{mm} = 2100\,\text{N mm}$$

$$= 2100\,\text{N mm} \times \frac{1\,\text{m}}{1000\,\text{mm}} = \mathbf{2.1\,N\,m}$$

(b) Turning moment, M is 2100 N mm and the effective length d becomes 100 mm (see Figure 3.34(b)). Applying $M = \mathbf{F} \times d$ gives:

$$2100\,\text{N mm} = \mathbf{F} \times 100\,\text{mm}$$

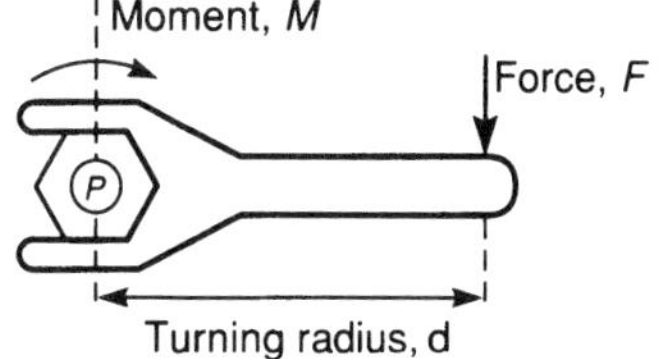

Figure 3.33

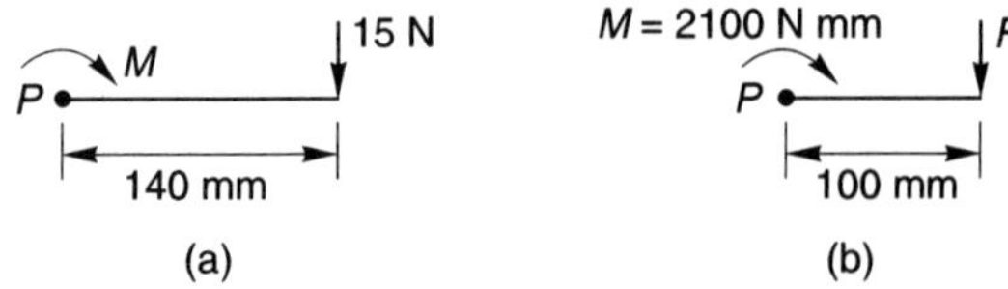

Figure 3.34

from which,

$$\text{force } \mathbf{F} = \frac{2100\,\text{N mm}}{100\,\text{mm}} = \mathbf{21\,N}$$

Problem 3.125 A moment of 25 N m is required to operate a lifting jack. Determine the effective length of the handle of the jack if the force applied to it is 125 N.

From the above, moment $M = \mathbf{F} \times d$, where **F** is the force applied at right angles to the handle and d is the effective length of the handle. Thus:
25 N m = 125 N, from which

$$\text{effective length, } d = \frac{25\,\text{N m}}{125\,\text{N}} = \frac{1}{5}\,\text{m}$$
$$= \frac{1000}{5}\,\text{mm}$$
$$= \mathbf{200\,mm}$$

Equilibrium and the principle of moments

If more than one force is acting on an object and the forces do not act at a single point, then the turning effect of the forces, that is, the moment of the forces, must be considered.

Figure 3.35 shows a beam with its support (known as its pivot or fulcrum) at P, acting vertically upwards, and forces $\mathbf{F}_1$ and $\mathbf{F}_2$ acting vertically downwards at distances a and b respectively, from the fulcrum.

A beam is said to be in **equilibrium** when there is no tendency for it to move. There are two conditions for equilibrium:

(i) The sum of the forces acting vertically downwards must be equal to the sum of the forces acting vertically upwards, i.e. for Figure 3.35,

$$\mathbf{R}_P = \mathbf{F}_1 + \mathbf{F}_2$$

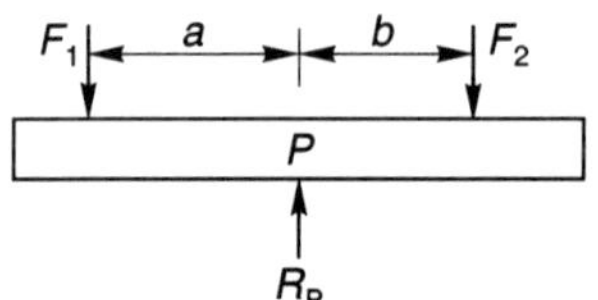

Figure 3.35

(ii) The total moment of the forces acting on a beam must be zero; for the total moment to be zero:

the sum of the clockwise moments about any point must be equal to the sum of the anticlockwise moments about that point.

This statement is known as the **principle of moments**. Hence, taking moments about P in Figure 3.35,

$$\mathbf{F}_2 \times b = \text{the clockwise moment, and}$$
$$\mathbf{F}_1 \times a = \text{the anticlockwise moment.}$$

Thus for equilibrium:

$$\mathbf{F}_1 \boldsymbol{a} = \mathbf{F}_2 \boldsymbol{b}$$

Problem 3.126 A system of forces is as shown in Figure 3.36.

(a) If the system is in equilibrium find the distance d.

(b) If the point of application of the 5 N force is moved to point P, distance 200 mm from the support, find the new value of **F** to replace the 5 N force for the system to be in equilibrium.

(a) From above, the clockwise moment M_1 is due to a force of 7 N acting at a distance d from the support, called the **fulcrum**, i.e.

$$M_1 = 7\,\text{N} \times d$$

The anticlockwise moment M_2 is due to a force of 5 N acting at a distance of 140 mm from the fulcrum, i.e.

$$M_2 = 5\,\text{N} \times 140\,\text{mm}$$

Applying the principle of moments, for the system to be in equilibrium about the fulcrum:
clockwise moment = anticlockwise moment, i.e.

$$7\,\text{N} \times d = 5 \times 140\,\text{N mm}$$

Hence

$$\text{distance, } d = \frac{5 \times 140\,\text{N mm}}{7\,\text{N}} = \mathbf{100\,mm}$$

(b) When the 5 N force is replaced by force **F** at a distance of 200 mm from the fulcrum, the new value of the anticlockwise moment is **F** × 200. For the system to be in equilibrium: clockwise moment = anticlockwise moment, i.e. (7 × 100) N mm = **F** × 200 mm

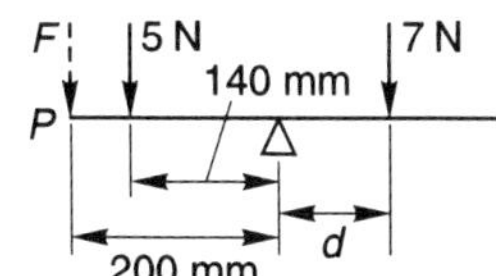

Figure 3.36

Hence,

$$\text{new value of force, } \mathbf{F} = \frac{700\,\text{N mm}}{200\,\text{mm}} = \mathbf{3.5\,N}$$

Problem 3.127 A beam is supported at its centre on a fulcrum and forces act as shown in Figure 3.37. Calculate (a) force **F** for the beam to be in equilibrium, (b) the new position of the 23 N force when **F** is decreased to 21 N, if equilibrium is to be maintained.

(a) The clockwise moment, M_1, is due to the 23 N force acting at a distance of 100 mm from the fulcrum, i.e.

$$M_1 = 23 \times 100 = 2300\,\text{N mm}$$

There are two forces giving the anticlockwise moment M_2. One is the force **F** acting at a distance of 20 mm from the fulcrum and the other a force of 12 N acting at a distance of 80 mm. Thus

$$M_2 = (\mathbf{F} \times 20) + (12 \times 80)\,\text{N mm}$$

Applying the principle of moments about the fulcrum: clockwise moment = anticlockwise moment, i.e.

$$2300 = (\mathbf{F} \times 20) + (12 \times 80)$$

Hence

$$\mathbf{F} \times 20 = 2300 - 960$$

i.e.

$$\text{force } \mathbf{F} = \frac{1340}{20} = \mathbf{67\,N}$$

(b) The clockwise moment is now due to a force of 23 N acting at a distance of, say, d from the fulcrum. Since the value of **F** is decreased to 21 N, the anticlockwise moment is

$(21 \times 20) + (12 \times 80)$ N mm.
Applying the principle of moments,

$$23 \times d = (21 \times 20) + (12 \times 80)$$

Test your knowledge 3.24

1 A moment of 7.5 N m is required to turn a wheel. If a force of 37.5 N is applied to the rim of the wheel, calculate the effective distance from the rim to the hub of the wheel.

2 For the centrally supported uniform beam shown in Figure 3.38 determine the values of forces $\mathbf{F}_1$ and $\mathbf{F}_2$ when the beam is in equilibrium.

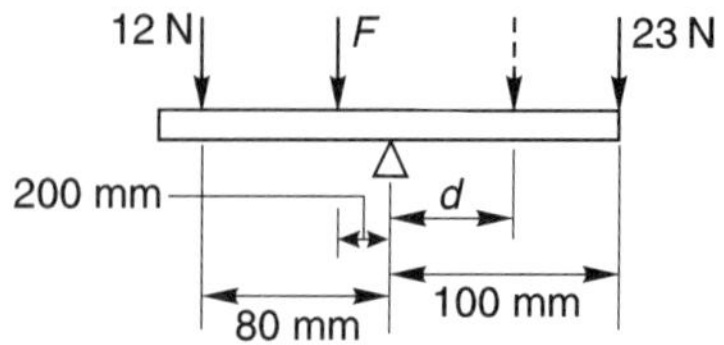

Figure 3.37

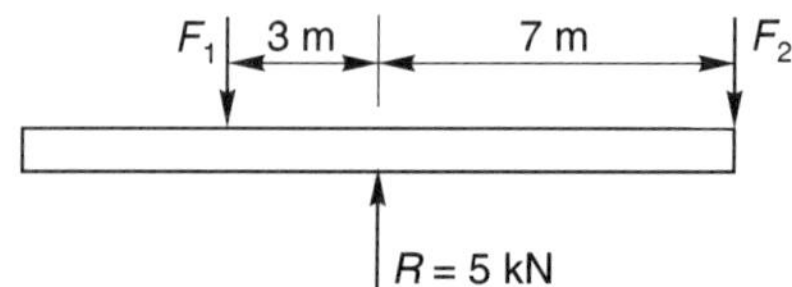

Figure 3.38

i.e.

$$\text{distance } d = \frac{420 + 960}{23} = \frac{1380}{23} = \mathbf{60\,mm}$$

Simply supported beams having point loads

A simply supported beam is one which rests on two supports and is free to move horizontally. Two typical simply supported beams having loads acting at given points on the beam (called **point loading**) are shown in Figure 3.39.

A man whose mass exerts a force **F** vertically downwards, standing on a wooden plank which is simply supported at its ends, may, for example, be represented by the beam diagram of Figure 3.39(a) if the mass of the plank is neglected.

The forces exerted by the supports on the plank, $\mathbf{R}_A$ and $\mathbf{R}_B$, act vertically upwards, and are called **reactions**. When the forces acting are all in one plane, the algebraic sum of the moments can be taken about any point.

For the beam in Figure 3.39(a) at equilibrium:

(i) $\mathbf{R}_A + \mathbf{R}_B = \mathbf{F}$, and

(ii) taking moments about A, $\mathbf{F}a = \mathbf{R}_B(a + b)$.

(Alternatively, taking moments about C, $\mathbf{R}_A a = \mathbf{R}_B b$).

For the beam in Figure 3.39(b), at equilibrium

(i) $\mathbf{R}_A + \mathbf{R}_B = \mathbf{F}_1 + \mathbf{F}_2$, and

(ii) taking moments about B, $\mathbf{R}_A(a + b) + \mathbf{F}_2 c = \mathbf{F}_1 b$.

Typical practical applications of simply supported beams with point loadings include bridges, beams in buildings, and beds of machine tools.

Problem 3.128 A beam is loaded as shown in Figure 3.40.

Determine (a) the force acting on the beam support at B, (b) the force acting on the beam support at A, neglecting the mass of the beam.

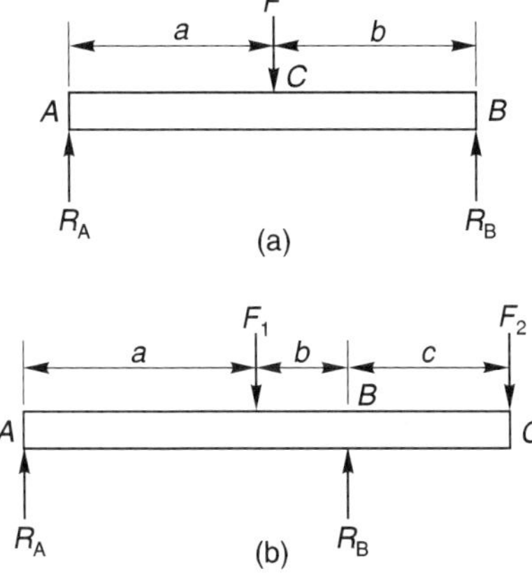

Figure 3.39

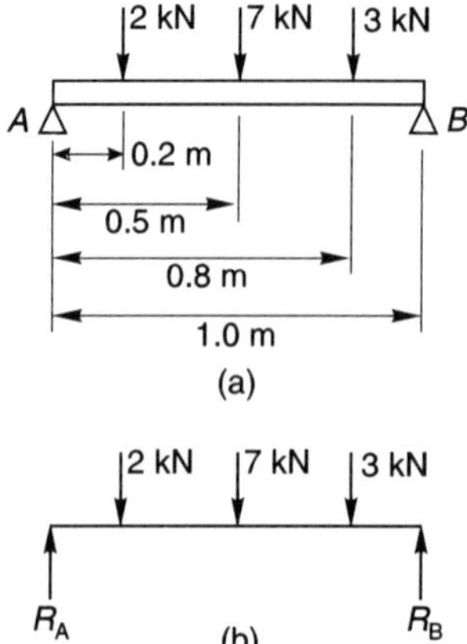

Figure 3.40

A beam supported as shown in Figure 3.40 is called a simply supported **beam**.

(a) Taking moments about point A and applying the principle of moments gives:
clockwise moments = anticlockwise moments
$(2 \times 0.2) + (7 \times 0.5) + (3 \times 0.8)$ kN m $= \mathbf{R_B} \times 1.0$ m, where $\mathbf{R_B}$ is the force supporting the beam at B, as shown in Figure 3.40(b).
Thus $(0.4 + 3.5 + 2.4)$ kN m $= \mathbf{R_B} \times 1.0$ m, i.e.

$$\mathbf{R_B} = \frac{6.3\,\text{kN m}}{1.0\,\text{m}} = \mathbf{6.3\,kN}$$

(b) For the beam to be in equilibrium, the forces acting upwards must be equal to the forces acting downwards, thus

$$\mathbf{R_A} + \mathbf{R_B} = (2 + 7 + 3)\,\text{kN}$$
$$\mathbf{R_B} = 6.3\text{kN, thus } \mathbf{R_A} = 12 - 6.3 = \mathbf{5.7\,kN}$$

Problem 3.129 For the beam shown in Figure 3.41 calculate (a) the force acting on support A, (b) distance *d*, neglecting any forces arising from the mass of the beam.

(a) From above,

(the forces actring in an upward direction) = (the forces acting in a downward direction)

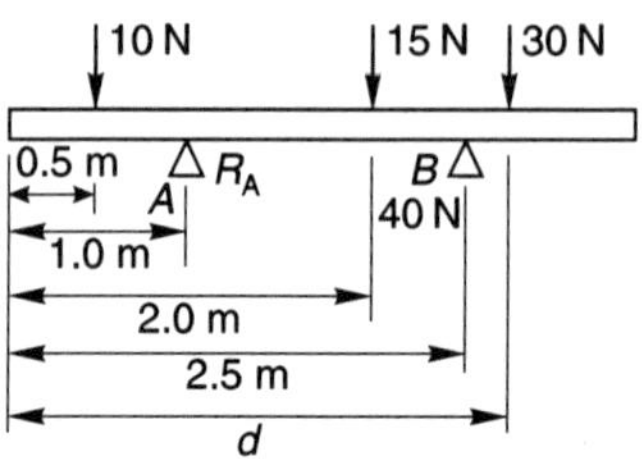

Figure 3.41

Test your knowledge 3.25

1. A metal bar AB is 4.0 m long and is supported at each end in a horizontal position. It carries loads of 2.5 kN and 5.5 kN at distances of 2.0 m and 3.0 m, respectively, from A. Neglecting the mass of the beam, determine the reactions of the supports when the beam is in equilibrium.

2. A simply supported beam AB is loaded as shown in Figure 3.42. Determine the load **F** in order that the reaction at A is zero.

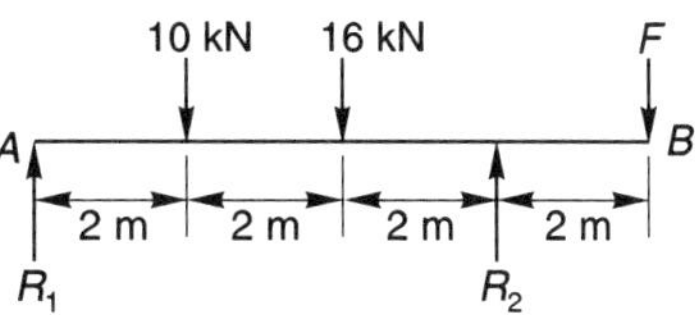

Figure 3.42

Hence

$$(\mathbf{R_A} + 40)\,\text{N} = (10 + 15 + 30)\,\text{N}$$
$$\mathbf{R_A} = 10 + 15 + 30 - 40 = \mathbf{15\,N}$$

(b) Taking moments about the left-hand end of the beam and applying the principle of moments gives:
clockwise moments = anticlockwise moments

$$(10 \times 0.5) + (15 \times 2.0) + 30\,\text{N} \times d = (15 \times 1.0) + (40 \times 2.5)$$

i.e.

$$35\,\text{N m} + 30\,\text{N} \times d = 115\,\text{N m}$$

from which

$$\text{distance, } d = \frac{(115 - 35)\,\text{N m}}{30\,\text{N}} = \mathbf{2.67\,m}$$

Introduction to friction

When an object, such as a block of wood, is placed on a floor and sufficient force is applied to the block, the force being parallel to the floor, the block slides across the floor. When the force is removed, motion of the block stops; thus there is a force which resists sliding. This force is called **dynamic** or **sliding friction**. A force may be applied to the block which is insufficient to move it. In this case, the force resisting motion is called the **static friction** or **striction**. Thus there are two categories into which a frictional force may be split:

(i) dynamic or sliding friction force which occurs when motion is taking place, and

(ii) static friction which occurs before motion takes place.

There are three factors which affect the size and direction of frictional forces.

(i) The size of the frictional force depends on the type of surface (a block of wood slides more easily on a polished metal surface than on a rough concrete surface).

(ii) The size of the frictional force depends on the size of the force acting at right angles to the surfaces in contact, called the **normal force**. Thus, if the weight of a block of wood is doubled, the frictional force is doubled when it is sliding on the same surface.

(iii) The direction of the frictional force is always opposite to the direction of motion. Thus the frictional force opposes motion, as shown in Figure 3.43.

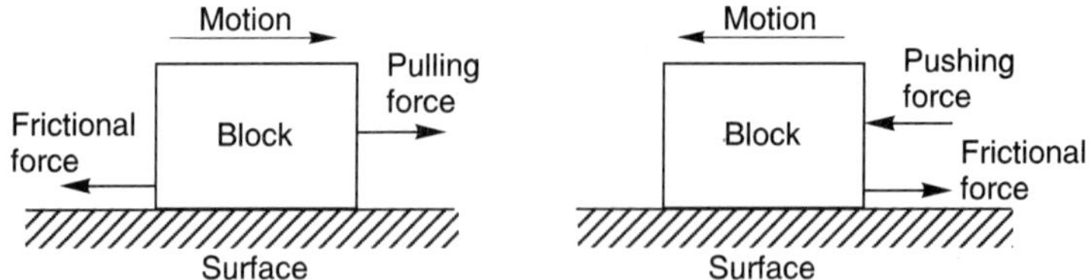

Figure 3.43

Coefficient of friction

The coefficient of friction μ, is a measure of the amount of friction existing between two surfaces. A low value of the coefficient of friction indicates that the force required for sliding to occur is less than the force required when the coefficient of friction is high. The value of the coefficient of friction is given by

$$\mu = \frac{\text{frictional force } (\mathbf{F})}{\text{normal force } (\mathbf{N})}$$

Transposing gives: frictional force $= \mu \times$ normal force,

$$\text{i.e. } \mathbf{F} \times \mu\mathbf{N}$$

The direction of the forces given in this equation is as shown in Fgure 3.44.

The coefficient of friction is the ratio of a force to a force, and hence has no units. Typical values for the coefficient of friction when sliding is occurring, i.e. the dynamic coefficient of friction, are:

For polished oiled metal surfaces	less than 0.1
For glass on glass	0.4
For rubber on tarmac	close to 1.0

Problem 3.130 A block of steel requires a force of 10.4 N applied parallel to a steel plate to keep it moving with constant velocity across the plate. If the normal force between the block and the plate is 40 N, determine the dynamic coefficient of friction.

As the block is moving at constant velocity, the force applied must be that required to overcome frictional forces, i.e.

$$\text{frictional force, } \mathbf{F} = 10.4\,\text{N}$$

The normal force is 40 N, and since $\mathbf{F} = \mu\mathbf{N}$,

$$\mu = \frac{\mathbf{F}}{\mathbf{N}} = \frac{10.4}{40} = 0.26$$

i.e. the dynamic coefficient of friction is 0.26

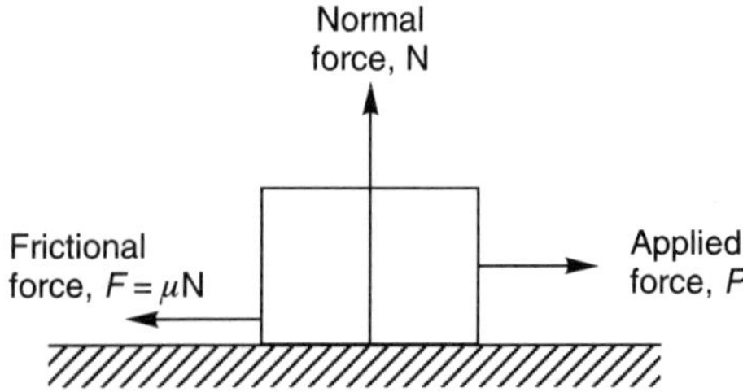

Figure 3.44

Problem 3.131 The surface between the steel block and plate of problem 3.130 is now lubricated and the dynamic coefficient of friction falls to 0.12. Find the new value of force required to push the block at a constant speed.

The normal force depends on the weight of the block and remains unaltered at 40 N. The new value of the dynamic coefficient of friction is 0.12 and since the frictional force $\mathbf{F} = \mu\mathbf{N}$, $\mathbf{F} = 0.12 \times 40 = 4.8$ N. The block is sliding at constant speed, thus the force required to overcome the frictional force is also 4.8 N, i.e. **the required applied force is 4.8 N**.

Applications of friction

In some applications, a low coefficient of friction is desirable, for example, in bearings, pistons moving within cylinders, on ski runs, and so on. However, for such applications as force being transmitted by belt drives and braking systems, a high value of the coefficient is necessary. Instances where frictional forces are an advantage include:

(i) Almost all fastening devices rely on frictional forces to keep them in place once secured, examples being screws, nails, nuts, clips and clamps.

(ii) Satisfactory operation of brakes and clutches relies on frictional forces being present.

(iii) In the absence of frictional forces, most accelerations along a horizontal surface are impossible. For example, a person's shoes just slip when walking is attempted and the tyres of a car just rotate with no forward motion of the car being experienced.

Test your knowledge 3.26

The material of a brake is being tested and it is found that the dynamic coefficient of friction between the material and steel is 0.91. Calculate the normal force when the frictional force is 0.728 kN.

Disadvantages of frictional forces include:

(i) Energy is wasted in the bearings associated with shafts, axles and gears due to heat being generated.

(ii) Wear is caused by friction, for example, in shoes, brake lining materials and bearings.

(iii) Energy is wasted when motion through air occurs (it is much easier to cycle with the wind rather than against it).

Work

If a body moves as a result of a force being applied to it, the force is said to do work on the body. The amount of work done is the product of the applied force and the distance, i.e.

work done = force × distance moved in the direction of the force

The unit of work is the **joule, J**, which is defined as the amount of work done when a force of 1 N acts for a distance of 1 m in the direction of the force. Thus,

$$1\,\text{J} = 1\,\text{N m}$$

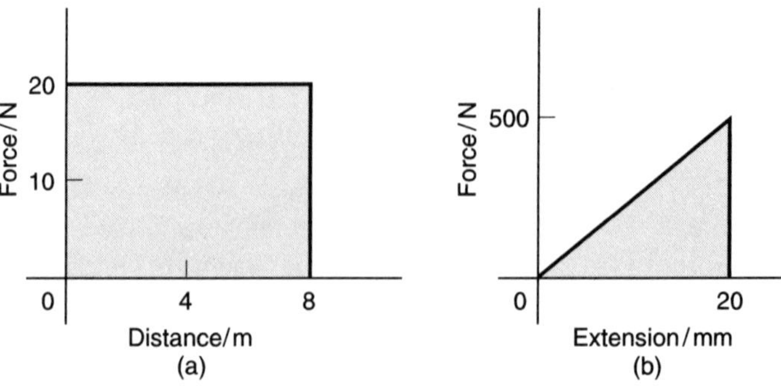

Figure 3.45

If a graph is plotted of experimental values of force (on the vertical axis) against distance moved (on the horizontal axis) a force/distance graph or work diagram is produced. **The area under the graph represents the work done**. For example, a constant force of 20 N used to raise a load a height of 8 m may be represented on a force/distance graph as shown in Figure 3.45(a). The area under the graph, shown shaded, represents the work done. Hence

$$\text{work done} = 20\,\text{N} \times 8\,\text{m} = 160\,\text{J}$$

Similarly, a spring extended by 20 mm by a force of 500 N may be represented by the work diagram shown in Figure 3.45(b).

$$\begin{aligned}\text{work done} &= \text{shaded area}\\ &= \frac{1}{2} \times \text{base} \times \text{height}\\ &= \frac{1}{2} \times (20 \times 10^{-3})\,\text{m} \times 500\,\text{N} = 5\,\text{J}\end{aligned}$$

Problem 3.132 Calculate the work done when a force of 40 N pushes an object a distance of 500 m in the same direction as the force.

$$\begin{aligned}\text{work done} &= \text{force} \times \text{distance moved in the direction of the force}\\ &= 40\,\text{N} \times 500\,\text{m} = 20\,000\,\text{J (since } 1\,\text{J} = 1\,\text{N m)}\end{aligned}$$

i.e. work done = **20 kJ**

Problem 3.133 A motor supplies a constant force of 1 kN which is used to move a load a distance of 5 m. The force is then changed to a constant 500 N and the load is moved a further 15 m. Draw the force/distance graph for the operation and from the graph determine the work done by the motor.

The force/distance graph of work diagram is shown in Figure 3.46. Between points A and B a constant force of 1000 N moves the load 5 m; between points C and D a constant force of 500 N moves the load from 5 m to 20 m.

Total work done
= area under the force/distance graph
= area ABFE = area CDGF

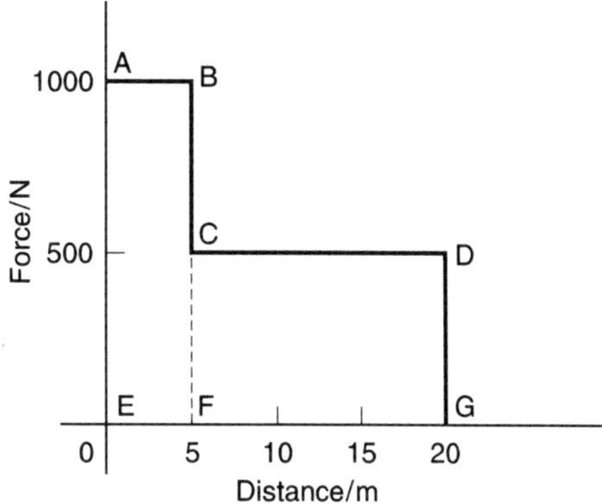

Figure 3.46

$= (1000\,\text{N} \times 5\,\text{m}) + (500\,\text{N} \times 15\,\text{m})$
$= 5000\,\text{J} + 7500\,\text{J} = 12\,500\,\text{J} = \mathbf{12.5\,kJ}$.

Problem 3.134 A spring, initially in a relaxed state, is extended by 100 mm. Determine the work done by using a work diagram if the spring requires a force of 0.6 N per mm of stretch.

Force required for a 100 mm extension $= 100\,\text{mm} \times 0.6\,\text{N}\,\text{mm}^{-1} = 60\,\text{N}$. Figure 3.47 shows the force/extension graph or work diagram representing the increase in extension in proportion to the force, as the force is increased from 0 to 60 N. The work done is the area under the graph (shown shaded). Hence

$$\begin{aligned} \text{work done} &= \frac{1}{2} \times \text{base} \times \text{height} \\ &= \frac{1}{2} \times 100\,\text{mm} \times 60\,\text{N} \\ &= \frac{1}{2} \times 100 \times 10^{-3}\,\text{m} \times 60\,\text{N} = 3\,\text{J}. \end{aligned}$$

(Alternatively, average force during extension $= (60 - 0)/2 = 30\,\text{N}$ and total extension $= 100\,\text{mm} = 0.1\,\text{m}$. Hence work done = average force × extension $= 30\,\text{N} \times 0.1\,\text{m} = 3\,\text{J}$.)

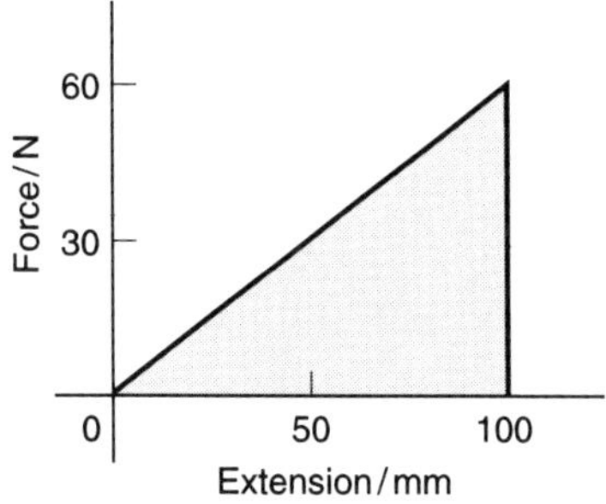

Figure 3.47

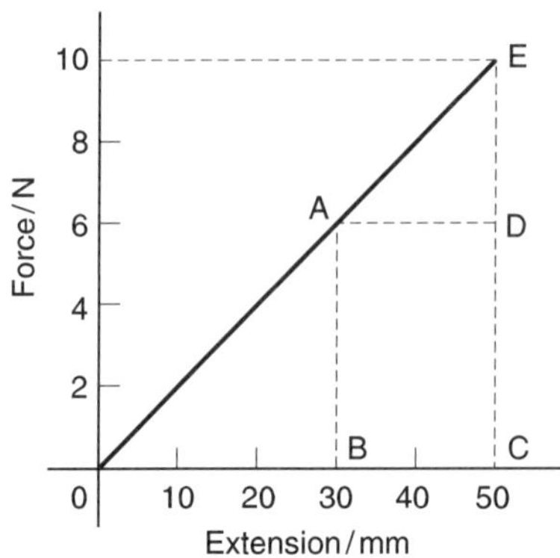

Figure 3.48

Problem 3.135 A spring requires a force of 10 N to cause an extension of 50 mm. Determine the work done in extending the spring (a) from zero to 30 mm, and (b) from 30 mm to 50 mm.

Figure 3.48 shows the force/extension graph for the spring.

(a) Work done in extending the spring from zero to 30 mm is given by area ABO of Figure 3.48, i.e.

$$\begin{aligned}\text{work done} &= \frac{1}{2} \times \text{base} \times \text{height}\\ &= \frac{1}{2} \times 30 \times 10^{-3}\,\text{m} \times 6\,\text{N}\\ &= 90 \times 10^{-3}\,\text{J} = \mathbf{0.09\,J}.\end{aligned}$$

(b) Work done in extending the spring from 30 mm to 50 mm is given by area ABCE of Figure 3.48, i.e.

$$\begin{aligned}\text{work done} &= \text{area ABCD} + \text{area ADE}\\ &= (20 \times 10^{-3}\,\text{m} \times 6\,\text{N}) + \frac{1}{2}(20 \times 10^{-3}\,\text{m})(4\,\text{N})\\ &= 0.12\,\text{J} + 0.04\,\text{J} = 0.16\,\text{J}\end{aligned}$$

Problem 3.136 Calculate the work done when a mass of 20 kg is lifted vertically through a distance of 5.0 m.

The force to be overcome when lifting a mass of 20 kg vertically upwards is mg, i.e. $20 \times 9.81 = 196.2$ N
Work done = force × distance = $196.2 \times 5.0 =$ **981 J**.

Problem 3.137 Water is pumped vertically upwards through a distance of 50.0 m and the work done is 294.3 kJ. Determine the number of litres of water pumped. (1 litre of water has a mass of 1 kg.)

Work done = force × distance, i.e. 294 300 = force × 50.0, from which force = 294 300/50.0 = 5886 N. The force to be overcome when lifting a mass m kg vertically upwards is mg, i.e. $(m \times 9.81)$ N. Thus $5886 = m \times 9.81$, from which mass $m = 5886/9.81 = 600$ kg. Since 1 litre of water has a mass of 1 kg, **600 litres of water are pumped**.

Test your knowledge 3.27

1 Calculate the work done when a mass is lifted vertically by a crane to a height of 5 m, the force required to lift the mass being 98 N.

2 A spring requires a force of 50 N to cause an extension of 100 mm. Determine the work done in extending the spring (a) from 0 to 100 mm, and (b) from 40 mm to 100 mm.

3 Calculate the work done when a mass of 50 kg is lifted vertically through a distance of 30 m.

Energy

Energy is the capacity, or ability, to do work. The unit of energy is the joule, the same as for work. Energy is expended when work is done. There are several forms of energy and these include:

(i) mechanical energy
(ii) heat or thermal energy
(iii) electrical energy
(iv) chemical energy
(v) nuclear energy
(vi) light energy
(vii) sound energy.

Energy may be converted from one form to another. The **principle of conservation of energy** states that the total amount of energy remains the same in such conversions, i.e. energy cannot be created or destroyed. Some examples of energy conversions include:

(i) Mechanical energy is converted to electrical energy by a generator.

(ii) Electrical energy is converted to mechanical energy by a motor.

(iii) Heat energy is converted to mechanical energy by a steam engine.

(iv) Mechanical energy is converted to heat energy by friction.

(v) Heat energy is converted to electrical energy by a solar cell.

(vi) Electrical energy is converted to heat energy by an electric fire.

(vii) Heat energy is converted to chemical energy by living plants.

(viii) Chemical energy is converted to heat energy by burning fuels.

(ix) Heat energy is converted to electrical energy by a thermocouple.

(x) Chemical energy is converted to electrical energy by batteries.

(xi) Electrical energy is converted to light energy by a light bulb.

(xii) Sound energy is converted to electrical energy by a microphone.

(xiii) Electrical energy is converted to chemical energy by electrolysis.

Efficiency is defined as the ratio of the useful output energy to the input energy. The symbol for efficiency is η (Greek letter eta). Hence

$$\text{efficiency, } \eta = \frac{\text{useful output energy}}{\text{input energy}}$$

Efficiency has no units and is often stated as a percentage. A perfect machine would have an efficiency of 100%. However, all machines have an efficiency lower than this due to friction and other losses. Thus, if the input energy to a motor is 1000 J and the output energy is 800 J then the efficiency is

$$\frac{800}{1000} \times 100\%, \text{ i.e. } 80\%.$$

Problem 3.138 A machine exerts a force of 200 N in lifting a mass through a height of 6 m. If 2 kJ of energy are supplied to it, what is the efficiency of the machine?

Work done in lifting mass
= force × distance moved
= weight of body × distance moved
= 200 N × 6 m = 1200 J = useful energy output

$$\text{Energy output} = 2\,\text{kJ} = 2000\,\text{J}$$

$$\text{Efficiency, } \eta = \frac{\text{useful output energy}}{\text{input energy}}$$

$$= \frac{1200}{2000} = \mathbf{0.6 \text{ or } 60\%}$$

Problem 3.139 4 kJ of energy are supplied to a machine used for lifting a mass. The force required is 800 N. If the machine has an efficiency of 50%, to what height will it lift the mass?

$$\text{Efficiency, } \eta = \frac{\text{output energy}}{\text{input energy}}$$

i.e.

$$\frac{50}{100} = \frac{\text{output energy}}{4000\,\text{J}}$$

from which,

$$\text{output energy} = \frac{50}{100} \times 4000 = 2000\,\text{J}$$

Work done = force × distance moved, hence 2000 J = 800 N × height, from which,

$$\text{height} = \frac{2000\,\text{J}}{800\,\text{N}} = \mathbf{2.5\,m}$$

Problem 3.140 A hoist exerts a force of 500 N in raising a load through a height of 20 m. The efficiency of the hoist gears is 75% and the efficiency of the motor is 80%. Calculate the input energy to the hoist.

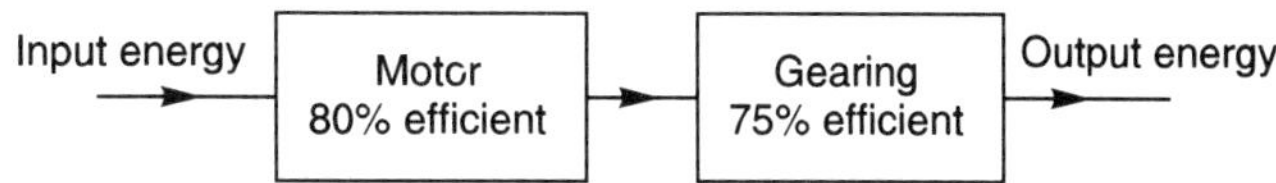

Figure 3.49

The hoist system is shown diagrammatically in Figure 3.49.

$$\begin{aligned}\text{Output energy} &= \text{work done}\\ &= \text{force} \times \text{distance} = 500\,\text{N} \times 20\,\text{m}\\ &= 10\,000\,\text{J}\end{aligned}$$

For the gearing,

$$\text{efficiency} = \frac{\text{output energy}}{\text{input energy}}$$

$$\text{i.e.}\quad \frac{75}{100} = \frac{10\,000}{\text{input energy}}$$

from which, the input energy to the gears $= 10\,000 \times (100/75) = 13\,333\,\text{J}$. The input energy to the gears is the same as the output energy of the motor. Thus, for the motor,

$$\text{efficiency} = \frac{\text{output energy}}{\text{input energy}} = \frac{80}{100} = \frac{13\,333}{\text{input energy}}$$

Hence, input energy to the system

$$= 13\,333 \times \frac{100}{80} = 16\,670\,\text{J} = \mathbf{16.67\,kJ}$$

Test your knowledge 3.28

1 Calculate the useful output of energy of an electric motor which is 70% efficient if it uses 600 J of electrical energy.

2 A machine which is used for lifting a particular mass is supplied with 5 kJ of energy. If the machine has an efficiency of 65% and exerts a force of 812.5 N, to what height will it lift the mass?

Power

Power is a measure of the rate at which work is done or at which energy is converted from one form to another.

$$\text{Power } P = \frac{\text{energy used}}{\text{time taken}}$$

$$\left(\text{or } P = \frac{\text{work done}}{\text{time taken}}\right)$$

The unit of power is the **watt, W**, where 1 W is equal to 1 joule per second. The watt is a small unit for many purposes and a larger unit called the kilowatt, kW, is used, where 1 kW = 1000 W. The power output of a motor which does 120 kJ of work in 30 s is thus given by

$$P = \frac{120\,\text{kJ}}{30\,\text{s}} = 4\,\text{kW}$$

Since work done = force × distance, then

$$\begin{aligned}\text{Power} &= \frac{\text{work done}}{\text{time taken}} = \frac{\text{force} \times \text{distance}}{\text{time taken}}\\ &= \text{force} \times \frac{\text{distance}}{\text{time taken}}\end{aligned}$$

$$\text{However,}\ \frac{\text{distance}}{\text{time taken}} = \text{velocity}$$

Hence **power = force × velocity**

Problem 3.141 The output power of a motor is 8 kW. How much work does it do in 30 s?

Power = (work done)/(time taken), from which,

$$\text{work done} = \text{power} \times \text{time}$$
$$= 8000\,\text{W} \times 30\,\text{s} = 240\,000\,\text{J}$$
$$= \mathbf{240\,kJ}$$

Problem 3.142 Calculate the power required to lift a mass through a height of 10 m in 20 s if the force required is 3924 N.

Work done = force × distance moved = 3924 N × 10 m = 39 240 J

$$\text{Power} = \frac{\text{work done}}{\text{time taken}} = \frac{39\,240\,\text{J}}{20\,\text{s}} = \mathbf{1962\,W\ or\ 1.962\,kW}$$

Problem 3.143 A car hauls a trailer at 90 km/h when exerting a steady pull of 600 N. Calculate (a) the work done in 30 min and (b) the power required.

(a) Work done = force × distance moved.

Distance moved in 30 min, i.e. $\frac{1}{2}$ h, at 90 km/h = 45 km

Hence work done = 600 N × 45 000 m = **27 000 kJ or 27 MJ**

(b) $$\text{Power required} = \frac{\text{work done}}{\text{time taken}} = \frac{27 \times 10^6\,\text{J}}{30 \times 60\,\text{s}}$$
$$= \mathbf{15\,000\,W\ or\ 15\,kW}$$

Problem 3.144 To what height will a mass of weight 981 N be raised in 40 s by a machine using a power of 2 kW?

tf = "times" Work done = force × distance. Hence, work done = 981 N × height

Power = (work done)/(time taken), from which,

$$\text{work done} = \text{power} \times \text{time taken}$$
$$= 2000\,\text{W} \times 40\,\text{s} = 80\,000\,\text{J}$$

Hence 80 000 = 981 N × height, from which,

$$\text{height} = \frac{80\,000\,\text{J}}{981\,\text{N}} = \mathbf{81.55\,m}$$

Test your knowledge 3.29

1 10 kJ of work is done by a force in moving a body uniformly through 125 m in 50 s. Determine (a) the value of the force and (b) the power.

2 A planing machine has a cutting stroke of 2 m and the stroke takes 4 s. If the constant resistance to the cutting tool is 900 N calculate for each cutting stroke (a) the power consumed at the tool point, and (b) the power input to the system if the efficiency of the system is 75%.

3 An electric motor provides power to a winding machine. The input power to the motor is 2.5 kW and the overall efficiency is 60%. Calculate (a) the output power of the machine, (b) the rate at which it can raise a 300 kg load vertically upwards.

The effects of forces on materials

A force exerted on a body can cause a change in either the shape or the motion of the body. The unit of force is the **newton, N**.

No solid body is perfectly rigid and when forces are applied to it, changes in dimensions occur. Such changes are not always perceptible to the human eye since they are so small. For example, the span of a bridge will sag under the weight of a vehicle and a spanner will bend slightly when tightening a nut. It is important for engineers and designers to appreciate the effects of forces on materials, together with their mechanical properties.

The three main types of mechanical force that can act on a body are (i) tensile, (ii) compressive, and (iii) shear.

Tensile force

Tension is a force which tends to stretch a material, as shown in Figure 3.50(a). Examples include:

(i) the rope or cable of a crane carrying a load is in tension;

(ii) rubber bands, when stretched, are in tension;

(iii) a bolt; when a nut is tightened, a bolt is under tension.

A tensile force, i.e. one producing tension, increases the length of the material on which it acts.

Compressive force

Compression is a force which tends to squeeze or crush a material, as shown in Figure 3.50(b). Examples include:

(i) a pillar supporting a bridge is in compression;

(ii) the sole of a shoe is in compression;

(iii) the jib of a crane is in compression.

A compressive force, i.e. one producing compression, will decrease the length of the material on which it acts.

Shear force

Shear is a force which tends to slide one face of the material over an adjacent face. Examples include:

(i) a rivet holding two plates together is in shear if a tensile force is applied between the plates (as shown in Figure 3.51);

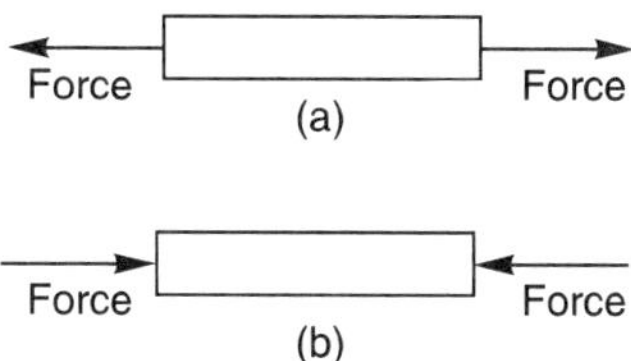

Figure 3.50

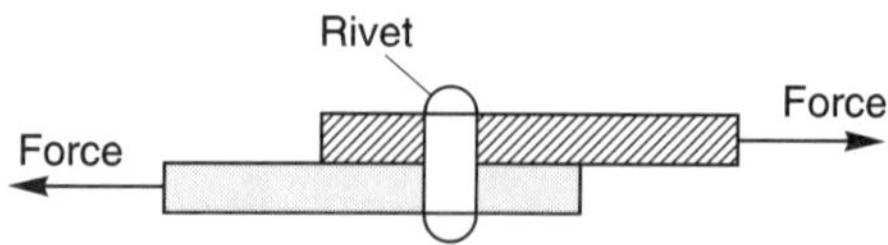

Figure 3.51

(ii) a guillotine cutting sheet metal, or garden shears, each provide a shear force;

(iii) a horizontal beam is subject to shear force;

(iv) transmission joints on cars are subject to shear forces.

A shear force can cause a material to bend, slide or twist.

Problem 3.145 Figure 3.52(a) represents a crane and Figure 3.52(b) a transmission joint. State the types of forces acting labelled A to F.

(a) For the crane, A, a supporting member, is in **compression**, B, a horizontal beam, is in **shear**, and C, a rope, is in **tension**.

(b) For the transmission joint, parts D and F are in **tension**, and E, the rivet or bolt, is in **shear**.

Stress

Forces acting on a material cause a change in dimensions and the material is said to be in a state of stress. Stress is the ratio of the applied force **F** to the cross-sectional area A of the material. The symbol used for tensile or compressive stress is σ (Greek letter sigma). The unit of stress is the **Pascal, Pa**, where $1\,\text{Pa} = 1\,\text{N/m}^2$. Hence

$$\sigma = \frac{\mathbf{F}}{A}\ \mathbf{Pa}$$

where **F** is the force in newtons and A is the cross-sectional area in square metres. For tensile and compressive forces, the cross-sectional area is that which is at right angles to the direction of the force. For a shear force the shear stress is equal to $\mathbf{F}/A$, where the cross-sectional area A is that which is parallel to the direction of the force. The symbol used for shear stress is the Greek letter tau, τ.

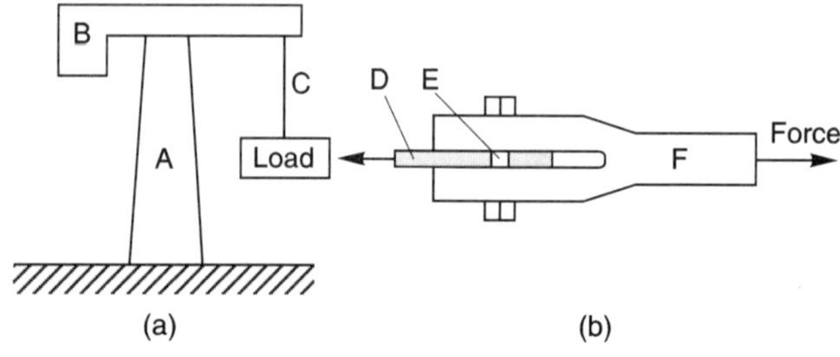

Figure 3.52

Problem 3.146 A rectangular bar having a cross-sectional area of $75\,\text{mm}^2$ has a tensile force of 15 kN applied to it. Determine the stress in the bar.

Cross-sectional area $A = 75\,\text{mm}^2 = 75 \times 10^{-6}\,\text{m}^2$; force $\mathbf{F} = 15\,\text{kN} = 15 \times 10^3\,\text{N}$. Stress in bar,

$$\sigma = \frac{\mathbf{F}}{A} = \frac{15 \times 10^3\,\text{N}}{75 \times 10^{-6}\,\text{m}^2} = 0.2 \times 10^9\,\text{Pa} = \mathbf{200\,MPa}$$

Problem 3.147 A circular wire has a tensile force of 60.0 N applied to it and this force produces a stress of 3.06 MPa in the wire. Determine the diameter of the wire.

Force $\mathbf{F} = 60.0\,\text{N}$; stress $\sigma = 3.06\,\text{MPa} = 3.06 \times 10^6\,\text{Pa}$.
Since $\sigma = \mathbf{F}/A$, then

$$\text{area } A = \frac{\mathbf{F}}{\sigma} = \frac{60.0\,\text{N}}{3.06 \times 10^6\,\text{Pa}} = 19.61 \times 10^{-6}\,\text{m}^2 = 19.61\,\text{mm}^2$$

Cross-sectional area $A = \dfrac{\pi d^2}{4}$. Hence $19.61 = \dfrac{\pi d^2}{4}$, from which,

$$d^2 = \frac{4 \times 19.61}{\pi} \text{ and } d = \sqrt{\left(\frac{4 \times 19.61}{\pi}\right)}$$

i.e. **diameter of wire = 5.0 mm.**

Test your knowledge 3.30

1. A rectangular bar having a cross-sectional area of $80\,\text{mm}^2$ has a tensile force of 20 kN applied to it. Determine the stress in the bar.
2. A circular cable has a tensile force of 1 kN applied to it and the force produces a stress of 7.8 MPa in the cable. Calculate the diameter of the cable.

Strain

The fractional change in a dimension of a material produced by a force is called the **strain**. For a tensile or compressive force, strain is the ratio of the change in length to the original length. The symbol used for strain is ϵ (Greek epsilon). For a material of length l metres which changes in length by an amount x metres when subjected to stress

$$\epsilon = \frac{x}{l}$$

Strain is dimensionless and is often expressed as a percentage, i.e.

$$\text{percentage strain} = \frac{x}{l} \times 100$$

For a shear force, strain is denoted by the symbol γ (Greek letter gamma) and, with reference to Figure 3.53, is given by:

$$\gamma = \frac{x}{l}$$

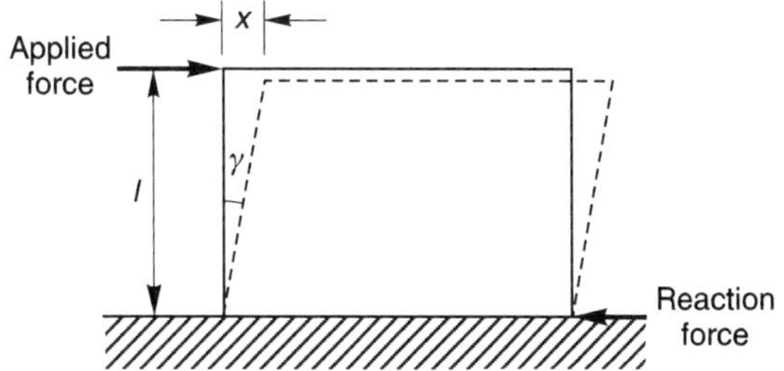

Figure 3.53

Problem 3.148 A bar 1.60 m long contracts to 0.1 mm when a compressive load is applied to it. Determine the strain and the percentage strain.

$$\text{Strain } \epsilon = \frac{\text{contraction}}{\text{original length}} = \frac{0.1\text{ mm}}{1.60 \times 10^3\text{ mm}} = \frac{0.1}{1600} = \mathbf{0.000\,062\,5}$$

$$\text{Percentage strain} = 0.000\,062\,5 \times 100 = \mathbf{0.006\,25\%}$$

Problem 3.149 (a) A rectangular metal bar has a width of 10 mm and can support a maximum compressive stress of 20 MPa. Determine the minimum breadth of the bar when loaded with a force of 3 kN. (b) If the bar in (a) is 2 m long and decreases in length by 0.25 mm when the force is applied, determine the strain and the percentage strain.

(a) Since stress $\sigma = \dfrac{\text{force } \mathbf{F}}{\text{area } A}$ then area $A = \dfrac{\mathbf{F}}{\sigma} = \dfrac{3000\text{ N}}{20 \times 10^6\text{ Pa}}$

$$= 150 \times 10^{-6}\text{ m}^2$$

$$= 150\text{ mm}^2$$

Cross-sectional area = width × breadth, hence

$$\text{breadth} = \frac{\text{area}}{\text{width}} = \frac{150}{10} = \mathbf{15\,mm}$$

(b) Strain $\epsilon = \dfrac{\text{contraction}}{\text{original length}} = \dfrac{0.25}{2000} = 0.000\,125$

$$\text{Percentage strain} = 0.000\,125 \times 100 = \mathbf{0.0125\%}$$

Problem 3.150 A rectangular block of plastic material 500 mm long by 20 mm wide by 300 mm high has its lower face glued to a bench and a force of 200 N is applied to the upper face and in line with it. The upper face moves 15 mm relative to the lower face. Determine (a) the shear stress, and (b) the shear strain in the upper face, assuming the deformation is uniform.

(a) Shear stress $\tau = \dfrac{\text{force}}{\text{area parallel to the force}}$

Area of any face parallel to the force $= 500\text{ mm} \times 20\text{ mm}$

$$= (0.5 \times 0.02)\text{ m}^2$$

$$= 0.01\text{ m}^2$$

$$\text{Hence shear stress } \tau = \frac{200\text{ N}}{0.01\text{ m}^2} = \mathbf{20\,000\,Pa} \text{ or } \mathbf{20\,kPa}$$

(b) Shear strain $= \dfrac{x}{l}$ (see side view of Figure 3.54)

$$= \frac{15}{300} = \mathbf{0.05\ (or\ 5\%)}$$

Test your knowledge 3.31

1 A wire of length 2.50 m has a percentage strain of 0.012% when loaded with a tensile force. Determine the extension of the wire.

2 A pipe has an outside diameter of 25 mm, an inside diameter of 15 mm and length 0.40 m and it supports a compressive load of 40 kN. The pipe shortens by 0.5 mm when the load is applied. Determine (a) the compressive stress, (b) the compressive strain in the pipe when supporting this load.

3 A circular hole of diameter 50 mm is to be punched out of a 2 mm thick metal plate. The shear stress needed to cause fracture is 500 MPa. Determine (a) the minimum force to be applied to the punch, (b) the compressive stress in the punch at this value.

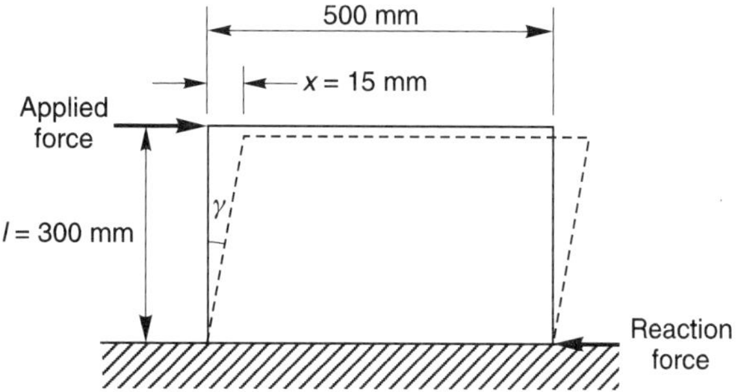

Figure 3.54

Elasticity and elastic limit

Elasticity is the ability of a material to return to its original shape and size on the removal of external forces.

Plasticity is the property of a material of being permanently deformed by a force without breaking. Thus if a material does not return to the original shape, it is said to be plastic.

Within certain load limits, mild steel, copper, polythene and rubber are examples of elastic materials; lead and Plasticine are examples of plastic materials. If a tensile force applied to a uniform bar of mild steel is gradually increased and the corresponding extension of the bar is measured, then provided the applied force is not too large, a graph depicting these results is likely to be as shown in Figure 3.55. Since the graph is a straight line, **extension is directly proportional to the applied force**.

If the applied force is large, it is found that the material no longer returns to its original length when the force is removed. The material is then said to have passed its elastic limit and the resulting graph of force/extension is no longer a straight line. Stress $= \sigma = \mathbf{F}/A$, and since, for a particular bar, A can be considered as a constant, then $\mathbf{F} \propto \sigma$. Strain $\epsilon = x/l$, and since for a particular bar l is constant, then $x \propto \epsilon$. Hence for stress applied to a material below the elastic limit a graph of stress/strain will be as shown in Figure 3.56, and is a similar shape to the force/extension graph of Figure 3.55.

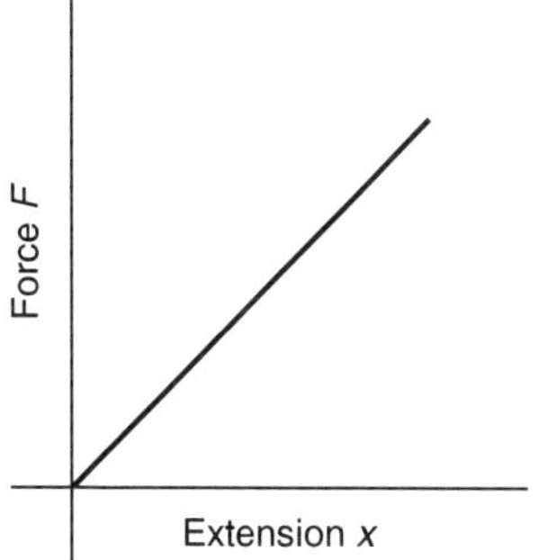

Figure 3.55

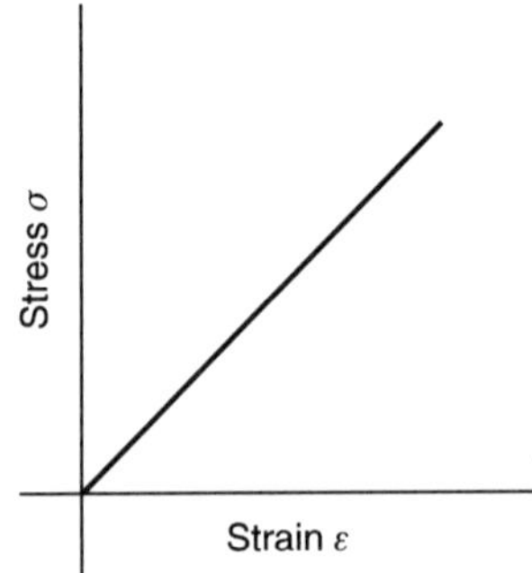

Figure 3.56

Hooke's Law

Hooke's law states:

> *Within the elastic limit, the extension of a material is proportional to the applied force.*

It follows that:

> *Within the elastic limit of a material, the strain produced is directly proportional to the stress producing it.*

Young's modulus of elasticity

Within the elastic limit, stress $\propto$ strain, hence stress = (a constant) $\times$ strain. This constant of proportionality is called **Young's modulus of elasticity** and is given the symbol E. The value of E may be determined from the gradient of the straight line portion of the stress/strain graph. The dimensions of E are pascals (the same as for stress, since strain is dimensionless).

$$E = \frac{\sigma}{\epsilon}\ \text{Pa}$$

Some typical values for Young's modulus of elasticity, E, include:

Aluminium	70 GPa (i.e. 70×10^9 Pa)		
Brass	100 GPa	Copper	96 GPa
Diamond	1200 GPa	Mild steel	210 GPa
Lead	18 GPa	Tungsten	410 GPa
Cast iron	110 GPa	Zinc	85 GPa

Stiffness

A material having a large value of Young's modulus is said to have a high value of stiffness, where stiffness is defined as:

$$\textbf{Stiffness} = \frac{\textbf{force F}}{\textbf{extension } \boldsymbol{x}}$$

For example, mild steel is much stiffer than lead.

Since $E = \sigma/\epsilon$ and $\sigma = \mathbf{F}/A$ and $\epsilon = x/l$, then

$$E = \frac{\mathbf{F}/A}{x/l}$$

$$\text{i.e. } E = \frac{Fl}{Ax} = \left(\frac{\mathbf{F}}{x}\right)\left(\frac{l}{A}\right)$$

$$\text{and } E = (\text{stiffness}) \times \left(\frac{l}{A}\right)$$

Stiffness ($= \mathbf{F}/x$) is also the gradient of the force/extension graph, hence

$$E = (\text{gradient of force/extension graph})\left(\frac{l}{A}\right)$$

Since l and A for a particular specimen are constant, the greater Young's modulus the greater the stiffness.

Problem 3.151 A wire is stretched 2 mm by a force of 250 N. Determine the force that would stretch the wire 5 mm, assuming that the elastic limit is not exceeded.

Hooke's law states that extension x is proportional to force $\mathbf{F}$, provided that the elastic limit is not exceeded, i.e. $x \propto \mathbf{F}$ or $x = k\mathbf{F}$ where k is a constant.

When $x = 2$ mm, $\mathbf{F} = 250$ N, thus $2 = k(250)$,
from which, constant $k = (2/250) = (1/125)$.

When $x = 5$ mm, then $5 = k\mathbf{F}$, i.e. $5 = (1/125)\mathbf{F}$
from which, force $\mathbf{F} = 5(125) = 625$ N

Thus to stretch the wire 5 mm a force of 625 N is required.

Problem 3.152 A force of 10 kN applied to a component produces an extension of 0.1 mm. Determine (a) the force needed to produce an extension of 0.12 mm, and (b) the extension when the applied force is 6 kN, assuming in each case that the elastic limit is not exceeded.

From Hooke's law, extension x is proportional to force $\mathbf{F}$ within the elastic limit, i.e. $x \propto \mathbf{F}$ or $x = k\mathbf{F}$, where k is a constant. If a force of 10 kN produces an extension of 0.1 mm, then $0.1 = k(10)$ from which, constant $k = 0.1/10 = 0.01$.

(a) When extension $x = 0.12$ mm, then $0.12 = k(\mathbf{F})$, i.e. $0.12 = 0.01\,\mathbf{F}$, from which

$$\text{force } \mathbf{F} = \frac{0.12}{0.01} = \mathbf{12\,kN}$$

(b) When force $\mathbf{F} = 6$ kN, then **extension** $\boldsymbol{x} = k(6) = (0.01)(6)$ $=$ **0.06 mm**

Problem 3.153 In an experiment to determine the modulus of elasticity of a sample of mild steel, a wire is loaded and the corresponding extension noted. The results of the experiment are as shown:

Load (N)	0	40	110	160	200	250	290	340
Extension (mm)	0	1.2	3.3	4.8	6.0	7.5	10.0	16.2

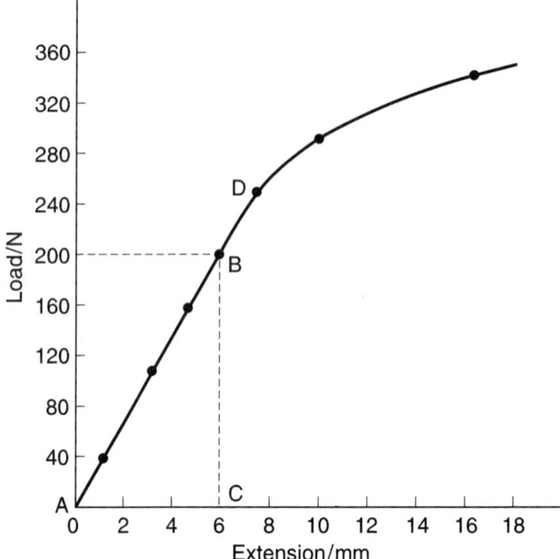

Figure 3.57

Draw the load/extension graph. The mean diameter of the wire is 1.3 mm and its length is 8.0 m. Determine the modulus of elasticity E of the sample, and the stress at the elastic limit.
A graph of load/extension is shown in Figure 3.57.

$$E = \frac{\sigma}{\epsilon} = \frac{\mathbf{F}/A}{x/l} = \left(\frac{\mathbf{F}}{x}\right)\left(\frac{l}{A}\right)$$

$(\mathbf{F}/x)$ is the gradient of the straight line part of the load/extension graph.

$$\text{Gradient} = \frac{\mathbf{F}}{x} = \frac{\text{BC}}{\text{AC}} = \frac{200\ \text{N}}{6 \times 10^{-3}\ \text{m}}$$
$$= 33.33 \times 10^3\ \text{N/m}$$

Modulus of elasticity $=$ (gradient of graph) (l/A)
Length of specimen, $l = 8.0$ m

$$\text{Cross-sectional area } A = \frac{\pi d^2}{4} = \frac{\pi(0.0013)^2}{4} = 1.327 \times 10^{-6}$$

$$\text{Hence modulus of elasticity} = (33.33 \times 10^3)\ \frac{8.0}{1.327 \times 10^{-6}}$$
$$= 201\ \text{GPa}$$

The elastic limit is at point D in Figure 3.57 where the graph no longer follows a straight line. This point corresponds to a load of 250 N as shown.

$$\text{Stress at elastic limit} = \frac{\text{force}}{\text{area}} = \frac{250}{1.327 \times 10^{-6}} = 188.4 \times 10^6\ \text{Pa}$$
$$= \mathbf{188.4\ MPa}$$

Ductility, brittleness and malleability

Ductility is the ability of a material to be plastically deformed by elongation, without fracture. This is a property which enables a material to be drawn out into wires. For ductile materials such as mild steel, copper and gold, large extensions can result before fracture occurs with increasing tensile force. Ductile materials usually have a percentage elongation value of about 15% or more.

Brittleness is the property of a material manifested by fracture without appreciable prior plastic deformation. Brittleness is a lack of ductility, and brittle materials such as cast iron, glass, concrete, brick and ceramics, have virtually no plastic stage, the elastic stage being followed by immediate fracture. Little or no 'waist' occurs before fracture in a brittle material undergoing a tensile test.

Malleability is the property of a material whereby it can be shaped when cold by hammering or rolling. A malleable material is capable of undergoing plastic deformation without fracture. Examples of malleable materials include lead, gold, putty and mild steel.

Problem 3.154 Sketch typical load/extension curves for (a) an elastic non-metallic material, (b) a brittle material and (c) a ductile material. Give a typical example of each type of material.

(a) A typical load/extension curve for an elastic non-metallic material is shown in Figure 3.58(a), and an example of such a material is **polythene**.

(b) A typical load/extension curve for a brittle material is shown in Figure 3.58(b), and an example of such a material is **cast iron**.

(c) A typical load/extension curve for a ductile material is shown in Figure 3.58(c), and an example of such a material is **mild steel**.

Test your knowledge 3.32

1 A rubber band extends 50 mm when a force of 300 N is applied to it. Assuming the band is within the elastic limit, determine the extension produced by a force of 60 N.

2 A copper rod of diameter 20 mm and length 2.0 m has a tensile force of 5 kN applied to it. Determine (a) the stress in the rod, (b) by how much the rod extends when the load is applied. Take the modulus of elasticity for copper as 96 GPa.

3 An aluminium rod has a length of 200 mm and a diameter of 10 mm. When subjected to a compressive force the length of the rod is 199.6 mm. Determine (a) the stress in the rod when loaded, and (b) the magnitude of the force. Take the modulus of elasticity for aluminium as 70 GPa.

Pressure

The pressure acting on a surface is defined as the perpendicular force per unit area of surface. The unit of pressure is the **pascal, Pa**, where 1 pascal is equal to 1 newton per square metre. Thus pressure,

$$p = \frac{\mathbf{F}}{A}\ \textbf{pascals}$$

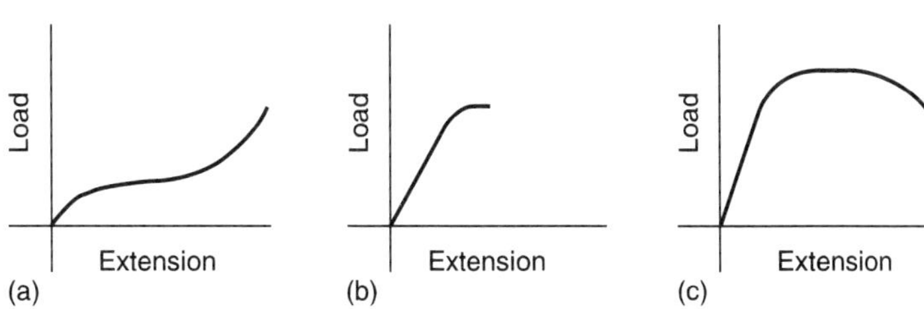

Figure 3.58

where **F** is the force in newtons acting at right angles to a surface of area A square metres.

When a force of 20 N acts uniformly over, and perpendicular to, an area of $4\,m^2$, then the pressure on the area, p, is given by

$$p = \frac{20\,N}{4\,m^2} = 5\,Pa$$

Problem 3.155 A table loaded with books has a force of 250 N acting in each of its legs. If the contact area between each leg and the floor is $50\,mm^2$, find the pressure each leg exerts on the floor.

From above, pressure $p =$ (force)/(area). Hence

$$p = \frac{250\,N}{50\,mm^2} \times \frac{10^6\,mm^2}{1\,m^2}$$
$$= 5 \times 10^6\,N/m^2 = \mathbf{5\,MPa}$$

That is, **the pressure exerted by each leg on the floor is 5 MPa.**

Problem 3.156 A circular piston exerts a pressure of 80 kPa on a fluid, when the force applied to the piston is 0.2 kN. Find the diameter of the piston. From above, pressure = (force)/(area). Hence, area = (force)/(pressure). Force in newtons is

$$0.2\,kN \times \frac{1000\,N}{1\,kN} = 200\,N$$

Pressure in pascals is 80 kPa = 80 000 Pa = 80 000 N/m^2. Hence

$$\text{area} = \frac{200\,N}{80\,000\,N/m^2} = 0.0025\,m^2$$

Since the piston is circular, its area is given by $\pi d^2/4$, where d is the diameter of the piston. Hence,

$$\text{area} = \frac{\pi d^2}{4} = 0.0025$$
$$d^2 = 0.0025 \times \frac{4}{\pi} = 0.003\,183$$
$$\text{i.e. } d = 0.0564\,m, \text{ i.e. } 56.4\,mm$$

Hence, **the diameter of the piston is 56.4 mm**.

Fluid pressure

A fluid is either a liquid or a gas and there are four basic factors governing the pressure within fluids.

(a) The pressure at a given depth in a fluid is equal in all directions, see Figure 3.59(a).

(b) The pressure at a given depth in a fluid is independent of the shape of the container in which the fluid is held. In Figure 3.59(b), the pressure at X is the same as the pressure at Y.

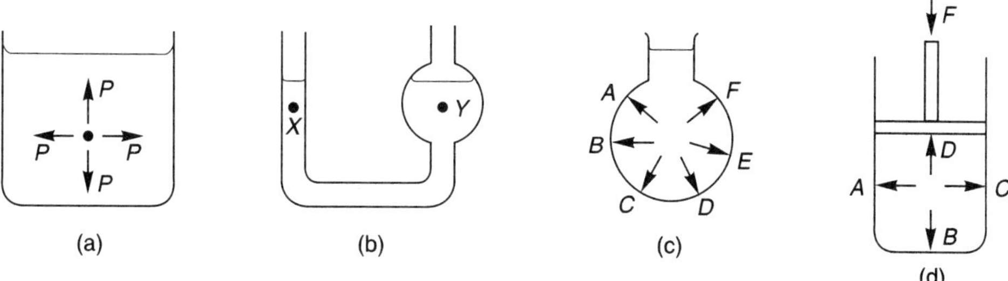

Figure 3.59

(c) Pressure acts at right angles to the surface containing the fluid. In Figure 3.59(c), the pressures at points A–F all act at right angles to the container.

(d) When a pressure is applied to a fluid, this pressure is transmitted equally in all directions. In Figure 3.59(d), if the mass of the fluid is neglected, the pressures at points A–D are all the same.

The pressure, p, at any point in a fluid depends on three factors:

(a) the density of the fluid, ρ, in kg/m^3

(b) the gravitational acceleration, g, taken as approximately $9.8\,m/s^2$ (or the gravitational field force in N/kg); and

(c) the height of fluid vertically above the point, h metres. The relationship connecting these quantities is:

$$\mathbf{p = \rho gh\ Pa}$$

When the container shown in Figure 3.60 is filled with water of density $1000\,kg/m^3$, the pressure due to the water at a depth of 0.03 m below the surface is given by:

$$\begin{aligned} p &= \rho gh \\ &= (1000 \times 9.8 \times 0.03)\ \text{Pa} \\ &= \mathbf{294\ Pa} \end{aligned}$$

Problem 3.157 A tank contains water to a depth of 600 mm. Calculate the water pressure (a) at a depth of 350 mm and (b) at the base of the tank. Take the density of water as $1000\,kg/m^3$ and the gravitational acceleration as $9.8\,m/s^2$.

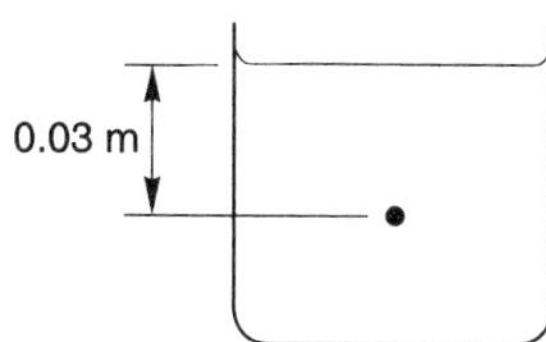

Figure 3.60

From above, pressure p at any point in a fluid is given by $p = \rho gh$ pascals, where ρ is the density in kg/m^3, g is the gravitational acceleration in m/s^2 and h is the height of fluid vertically above the point.

(a) At a depth of 350 mm, i.e. 0.35 m

$$p = 1000 \times 9.8 \times 0.35 = \mathbf{3430\ Pa} = \mathbf{3.43\ kPa}$$

(b) At the base of the tank, the vertical height of the water is 600 mm, that is, 0.6 m. Hence

$$p = 1000 \times 9.8 \times 0.6 = \mathbf{5880\ Pa} = \mathbf{5.88\ kPa}$$

Problem 3.158 A storage tank contains petrol to a height of 4.7 m. If the pressure at the base of the tank is 32.3 kPa, determine the density of the petrol. Take the gravitational field force as $9.8\ m/s^2$.

From above, pressure $p = \rho gh$ pascals, where ρ is the density in kg/m^3, g is the gravitational acceleration in m/s^2 and h is the vertical height of the petrol. Transposing gives:

$$\rho = \frac{p}{gh}$$

The pressure p is 32.2 kPa, that is, 32 200 Pa, hence,

$$\text{density, } \rho = \frac{32\,200}{9.8 \times 4.7} = 699\ kg/m^3$$

That is, the **density of the petrol is 699 kg/m³**.

Test your knowledge 3.33

1. Determine the pressure acting at the base of a dam, when the surface of the water is 35 m above base level.
2. Calculate the force exerted by the atmosphere on a pool of water which is 30 m long by 10 m wide, when the atmospheric pressure is 100 kPa.
3. A vertical tube is partly filled with mercury of density 13 600 kg/m³. Find the height, in millimetres, of the column of mercury, when the pressure at the base of the tube is 101 kPa. Take the gravitational field force as $9.8\ m/s^2$.

Atmospheric pressure

The air above the earth's surface is a fluid, having a density, ρ, which varies from approximately $1.225\ kg/m^3$ at sea level to zero in outer space. Since $p = \rho gh$, where height h is several thousands of metres, the air exerts a pressure on all points on the earth's surface. This pressure, called **atmospheric pressure**, has a value of approximately 100 kPa. Two terms are commonly used when measuring pressures:

(a) **absolute pressure**, meaning the pressure above that of an absolute vacuum (i.e. zero pressure); and

(b) **gauge pressure**, meaning the pressure above that normally present due to the atmosphere. Thus:

absolute pressure = atmospheric pressure + gauge pressure

Thus, a gauge pressure of 50 kPa is equivalent to an absolute pressure of (100 + 50) kPa, i.e. 150 kPa, since the atmospheric pressure is approximately 100 kPa.

Problem 3.159 Calculate the absolute pressure at a point on a submarine at a depth of 30 m below the surface of the sea, when the atmospheric pressure is 101 kPa. Take the density of sea water as $1030\ kg/m^3$ and the gravitational acceleration as $9.8\ m/s^2$.

Test your knowledge 3.34

A Bourdon pressure gauge shows a pressure of 1.151 MPa. If the absolute pressure is 1.25 MPa, find the atmospheric pressure in millimetres of mercury.

The pressure due to the sea, that is, the gauge pressure (p_g) is given by $p_g = gh$ pascals, i.e.

$$p_g = 1030 \times 9.8 \times 30 = 302\,820\,\text{Pa} = 302.82\,\text{kPa}$$

From the above,

$$\begin{aligned}\text{absolute pressure} &= \text{atmospheric pressure} + \text{gauge pressure} \\ &= (101 + 302.82)\,\text{kPa} = 403.82\,\text{kPa}\end{aligned}$$

that is, the **absolute pressure at a depth of 30 m is 403.82 kPa**.

Measurement of pressure

Pressure indicating instruments are made in a wide variety of forms because of their many different applications. Apart from the obvious criteria such as pressure range, accuracy and response, many measurements also require special attention to material, sealing and temperature effects. The fluid whose pressure is being measured may be corrosive or may be at high temperatures. Pressure indicating devices used in science and industry include:

(i) barometers – Fortin and aneroid types,

(ii) manometers – U-tube and inclined types,

(iii) Bourdon pressure gauge, and

(iv) McLeod and Pirani vacuum gauges

(i) Barometers

Introduction

A barometer is an instrument for measuring atmospheric pressure. It is affected by seasonal changes of temperature. Barometers are therefore also used for the measurement of altitude and also as one of the aids in weather forecasting. The value of atmospheric pressure will thus vary with climatic conditions, although not usually by more than about 10% of standard atmospheric pressure.

Construction and principle of operation

A simple barometer consists of a glass tube, just under 1 m in length, sealed at one end, filled with mercury and then inverted into a trough containing more mercury. Care must be taken to ensure that no air enters the tube during this latter process. Such a barometer is shown in Figure 3.61(a) and it is seen that the level of the mercury column falls, leaving an empty space, called a vacuum. Atmospheric pressure acts on the surface of the mercury in the trough as shown and this pressure is equal to the pressure at the base of the column of mercury in the inverted tube, i.e. the pressure of the atmosphere is supporting the column of mercury. If the atmospheric pressure falls the barometer height h decreases. Similarly, if the atmospheric pressure rises then h increases. Thus atmospheric pressure can be measured

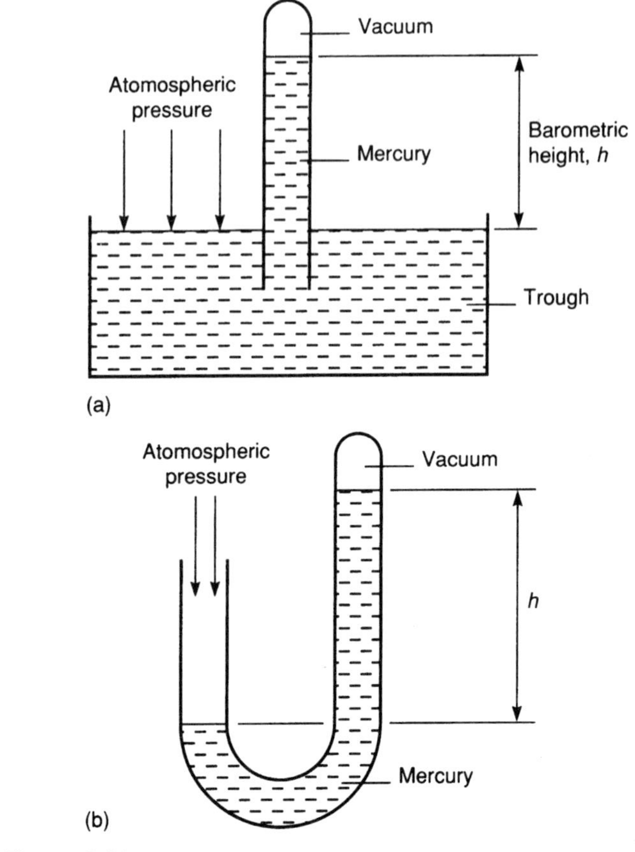

Figure 3.61

in terms of the height of the mercury column. It may be shown that for mercury the height h is 760 mm at standard atmospheric pressure, i.e. a vertical column of mercury 760 mm high exerts a pressure equal to the standard value of atmospheric pressure.

There are thus several ways in which atmospheric pressure can be expressed:

Standard atmospheric pressure = 101 325 Pa or 101.325 kPa
= 101 325 N/m^2 or 101.325 kN/m^2
= 1.013 25 bars or 1013.25 mbars
= 760 mm of mercury

Another arrangement of a typical barometer is shown in Figure 3.61(b) where a U-tube is used instead of an inverted tube and trough, the principle being similar. If, instead of mercury, water was used as the liquid in a barometer, then the barometric height h at standard atmospheric pressure would be 13.6 times more than for mercury, i.e. about 10.4 m high, which is not very practicable. This is because the relative density of mercury is 13.6.

Types of barometer

The **Fortin barometer** is an example of a mercury barometer which enables barometric heights to be measured to a high degree of accuracy (of the order of one-tenth of a millimetre or less). Its construction is merely a more sophisticated arrangement of the inverted tube and trough shown in Figure 3.61(a), with the addition of a vernier scale to measure the barometric height with great accuracy. A disadvantage of this type of barometer is that it is not portable.

A **Fortin barometer** is shown in Figure 3.62. Mercury is contained in a leather bag at the base of the mercury reservoir, and height H, of the mercury in the reservoir can be adjusted using the screw at the base of the barometer to depress or release the leather bag. To measure the atmospheric pressure the screw is adjusted until the pointer at H is just touching the surface of the mercury and the height of the mercury column is then read using the main and vernier scales. The measurement of atmospheric pressure using a Fortin barometer is achieved much more accurately than by using a simple barometer.

A portable type often used is the **aneroid barometer**. Such a barometer consists basically of a circular, hollow, sealed vessel, S, usually made from thin flexible metal. The air pressure in the vessel is reduced to nearly zero before sealing, so that a change in atmospheric pressure will cause the shape of the vessel to expand or contract. These small changes can be magnified by means of a lever and be made to move a pointer over a calibrated scale. Figure 3.63 shows a typical arrangement of an aneroid barometer. The scale is usually circular and calibrated in millimetres of mercury. These instruments require frequent calibration.

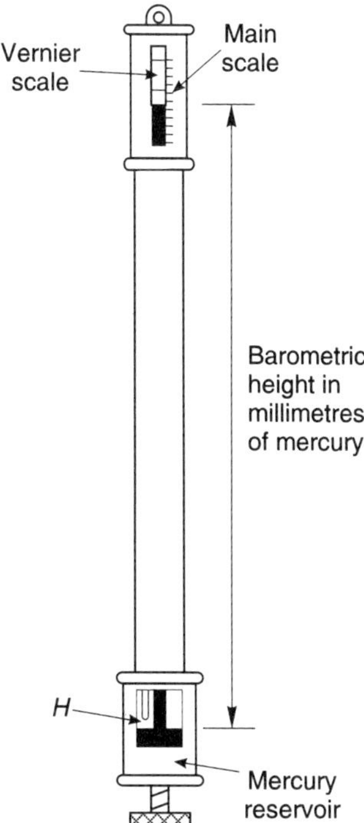

Figure 3.62

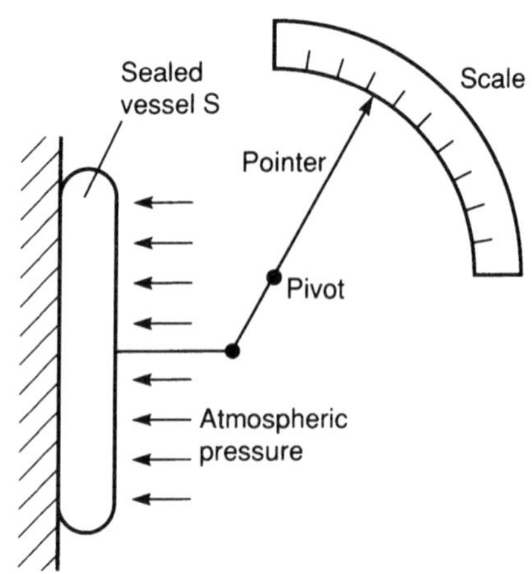

Figure 3.63

(ii) The manometer

A manometer is a device for measuring or comparing fluid pressures, and is the simplest method of indicating such pressures.

U-tube manometer

A U-tube manometer consists of a glass tube bent into a U-shape and containing a liquid such as mercury. A U-tube manometer is shown in Figure 3.64(a). If limb A is connected to a container of

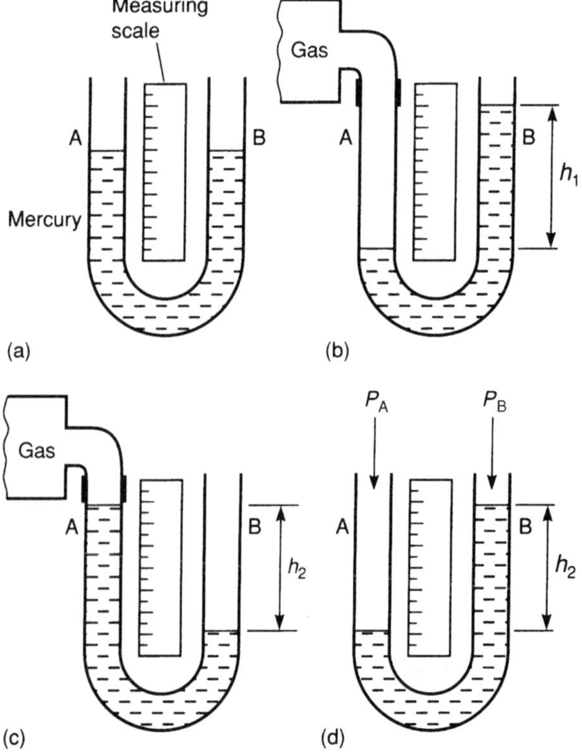

Figure 3.64

gas whose pressure is above atmospheric, then the pressure of the gas will cause the levels of mercury to move, as shown in Figure 3.64(b), such that the difference in height is h_1. The measuring scale can be calibrated to give the gauge pressure of the gas as h_1 mm of mercury. If limb A is connected to a container of gas whose pressure is below atmospheric then the levels of mercury will move, as shown in Figure 3.64(c), such that their pressure difference is h_2 mm of mercury.

It is also possible merely to compare two pressures, say P_A and P_B, using a U-tube manometer. Figure 3.64(d) shows such an arrangement with $(P_B - P_A)$ equivalent to h mm of mercury. One application of this differential pressure-measuring device is in determining the velocity of fluid flow in pipes. For the measurement of lower pressures, water or paraffin may be used instead of mercury in the U-tube to give larger values of h and thus greater sensitivity.

Inclined manometers

For the measurement of very low pressures, greater sensitivity is achieved by using an inclined manometer, a typical arrangement of which is shown in Figure 3.65. With the inclined manometer the liquid used is water and the scale attached to the inclined tube is calibrated in terms of the vertical height h. Thus when a vessel containing gas under pressure is connected to the reservoir, movement of the liquid levels of the manometer occurs. Since small bore tubing is used the movement of the liquid in the reservoir is very small compared with the movement in the inclined tube and is thus neglected. Hence the scale on the manometer is usually used in the range 0.2 mbar to 2 mbar.

The pressure of a gas that a manometer is capable of measuring is naturally limited by the length of tube used. Most manometer tubes are less than 2 m in length and this restricts measurement to a maximum pressure of about 2.5 bar (or 250 kPa) when mercury is used.

(iii) The Bourdon pressure gauge

Pressures many times greater than atmospheric can be measured by the Bourdon pressure gauge, which is the most extensively

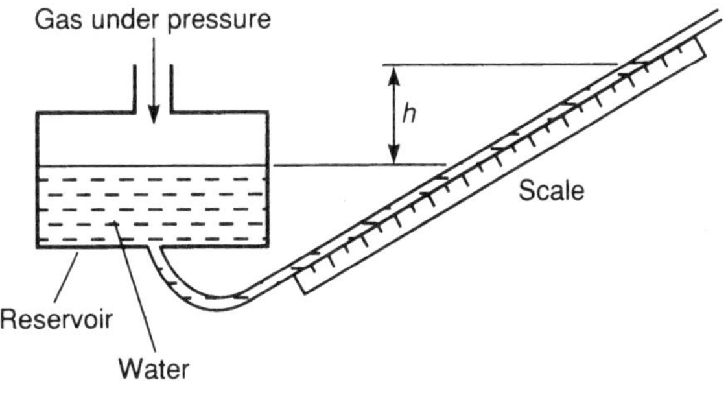

Figure 3.65

used of all pressure-indicating instruments. It is a robust instrument. Its main component is a piece of metal tube (called the Bourdon tube), usually made of phosphor bronze or alloy steel, of oval or elliptical cross-section, sealed at one end and bent into an arc. In some forms the tube is bent into a spiral for greater sensitivity. A typical arrangement is shown in Figure 3.66(a). One end, E, of the Bourdon tube is fixed and the fluid whose pressure is to be measured is connected to this end. The pressure acts at right angles to the metal tube wall as shown in the cross-section of the tube in Figure 3.66(b). Because of its elliptical shape it is clear that the sum of the pressure components, i.e. the total force acting on the sides A and C, exceeds the sum of the pressure components acting on ends B and D. The result is that sides A and C tend to move outwards and B and D inwards tending to form a circular cross-section. As the pressure in the tube is increased the tube tends to uncurl, or if the pressure is reduced the tube curls up further.

The movement of the free end of the tube is, for practical purposes, proportional to the pressure applied to the tube, this pressure, of course, being the gauge pressure (i.e. the difference between atmospheric pressure acting on the outside of the tube and the applied pressure acting on the inside of the tube). By

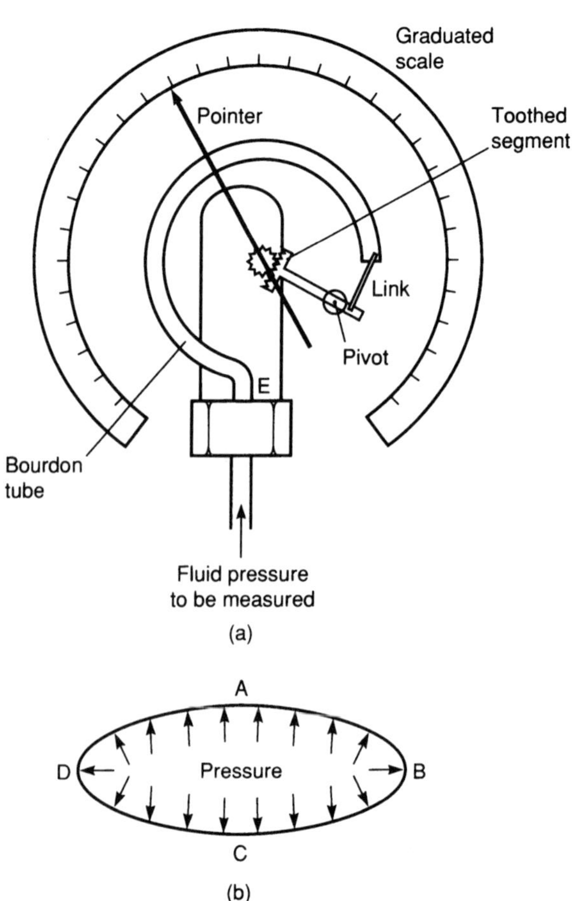

Figure 3.66

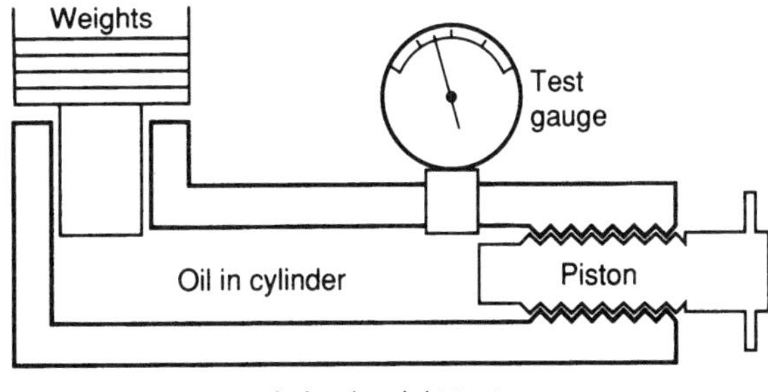

A dead weight tester

Figure 3.67

using a link, a pivot and a toothed segment as shown in Figure 3.66(a), the movement can be converted into the rotation of a pointer over a graduated calibrated scale. The Bourdon tube pressure is capable of measuring high pressures up to 10^4 bar (i.e. 7600 m of mercury) with the addition of special safety features.

A pressure gauge must be calibrated, and this is done either by a manometer, for low pressures, or by a piece of equipment called a '**dead weight tester**'. This tester consists of a piston operating in an oil-filled cylinder of known bore, and carrying accurately known weights, as shown in Figure 3.67. The gauge under test is attached to the tester and the required pressure is applied by a screwed piston or ram, until the weights are just lifted. While the gauge is being read, the weights are turned to reduce friction effects.

(iv) Vacuum gauges

Vacuum gauges are instruments for giving a visual indication, by means of a pointer, of the amount by which the pressure of a fluid applied to the gauge is less than the pressure of the surrounding atmosphere. Two examples of vacuum gauges are the McLeod gauge and the Pirani gauge.

McLeod gauge

The McLeod gauge is normally regarded as a standard and is used to calibrate other forms of vacuum gauges. The basic principle of this gauge is that it takes a known volume of gas at a pressure so low that it cannot be measured, then compresses the gas in a known ratio until the pressure becomes large enough to be measured by an ordinary manometer. This device is used to measure low pressures, often in the range 10^{-6}–1.0 mm of mercury. A disadvantage of the McLeod gauge is that it does not give a continuous reading of pressure and is not suitable for registering rapid variations in pressure.

Pirani gauge

The Pirani gauge measures the resistance and thus the temperature of a wire through which current is flowing. The thermal conductvity decreases with the pressure in the range

Test your knowledge 3.35

State the most suitable pressure indicating device for the following:

(a) A robust device to measure high pressures in the range 0–30 MPa.

(b) Calibration of a Pirani gauge.

(c) Measurement of gas pressures comparable with atmospheric pressure.

(d) To measure pressures of the order of 200 kPa.

(e) Measurement of atmospheric pressure to a high degree of accuracy.

10^{-1}–10^{-4} mm of mercury so that the increase in resistance can be used to measure pressure in this region. The Pirani gauge is calibrated by comparison with a McLeod gauge.

Heat energy

Heat is a form of energy and is measured in joules. **Temperature** is the degree of hotness or coldness of a substance. Heat and temperature are thus not the same thing. For example, twice the heat energy is needed to boil a full container of water than half a container – that is, different amounts of heat energy are needed to cause an equal rise in the temperature of different amounts of the same substance.

Temperature is measured either (i) on the **Celsius (°C) scale** (formerly Centigrade), where the temperature at which ice melts, i.e. the freezing point of water, is taken as 0°C and the point at which water boils under normal atmospheric pressure is taken as 100°C, or (ii) on the **thermodynamic scale**, in which the unit of temperature is the kelvin (K). The kelvin scale uses the same temperature interval as the Celsius scale but as its zero takes the 'absolute zero of temperature' which is at about −273°C. Hence,

$$\text{kelvin temperature} = \text{degree Celsius} + 273$$

i.e.

$$K = (°C) + 273$$

Thus, for example, 0°C = 273 K, 25°C = 298 K and 100°C = 373 K.

Problem 3.160 Convert the following temperatures into the kelvin scale: (a) 37°C (b) −28°C

From above, kelvin temperature = degree Celsius + 273

(a) 37°C corresponds to a kelvin temperature of 37 + 273, i.e. **310 K**

(b) −28°C corresponds to a kelvin temperature of −28 + 273, i.e. **245 K**.

Problem 3.161 Convert the following temperatures into the Celsius scale: (a) 365 K (b) 213 K

From the above, K = (°C) + 273
Hence, degree Celsius = kelvin temperature − 273

(a) 365 K corresponds to 365 − 273, i.e. **92°C**

(b) 213 K corresponds to 213 − 273, i.e. **−60°C**.

Specific heat capacity

The **specific heat capacity** of a substance is the quantity of heat energy required to raise the temperature of 1 kg of the substance by 1°C. The symbol used for specific heat capacity is c and the units are J/(kg °C) or J(kg K). (Note that these units may also be written as $\text{J kg}^{-1}\,{}^\circ\text{C}^{-1}$ or $\text{J kg}^{-1}\,\text{K}^{-1}$.)

Some typical values of specific heat capacity for the range of temperature 0°C to 100°C include:

Water	4190 J(kg °C)	Ice	2100 J(kg °C)
Aluminium	950 J/(kg °C)	Copper	390 J/(kg °C)
Iron	500 J(kg °C)	Lead	130 J/(kg °C)

Hence to raise the temperature of 1 kg of iron by 1°C requires 500 J of energy, to raise the temperature of 5 kg of iron by 1°C requires (500 × 5) J of energy, and to raise the temperature of 5 kg of iron by 40°C requires (500 × 5 × 40) J of energy, i.e. 100 kJ.

In general, the quantity of heat energy, Q, required to raise a mass m kg of a substance with a specific heat capacity c J/(kg °C) from temperature t_1 °C to t_2 °C is given by:

$$\boldsymbol{Q = mc(t_2 - t_1)\ \mathbf{J}}$$

Problem 3.162 Calculate the quantity of heat required to raise the temperature of 5 kg of water from 0°C to 100°C. Assume the specific heat capacity of water is 4200 J/(kg °C)

$$\begin{aligned}\text{Quantity of heat energy, } Q &= mc(t_2 - t_1)\\ &= 5\text{ kg} \times 4200\text{ J/(kg °C)}\\ &\quad \times (100 - 0)\text{°C}\\ &= 5 \times 4200 \times 100\\ &= \mathbf{2\,100\,000\ J\ or\ 2100\ kJ}\\ &\quad \mathbf{or\ 2.1\ MJ}\end{aligned}$$

Proboem 3.163 A block of cast iron having a mass of 10 kg cools from a temperature of 150°C to 50°C. How much energy is lost by the cast iron? Assume the specific heat capacity of iron is 500 J/(kg °C).

$$\begin{aligned}\text{Quantity of heat energy, } Q &= mc(t_2 - t_1)\\ &= 10\text{ kg} \times 500\text{ J/(kg °C)}\\ &\quad \times (50 - 150)\text{ °C}\\ &= 10 \times 500 \times (-100)\\ &= \mathbf{-500\,000\ J\ or\ -500\ kJ}\\ &\quad \mathbf{or\ -0.5\ MJ}\end{aligned}$$

(Note that the minus sign indicates that heat is given out or lost.)

Problem 3.164 Some lead having a specific heat capacity of 130 J/(kg °C) is heated from 27°C to its melting point at 327°C. If the quantity of heat required is 780 kJ determine the mass of the lead.

Quantity of heat, $Q = mc(t_2 - t_1)$,
hence $780 \times 10^3\text{ J} = m \times 130\text{ J/(kg °C)} \times (327 - 27)\text{ °C}$, i.e.

$$780\,000 = m \times 130 \times 300$$

Test your knowledge 3.36

1 (a) Convert −63°C into the kelvin scale
(b) Change 225 K into the Celsius scale.

2 20.8 kJ of heat energy is required to raise the temperature of 2 kg of lead from 16°C to 96°C. Determine the specific heat capacity of lead.

3 5.7 MJ of heat energy are supplied to 30 kg of aluminium which is initially at a temperature of 20°C. If the specific heat capacity of aluminium is 950 J/(kg °C), determine its final temperature.

from which,

$$\text{mass } m = \frac{780\,000}{130 \times 300}\ \text{kg} = \mathbf{20\ kg}$$

Change of state

A material may exist in any one of three states – solid, liquid or gas. If heat is supplied at a constant rate to some ice initially at, say, −30°C, its temperature rises, as shown in Figure 3.68. Initially the temperature increases from −30°C to 0°C as shown by the line AB. It then remains constant at 0°C for the time BC required for the ice to melt into water.

When melting commences, the energy gained by continuous heating is offset by the energy required for the change of state and the temperature remains constant even though heating is continued. When the ice is completely melted to water, continuous heating raises the temperature to 100°C, as shown by CD in Figure 3.68. The water then begins to boil and the temperature again remains constant at 100°C, shown as DE, until all the water has vaporized. Continuous heating raises the temperature of the steam, as shown by EF in the region where the steam is termed superheated.

Changes of state from solid to liquid or liquid to gas occur without change of temperature and such changes are reversible processes. When heat energy flows to or from a substance and causes a change of temperature, such as between A and B, between C and D and between E and F in Figure 3.68, it is called **sensible heat** (since it can be 'sensed' by a thermometer).

Heat energy which flows to or from a substance while the temperature remains constant, such as between B and C and between D and E in Figure 3.68, is called **latent heat** (latent means concealed or hidden).

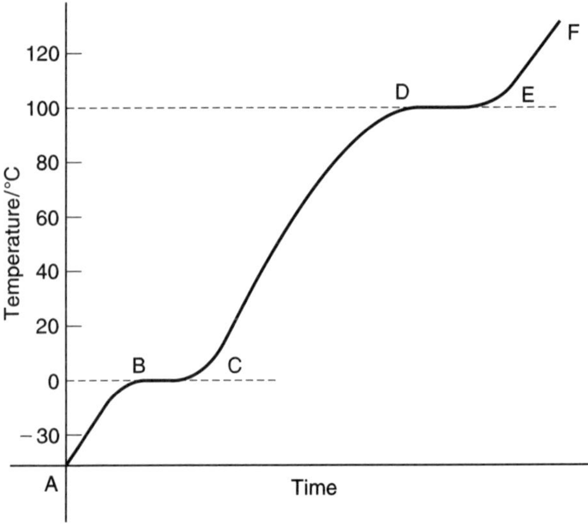

Figure 3.68

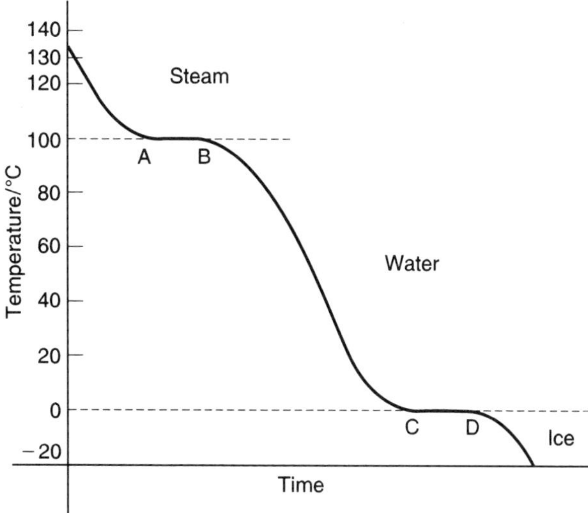

Figure 3.69

Steam, initially at a temperature of 130°C, is cooled to a temperature of 20°C below the freezing point of water, the loss of heat energy being at a constant rate. A temperature/time graph representing the change is shown in Figure 3.69. Initially steam cools until it reaches the boiling point of water at 100°C. The temperature then remains constant, i.e. between A and B, even though it is still giving off heat (i.e. latent heat). When all the steam at 100°C has changed to water at 100°C it starts to cool again until it reaches the freezing point of water at 0°C. From C to D the temperature again remains constant until all the water is converted to ice. The temperature of the ice then decreases as shown.

Latent heats of fusion and vaporization

The **specific latent heat of fusion** is the heat required to change 1 kg of a substance from the solid state to the liquid state (or vice versa) at constant temperature. The **specific latent heat of vaporization** is the heat required to change 1 kg of a substance from a liquid to a gaseous state (or vice versa) at constant temperature. The units of the specific latent heats of fusion and vaporization are J/kg, or more often kJ/kg, and some typical values are shown in Table 3.1. The quantity of heat Q supplied or given out during a change of state is given by:

$$Q = mL$$

where m is the mass in kilograms and L is the specific latent heat. Thus, for example, the heat required to convert 10 kg of ice at 0°C to water at 0°C is given by 10 kg × 335 kJ/kg, i.e. 3350 kJ or 3.35 MJ.

Besides changing temperature, the effects of supplying heat to a material can involve changes in dimensions, as well as in

Table 3.1

	Latent heat of fusion (kJ/kg)	*Melting point (°C)*
Mercury	11.8	−39
Lead	22	327
Silver	100	957
Ice	335	0
Aluminium	387	660
	Latent heat of vaporization (kJ/kg)	*Boiling point (°C)*
Oxygen	214	−183
Mercury	286	357
Ethyl alcohol	857	79
Water	2257	100

colour, state and electrical resistance. Most substances expand when heated and contract when cooled, and there are many practical applications and design implications of thermal movement.

Problem 3.165 How much heat is needed to melt completely 12 kg of ice at 0°C? Assume the latent heat of fusion of ice is 335 kJ/kg.

$$\begin{aligned}\text{Quantity of heat required, } Q &= mL \\ &= 12\,\text{kg} \times 335\,\text{kJ/kg} \\ &= \mathbf{4020\,kJ\ or\ 4.02\,MJ}\end{aligned}$$

Probloem 3.166 Calculate the heat required to convert 5 kg of water at 100°C to superheated steam at 100°C. Assume the latent heat of vaporization of water is 2260 kJ/kg.

$$\begin{aligned}\text{Quantity of heat required, } Q &= mL \\ &= 5\,\text{kg} \times 2260\,\text{kJ/kg} \\ &= \mathbf{11\,300\,kJ\ or\ 11.3\,MJ}\end{aligned}$$

Problem 3.1672 Determine the heat energy needed to convert 5 kg of ice initially at −20°C completely to water at 0°C. Assume the specific heat capacity of ice is 2100 J/(kg °C) and the specific latent heat of fusion of ice is 335 kJ/kg.

Quantity of heat energy needed, Q = sensible heat + latent heat. The quantity of heat needed to raise the temperature of ice from −20°C to 0°C, i.e. sensible heat, is given by

$$\begin{aligned}Q_1 &= mc(t_2 - t_1) \\ &= 5\,\text{kg} \times 2100\,\text{J/(kg °C)} \times (0 - -20)\text{°C} \\ &= (5 \times 2100 \times 20)\,\text{J} = 210\,\text{kJ}\end{aligned}$$

Test your knowledge 3.37

1. Determine the heat energy required to change 8 kg of water at 100°C to superheated steam at 100°C. Assume the specific latent heat of vaporization of water is 2260 kJ/kg.
2. Calculate the heat energy required to convert completely 10 kg of water at 50°C into steam at 100°C, given that the specific heat capacity of water is 4200 J/(kg °C) and the specific latent heat of vaporization of water is 2260 kJ/kg.

The quantity of heat needed to melt 5 kg of ice at 0°C, i.e. the latent heat, $Q_2 = mL = 5\,\text{kg} \times 335\,\text{kJ/kg} = 1675\,\text{kJ}$.
Total heat energy needed, $Q = Q_1 + Q_2 = 210 + 1675 =$ **1885 kJ**.

A simple refrigerator

The boiling point of most liquids may be lowered if the pressure is lowered. In a simple refrigerator a working fluid, such as ammonia or freon, has the pressure acting on it reduced. The resulting lowering of the boiling point causes the liquid to vaporize. In vaporizing, the liquid takes in the necessary latent heat from its surroundings, i.e. the freezer, which thus becomes cooled. The vapour is immediately removed by a pump to a condenser which is outside of the cabinet, where it is compressed and changed back into a liquid, giving out latent heat. The cycle is repeated when the liquid is pumped back to the freezer to be vaporized.

Measurement of temperature

A change in temperature of a substance can often result in a change in one or more of its physical properties. Thus, although temperature cannot be measured directly, its effects can be measured. Some properties of substances used to determine changes in temperature include changes in dimensions, electrical resistance, state, type and volume of radiation and colour.

Temperature measuring devices available are many and varied. Two of those most often used in science and industry are:

1. Liquid-in-glass thermometers
2. Thermocouples.

1. Liquid-in-glass thermometer

A **liquid-in-glass thermometer** uses the expansion of a liquid with increase in temperature as its principle of operation.

Construction

A typical liquid-in-glass thermometer is shown in Figure 3.70 and consists of a sealed stem of uniform small-bore tubing, called a capillary tube, made of glass, with a cylindrical glass bulb formed at one end. The bulb and part of the stem are filled with a liquid such as mercury or alcohol and the remaining part of the tube is evacuated. A temperature scale is formed by etch-

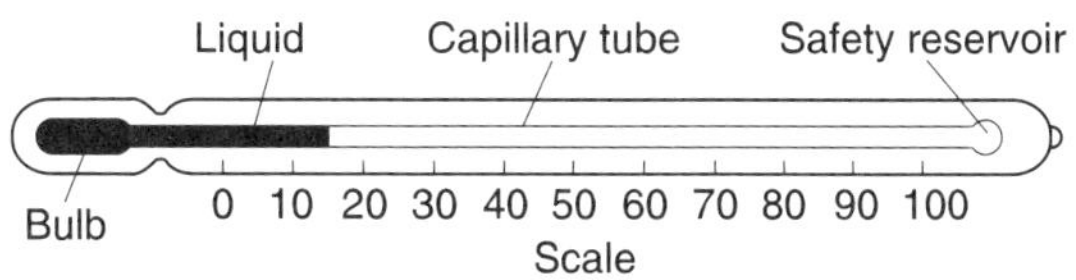

Figure 3.70

ing graduations on the stem. A safety reservoir is usually provided, into which the liquid can expand without bursting the glass if the temperature is raised beyond the upper limit of the scale.

Principle of operation

The operation of a liquid-in-glass thermometer depends on the liquid expanding with increase in temperature and contracting with decrease in temperature. The position of the end of the column of liquid in the tube is a measure of the temperature of the liquid in the bulb – shown as 15°C in Figure 3.70, which is about room temperature. Two fixed points are needed to calibrate the thermometer, with the interval between these points being divided into 'degrees'. In the first thermometer, made by Celsius, the fixed points chosen were the temperature of melting ice (0°C) and that of boiling water at standard atmospheric pressure (100°C), in each case the blank stem being marked at the liquid level. The distance between these two points is called the fundamental interval, and was divided into 100 equal parts, each equivalent to 1°C, thus forming the scale.

The **clinical thermometer**, with a limited scale around body temperature, the **maximum and/or minimum thermometer**, recording the maximum day temperature and minimum night temperature, and the **Beckman thermometer**, which is used only in accurate measurement of temperature change and has no fixed points, are particular types of liquid-in-glass thermometer which all operate on the same principle.

Advantages

The liquid-in-glass thermometer is simple in construction, relatively inexpensive, easy to use and portable, and is the most widely used method of temperature measurement having industrial, chemical, clinical and meteorological applications.

Disadvantages

Liquid-in-glass thermometers tend to be fragile and hence easily broken, can only be used where the liquid column is visible, cannot be used for surface temperature measurements, cannot be read from a distance and are unsuitable for high temperature measurements.

Advantages of mercury

The use of mercury in a thermometer has many advantages, for mercury:

(i) is clearly visible

(ii) has a fairly uniform rate of expansion

(iii) is readily obtainable in the pure state

(iv) does not 'wet' the glass

(v) is a good conductor of heat.

Mercury has a freezing point of −39°C and cannot be used in a thermometer below this temperature. Its boiling point is 357°C but before this temperature is reached some distillation of the mercury occurs if the space above the mercury is a vacuum. To prevent this, and to extend the upper temperature limits to over 500°C, an inert gas such as nitrogen under pressure is used to fill the remainder of the capillary tube. Alcohol, often dyed red to be seen in the capillary tube, is considerably cheaper than mercury and has a freezing point of −113°C, which is considerably lower than for mercury. However it has a low boiling point at about 79°C.

Errors

Typical errors in liquid-in-glass thermometers may occur due to:

(i) the slow cooling rate of glass

(ii) incorrect positioning of the thermometer

(iii) a delay in the thermometer becoming steady (i.e. slow response time)

(iv) non-uniformity of the bore of the capillary tube, which means that equal intervals marked on the stem do not correspond to equal temperature intervals.

2. *Thermocouples*

Thermocouples use the e.m.f. set up when the junction of two dissimilar metals is heated.

Principle of operation

At the junction between two different metals, say, copper and constantan, there exists a difference in electrical potential, which varies with the temperature of the junction. This is known as the 'thermoelectric effect'. If the circuit is completed with a second junction at a different temperature, a current will flow round the circuit. This principle is used in the thermocouple. Two different metal conductors having their ends twisted together are shown in Figure 3.71. If the two junctions are at different temperatures, a current I flows round the circuit.

The deflection on the galvanometer G depends on the difference in temperature between junctions X and Y and is caused by the difference between voltages V_X and V_Y. The higher

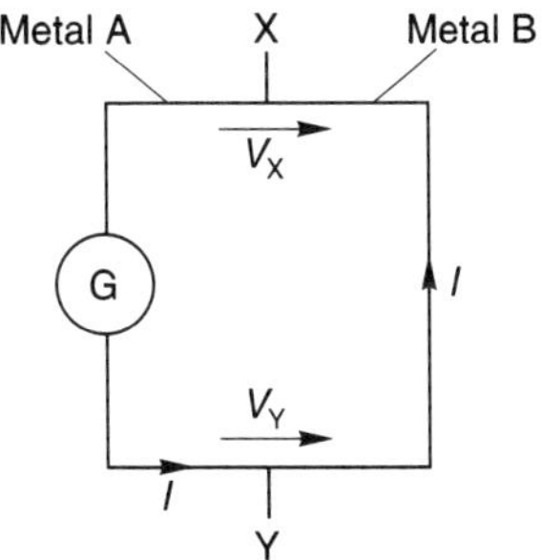

Figure 3.71

temperature junction is usually called the 'hot junction' and the lower temperature junction the 'cold junction'. If the cold junction is kept at a constant known temperature, the galvanometer can be calibrated to indicate the temperature of the hot junction directly. The cold junction is then known as the reference junction.

In many instrumentation situations, the measuring instrument needs to be located far from the point at which the measurements are to be made. Extension leads are then used, usually made of the same material as the thermocouple but of smaller gauge. The reference junction is then effectively moved to their ends. The thermocouple is used by positioning the hot junction where the temperature is required. The meter will indicate the temperature of the hot junction only if the reference junction is at 0°C, for:

(temperature of hot junction) = (temperature of the cold junction) + (temperature difference)

In a laboratory the reference junction is often placed in melting ice, but in industry it is often positioned in a thermostatically controlled oven or buried underground where the temperature is constant.

Construction

Thermocouple junctions are made by twisting together two wires of dissimilar metals before welding them. The construction of a typical copper–constantan thermocouple for industrial use is shown in Figure 3.72. Apart from the actual junction the two conductors used must be insulated electrically from each other with appropriate insulation and is shown in Figure 3.72 as twin-holed tubing. The wires and insulation are usually inserted into a sheath for protection from environments in which they might be damaged or corroded. A copper–constantan thermocouple can measure temperatures from −250°C up to about 400°C, and is used typically with boiler flue gases, food processing and with sub-zero temperature measurement.

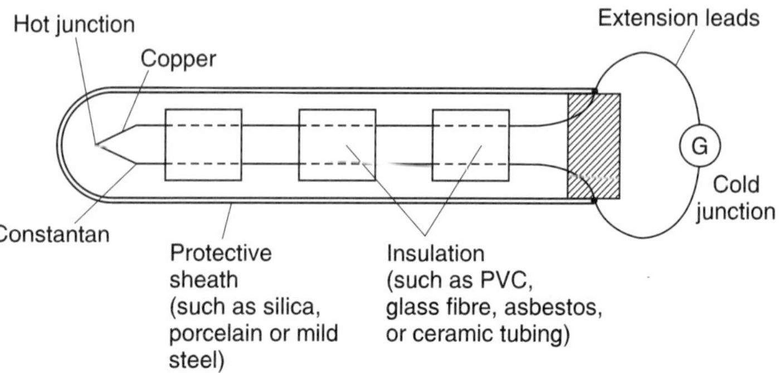

Figure 3.72

Applications

An iron–constantan thermocouple can measure temperatures from −200°C to about 850°C, and is used typically in paper and pulp mills, re-heat and annealing furnaces and in chemical reactors. A chromel–alumel thermocouple can measure temperatures from −200°C to about 1100°C and is used typically with blast furnace gases, brick kilns and in glass manufacture.

For the measurement of temperatures above 1100°C radiation pyrometers are normally used. However, thermocouples are available made of platinum–platinum/rhodium, capable of measuring temperatures up to 1400°C, or tungsten–molybdenum which can measure up to 2600°C.

Advantages

A thermocouple:

(i) has a very simple, relatively inexpensive construction

(ii) can be made very small and compact

(iii) is robust

(iv) is easily replaced if damaged

(v) has a small response time

(vi) can be used at a distance from the actual measuring instrument and is thus ideal for use with automatic and remote-control systems.

Sources of error

Sources of error in the thermocouple which are difficult to overcome include:

(i) voltage drops in leads and junctions

(ii) possible variations in the temperature of the cold junction

(iii) stray thermoelectric effects, which are caused by the addition of further metals into the 'ideal' two-metal thermocouple circuit. Additional leads are frequently necessary for extension leads or voltmeter terminal connections.

A thermocouple may be used with a battery- or mains-operated electronic thermometer instead of a millivoltmeter. These devices amplify the small e.m.f.s from the thermocouple before feeding them to a multi-range voltmeter calibrated directly with temperature scales. These devices have great accuracy and are almost unaffected by voltage drops in the leads and junctions.

Test your knowledge 3.38

State which device would be most suitable to measure the following:

(a) the air in an office in the range 0°C–40°C

(b) boiler flue gas in the range 15°C–300°C

(c) brick in a kiln up to 900°C

(d) an inexpensive method for food processing applications in the range −25°C to −75°C.

Thermal expansion

When heat is applied to most materials, **expansion** occurs in all directions. Conversely, if heat energy is removed from a material (i.e. the material is cooled) **contraction** occurs in all directions.

The effects of expansion and contraction each depend on the **change of temperature** of the material.

Practical applications of thermal expansion

Some practical applications where expansion and contraction of solid materials must be allowed for include:

(i) Overhead electrical transmission lines are hung so that they are slack in summer, otherwise their contraction in winter may snap the conductors or bring down pylons.

(ii) Gaps need to be left in lengths of railway lines to prevent buckling in hot weather (except where these are continuously welded).

(iii) Ends of large bridges are often supported on rollers to allow them to expand and contract freely.

(iv) Fitting a metal collar to a shaft or a steel tyre to a wheel is often achieved by first heating them so that they expand, fitting them in position, and then cooling them so that the contraction holds them firmly in place. This is known as a 'shrink-fit'. By a similar method hot rivets are used for joining metal sheets.

(v) The amount of expansion varies with different materials. Fgure 3.73(a) shows a bimetallic strip at room temperature (i.e. two strips of different metals riveted together). When heated, brass expands more than steel, and since the two metals are riveted together the bimetallic strip is forced into an arc, as shown in Figure 3.73(b). Such a movement can be arranged to make or break an electric circuit and bimetallic strips are used, in particular, in thermostats (which are temperature-operated switches) used to control central heating systems, cookers, refrigerators, toasters, irons, hot-water and alarm systems.

(vi) Motor engines use the rapid expansion of heated gases to force a piston to move.

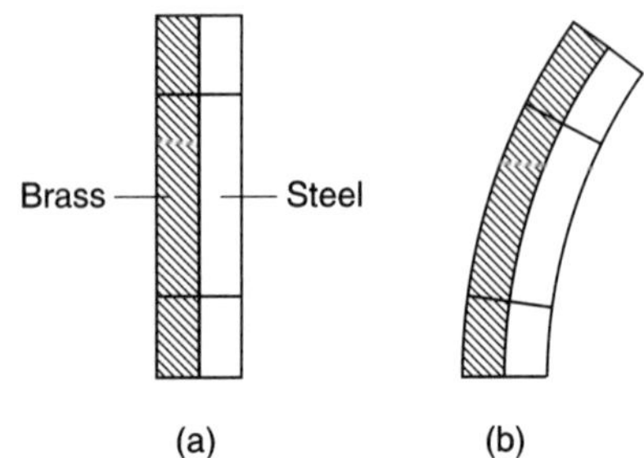

Figure 3.73

(vii) Designers must predict, and allow for, the expansion of steel pipes in a steam-raising plant so as to avoid damage and consequent danger to health.

Expansion and contraction of water

Water is a liquid which at low temperature displays an unusual effect. If cooled, contraction occurs until, at about 4°C, the volume is at a minimum. As the temperature is further decreased from 4°C to 0°C expansion occurs, i.e. the volume increases. When ice is formed, considerable expansion occurs and it is this expansion which often causes frozen water pipes to burst.

A practical application of the expansion of a liquid is with thermometers, where the expansion of a liquid, such as mercury or alcohol, is used to measure temperature.

Coefficient of linear expansion

The amount by which unit length of a material expands when the temperature is raised one degree is called the coefficient of linear expansion of the material and is represented by α (Greek alpha). The units of the coefficient of linear expansion are m/(mK), although it is usually quoted as just /K or K^{-1}. For example, copper has a coefficient of linear expansion value of $17 \times 10^{-6}\,K^{-1}$, which means that a 1 m long bar of copper expands by 0.000 017 m if its temperature is increased by 1 K (or 1°C). If a 6 m long bar of copper is subjected to a temperature rise of 25 K then the bar will expand by $(6 \times 0.000\,017 \times 25)$ m, i.e. 0.002 55 m or 2.55 mm. (Since the kelvin scale uses the same temperature interval as the Celsius scale, a change of temperature of, say, 50°C, is the same as a change of temperature of 50 K.) If a material, initially of length l_1 and at a temperature of t_1 and having a coefficient of linear expansion α, has its temperature increased to t_2, then the new length l_2 of the material is given by:

$$\text{new length} = \text{original length} + \text{expansion}$$

i.e. $l_2 = l_1 + l_1\alpha(t_2 - t_1)$
i.e. $l_2 = l_1[1 + \alpha(t_2 - t_1)]$

Some typical values for the coefficient of linear expansion include:

Aluminium	$23 \times 10^{-6}\,K^{-1}$	Brass	$18 \times 10^{-6}\,K^{-1}$
Concrete	$12 \times 10^{-6}\,K^{-1}$	Copper	$17 \times 10^{-6}\,K^{-1}$
Gold	$14 \times 10^{-6}\,K^{-1}$	Invar (nickel–steel alloy)	$0.9 \times 10^{-6}\,K^{-1}$
Iron	$11–12 \times 10^{-6}\,K^{-1}$	Nylon	$100 \times 10^{-6}\,K^{-1}$
Steel	$15–16 \times 10^{-6}\,K^{-1}$	Tungsten	$4.5 \times 10^{-6}\,K^{-1}$
Zinc	$31 \times 10^{-6}\,K^{-1}$		

Problem 3.168 The length of an iron steam pipe is 20.0 m at a temperature of 18°C. Determine the length of the pipe under working conditions

when the temperature is 300°C. Assume the coefficient of linear expansion of iron is $12 \times 10^{-6}\,K^{-1}$.

Length $l_1 = 20\,m$; temperature $t_1 = 18°C$; $t_2 = 300°C$; $\alpha = 12 \times 10^{-6}\,K^{-1}$

Length of pipe at 300°C is given by

$$\begin{aligned} l_2 &= l_1[1 + \alpha(t_2 - t_1)] \\ &= 20.0[1 + (12 \times 10^{-6})(300 - 18)] \\ &= 20.0[1 + 0.003\,384] = 20.0[1.003\,384] \\ &= \mathbf{20.067\,68\,m} \end{aligned}$$

i.e. an increase in length of 0.067 68 m, i.e. 67.68 mm.

In practice, allowances are made for such expansions. U-shaped expansion joints are connected into pipelines carrying hot fluids to allow some 'give' to take up the expansion.

Problem 3.169 An electrical overhead transmission line has a length of 80.0 m between its supports at 15°C. Its length increases by 92 mm at 65°C. Determine the coefficient of linear expansion of the material of the line.

Length $l_1 = 80.0\,m$; $l_2 = 80.0 + 92\,mm = 80.092\,m$; temperature $t_1 = 15°C$; temperature $t_2 = 65°C$

Length $l_2 = l_1[1 + \alpha(t_2 - t_1)]$, i.e.

$$80.092 = 80.0[1 + \alpha(65 - 15)]$$

$$80.092 = 80.0 + (80.0)(\alpha)(50)$$

i.e.

$$80.092 - 80.0 = (80.0)(\alpha)(50)$$

Hence the coefficient of linear expansion, $\alpha = \dfrac{0.092}{(80.0)(50)}$

$$= 0.000\,023$$

i.e. $\alpha = \mathbf{23 \times 10^{-6}\,K^{-1}}$ (which is aluminium – see above).

Test your knowledge 3.39

1 A length of lead piping is 50.0 m long at a temperature of 16°C. When hot water flows through it the temperature of the pipe rises to 80°C. Determine the length of the hot pipe if the coefficient of linear expansion of lead is $29 \times 10^{-6}\,K^{-1}$.

2 A measuring tape made of copper measures 5.0 m at a temperature of 288 K. Calculate the percentage error in measurement when the temperature has increased to 313 K. Take the coefficient of linear expansion of copper as $17 \times 10^{-6}\,K^{-1}$.

3 The copper tubes in a boiler are 4.20 m long at a temperature of 20°C. Determine the length of the tubes (a) when surrounded only by feed water at 10°C, (b) when the boiler is operating and the mean temperature of the tubes is 320°C. Assume the coefficient of linear expansion of copper to be $17 \times 10^{-6}\,K^{-1}$.

Standard symbols for electrical components

Symbols are used for components in electrical circuit diagrams and some of the more common ones are shown in Figure 3.74.

Electrical current and quantity of electricity

All **atoms** consist of **protons**, **neutrons** and **electrons**. The protons, which have positive electrical charges, and the neutrons, which have no electrical charge, are contained within the **nucleus**. Removed from the nucleus are minute negatively charged particles called electrons. Atoms of different materials differ from one another by having different numbers of protons, neutrons and electrons. An equal number of protons and electrons exist within an atom and it is said to be electrically balanced, as the positive and negative charges cancel each other out. When there

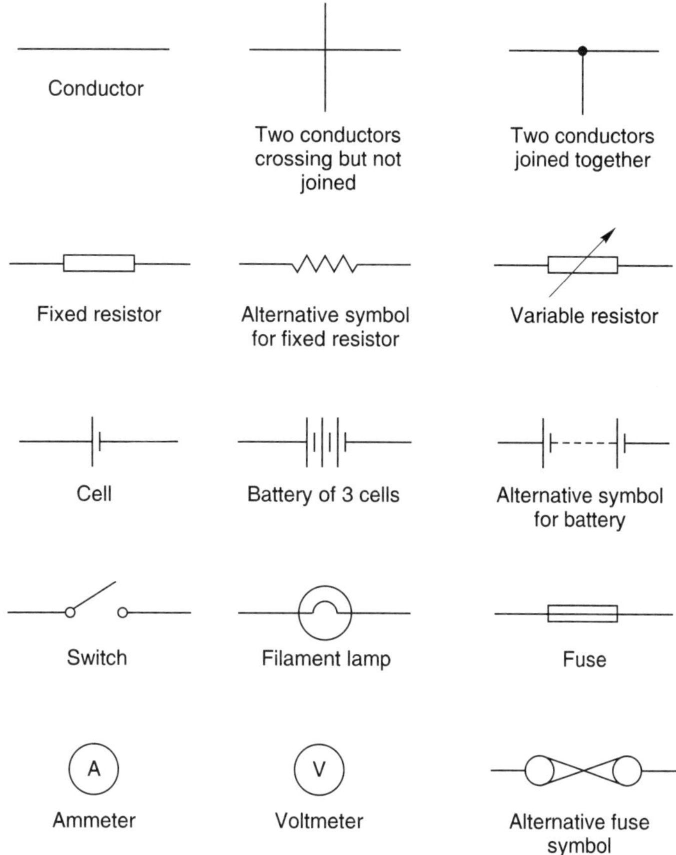

Figure 3.74

are more than two electrons in an atom the electrons are arranged into **shells** at various distances from the nucleus.

All atoms are bound together by powerful forces of attraction existing between the nucleus and its electrons. Electrons in the outer shell of an atom, however, are attracted to their nucleus less powerfully than are electrons whose shells are nearer the nucleus.

It is possible for an atom to lose an electron; the atom, which is now called an **ion**, is not now electrically balanced, but is positively charged and is thus able to attract an electron to itself from another atom. Electrons that move from one atom to another are called free electrons and such random motion can continue indefinitely. However, if an electric pressure or **voltage** is applied across any material there is a tendency for electrons to move in a particular direction. This movement of free electrons, known as **drift**, constitutes an electric current flow.

Thus, current is the rate of movement of charge.

Conductors are materials that have electrons that are loosely connected to the nucleus and can easily move through the material from one atom to another. **Insulators** are materials whose electrons are held firmly to their nucleus. The unit used to measure the **quantity of electrical charge** Q is called the **coulomb C** (where 1 coulomb $= 6.24 \times 10^{18}$ electrons). If the drift of electrons in a conductor takes place at the rate of one coulomb per second the resulting current is said to be a current of one ampere. Thus

$$\mathbf{1\ ampere = 1\ coulomb\ per\ second}$$

or $1\,\text{A} = 1\,\text{C/s}$

Hence

$$1\ \text{coulomb} = 1\ \text{ampere second or } 1\,\text{C} = 1\,\text{A s}$$

Generally, if I is the current in amperes and t the time in seconds during which the current flows, then $I \times t$ represents the quantity of electrical charge in coulombs, i.e. quantity of electrical charge transferred.

$$\boldsymbol{Q = I \times t}\ \textbf{coulombs}$$

Problem 3.170 If a current of 10 A flows for 4 min, find the quantity of electricity transferred.

Quantity of electricity, $Q = It$ coulombs. $I = 10\,\text{A}$, $t = 4 \times 60 = 240\,\text{s}$
Hence $Q = 10 \times 240 = \mathbf{2400\,C}$.

Potential difference and resistance

For a continuous current to flow between two points in a circuit a **potential difference (p.d.)** or **voltage**, V, is required between them; a complete conducting path is necessary to and from the source of electrical energy. The unit of p.d. is the **volt, V**.

Figure 3.75 shows a cell connected across a filament lamp. Current flow, by convention, is considered as flowing from the positive terminal of the cell around the circuit to the negative terminal.

The flow of electric current is subject to friction. This friction, or opposition, is called **resistance** $\boldsymbol{R}$ and is the property of a conductor that limits current. The unit of resistance is the **ohm Ω**. 1 Ω is defined as the resistance which will have a current of 1 A flowing through it when 1 V is connected across it, i.e.

$$\textbf{resistance } \boldsymbol{R} = \frac{\textbf{potential difference}}{\textbf{current}}$$

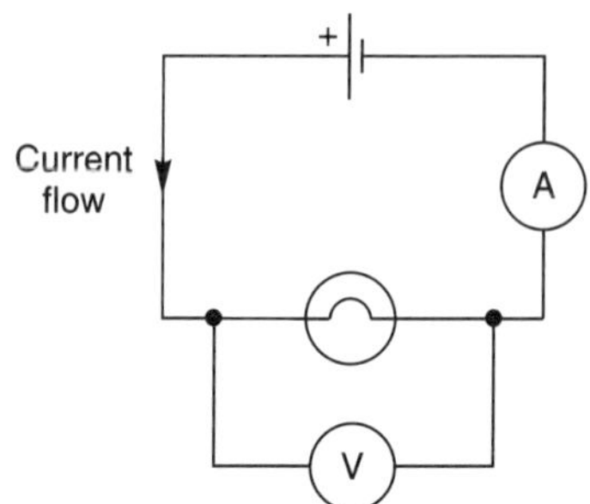

Figure 3.75

Basic electrical measuring instruments

An **ammeter** is an instrument used to measure current and must be connected in **series** with the circuit. Figure 3.75 shows an ammeter connected in series with a lamp to measure the current flowing through it. Since all the current in the circuit passes through the ammeter it must have a very **low resistance**.

A **voltmeter** is an instrument used to measure p.d. and must be connected in **parallel** with the part of the circuit whose p.d. is required. In Figure 3.75, a voltmeter is connected in parallel with the lamp to measure the p.d. across it. To avoid a significant current flowing through it a voltmeter must have a very **high resistance**.

An **ohmmeter** is an instrument for measuring resistance.

A **multimeter**, or universal instrument, may be used to measure voltage, current and resistance. An 'Avometer' is a typical example.

The **cathode ray oscilloscope** (CRO) may be used to observe waveforms and to measure voltages and currents. The display of a CRO involves a spot of light moving across a screen. The amount by which the spot is deflected from its initial position depends on the p.d. applied to the terminals of the CRO and the range selected. The displacement is calibrated in 'volts per cm'. For example, if the spot is deflected 3 cm and the volts/cm switch is on 10 V/cm then the magnitude of the p.d. is 3 cm × 10 V/cm, i.e. 30 V.

Ohm's law

Ohm's law states that the current I flowing in a circuit is directly proportional to the applied voltage V and inversely proportional to the resistance R, provided the temperature remains constant. Thus,

$$\boldsymbol{I = \frac{V}{R}} \text{ or } \boldsymbol{V = IR} \text{ or } \boldsymbol{R = \frac{V}{I}}$$

Problem 3.171 The current flowing through a resistor is 0.8 A when a p.d. of 20 V is applied. Determine the value of the resistance.

From Ohm's law,

$$\text{resistance } R = \frac{V}{I} = \frac{20}{0.8} = \frac{200}{8} = \mathbf{25\,\Omega}$$

Test your knowledge 3.40

1. What current must flow if 0.24 coulombs is to be transferred in 15 ms?
2. A p.d. of 50 V is applied across a heating element. If the resistance of the element is 12.5 Ω, find the current flowing through it.

Multiples and sub-multiples

Currents, voltages and resistances can often be very large or very small. Thus **multiples** and **sub-multiples** of units are often used. The most common ones, with an example of each, are listed in Table 3.2.

Table 3.2

Prefix	Name	Meaning	Example
M	mega	multiply by 1 000 000 (i.e. $\times 10^6$)	$2\,M\Omega = 2\,000\,000\,\Omega$
k	kilo	multiply by 1000 (i.e. $\times 10^3$)	$10\,kV = 10\,000\,V$
m	milli	divide by 1000 (i.e. $\times 10^{-3}$)	$25\,mA = \frac{25}{1000}\,A = 0.025\,A$
μ	micro	divide by 1 000 000 (i.e. $\times 10^{-6}$)	$50\,\mu V = \frac{50}{1\,000\,000}\,V = 0.000\,05\,V$

Problem 3.172 Determine the p.d. which must be applied to a 2 kΩ resistor in order that a current of 10 mA may flow.

Resistance $R = 2\,k\Omega = 2 \times 10^3 = 2000\,\Omega$

$$\text{Current } I = 10\,mA = 10 \times 10^{-3}\,A \text{ or } \frac{10}{10^3}\,A \text{ or } \frac{10}{1000}\,A = 0.01\,A.$$

From Ohm's law, potential difference, $V = IR = (0.01)(2000)$ $= \mathbf{20\,V}$

Problem 3.173 A coil has a current of 50 mA flowing through it when the applied voltage is 12 V. What is the resistance of the coil?

$$\text{Resistance, } R = \frac{V}{I} = \frac{12}{50 \times 10^{-3}} = \frac{12 \times 10^3}{50} = \frac{12\,000}{50} = \mathbf{240\,\Omega}$$

Test your knowledge 3.41

1. A 100 V battery is connected across a resistor and causes a current of 5 mA to flow. Determine the resistance of the resistor. If the voltage is now reduced to 25 V, what will be the new value of the current flowing?
2. What is the resistance of a coil which draws a current of (a) 50 mA and (b) 200 {μA from a 120 V supply?

Conductors and insulators

A **conductor** is a material having a low reistance which allows electric current to flow in it. All metals are conductors and some examples include copper, aluminium, brass, platinum, silver, gold and also carbon.

An **insulator** is a material having a high resistance which does not allow electric current to flow in it. Some examples of insulators include plastic, rubber, glass, porcelain, air, paper, cork, mica, ceramics and certain oils.

Electrical power

Power *P* in an electrical circuit is given by the product of potential difference V and current I. The unit of power is the **watt, W**. Hence

$$\boldsymbol{P = V \times I}\ \text{W} \qquad (1)$$

From Ohm's law, $V = IR$. Substituting for V in equation (1) gives:

$$P = (IR) \times I$$

$$\text{i.e. } \boldsymbol{P = I^2R} \text{ W}$$

Also, from Ohm's law, $I = V/R$. Substituting for I in equation (1) gives:

$$P = V \times \frac{V}{R}$$

$$\text{i.e. } \boldsymbol{P = \frac{V^2}{R}} \text{ W}$$

There are thus three possible formulae which may be used for calculating power.

Problem 3.174 A 100 W electric light bulb is connected to a 250 V supply. Determine (a) the current flowing in the bulb, and (b) the resistance of the bulb.

Power $P = V \times I$, from which, current $I = P/V$

$$\text{(a) Current } I = \frac{100}{250} = \frac{10}{25} = \frac{2}{5} = \mathbf{0.4\,A}$$

$$\text{(b) Resistance } R = \frac{V}{I} = \frac{250}{0.4} = \frac{2500}{4} = \mathbf{625\,\Omega}$$

Problem 3.175 An electric kettle has a resistance of 30 Ω. What current will flow when it is connected to a 240 V supply? Find also the power rating of the kettle.

$$\text{Current, } I = \frac{V}{R} = \frac{240}{30} = \mathbf{8\,A}$$

Power, $P = VI = 240 \times 8 = 1920\text{ W} = \mathbf{1.92\,kW}$ = power rating of kettle.

Problem 3.176 The current/voltage relationship for two resistors A and B is as shown in Figure 3.76. Determine the value of the resistance of each resistor.

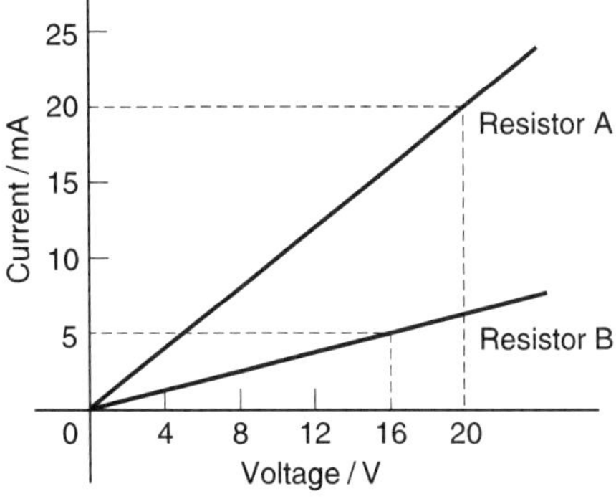

Figure 3.76

For resistor A,

$$R = \frac{V}{I} = \frac{20\,\text{V}}{20\,\text{mA}} = \frac{20}{0.02} = \frac{2000}{2} = \mathbf{1000\,\Omega \text{ or } 1\,k\Omega}$$

For resistor B,

$$R = \frac{V}{I} = \frac{16\,\text{V}}{5\,\text{mA}} = \frac{16}{0.005} = \frac{16\,000}{5} = \mathbf{3200\,\Omega \text{ or } 3.02\,k\Omega}$$

Problem 3.177 The hot resistance of a 240 V filament lamp is 960 Ω. Find the current taken by the lamp and its power rating.

From Ohm's law,

$$\text{current, } I = \frac{V}{R} = \frac{240}{960} = \frac{24}{96} = \mathbf{\frac{1}{4}\,A \text{ or } 0.25\,A}$$

$$\text{Power rating } P = VI = (240)\left(\frac{1}{4}\right) = \mathbf{60\,W}$$

Electrical energy

$$\text{Electrical energy} = \text{power} \times \text{time}$$

If the power is measured in watts and the time in seconds then the unit of energy is watt-seconds or **joules**. If the power is measured in kilowatts and the time in hours then the unit of energy is **kilowatt-hours**, often called the '**unit of electricity**'. The 'electricity meter' in the home records the number of kilowatt-hours used and is thus an energy meter.

Problem 3.178 A 12 V battery is connected across a load having a resistance of 40 Ω. Determine the current flowing in the load, the power consumed and the energy dissipated in 2 min.

$$\text{Current } I = \frac{V}{R} = \frac{12}{40} = \mathbf{0.3\,A}$$

Power consumed, $P = VI = (12)(0.3) = \mathbf{3.6\,W}$
Energy dissipated = power × time = $(3.6\,\text{W})(2 \times 60\,\text{s}) = \mathbf{432\,J}$ (since $1\,\text{J} = 1\,\text{W s}$).

Problem 3.179 A source of 15 V supplies a current of 2 A for 6 min. How much energy is provided in this time?

Energy = power × time, and power = voltage × current. Hence

$$\begin{aligned} \text{energy} = VIt &= 15 \times 2 \times (6 \times 60) \\ &= 10\,800\,\text{W s or J} \\ &= \mathbf{10.8\,kJ} \end{aligned}$$

Problem 3.180 Electrical equipment in an office takes a current of 13 A from a 240 V supply. Estimate the cost per week of electricity if the equipment is used for 30 hours each week and 1 kWh of energy costs 7p.

Power $= VI$ W $= 240 \times 13 = 3120$ W $= 3.12$ kW
Energy used per week = power × time = (3.12 kW) × (30 h) = 93.6 kW h
Cost at 7p per kWh $= 93.6 \times 7 = 655.2$ p. Hence

weekly cost of electricity = £6.55

Problem 3.181 An electric heater consumes 3.6 MJ when connected to a 240 V supply for 40 min. Find the power rating of the heater and the current taken from the supply.

$$\text{Power} = \frac{\text{energy}}{\text{time}} = \frac{3.6 \times 10^6\,\text{J}}{40 \times 60\,\text{s}} \text{ (or W)}$$
$$= 1500\,\text{W}$$

i.e. power rating of heater = **1.5 kW**
Power $P = VI$, thus

$$I = \frac{P}{V} = \frac{1500}{250} = 6\,\text{A}$$

Hence the current taken from the supply is **6 A**.

Test your knowledge 3.42

1 Calculate the power dissipated when a current of 4 mA flows through a resistance of 5 kΩ.

2 A current of 5 A flows in the winding of an electric motor, the resistance of the winding being 100 Ω. Determine (a) the p.d. across the winding, and (b) the power dissipated by the coil.

3 Determine the power dissipated by the element of an electric fire of resistance 20 Ω when a current of 10 A flows through it. If the fire is on for 6 hours determine the energy used and the cost if 1 unit of electricity costs 7p.

4 A business uses two 3 kW fires for an average of 20 hours each per week, and six 150 W lights for 30 hours each per week. If the cost of electricity is 7p per unit, determine the weekly cost of electricity to the business.

Test your knowledge 3.43

If 5 A, 10 A and 13 A fuses are available, state which is most appropriate for the following appliances which are both connected to a 240 V supply.

(a) Electric toaster having a power rating of 1 kW

(b) Electric fire having a power rating of 3 kW.

Main effects of electric current

The three main effects of an electric current are:

(a) magnetic effect

(b) chemical effect

(c) heating effect.

Some practical applications of the effects of an electric current include:

Magnetic effect: bells, relays, motors, generators, transformers, telephones, car ignition and lifting magnets

Chemical effect: primary and secondary cells and electroplating

Heating effect: cookers, water heaters, electric fires, irons, furnaces, kettles and soldering irons.

Fuses

A **fuse** is used to prevent overloading of electrical circuits. The fuse, which is made of material having a low melting point, utilizes the heating effect of an electric current. A fuse is placed in an electrical circuit and if the current becomes too large the fuse wire melts and so breaks the circuit. A circuit diagram symbol for a fuse is shown in Figure 3.74.

Resistance and resistivity

The resistance of an electrical conductor depends on four factors, these being: (a) the length of the conductor, (b) the cross-sectional area of the conductor, (c) the type of material and (d) the temperature of the material.

Resistance, R, is directly proportional to length, l, of a conductor, i.e. $R \propto l$. Thus, for example, if the length of a piece of wire is doubled, then the resistance is doubled.

Resistance, R, is inversely proportional to cross-sectional area, a, of a conductor, i.e. $R \propto 1/a$. Thus, for example, if the cross-sectional area of a piece of wire is doubled then the resistance is halved.

Since $R \propto l$ and $R \propto 1/a$ then $R \propto l/a$. By inserting a constant of proportionality into this relationship the type of material used may be taken into account. The constant of proportionality is known as the **resistivity** of the material and is given the symbol ρ (rho). Thus

$$\text{resistance } \boldsymbol{R} = \frac{\boldsymbol{\rho l}}{\boldsymbol{a}}\ \Omega$$

ρ is measured in ohm metres (Ω m). The value of the resistivity is that resistance of a unit cube of the material measured between opposite faces of the cube.

Resistivity varies with temperature and some typical values of resistivities measured at about room temperature are given below:

Copper	$1.7 \times 10^{-8}\ \Omega\,\text{m}$ (or $0.017\ \mu\Omega\,\text{m}$)
Aluminium	$2.6 \times 10^{-8}\ \Omega\,\text{m}$ (or $0.026\ \mu\Omega\,\text{m}$)
Carbon (graphite)	$10 \times 10^{-8}\ \Omega\,\text{m}$ (or $0.10\ \mu\Omega\,\text{m}$)
Glass	$1 \times 10^{10}\ \Omega\,\text{m}$
Mica	$1 \times 10^{13}\ \Omega\,\text{m}$

Note that good conductors of electricity have a low value of resistivity and good insulators have a high value of resistivity.

Problem 3.182 The resistance of a 5 m length of wire is 600 Ω. Determine (a) the resistance of an 8 m length of the same wire, and (b) the length of the same wire when the resistance is 420 Ω.

(a) Resistance, R, is directly proportional to length, l, i.e. $R \propto l$. Hence, $600\ \Omega \propto 5\ \text{m}$ or $600 = (k)(5)$, where k is the coefficient of proportionality. Hence

$$k = \frac{600}{5} = 120$$

When the length l is 8 m, then

$$\text{resistance } R = kl = (120)(8) = \mathbf{960\ \Omega}$$

(b) When the resistance is 420 Ω, $420 = kl$, from which

$$\text{length } l = \frac{420}{k} = \frac{420}{120} = \mathbf{3.5\ m}$$

Problem 3.183 A piece of wire of cross-sectional area $2\,\text{mm}^2$ has a resistance of $300\,\Omega$. Find (a) the resistance of a wire of the same length and material if the cross-sectional area is $5\,\text{mm}^2$, (b) the cross-sectional area of a wire of the same length and material of resistance $750\,\Omega$.

Resistance R is inversely proportional to cross-sectional area, a, i.e. $R \propto 1/a$. Hence

$$300\,\Omega \propto \frac{1}{2\,\text{mm}^2} \text{ or } 300 = (k)\left(\frac{1}{2}\right)$$

from which the coefficient of proportionality, $k = 300 \times 2 = 600$.

(a) When the cross-sectional area $a = 5\,\text{mm}^2$ then

$$R = (k)\left(\frac{1}{5}\right)$$
$$= (600)\left(\frac{1}{5}\right) = \mathbf{120\,\Omega}$$

(Note that resistance has decreased as the cross-sectional area is increased.)

(b) When the resistance is $750\,\Omega$ then $750 = (k)(1/a)$, from which

$$\text{cross-sectional area, } a = \frac{k}{750} = \frac{600}{750} = \mathbf{0.8\,mm^2}$$

Problem 3.184 Calculate the resistance of a 2 km length of aluminium over-head power cable if the cross-sectional area of the cable is $100\,\text{mm}^2$. Take the resistivity of aluminium to be $0.03 \times 10^{-6}\,\Omega\,\text{m}$.

Length $l = 2\,\text{km} = 2000\,\text{m}$; area, $a = 100\,\text{mm}^2 = 100 \times 10^{-6}\,\text{m}^2$; resistivity $= \rho = 0.03 \times 10^{-6}\,\Omega\,\text{m}$

$$\text{Resistance } R = \frac{\rho l}{a} = \frac{(0.03 \times 10^{-6}\,\Omega\,\text{m})(2000\,\text{m})}{(100 \times 10^{-6}\,\text{m}^2)}$$
$$= \frac{0.03 \times 2000}{100}\,\Omega = \mathbf{0.6\,\Omega}$$

Test your knowledge 3.44

1 A wire of length 8 m and cross-sectional area $3\,\text{mm}^2$ has a resistance of $0.16\,\Omega$. If the wire is drawn out until its cross-sectional area is $1\,\text{mm}^2$, determine the resistance of the wire.

2 The resistance of 1.5 km of wire of cross-sectional area $0.17\,\text{mm}^2$ is $150\,\Omega$. Determine the resistivity of the wire.

3 Determine the resistance of 1200 m of copper cable having a diameter of 12 mm if the resistivity of copper is $1.7 \times 10^{-8}\,\Omega\,\text{m}$.

Temperature coefficient of resistance

In general, as the temperature of a material increases, most conductors increase in resistance, insulators decrease in resistance, while the resistance of some special alloys remains almost constant.

The **temperature coefficient of resistance** of a material is the increase in the resistance of a $1\,\Omega$ resistor of that material when it is subjected to a rise of temperature of 1°C. The symbol used for the temperature coefficient of resistance is α (alpha). Thus, if some copper wire of resistance $1\,\Omega$ is heated through 1°C and its resistance is then measured as $1.0043\,\Omega$ then $\alpha = 0.0043\,\Omega/\Omega\,°\text{C}$ for copper. The units are usually expressed only as 'per °C', i.e. $\alpha = 0.0043/°\text{C}$ for copper. If the $1\,\Omega$ resistor of copper is heated through 100°C then the resistance at 100°C would be

$$1 + 100 \times 0.0043 = 1.43\,\Omega.$$

Some typical values of the temperature coefficient of resistance measured at 0°C are given below:

Copper	0.0043/°C	Aluminium	0.0038/°C
Nickel	0.0062/°C	Carbon	−0.000 48/°C
Constantan	0	Eureka	0.000 01/°C

(Note that the negative sign for carbon indicates that its resistance falls with increase of temperature.)

If the resistance of a material at 0°C is known the resistance at any other temperature can be determined from:

$$R_\Theta = R_0(1 + \alpha_0\theta)$$

where R_0 = resistance at 0°C
R_θ = resistance at temperature θ°C
α_0 = temperature coefficient of resistance at 0°C.

Problem 3.185 A coil of copper wire has a resistance of 100 Ω when its temperature is 0°C. Determine its resistance at 100°C if the temperature coefficient of resistance of copper at 0°C is 0.0043/°C.

Resistance $R_\Theta = R_0(1 + \alpha_0\theta)$.
Hence resistance at 100°C is given by

$$R_{100} = 100[1 + (0.0043)(100)]$$
$$= 100[1 + 0.43] = 100(1.43) = \mathbf{143\,\Omega}$$

Problem 3.186 A carbon resistor has a resistance of 1 kΩ at 0°C. Determine its resistance at 80°C. Assume that the temperature coefficient of resistance for carbon at 0°C is −0.0005.

Resistance at temperature θ°C is given by

$$R_\Theta = R_0(1 + \alpha_0\theta)$$

i.e.

$$R_\Theta = 1000[(1 + (-0.0005)(80)]$$
$$= 1000(1 - 0.040) = 1000(0.96) = \mathbf{960\,\Omega}$$

Test your knowledge 3.45

An aluminium cable has a resistance of 27 Ω at a temperature of 35°C. Determine its resistance at 0°C. Take the temperature coefficient of resistance at 0°C to be 0.0038/°C.

The simple cell

The purpose of an **electric cell** is to convert chemical energy into electrical energy. A **simple cell** comprises two dissimilar conductors (electrodes) in an electrolyte. Such a cell is shown in Figure 3.77, comprising copper and zinc electrodes. An electric current is found to flow between the electrodes. Other possible electrode pairs exist, including zinc–lead and zinc–iron. The electrode potential (i.e. the p.d. measured between the electrodes) varies for each pair of metals. By knowing the e.m.f. of each metal with respect to some standard electrode the e.m.f. of any pair of metals may be determined.

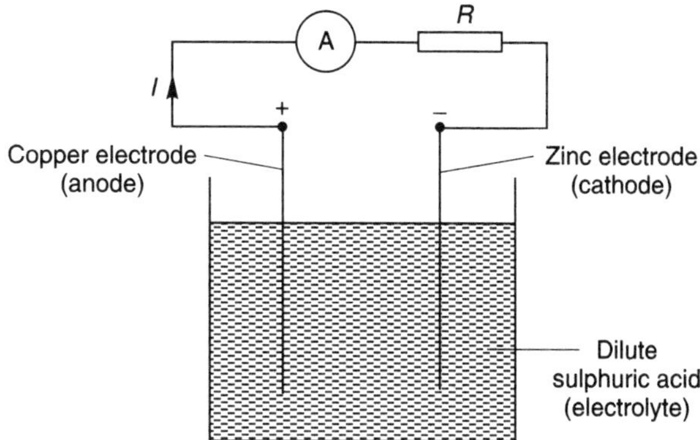

Figure 3.77

E.m.f. and internal resistance of a cell

The **electromotive force (e.m.f.)**, ***E***, of a cell is the p.d. between its terminals when it is not connected to a load (i.e. the cell is on 'no-load').

The e.m.f. of a cell is measured by using a **high resistance voltmeter** connected in parallel with the cell. The voltmeter must have a high resistance otherwise it will pass current and the cell will not be on no-load. For example, if the resistance of a cell is 1 Ω and that of a voltmeter 1 MΩ then the equivalent resistance of the circuit is 1 MΩ + 1 Ω, i.e. approximately 1 MΩ, hence no current flows and the cell is not loaded.

The voltage available at the terminals of a cell falls when a load is connected. This is caused by the **internal resistance** of the cell which is the opposition of the material of the cell to the flow of current. The internal resistance acts in series with other resistances in the circuit. Figure 3.78 shows a cell of e.m.f. E volts and internal resistance, R, and XY represents the terminals of the cell.

When a load (shown as resistance R) is not connected, no current flows and the terminal p.d., $V = E$. When R is connected a current I flows which causes a voltage drop in the cell, given by Ir. The p.d. available at the cell terminals is less than the e.m.f. of the cell and is given by:

$$\boldsymbol{V = E - Ir}$$

Thus if a battery of e.m.f. 12 volts and internal resistance 0.1 Ω delivers a current of 100 A, the terminal p.d. is given by:

$$V = 12 - (100)(0.01)$$
$$= 12 - 1 = 11\,\text{V}$$

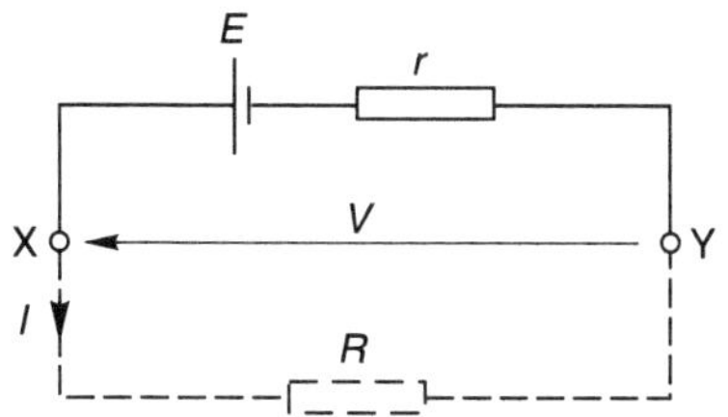

Figure 78

When a current is flowing in the direction shown in Figure 3.78 the cell is said to be **discharging** ($E > V$). When a current flows in the opposite direction to that shown in Figure 3.78 the cell is said to be **charging** ($V > E$).

A battery is a combination of more than one cell. The cells in a battery may be connected in series or in parallel.

(i) For cells connected in series:
Total e.m.f. = sum of cell e.m.f.s
Total internal resistance = sum of cell internal resistances

(ii) For cells connected in parallel:
If each cell has the same e.m.f. and internal resistance:
Total e.m.f. = e.m.f. of one cell
Total internal resistance of n cells
$= \frac{1}{n} \times$ internal resistance of one cell

Problem 3.187 Eight cells, each with an internal resistance of 0.2 Ω and an e.m.f. of 2.2 V are connected (a) in series, (b) in parallel. Determine the e.m.f. and the internal resistance of the batteries so formed.

(a) When connected in series, total e.m.f. = sum of cell e.m.f.s
$= 2.2 \times 8 = \mathbf{17.6\,V}$

Total internal resistance = sum of cell internal resistances
$= 0.2 \times 8 = \mathbf{1.6\,\Omega}$

(b) When connected in parallel, total e.m.f. = e.m.f. of one cell
$= \mathbf{2.2\,V}$

Total internal resistance of 8 cells

$$= \frac{1}{8} \times \text{internal resistance of one cell}$$
$$= \frac{1}{8} \times 0.2 = \mathbf{0.025\,\Omega}$$

Problem 3.188 A cell has an internal resistance of 0.02 Ω and an e.m.f. of 2.0 V. Calculate its terminal p.d. if it delivers (a) 5 A, (b) 50 A.

(a) Terminal p.d., $V = E - Ir$ where E = e.m.f. of cell, I = current flowing and r = internal resistance of cell. $E = 2.0$ V, $I = 5$ A and $r = 0.02\,\Omega$, hence

$$V = 2.0 - (5)\,(0.02) = 2.0 - 0.1 = \mathbf{1.9\,V}$$

(b) When the current is 50 A, terminal p.d. is given by

$$V = E - Ir = 2.0 - 50(0.02)$$

i.e. $V = 2.0 - 1.0 = \mathbf{1.0\,V}$

Thus the terminal p.d. decreases as the current drawn increases.

Problem 3.189 The p.d. at the terminals of a battery is 25 V when no-load is connected and 24 V when a load taking 10 A is connected. Determine the internal resistance of the battery.

Test your knowledge 3.46

1 A cell has an internal resistance of 0.03 Ω and an e.m.f. of 2.2 V. Calculate its terminal p.d. if it delivers (a) 1 A, (b) 20 A, (c) 50 A.

2 Ten 1.5 V cells, each having an internal resistance of 0.2 Ω, are connected in series to a load of 58 Ω. Determine (a) the current flowing in the circuit and (b) the p.d. at the battery terminals.

When no-load is connected the e.m.f. of the battery, E, is equal to the terminal p.d., V, i.e. $E = 25\,\text{V}$. When current $I = 10\,\text{A}$ and terminal p.d. $V = 24\,\text{V}$, then $V = E - Ir$, i.e. $25 = 25 - (10)r$ Hence, rearranging, gives

$$10r = 25 - 24 = 1$$

$$\text{and the internal resistance } r = \frac{1}{10} = \mathbf{0.1\,\Omega}$$

Primary cells

Primary cells cannot be recharged, that is, the conversion of chemical energy to electrical energy is irreversible and the cell cannot be used once the chemicals are exhausted. Examples of primary cells include the Leclanché cell and the mercury cell.

(a) Leclanché cell

A typical dry Leclanché cell is shown in Figure 3.79. Such a cell has an e.m.f. of about 1.5 V when new, but this falls rapidly if in continuous use due to polarization. The hydrogen film on the carbon electrode forms faster than can be dissipated by the depolarizer. The Leclanché cell is suitable only for intermittent use, applications including torches, transistor radios, bells, indicator circuits, gas lighters, controlling switch-gear, and so on. The cell is the most commonly used of primary cells, is cheap, requires little maintenance and has a shelf life of about 2 years.

(b) Mercury cell

A typical mercury cell is shown in Figure 3.80. Such a cell has an e.m.f. of about 1.3 V which remains constant for a relatively long time. Its main advantages over the Leclanché cell is its smaller

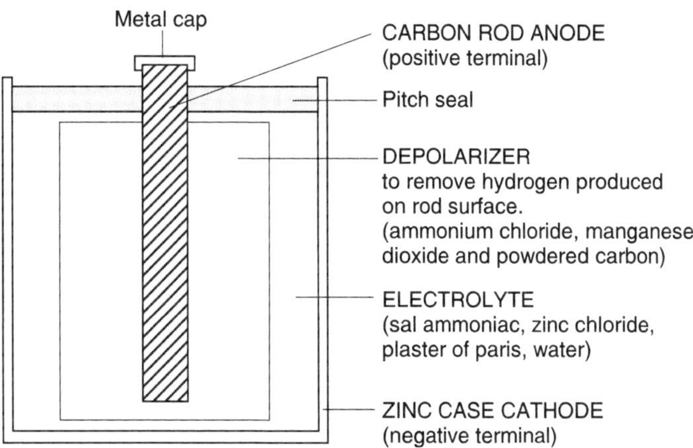

Figure 3.79 *Dry Leclanché cell*

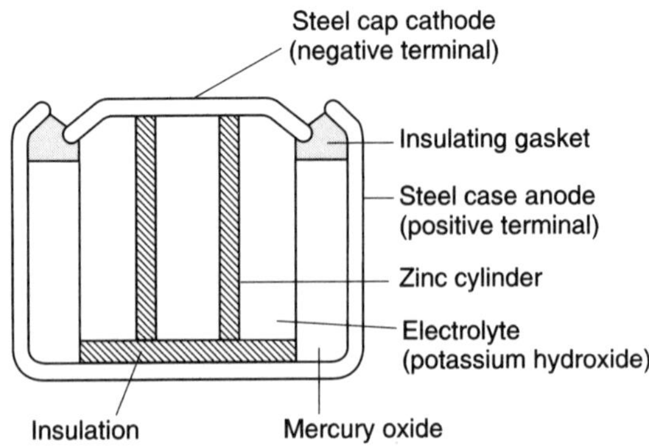

Figure 3.80 *Mercury cell*

size and its long shelf life. Typical practical applications include hearing aids, medical electronics and guided missiles.

Secondary cells

Secondary cells can be recharged after use, that is, the conversion of chemical energy to electrical energy is reversible and the cell may be used many times. Examples of secondary cells include the lead–acid cell and alkaline cells. Practical applications of such cells include car batteries, telephone circuits and for traction purposes – such as milk delivery vans and fork lift trucks.

(a) Lead–acid cell

A typical lead–acid cell is constructed of:

(i) A container made of glass, ebonite or plastic.

(ii) Lead plates
 (a) the negative plate (cathode) consists of spongy lead
 (b) the positive plate (anode) is formed by pressing lead peroxide into the lead grid.

The plates are interleaved, as shown in the plan view of Figure 3.81, to increase their effective cross-sectional area and to minimize internal resistance.

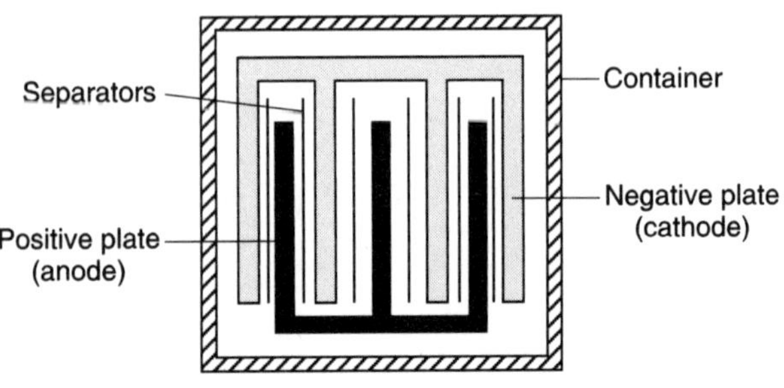

Figure 3.81 *Plan view of lead–acid cell*

(iii) **Separators** made of glass, celluloid or wood.

(iv) An **electrolyte** which is a mixture of sulphuric acid and distilled water.

The relative density of a lead–acid cell, which may be measured using a hydrometer, varies between about 1.26 when the cell is fully charged to about 1.19 when discharged. The terminal p.d. of a lead–acid cell is about 2 V.

When a cell supplies current to a load it is said to be **discharging**. During discharge:

(i) the lead peroxide (positive plate) and the spongy lead (negative plate) are converted into lead sulphate, and

(ii) the oxygen in the lead peroxide combines with hydrogen in the electrolyte to form water. The electrolyte is therefore weakened and the relative density falls.

The terminal p.d. of a lead–acid cell when fully discharged is about 1.8 V. A cell is **charged** by connecting a d.c. supply to its terminals, the positive terminal of the cell being connected to the positive terminal of the supply. The charging current flows in the reverse direction to the discharge current and the chemical action is reversed. During charging:

(i) the lead sulphate on the positive and negative plates is converted back to lead peroxide and lead respectively, and

(ii) the water content of the electrolyte decreases as the oxygen released from the electrolyte combines with the lead of the positive plate. The relative density of the electrolyte thus increases.

The colour of the positive plate when fully charged is dark brown and when discharged is light brown. The colour of the negative plate when fully charged is grey and when discharged is light grey.

(b) Alkaline cell

There are two main types of alkaline cell – the nickel–iron cell and the nickel–cadmium cell. In both types the positive plate is made of nickel hydroxide enclosed in finely perforated steel tubes, the resistance being reduced by the addition of pure nickel or graphite. The tubes are assembled into nickel–steel plates.

In the nickel–iron cell (sometimes called the Edison cell or nife cell), the negative plate is made of iron oxide, with the resistance being reduced by a little mercuric oxide, the whole being enclosed in perforated steel tubes and assembled in steel plates. In the nickel–cadmium cell the negative plate is made of cadmium. The electrolyte in each type of cell is a solution of potassium hydroxide which does not undergo any chemical change and thus the quantity can be reduced to a minimum. The plates are separated by insulating rods and assembled in steel containers which are then enclosed in a non-metallic crate to insulate the cells from one another. The average discharge p.d. of an alkaline cell is about 1.2 V.

Advantages of an alkaline cell (for example, a nickel–cadmium cell or a nickel–iron cell) over a lead–acid cell include:

(i) more robust construction

(ii) capable of withstanding heavy charging and discharging currents without damage

(iii) has a longer life

(iv) for a given capacity is lighter in weight

(v) can be left indefinitely in any state of charge or discharge without damage

(vi) is not self-discharging.

Disadvantages of an alkaline cell over a lead–acid cell include:

(i) is relatively more expensive

(ii) requires more cells for a given e.m.f.

(iii) has a higher internal resistance

(iv) must be kept sealed

(v) has a lower efficiency.

Alkaline cells may be used in extremes of temperature, in conditions where vibration is experienced or where duties require long idle periods or heavy discharge currents. Practical examples include traction and marine work, lighting in railway carriages, military portable radios and for starting diesel and petrol engines. However, the lead–acid cell is the most common one in practical use.

Series circuits

Figure 3.82 shows three resistors, R_1, R_2 and R_3 connected end to end, i.e. in series, with a battery source of V volts. Since the circuit is closed a current I will flow and the p.d. across each

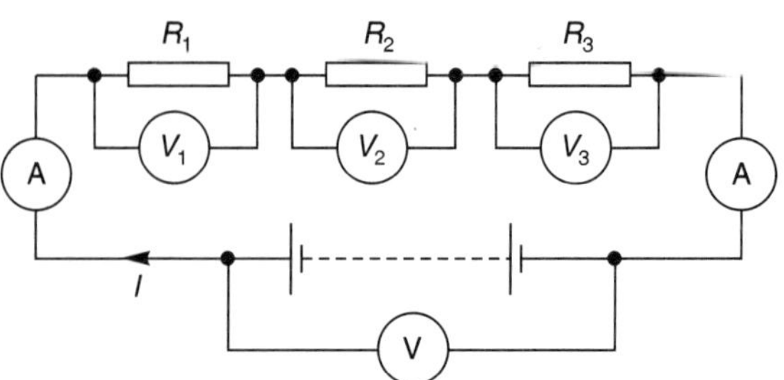

Figure 3.82

resistor may be determined from the voltmeter readings V_1, V_2 and V_3.

In a series circuit:

(a) the current I is the same in all parts of the circuit and hence the same reading is found on each of the ammeters shown, and

(b) the sum of the voltages V_1, V_2 and V_3 is equal to the total applied voltage, V, i.e.

$$\boldsymbol{V = V_1 + V_2 + V_3}$$

From Ohm's law:

$$V_1 = IR_1,\ V_2 = IR_2,\ V_3 = IR_3 \text{ and } V = IR$$

where R is the total circuit resistance.

Since $V = V_1 + V_2 + V_3$, then $IR = IR_1 + IR_2 + IR_3$. Dividing throughout by I gives

$$\boldsymbol{R = R_1 + R_2 + R_3}$$

Thus for a series circuit, the total resistance is obtained by adding together the values of the separate resistances.

Problem 3.190 For the circuit shown in Figure 3.83 determine (a) the battery voltage V, (b) the total resistance of the circuit, and (c) the values of resistance of resistors R_1, R_2 and R_3, given that the p.d.s across R_1, R_2 and R_3 are 5 V, 2 V and 6 V, respectively.

(a) Battery voltage $V = V_1 + V_2 + V_3 = 5 + 2 + 6 = \mathbf{13\,V}$

(b) Total circuit resistance $R = \dfrac{V}{I} = \dfrac{13}{4} = \mathbf{3.25\,\Omega}$

(c) Resistance $R_1 = \dfrac{V_1}{I} = \dfrac{5}{4} = \mathbf{1.25\,\Omega}$

Resistance $R_2 = \dfrac{V_2}{I} = \dfrac{2}{4} = \mathbf{0.5\,\Omega}$

Resistance $R_3 = \dfrac{V_3}{I} = \dfrac{6}{4} = \mathbf{1.5\,\Omega}$

(Check: $R_1 + R_2 + R_3 = 1.25 + 0.5 + 1.5 = 3.25\,\Omega = R$)

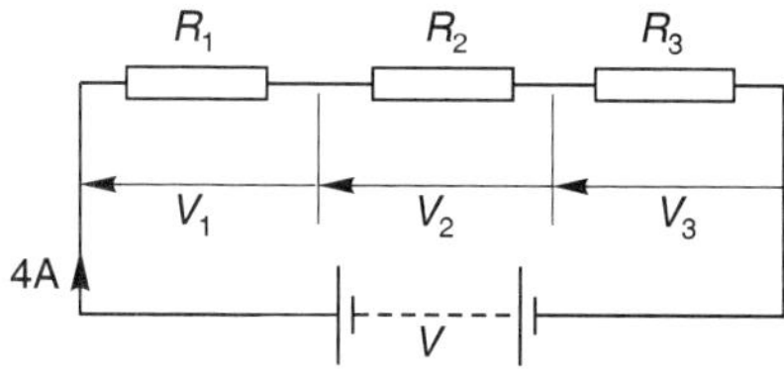

Figure 3.83

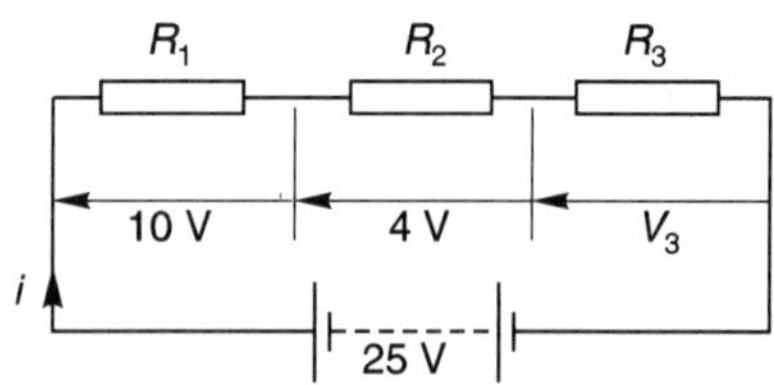

Figure 3.84

Problem 3.191 For the circuit shown in Figure 3.84, determine the p.d. across resistor R_3. If the total resistance of the circuit is 100 Ω, determine the current flowing through resistor R_1. Find also the value of resistor R_2.

p.d. across R_3, $V_3 = 25 - 10 - 4 = \mathbf{11\,V}$

$$\text{Current } I = \frac{V}{R} = \frac{25}{100} = \mathbf{0.25\,A}$$

which is the current flowing in each resistor.

$$\text{Resistance } R_2 = \frac{V_2}{I} = \frac{4}{0.25} = \mathbf{16\,\Omega}$$

Test your knowledge 3.47

A 12 V battery is connected in a circuit having three series-connected resistors with resistances of 4 Ω, 9 Ω and 11 Ω. Determine the current flowing through, and the p.d. across, the 9 Ω resistor. Find also the power dissipated in the 11 Ω resistor.

Parallel networks

Figure 3.85 shows three resistors R_1, R_2 and R_3 connected across each other, i.e. in parallel, across a battery source of V volts.

In a parallel circuit:

(a) the sum of the currents I_1, I_2 and I_3 is equal to the total circuit current, I, i.e.

$$I = I_1 + I_2 + I_3$$

and

(b) the source p.d., V volts, is the same across each of the resistors.

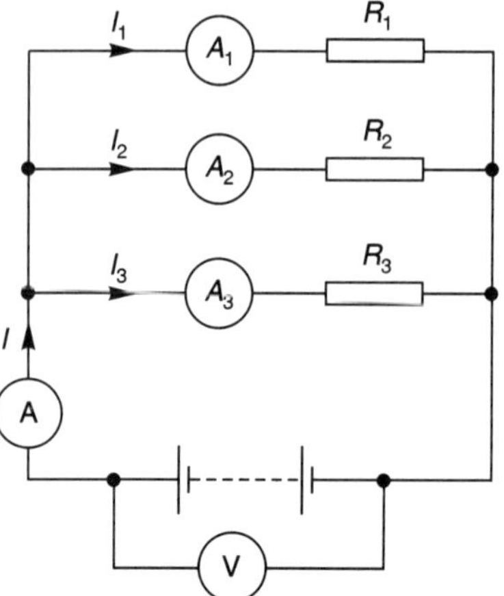

Figure 3.85

From Ohm's law:

$$I_1 = \frac{V}{R_1},\ I_2 = \frac{V}{R_2},\ I_3 = \frac{V}{R_3} \text{ and } I = \frac{V}{R}$$

where R is the total circuit resistance. Since

$$I = I_1 + I_2 + I_3$$

then

$$\frac{V}{R} = \frac{V}{R_1} + \frac{V}{R_2} + \frac{V}{R_3}$$

Dividing throughout by V gives

$$\frac{1}{R} = \frac{1}{R_1} + \frac{1}{R_2} + \frac{1}{R_3}$$

This equation must be used when finding the total resistance R of a parallel circuit. For the special case of **two resistors in parallel**

$$\frac{1}{R} = \frac{1}{R_1} + \frac{1}{R_2} = \frac{R_2 + R_1}{R_1 R_2}$$

Hence

$$R = \frac{R_1 R_2}{R_1 + R_2}\left(\text{i.e. } \frac{\text{product}}{\text{sum}}\right)$$

Problem 3.192 For the circuit shown in Figure 3.86, determine (a) the reading on the ammeter, and (b) the value of resistor R_2.

P.d. across R_1 is the same as the supply voltage V.
Hence supply voltage, $V = 8 \times 5 = 40\,\text{V}$.

(a) Reading on ammeter

$$I = \frac{V}{R_3} = \frac{40}{20} = \mathbf{2\,A}$$

(b) Current flowing through $R_2 = 11 - 8 - 2 = 1\,\text{A}$, hence

$$R_2 = \frac{V}{I_2} = \frac{40}{1} = \mathbf{40\,\Omega}$$

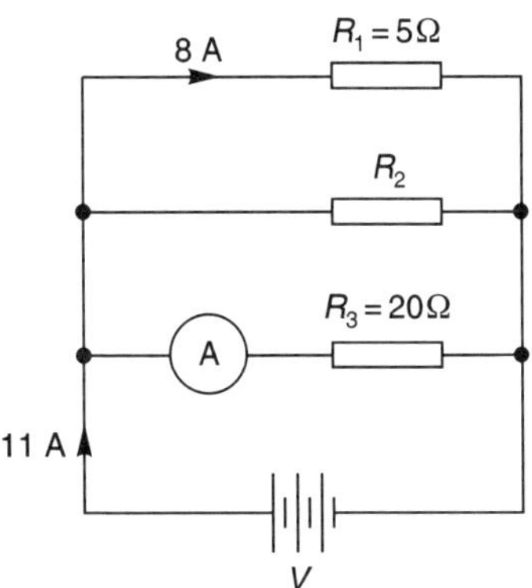

Figure 3.86

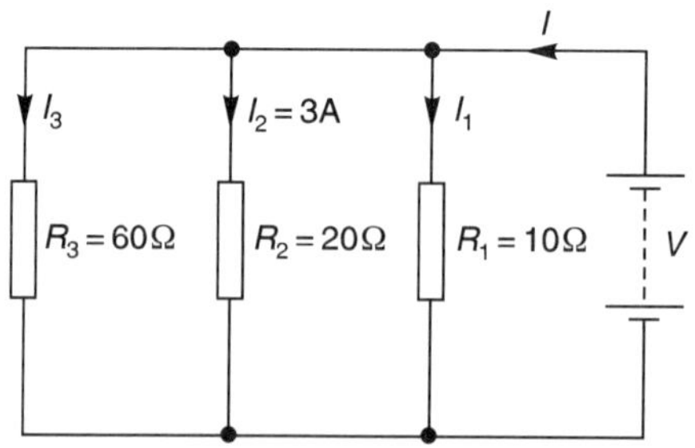

Figure 3.87

Problem 3.193 For the circuit shown in Figure 3.87, find (a) the value of the supply voltage V and (b) the value of current I.

(a) P.d. across $20\,\Omega$ resistor $= I_2R_2 = 3 \times 20 = 60\,\text{V}$, hence supply voltage $\mathbf{V = 60\,V}$ since the circuit is connected in parallel.

(b) Current $I_1 = \dfrac{V}{R_1} = \dfrac{60}{10} = 6\,\text{A}$; $I_2 = \mathbf{3\,A}$

$$I_3 = \frac{V}{R_3} = \frac{60}{60} = 1\,\text{A}$$

Current $I = I_1 + I_2 + I_3$ and hence $I = 6 + 3 + 1 = \mathbf{10\,A}$

Alternatively

$$\frac{1}{R} = \frac{1}{60} + \frac{1}{20} + \frac{1}{10} = \frac{1+3+6}{60} = \frac{10}{60}$$

Hence total resistance $R = \dfrac{60}{10} = 6\,\Omega$

$$\text{Current } I = \frac{V}{R} = \frac{60}{6} = \mathbf{10\,A}$$

Problem 3.194 Find the equivalent resistance for the circuit shown in Figure 3.88.

R_3, R_4 and R_5 are connected in parallel and their equivalent resistance R is given by:

$$\frac{1}{R} = \frac{1}{3} + \frac{1}{6} + \frac{1}{18} = \frac{6+3+1}{18} = \frac{10}{18}$$

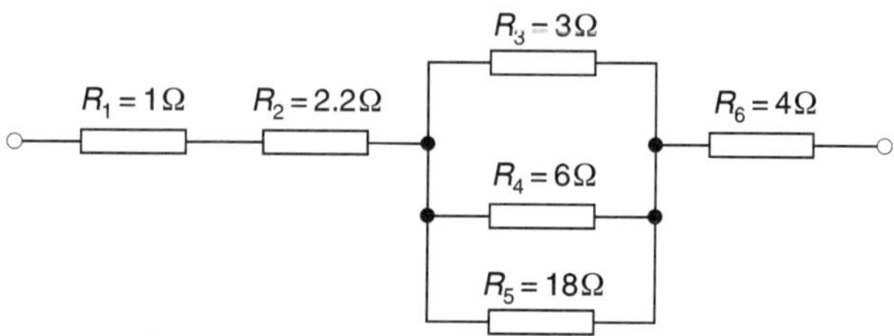

Figure 3.88

Test your knowledge 3.48

1. Two resistors, of resistance 3 Ω and 6 Ω, are connected in parallel across a battery having a voltage of 12 V. Determine (a) the total circuit resistance and (b) the current flowing in the 3 Ω resistor.
2. Given four 1 Ω resistors, state how they must be connected to give an overall resistance of (a) $\frac{1}{4}$ Ω, (b) 1 Ω, (c) $1\frac{1}{3}$ Ω, (d) $2\frac{1}{2}$ Ω, all four resistors being connected in each case.
3. Resistances of 10 Ω, 20 Ω and 30 Ω are connected (a) in series and (b) in parallel to a 240 V supply. Calculate the supply current in each case.

Hence $R = (18/10) = 1.8\,\Omega$. The circuit is now equivalent to four resistors in series and the equivalent circuit resistance $= 1 + 2.2 + 1.8 + 4 = \mathbf{9\,\Omega}$.

Wiring lamps in series and in parallel

Series connection

Figure 3.89 shows three lamps, each rated at 240 V, connected in series across a 240 V supply.

(i) Each lamp has only (240/3) V, i.e. 80 V across it and thus each lamp glows dimly.

(ii) If another lamp of similar rating is added in series with the other three lamps, then each lamp now has (240/4) V, i.e. 60 V across it and each now glows even more dimly.

(iii) If a lamp is removed from the circuit or if a lamp develops a fault (i.e. an open circuit) or if the switch is opened then the circuit is broken, no current flows, and the remaining lamps will not light up.

(iv) Less cable is required for a series connection than for a parallel one.

The series connection of lamps is usually limited to decorative lighting such as for Christmas tree lights.

Parallel connection

Figure 3.90 shows three similar lamps, each rated at 240 V, connected in parallel across a 240 V supply.

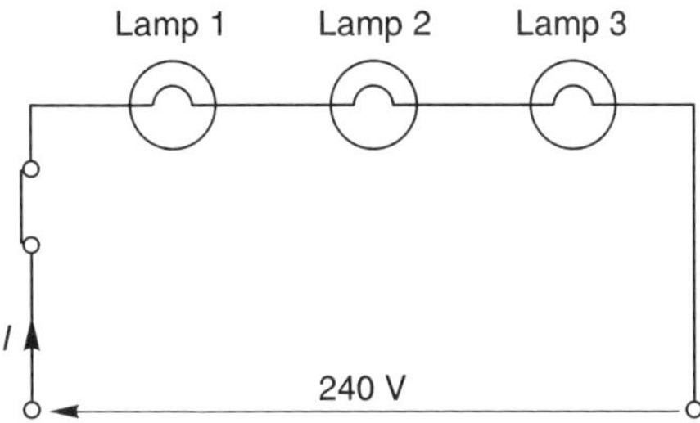

Figure 3.89

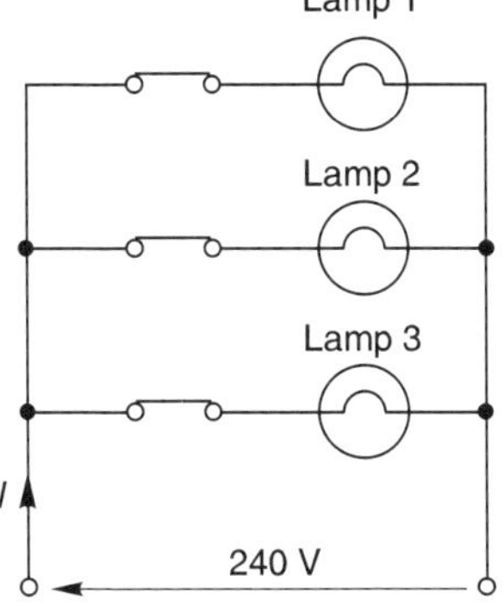

Figure 3.90

(i) Each lamp has 240 V across it and thus each will glow brilliantly at their rated voltage.

(ii) If any lamp is removed from the circuit or develops a fault (open circuit) or a switch is opened, the remaining lamps are unaffected.

(iii) The addition of further similar lamps in parallel does not affect the brightness of the other lamps.

(iv) More cable is required for parallel connection than for a series one.

The parallel connection of lamps is the most widely used in electrical installations.

Problem 3.195 If three identical lamps are connected in parallel and the combined resistance is 150 Ω, find the resistance of one lamp.

Let the resistance of one lamp be R, then

$$\frac{1}{150} = \frac{1}{R} + \frac{1}{R} + \frac{1}{R} = \frac{3}{R}$$

from which, $R = 3 \times 150 =$ **450 Ω**

Test your knowledge 3.49

Three identical lamps A, B and C are connected in series across a 150 V supply. State (a) the voltage across each lamp, and (b) the effect of lamp C failing.

Introduction to Kirchhoff's laws

More complex d.c. circuits cannot be solved by Ohm's law and the formulae for series and parallel resistors alone. Kirchhoff (a German physicist) developed two laws which further help the determination of unknown currents and voltages in d.c. series/parallel networks.

Kirchhoff's current and voltage laws

Current law

At any junction in an electric circuit the total current flowing towards that junction is equal to the total current flowing away from the junction, i.e. $\sum I = 0$.

Thus referring to Figure 3.91

$$I_1 + I_2 + I_3 = I_4 + I_5$$

or

$$I_1 + I_2 + I_3 - I_4 - I_5 = 0$$

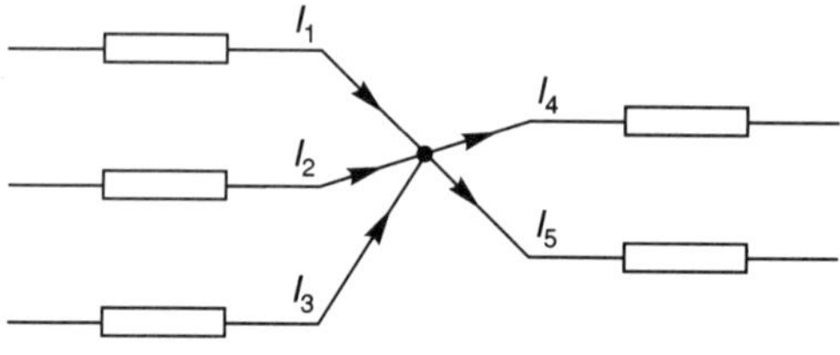

Figure 3.91

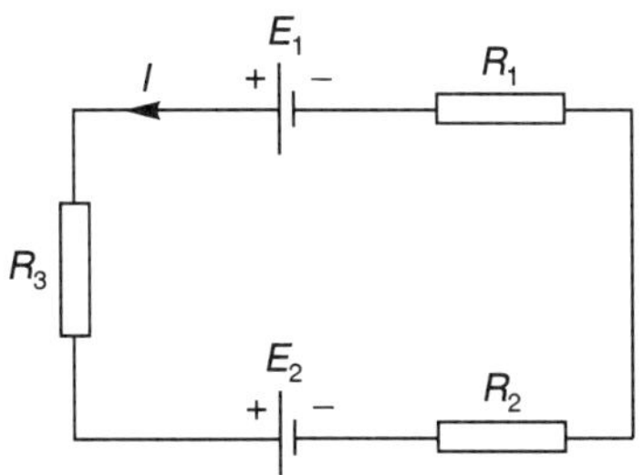

Figure 3.92

Voltage law

In any closed loop in a network, the algebraic sum of the voltage drops (i.e. products of current and resistance) taken around the loop is equal to the resultant e.m.f. acting in that loop. Thus referring to Figure 3.92:

$$E_1 - E_2 = IR_1 + IR_2 + IR_3$$

(Note that if current flows away from the positive terminal of a source, that source is considered by convention to be positive. Thus moving anticlockwise around the loop of Figure 3.92, E_1 is positive and E_2 is negative.)

Problem 3.196 Determine the value of the unknown currents marked in Figure 3.93.

Applying Kirchhoff's current law to each junction in turn gives:

For junction A: $15 = 5 + I_1$

Hence $\mathbf{I_1 = 10\,A}$

For junction B: $5 + 2 = I_2$

Hence $\mathbf{I_2 = 7\,A}$

For junction C: $I_1 = 27 + I_3$

i.e. $10 = 27 + I_3$

Hence $\mathbf{I_3 = 10 - 27 = -17\,A}$

(i.e. in the opposite direction to that shown in Figure 3.93).

For junction D: $I_3 + I_4 = 2$

i.e. $-17 + I_4 = 2$

Hence $\mathbf{I_4 = 17 + 2 = 19\,A}$

For junction E: $27 = 6 + I_5$

Hence $\mathbf{I_5 = 27 - 6 = 21\,A}$

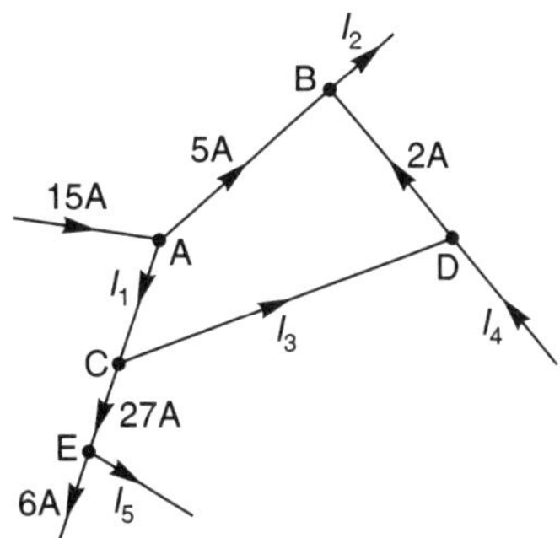

Figure 3.93

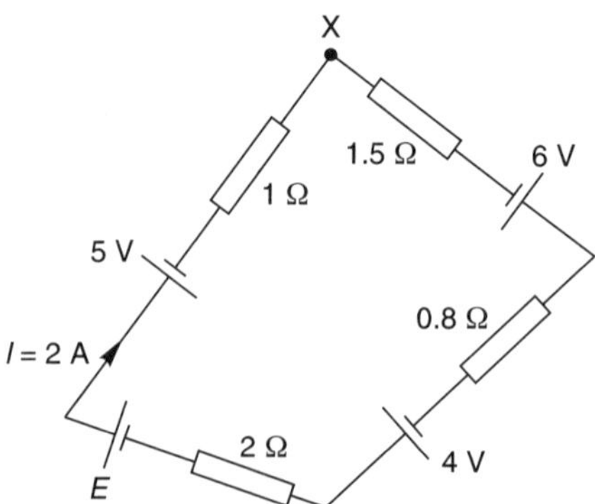

Figure 3.94

Problem 3.197 Determine the value of e.m.f. E in Figure 3.94.

Applying Kirchhoff's voltage law and moving clockwise around the loop of Figure 3.94 starting at point X gives:

$$6 + 4 + E - 5 = I(1.5) + I(0.8) + I(2) + I(1)$$

$$5 + E = I(5.3) = 2(5.3)$$

since current I is 2 A.
Hence

$$5 + E = 10.6$$

and

$$\textbf{e.m.f. } \boldsymbol{E} = 10.6 - 5 = \mathbf{5.6\,V}$$

Magnetic fields

A **permanent magnet** is a piece of ferromagnetic material (such as iron, nickel or cobalt) which has properties of attracting other pieces of these materials. The area around a magnet is called the **magnetic field** and it is in this area that the effects of the **magnetic force** produced by the magnet can be detected. The magnetic field of a bar magnet can be represented pictorially by the 'lines of force' (or lines of 'magnetic flux' as they are called) as shown in Figure 3.95. Such a field pattern can be produced by placing iron filings in the vicinity of the magnet.

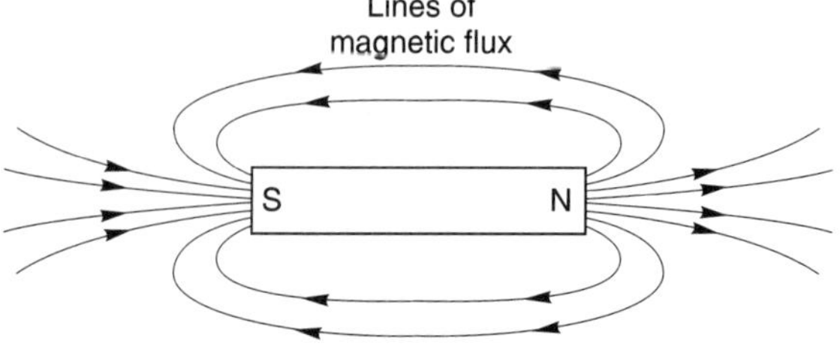

Figure 3.95

The field direction at any point is taken as that in which the north-seeking pole of a compass needle points when suspended in the field. External to the magnet the direction of the field is north to south.

The laws of magnetic attraction and repulsion can be demonstrated by using two bar magnets. In Figure 3.96(a), with **unlike poles** adjacent, **attraction** occurs.

In Figure 3.96(b), with **like poles** adjacent, **repulsion** occurs.

Magnetic fields are produced by electric currents as well as by permanent magnets. The field forms a circular pattern with the current-carrying conductor at the centre. The effect is portrayed in Figure 3.97 where the convention adopted is:

(a) current flowing **away** from the viewer is shown by X and can be thought of as the feathered end of a shaft of an arrow;

(b) current flowing **towards** the viewer is shown by · and can be thought of as the tip of an arrow.

The direction of the fields in Figure 3.97 is remembered by the **screw rule** which states: 'If a normal right-hand thread screw is screwed along the conductor in the direction of the current, the direction of rotation of the screw is in the direction of the magnetic field'.

A magnetic field produced by a long coil, or **solenoid**, is shown in Figure 3.98 and is seen to be similar to that of a bar magnet

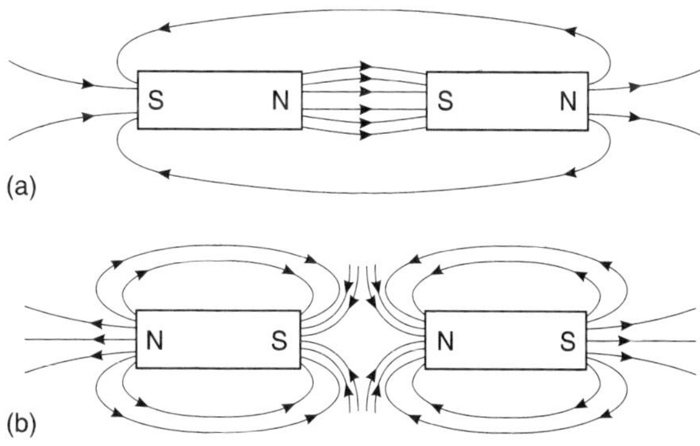

Figure 3.96

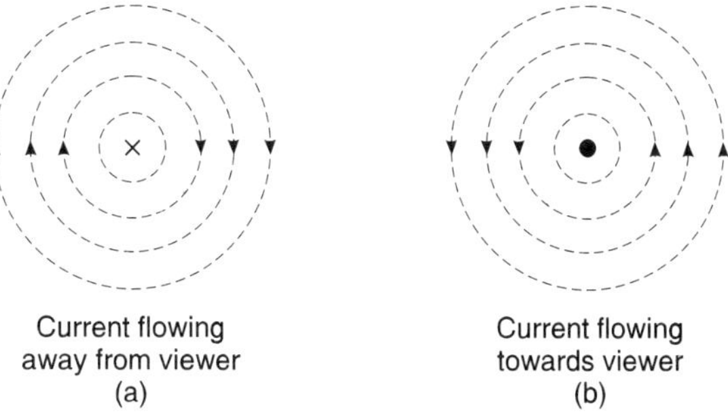

Figure 3.97

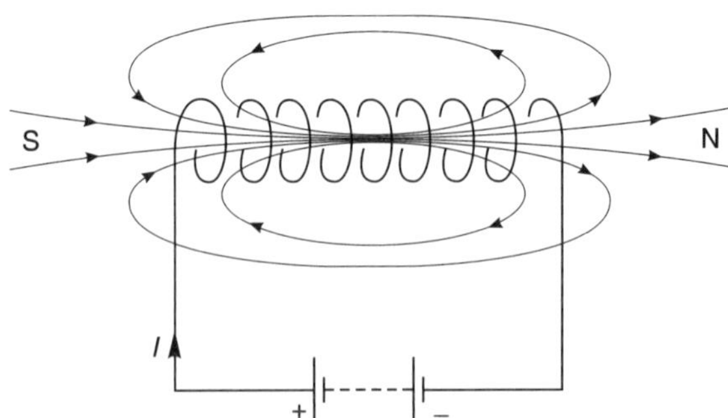

Figure 3.98

shown in Figure 3.95. If the solenoid is wound on an iron bar an even stronger field is produced. The **direction** of the field produced by current I is determined by a compass and is remembered by either:

(a) the **screw rule**, which states that if a normal right-hand thread screw is placed along the axis of the solenoid and is screwed in the direction of the current it moves in the direction of the solenoid (i.e. points in the direction of the north pole); or

(b) the **grip rule**, which states that if the coil is gripped with the right hand with the fingers pointing in the direction of the current, then the thumb, outstretched parallel to the axis of the solenoid, points in the direction of the magnetic field inside the solenoid (i.e. points in the direction of the north pole).

Problem 3.198 Figure 3.99 shows a coil of wire wound on an iron core connected to a battery. Sketch the magnetic field pattern associated with the current-carrying coil and determine the polarity of the field.

The magnetic field associated with the solenoid in Figure 3.99 is similar to the field associated with a bar magnet and is shown in Figure 3.100. The polarity of the field is determined either by the screw rule or by the grip rule. Thus the north pole is at the bottom and the south pole at the top.

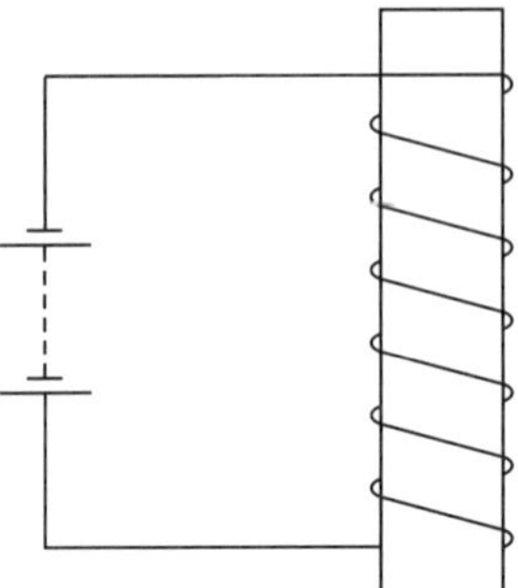

Figure 3.99

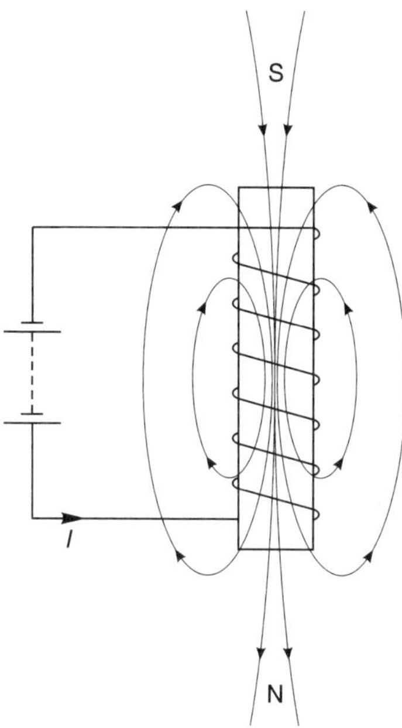

Figure 3.100

Electromagnets

An **electromagnet**, which is a solenoid wound on an iron core, provides the basis of many items of electrical equipment, examples including an electric bell, relays, lifting magnets and telephone receivers.

(a) Electric bell

There are various types of electric bell, including the single-strike bell, the trembler bell, the buzzer and a continuously ringing bell, but all depend on the attraction exerted by an electromagnet on a soft iron armature. A typical single-stroke bell circuit is shown in Figure 3.101. When the push button is operated a current passes

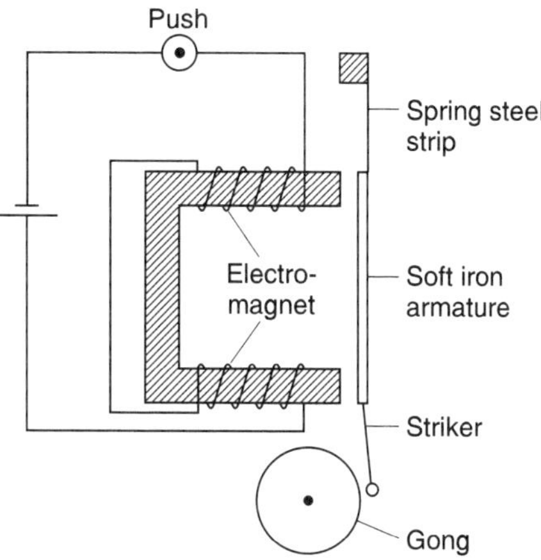

Figure 3.101

through the coil. Since the iron-cored coil is energized the soft iron armature is attracted to the electromagnet. The armature also carries a striker which hits the gong. When the circuit is broken the coil becomes demagnetized and the spring steel strip pulls the armature back to its original position. The striker will only operate when the push is operated.

(b) Relay

A relay is similar to an electric bell except that contacts are opened or closed by operation instead of a gong being struck. A typical simple relay is shown in Figure 3.102, which consists of a coil wound on a soft iron core. When the coil is energized the hinged soft iron armature is attracted to the electromagnet and pushes against two fixed contacts so that they are connected together, thus closing some other electrical circuit.

(c) Lifting magnet

Lifting magnets, incorporating large electromagnets, are used in iron and steel works for lifting scrap metal. A typical robust lifting magnet, capable of exerting large attractive forces, is shown in the elevation and plan view of Figure 3.103 where a coil, C, is wound round a central core, P, of the iron casting. Over the face of the electromagnet is placed a protective non-magnetic sheet of material R. The load, Q, which must be of magnetic material, is lifted when the coils are energized, the magnetic flux paths, M, being shown by the broken lines.

(d) Telephone receiver

Whereas a transmitter or microphone changes sound waves into corresponding electrical signals, a telephone receiver converts the electrical waves back into sound waves. A typical telephone receiver is shown in Figure 3.104 and consists of a permanent

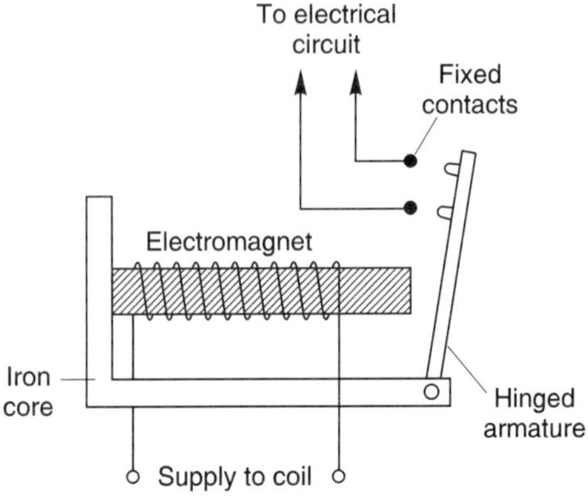

Figure 3.102

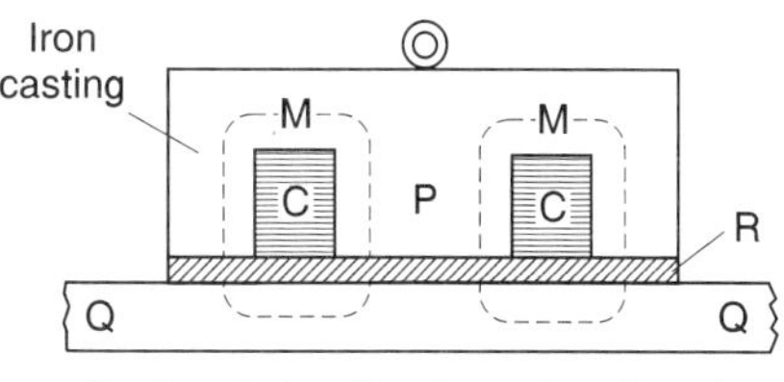

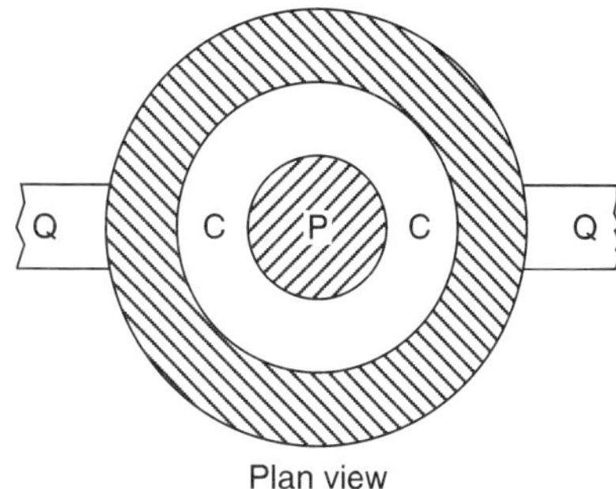

Figure 3.103

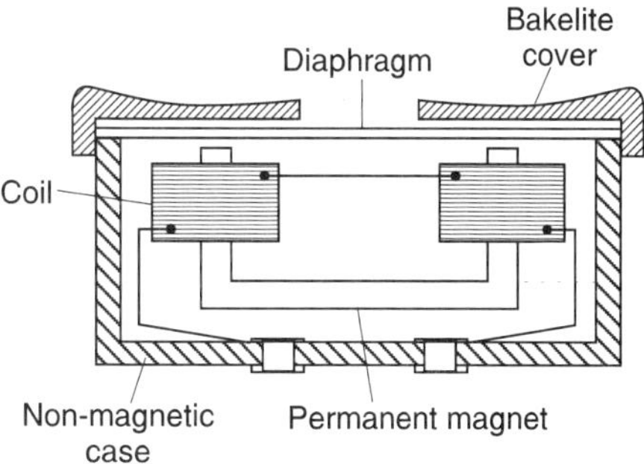

Figure 3.104

magnet with coils wound on its poles. A thin, flexible diaphragm of magnetic material is held in position near to the magnetic poles but not touching them. Variation in current from the transmitter varies the magnetic field and the diaphragm consequently vibrates. The vibration produces sound variations corresponding to those transmitted.

Magnetic flux and flux density

Magnetic flux is the amount of magnetic field (or the number of lines of force) produced by a magnetic source. The symbol for magnetic flux is Φ (Greek letter cap phi). The unit of magnetic flux is the **weber**, **Wb**.

Magnetic flux density is the amount of flux passing through a defined area that is perpendicular to the direction of the flux.

$$\textbf{Magnetic flux density} = \frac{\textbf{magnetic flux}}{\textbf{area}}$$

The symbol for magnetic flux density is B. The unit of magnetic flux density is the tesla, T, where $1\,T = 1\,Wb/m^2$. Hence

$$B = \frac{\Phi}{A} \text{ tesla}$$

where $A\,m^2$ is the area.

Problem 3.199 A magnetic pole face has a rectangular section having dimensions 200 mm by 100 mm. If the total flux emerging from the pole is 150 μWb, calculate the flux density.

Flux $\Phi = 150\,\mu Wb = 150 \times 10^{-6}$ Wb.
Cross-sectional area $A = 200 \times 100 = 20\,000\,mm^2 = 20\,000 \times 10^{-6}\,m^2$.

$$\text{Flux density } B = \frac{\Phi}{A} = \frac{150 \times 10^{-6}}{20\,356000 \times 10^{-6}}$$
$$= \mathbf{0.0075\,T \text{ or } 7.5\,mT}$$

Test your knowledge 3.50

1 Determine the flux density in a magnetic field of cross-sectional area 20 cm^2 having a flux of 3 mWb.

2 The maximum working flux density of a lifting electromagnet is 1.8 T and the effective area of a pole face is circular in cross-section. If the total magnetic flux produced is 353 mWb, determine the radius of the pole face.

Force on a current-carrying conductor

If a current-carrying conductor is placed in a magnetic field produced by permanent magnets, then the fields due to the current-carrying conductor and the permanent magnets interact and cause a force to be exerted on the conductor. The force on the current-carrying conductor in a magnetic field depends upon:

(a) the flux density of the field, B tesla

(b) the strength of the current, I amperes

(c) the length of the conductor perpendicular to the magnetic field, l m

(d) the directions of the field and the current.

When the magnetic field, the current and the conductor are mutually at right angles then:

$$\text{Force } \mathbf{F} = BIl \text{ N}$$

When the conductor and the field are at an angle θ° to each other then:

$$\text{Force } \mathbf{F} = BIl \sin\theta \text{ N}$$

Since when the magnetic field, current and conductor are mutually at right angles, $\mathbf{F} = BIl$, the magnetic flux density B may be defined by $B = \mathbf{F}/Il$, i.e. the flux density is 1 T if the force exerted on 1 m of a conductor when the conductor carries a current of 1 A is 1 N.

Problem 3.200 A conductor carries a current of 20 A and is at right angles to a magnetic field having a flux density of 0.9 T. If the length of the conductor in the field is 30 cm, calculate the force acting on the conductor. Determine also the value of the force if the conductor is inclined at an angle of 30° to the direction of the field.

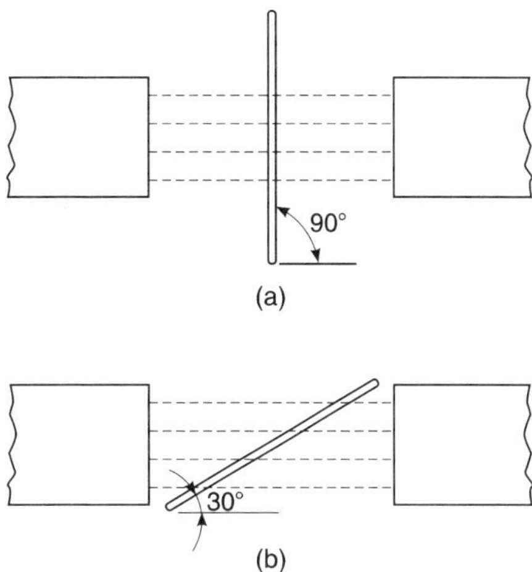

Figure 3.105

$B = 0.9\,\text{T}$; $I = 20\,\text{A}$; $l = 30\,\text{cm} = 0.30\,\text{m}$. Force $\mathbf{F} = BIl = (0.9)(20)(0.30)\,\text{N}$ when the conductor is at right angles to the field, as shown in Figure 3.105(a), i.e. $\mathbf{F} = \mathbf{5.4\,N}$. When the conductor is inclined at 30° to the field, as shown in Figure 3.105(b), then

$$\text{Force } \mathbf{F} = BIl \sin\theta$$
$$\mathbf{F} = (0.9)(20)(0.30)\sin 30^\circ$$
$$\mathbf{F} = \mathbf{2.7\,N}$$

If the current-carrying conductor shown in Figure 3.97(a) is placed in the magnetic field shown in Figure 3.106(a), then the two fields interact and cause a force to be exerted on the conductor, as shown in Figure 3.106(b). The field is strengthened above the conductor and weakened below, thus tending to move the conductor downwards. This is the basic principle of operation of the **electric motor** and the **moving-coil instrument.**

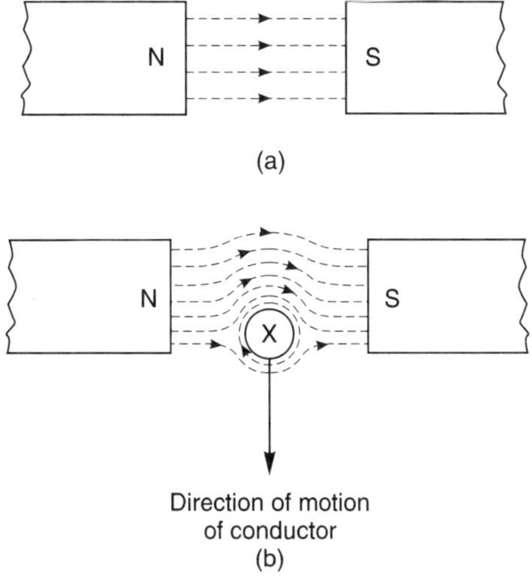

Figure 3.106

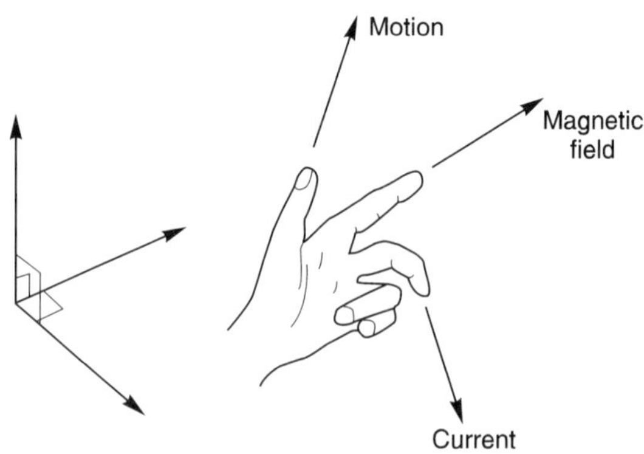

Figure 3.107

The direction of the force exerted on a conductor can be predetermined by using Fleming's left-hand rule (often called the motor rule) which states:

> Let the thumb, first finger and second finger of the left hand be extended such that they are all at right-angles to each other (as shown in Figure 3.107). If the first finger points in the direction of the magnetic field, the second finger points in the direction of the current, then the thumb will point in the direction of the motion of the conductor.

Summarizing:

*F*irst finger = *F*ield
Se*C*ond finger = *C*urrent
Thu*M*b = *M*otion.

Problem 3.201 Determine the current required in a 400 mm length of conductor of an electric motor, when the conductor is situated at right angles to a magnetic field of flux density 1.2 T, if a force of 1.92 N is to be exerted on the conductor. If the conductor is vertical, the current flowing downwards and the direction of the magnetic field is from left to right, what is the direction of the force?

$\mathbf{F} = 1.92\,\text{N}$; $l = 400\,\text{mm} = 0.40\,\text{m}$; $B = 1.2\,\text{T}$.
Since $\mathbf{F} = BIl$, $IO = \mathbf{F}/Bl$. Hence

$$\text{current } I = \frac{1.92}{(1.2)(0.4)} = \mathbf{4\,A}$$

If the current flows downwards, the direction of its magnetic field due to the current alone will be clockwise when viewed from above. The lines of flux will reinforce (i.e. strengthen) the main magnetic field at the back of the conductor and will be in opposition in the front (i.e. weaken the field). **Hence the force on the conductor will be from back to front (i.e. toward the viewer)**. This direction may also have been deduced using Fleming's left-hand rule.

Problem 3.202 A conductor 350 mm long carries a current of 10 A and is at right angles to a magnetic field lying between two circular pole faces

Test your knowledge 3.51

1. A conductor carries a current of 70 A at right angles to a magnetic field having a flux density of 1.5 T. If the length of the conductor in the field is 200 mm calculate the force on the conductor. What is the force when the conductor and field are at an angle of 45°?

2. With reference to Figure 3.108 determine (a) the direction of the force on the conductor in Figure 3.108(a), (b) the direction of the force on the conductor in Figure 3.108(b), (c) the direction of the current in Figure 3.108(c), (d) the polarity of the magnetic system in Figure 3.108(d).

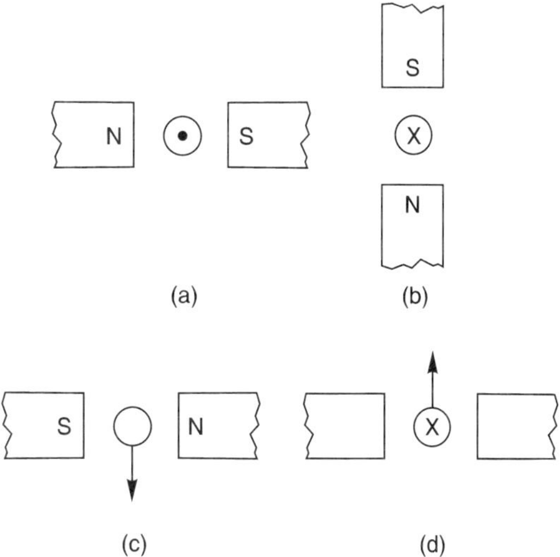

Figure 3.108

each of radius 60 mm. If the total flux between the pole faces is 0.5 mWb, calculate the magnitude of the force exerted on the conductor.

$l = 350\text{ mm} = 0.35\text{ m}$; $I = 10\text{ A}$; area of pole face $A = \pi r^2 = \pi(0.06)^2\text{ m}^2$; $\Phi = 0.5\text{ mWb} = 0.5 \times 10^{-3}\text{ Wb}$. Force $\mathbf{F} = BIl$, and $B = \Phi/A$, hence

$$\text{Force} = \left(\frac{\Phi}{A}\right) Il$$

$$= \frac{(0.5 \times 10^{-3})}{\pi(0.06)^2}(10)(0.35)\text{ N}$$

i.e. **force = 0.155 N**

Principle of operation of a simple d.c. motor

A rectangular coil which is free to rotate about a fixed axis is shown placed inside a magnetic field produced by permanent magnets in Figure 3.109. A direct current is fed into the coil via carbon brushes bearing on a commutator, which consists of a metal ring split into two halves separated by insulation. when current flows in the coil a magnetic field is set up around the coil which interacts with the magnetic field produced by the magnets. This causes a force **F** to be exerted on the current-carrying conductor which, by Fleming's left-hand rule, is downwards between points A and B and upward between C and D for the current direction shown. This causes a torque and the coil rotates anticlockwise. When the coil has turned through 90° from the position shown in Figure 3.109 the brushes connected to the positive and negative terminals of the supply make contact with different halves of the commutator ring, thus reversing the direction of the current flow in the conductor. If the current is not reversed and the coil rotates past this position the forces acting on it change direction and it rotates in the opposite

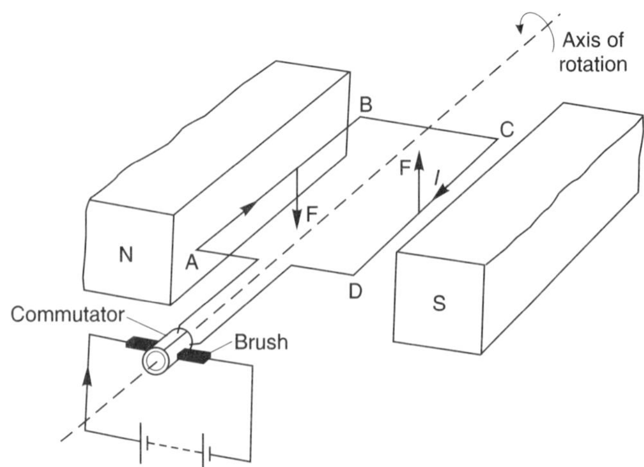

Figure 3.109

direction thus never making more than half a revolution. The current direction is reversed every time the coil swings through the vertical position and thus the coil rotates anticlockwise for as long as the current flows. This is the principle of operation of a d.c. motor, which is thus a device that takes in electrical energy and converts it into mechanical energy.

Principle of operation of a moving-coil instrument

A moving-coil instrument operates on the motor principle. When a conductor carrying current is placed in a magnetic field, a force **F** is exerted on the conductor, given by $\mathbf{F} = BIl$. If the flux density B is made constant (by using permanent magnets) and the conductor is a fixed length (say, a coil) then the force will depend only on the current flowing in the conductor.

In a moving-coil instrument a coil is placed centrally in the gap between shaped pole pieces as shown by the front elevation in Figure 3.110(a). (The airgap is kept as small as possible, although for clarity it is shown exaggerated in Figure 3.110(b).) The coil is supported by steel pivots, resting in jewel bearings, on a cylindrical iron core. Current is led into and out of the coil by two phosphor bronze spiral hairsprings which are wound in opposite directions to minimize the effect of temperature change and to limit the coil swing (i.e. to control the movement) and return the movement to zero position when no current flows. Current flowing in the coil produces forces as shown in Figure 3.110(b), the directions being obtained by Fleming's left-hand rule. The two forces, $\mathbf{F}_A$ and $\mathbf{F}_B$, produce a torque which will move the coil in a clockwise direction, i.e. move the pointer from left to right. Since force is proportional to current the scale is linear.

When the aluminium frame, on which the coil is wound, is rotated between the poles of the magnet, small currents (called eddy currents) are induced into the frame, and this provides

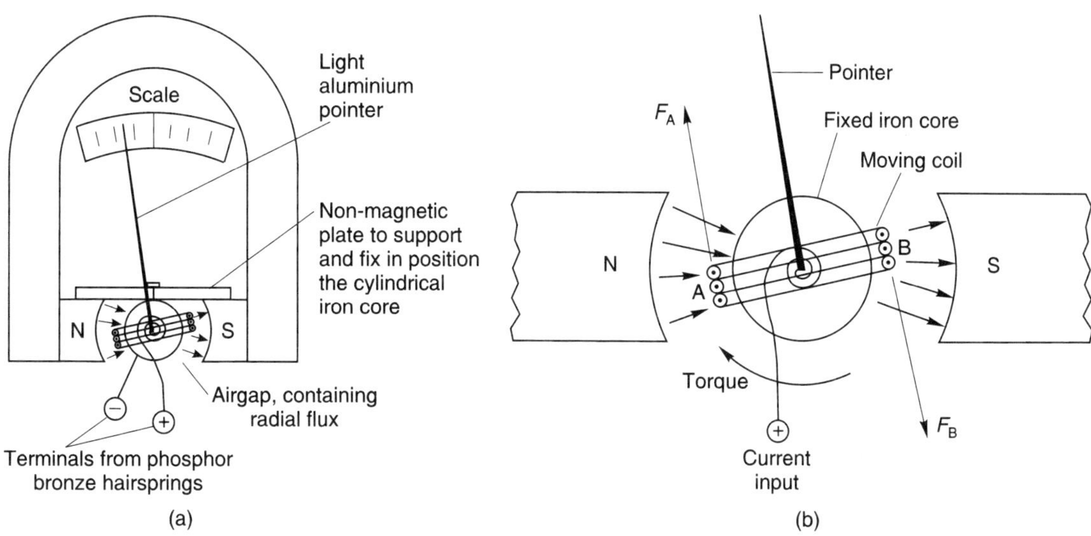

Figure 3.110

automatically the necessary **damping** of the system due to the reluctance of the former to move within the magnetic field.

The moving-coil instrument will measure only direct current or voltage and the terminals are marked positive and negative to ensure that the current passes through the coil in the correct direction to deflect the pointer 'up the scale'. The range of this sensitive instrument is extended by using shunts and multipliers.

Introduction to electromagnetic induction

When a conductor is moved across a magnetic field so as to cut through the lines of force (flux), an electromotive force (e.m.f.) is produced in the conductor. If the conductor forms part of a closed circuit then the e.m.f. produced causes an electric current to flow round the circuit. Hence, an e.m.f. (and thus current) is 'induced' in the conductor as a result of its movement across the magnetic field. This effect is known as '**electromagnetic induction**'.

Figure 3.111(a) shows a coil of wire connected to a centre-zero galvanometer, which is a sensitive ammeter with a zero-current position in the centre of the scale.

(a) When the magnet is moved at constant speed towards the coil (Figure 3.111(a)), a deflection is noted on the galvanometer showing that a current has been produced in the coil.

(b) When the magnet is moved at the same speed as in (a) but away from the coil the same deflection is noted but is in the opposite direction (see Figure 3.111(b)).

(c) When the magnet is held stationary even within the coil no deflection is recorded.

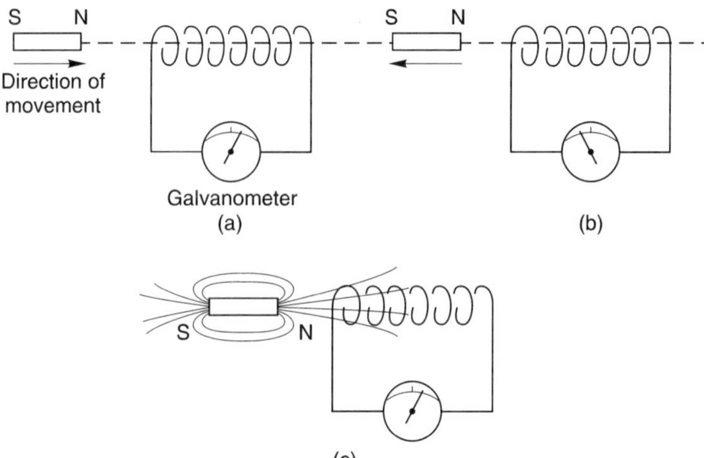

Figure 3.111

(d) When the coil is moved at the same speed as in (a) and the magnet held stationary the same galvanometer deflection is noted.

(e) When the relative speed is, say, doubled, the galvanometer deflection is doubled.

(f) When a stronger magnet is used, a greater galvanometer deflection is noted.

(g) When the number of turns of wire of the coil is increased, a greater galvanometer deflection is noted.

Figure 3.111(c) shows the magnetic field associated with the magnet. As the magnet is moved towards the coil, the magnetic flux of the magnet moves across, or cuts, the coil. **It is the relative movement of the magnetic flux and the coil that causes an e.m.f. and thus current to be induced in the coil**. This effect is known as **electromagnetic induction**. The laws of electromagnetic induction evolved from experiments such as those described above.

Laws of electromagnetic induction

Faraday's laws of electromagnetic induction state:

(i) An induced e.m.f. is set up whenever the magnetic field linking a circuit changes.

(ii) The magnitude of the induced e.m.f. in any circuit is proportional to the rate of change of the magnetic flux linking the circuit.

Lenz's law states:

The direction of an induced e.m.f. is always such that it tends to set up a current opposing the motion or the change of flux responsible for inducing that e.m.f.

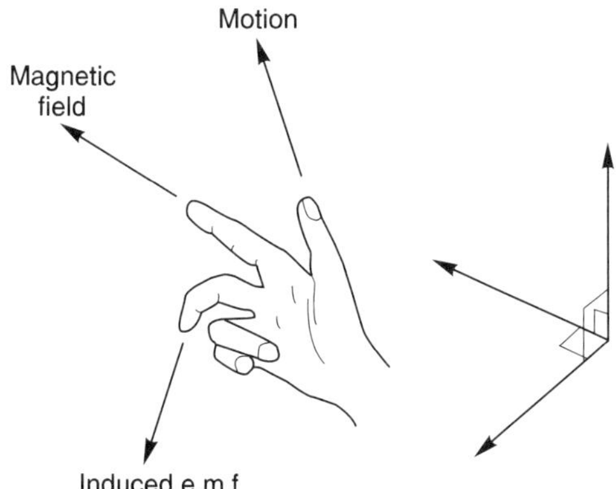

Figure 3.112

An alternative method to Lenz's law of determining relative directions is given by **Fleming's *R*ight-hand rule** (often called the gene*R*ator rule) which states:

> Let the thumb, first finger and second finger of the right hand be extended such that they are all at right angles to each other (as shown in Figure 3.112). If the first finger points in the direction of the magnetic field, the thumb points in the direction of motion of the conductor relative to the magnetic field, then the second finger will point in the direction of the induced e.m.f.

Summarizing:

*F*irst finger = *F*ield
Thu*M*b = *M*otion
S*E*cond finger = *E*.m.f.

In a generator, conductors forming an electric circuit are made to move through a magnetic field. By Farady's law an e.m.f. is induced in the conductors and thus a source of e.m.f. is created. A generator converts mechanical energy into electrical energy.

The induced e.m.f. E set up between the ends of the conductor shown in Figure 3.113 is given by:

$$E = Blv \text{ V}$$

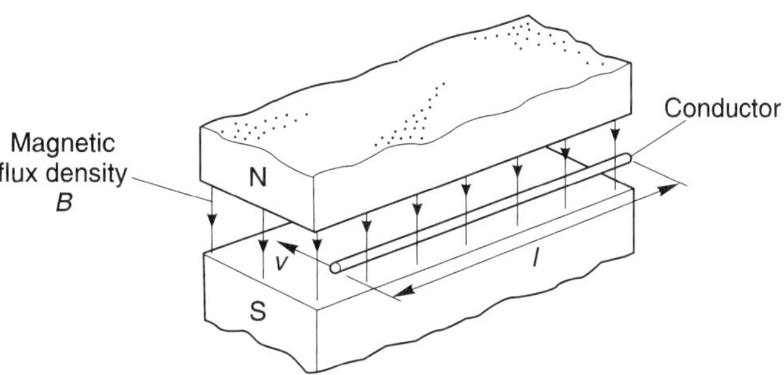

Figure 3.113

where B the flux density is measured in teslas, l the length of conductor in the magnetic field is measured in metres, and v the conductor velocity is measured in metres per second.

If the conductor moves at an angle θ°C to the magnetic field (instead of at 90° as assumed above) then

$$\boldsymbol{E = Blv \sin \theta}$$

Problem 3.203 A conductor 300 mm long moves at a uniform speed of 4 m/s at right angles to a uniform magnetic field of flux density 1.25 T. Determine the current flowing in the conductor when (a) its ends are open-circuited, (b) its ends are connected to a load of 20 Ω resistance.

When a conductor moves in a magnetic field it will have an e.m.f. induced in it but this e.m.f. can only produce a current if there is a closed circuit.

$$\text{Induced e.m.f. } E = Blv = (1.25)\left(\frac{300}{1000}\right)(4) = 1.5\,\text{V}$$

(a) If the ends of the conductor are open-circuited no current will flow even though 1.5 V has been induced.

(b) From Ohm's law

$$I = \frac{E}{R} = \frac{1.5}{20} = \mathbf{0.075\,A} \text{ or } \mathbf{75\,mA}$$

Problem 3.204 The wing span of a metal aeroplane is 36 m. If the aeroplane is flying at 400 km/h, determine the e.m.f. induced between its wing tips. Assume the vertical component of the earth's magnetic field is 40 μT.

Induced e.m.f. across wing tips, $E = Blv$. $B = 40\,\mu\text{T} = 40 \times 10^{-6}$ T; $l = 36$ m.

$$v = 400\,\frac{\text{km}}{\text{h}} \times 1000\,\frac{\text{m}}{\text{km}} \times \frac{1\,\text{h}}{60 \times 60\,\text{s}} = \frac{(400)(1000)}{3600} = \frac{4000}{36}\,\text{m/s}$$

Hence

$$E = (40 \times 10^{-6})(36)\left(\frac{4000}{36}\right) = \mathbf{0.16\,V}$$

Problem 3.205 The diagram shown in Figure 3.114 represents the generation of e.m.f.s. Determine (i) the direction in which the conductor has to be moved in Figure 3.114(a), (ii) the direction of the induced e.m.f. in Figure 3.114(b), (iii) the polarity of the magnetic system in Figure 3.114(c).

The direction of the e.m.f. and thus the current due to the e.m.f. may be obtained by either Lenz's rule or Fleming's *R*ight-hand rule (i.e. Gene*R*ator rule).

(i) Using Lenz's law: The field due to the magnet and the field due to the current-carrying conductor are shown in

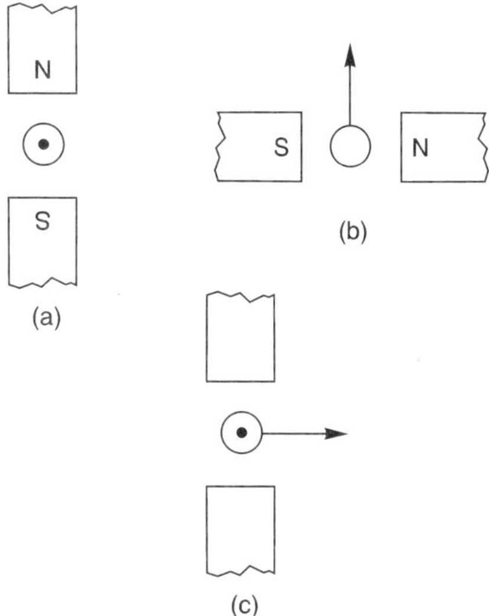

Figure 3.114

Figure 3.115(a) and are seen to reinforce to the left of the conductor. Hence the force on the conductor is to the right. However Lenz's law says that the direction of the induced e.m.f. is always such as to oppose the effect producing it. **Thus the conductor will have to be moved to the left**.

(ii) Using Fleming's right-hand rule:
*F*irst finger = *F*ield, i.e. N → S, or right to left;
Thu*M*b = *M*otion, i.e. upwards;
S*E*cond finger = *E*.m.f., i.e. **towards the viewer or out of the paper**, as shown in Figure 3.115(b).

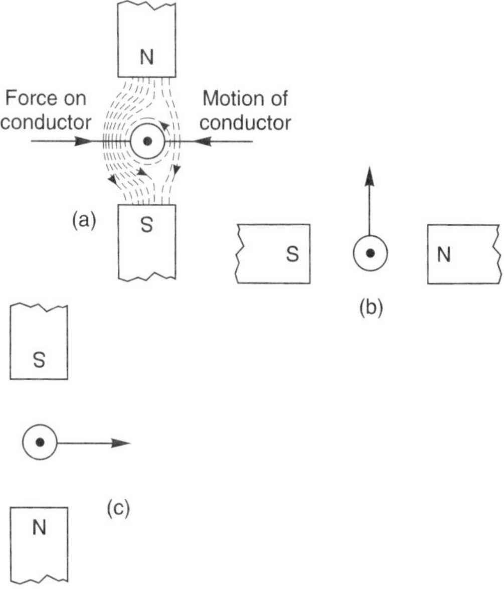

Figure 3.115

Test your knowledge 3.52

1 At what velocity must a conductor 75 mm long cut a magnetic field of flux density 0.6 T if an e.m.f. of 9 V is to be induced in it? Assume the conductor, the field and the direction of motion are mutually perpendicular.

2 A conductor moves with a velocity of 15 m/s at an angle of (a) 90°, (b) 60°, and (c) 30° to a magnetic field produced between two square-faced poles of side length 2 cm. If the flux leaving a pole face is 5 μWb, find the magnitude of the induced e.m.f. in each case.

(iii) The polarity of the magnetic system of Figure 3.114(c) is shown in Figure 3.115(c) and is obtained using Fleming's right-hand rule.

Introduction to alternating voltages and currents

Electricity is produced by generators at power stations and then distributed by a vast network of transmission lines (called the National Grid system) to industry and for domestic use. It is easier and cheaper to generate alternating current (a.c.) than direct current (d.c.) and a.c. is more conveniently distributed than d.c. since its voltage can be readily altered using transformers. Whenever d.c. is neded in preference to a.c., devices called rectifiers are used for conversion.

The a.c. generator

Let a single turn coil be free to rotate at constant angular velocity symmetrically between the poles of a magnet system as shown in Figure 3.116. An e.m.f. is generated in the coil (from Faraday's law) which varies in magnitude and reverses its direction at regular intervals. The reason for this is shown in Figure 3.117.

In positions (a), (e) and (i) the conductors of the loop are effectively moving along the magnetic field, no flux is cut and hence no e.m.f. is induced. In position (c) maximum e.m.f. is induced. In position (g), maximum flux is cut and hence maximum e.m.f. is again induced. However, using Fleming's right-hand rule, the induced e.m.f. is in the opposite direction to that in position (c) and is thus shown as $-E$. In positions (b), (d), (f) and (h) some flux is cut and hence some e.m.f. is induced. If all such positions of the coil are considered, in one revolution of the coil, one cycle of alternating e.m.f. is produced as shown. This is the principle of operation of the **a.c. generator** (i.e. the **alternator**).

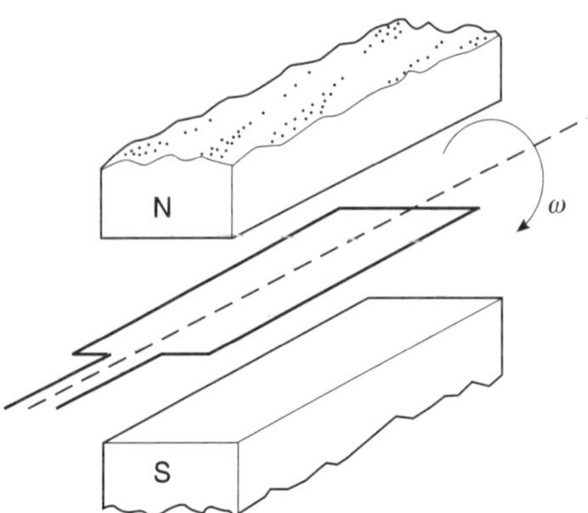

Figure 3.116

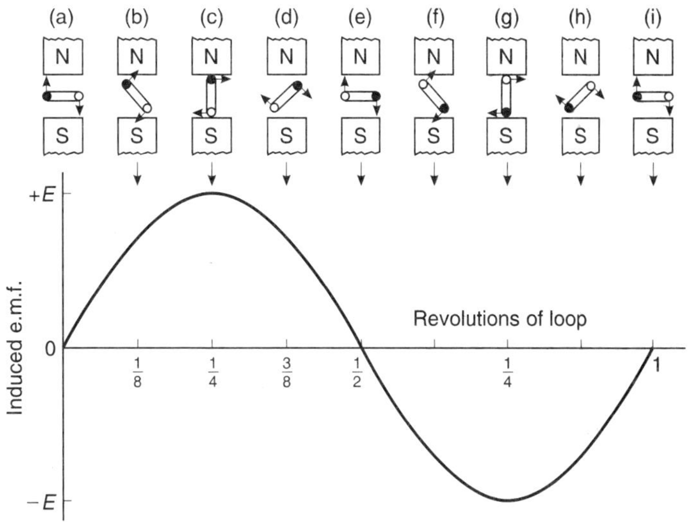

Figure 3.117

Waveforms

If values of quantities which vary with time t are plotted to a base of time, the resulting graph is called a **waveform**. Some typical waveforms are shown in Figure 3.118. Waveforms (a) and (b) are **unidirectional waveforms**, for, although they vary considerably with time, they flow in one direction only (i.e. they do not cross the time axis and become negative). Waveforms (c) to (g) are called **alternating waveforms** since their quantities are continually changing in direction (i.e. alternately positive and negative).

A waveform of the type shown in Figure 3.118(g) is called a **sine wave**. It is the shape of the waveform of e.m.f. produced by an alternator and thus the mains electricity supply is of 'sinusoidal' form.

One complete series of values is called a **cycle** (i.e. from O to P in Figure 3.118(g)). The time taken for an alternating quantity to complete one cycle is called the **period** or the **periodic time**, T, of the waveform. The number of cycles completed in one second is called the **frequency**, f, of the supply and is **measured in hertz, Hz**. The standard frequency of the electricity supply in Great Britain is 50 Hz.

$$\boldsymbol{T = \frac{1}{f} \text{ or } f = \frac{1}{T}}$$

Problem 3.206 Determine the periodic time for frequencies of (a) 50 Hz and (b) 20 kHz.

(a) Periodic time $T = \frac{1}{f} = \frac{1}{50} = \mathbf{0.02\,s \text{ or } 20\,ms}$

(b) Periodic time $T = \frac{1}{f} = \frac{1}{20\,000} = \mathbf{0.000\,05\,s \text{ or } 50\,\mu s}$

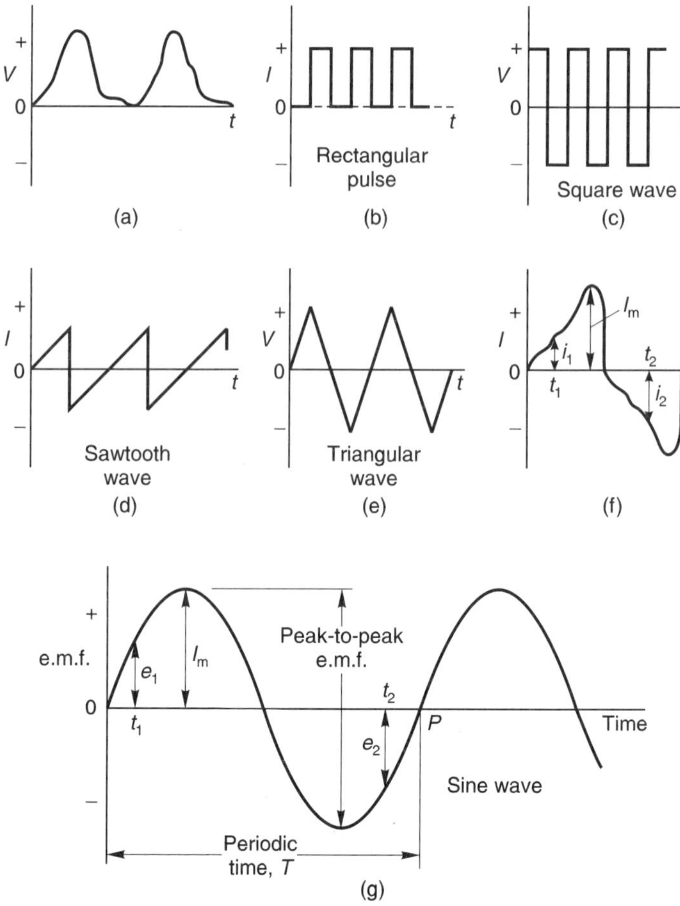

Figure 3.118

Problem 3.207 Determine the frequencies for periodic times of (a) 4 ms, (b) 4 μs.

(a) Frequency $f = \frac{1}{T} = \frac{1}{4 \times 10^{-3}} = \frac{1000}{4} = \mathbf{250\ Hz}$

(b) Frequency $f = \frac{1}{T} = \frac{1}{4 \times 10^{-6}} = \frac{1\,000\,000}{4}$

$= 250\,000$ Hz or 250 kHz or **0.25 MHz**

Test your knowledge 3.53

1 Determine the period times for the following frequencies:
(a) 2.5 Hz (b) 100 Hz
(c) 40 kHz.

2 An alternating current completes 5 cycles in 8 ms. What is its frequency?

A.c. values

Instantaneous values are the values of the alternating quantities at any instant of time. They are represented by small letters, i, v, e, etc. (see Figures 3.118(f) and (g)). The largest value reached in a half cycle is called the **peak value** or the **maximum value** or the **crest value** or the **amplitude** of the waveform. Such values are represented by V_m, I_m, E_m, etc. (see Figures 3.118(f) and (g)). A **peak-to-peak** value of e.m.f. is shown in Figure 3.118(g) and is the difference between the maximum and minimum values in a cycle.

The **average** or **mean value** of a symmetrical alternating quantity (such as a sine wave) is the average value measured over a half cycle (since over a complete cycle the average value is zero).

$$\textbf{average or mean value} = \frac{\textbf{area under the curve}}{\textbf{length of base}}$$

Average values are represented by V_{AV}, I_{AV}, etc.

For a sine wave,
average value $= \mathbf{0.637 \times}$ **maximum value**
(i.e. $\mathbf{2/\pi \times}$ **maximum value**)

The **effective value** of an alternating current is that current which will produce the same heating effect as an equivalent direct current. The effective value is called the **root mean square (r.m.s.) value** and whenever an alternating quantity is given, it is assumed to be the r.m.s. value. For example, the domestic mains supply in Great Britain is 240 V and is assumed to mean '240 V r.m.s.'. The symbols used for r.m.s. values are I, V, E, etc. For a non-sinusoidal waveform as shown in Figure 3.119 the r.m.s. value is given by:

$$I = \sqrt{\frac{i_1^2 + i_2^2 + \ldots + i_n^2}{n}}$$

where n is the number of intervals used.

For a sine wave,
r.m.s. value $= \mathbf{0.707 \times}$ **maximum value (i.e.** $\mathbf{1/\sqrt{2} \times}$ **maximum value)**

Problem 3.208 For the periodic waveform shown in Figure 3.120 determine (i) the frequency, (ii) the average value over half a cycle, and (iii) the r.m.s. value.

(i) Time for 1 complete cycle = 20 ms = periodic time T,

$$\text{hence frequency } F = \frac{1}{T} = \frac{1}{20 \times 10^{-3}} = \frac{1000}{210} = \mathbf{50\,Hz}$$

(ii) Area under the triangular waveform for a half cycle

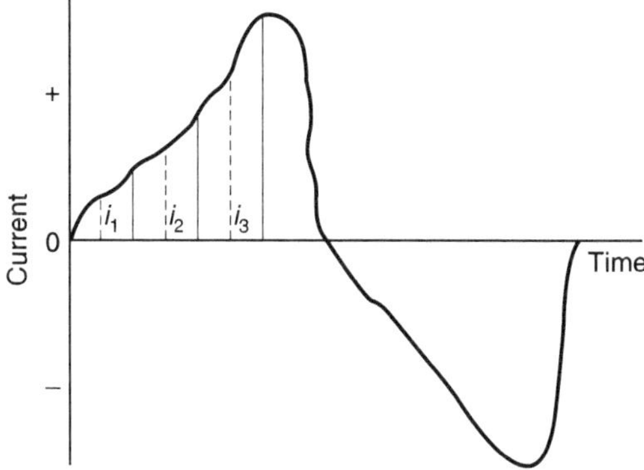

Figure 3.119

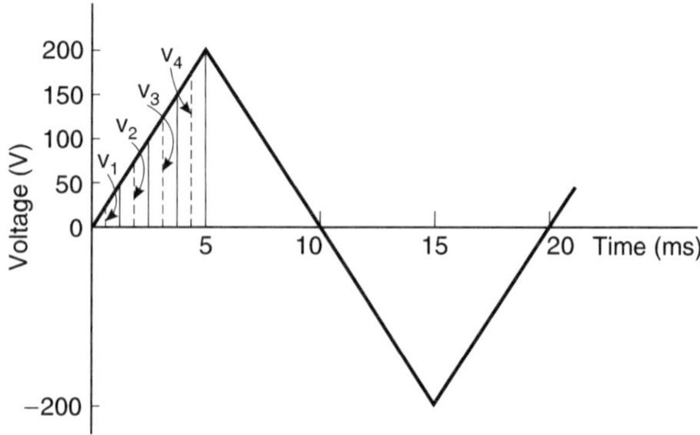

Figure 3.120

$$= \frac{1}{2} \times \text{base} \times \text{height} = \frac{1}{2} \times (10 \times 10^{-3}) \times 200$$
$$= 1\,\text{V s}$$

Average value of waveform

$$= \frac{\text{area under curve}}{\text{length of base}} = \frac{1\,\text{V s}}{10 \times 10^{-3}\,\text{s}} = \frac{1000}{10} = \mathbf{100\,V}$$

(iii) In Figure 3.120 the first 1/4 cycle is divided into 4 intervals. Thus

$$\text{r.m.s. value} = \sqrt{\left(\frac{v_1^2 + v_2^2 + v_3^2 + v_4^2}{4}\right)}$$
$$= \sqrt{\left(\frac{25^2 + 75^2 + 125^2 + 175^2}{4}\right)}$$
$$= \mathbf{114.6\,V}$$

(Note that the greater the number of intervals chosen, the greater the accuracy of the result. For example, if twice the number of ordinates as that chosen above are used, the r.m.s value is found to be 115.6 V.)

Problem 3.209 Calculate the r.m.s. value of a sinusoidal current of maximum value 20 A.

For a sine wave,

$$\text{r.m.s. value} = 0.707 \times \text{maximum value}$$
$$= 0.707 \times 20 = \mathbf{14.14\,A}$$

Problem 3.210 Determine the peak and mean values of a 240 V mains supply.

For a sine wave, r.m.s. value of voltage $V = 0.707 \times V_m$. A 240 V mains supply means that 240 V is the r.m.s. value. Hence

$$V_m = \frac{V}{0.707} = \frac{240}{0.707} = \mathbf{339.5\,V = peak\ value}$$

Mean value, $V_{AV} = 0.637\,V_m = 0.637 \times 339.5 = \mathbf{216.3\,V}$

Test your knowledge 3.54

1. For the period waveform shown in Figure 3.121 determine (i) its frequency, (ii) the average value over half a cycle, and (iii) its r.m.s. value.
2. A supply voltage has a mean value of 150 V. Determine its maximum value and its r.m.s. value.

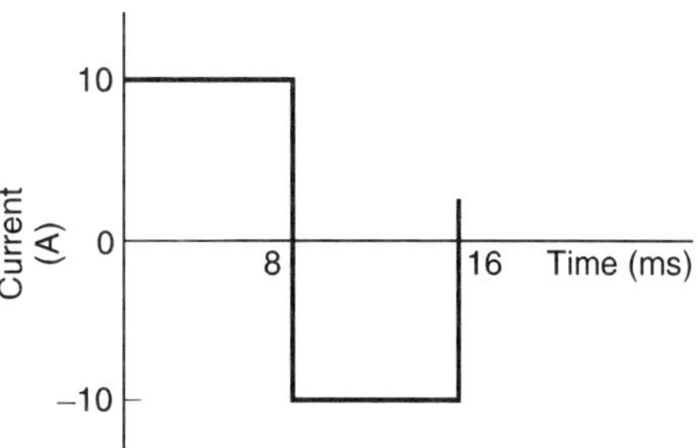

Figure 3.121

Introduction to electrical measuring instruments and measurements

Tests and measurements are important in designing, evaluating, maintaining and servicing electrical circuits and equipment. In order to detect electrical quantities such as current, voltage, resistance or power, it is necessary to transform an electrical quantity or condition into a visible indication. This is done with the aid of instruments (or meters) that indicate the magnitude of quantities either by the position of a pointer moving over a graduated scale (called an analogue instrument) or in the form of a decimal number (called a digital instrument).

Analogue instruments

All analogue electrical indicating instruments require three essential devices:

(a) **A deflecting or operating device**. A mechanical force is produced by the current or voltage which causes the pointer to deflect from its zero position.

(b) **A controlling device**. The controlling force acts in opposition to the deflecting force and ensures that the deflection shown on the meter is always the same for a given measured quantity. It also prevents the pointer always going to the maximum deflection. There are two main types of controlling device – spring control and gravity control.

(c) **A damping device**. The damping force ensures that the pointer comes to rest in its final position quickly and without undue oscillation. There are three main types of damping used – eddy-current damping, air-friction damping and fluid-friction damping.

There are basically two types of scale, linear and non-linear. A linear scale is shown in Figure 3.122(a), where the divisions or graduations are evenly spaced. The voltmeter shown has a range 0–100 V, i.e. a full-scale deflection (f.s.d.) of 100 V. A **non-linear scale** is shown in Figure 3.122(b). The scale is cramped at the

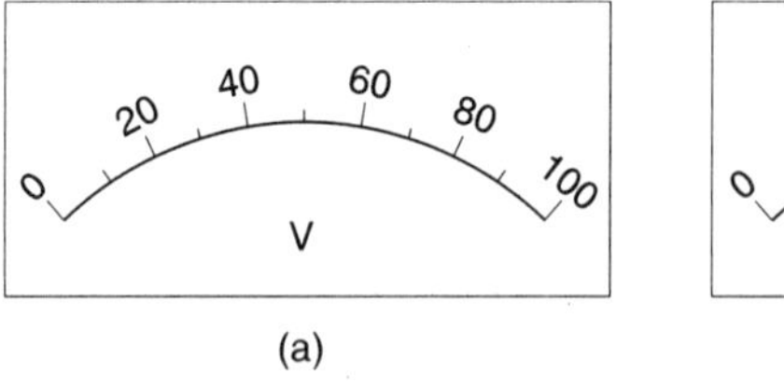

(a)

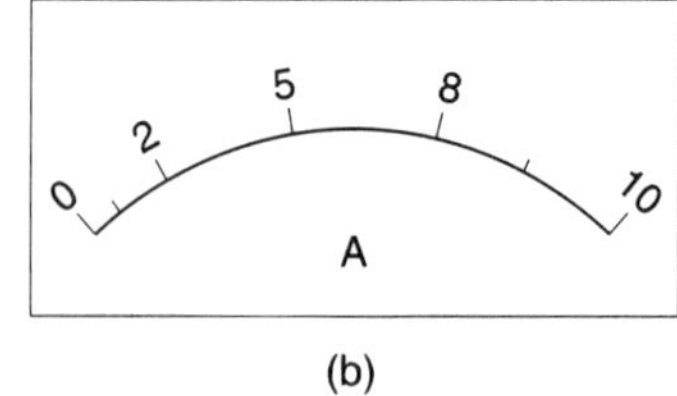

(b)

Figure 3.122

beginning and the graduations are uneven throughout the range. The ammeter shown has a f.s.d. of 10 A.

Moving-iron instrument

(i) An **attraction type** of moving-iron instrument is shown diagrammatically in Figure 3.123(a). When current flows in the solenoid, a pivoted soft-iron disc is attracted towards the solenoid and the movement causes a pointer to move across a scale.

(ii) In the **repulsion type** moving-iron instrument shown diagrammatically in Figure 3.123(b), two pieces of iron are placed inside the solenoid, one being fixed, and the other attached to the spindle carrying the pointer. When current passes through the solenoid, the two pieces of iron are magnetized in the same direction and therefore repel each other. The pointer thus moves across the scale. The

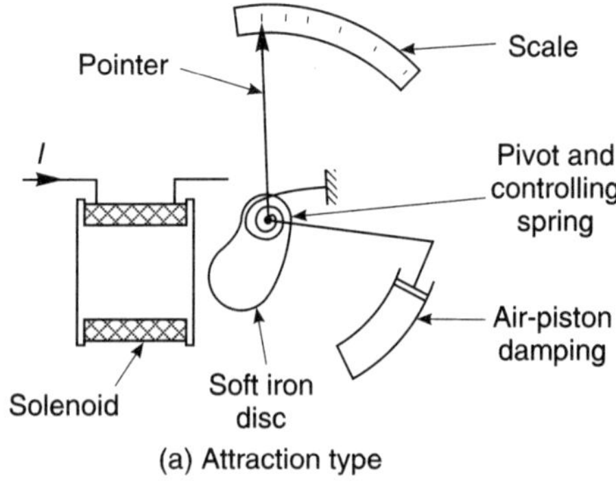

(a) Attraction type

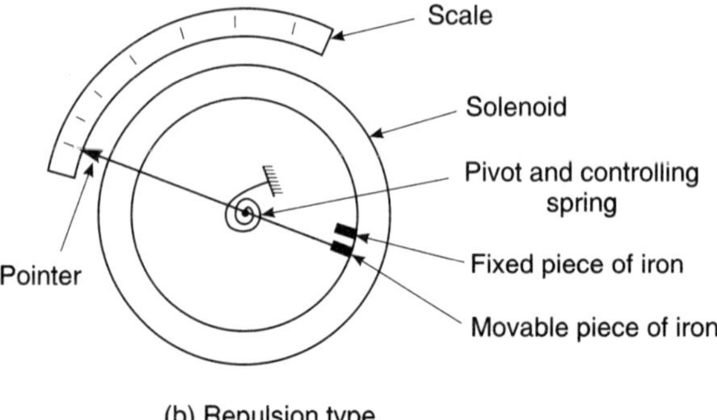

(b) Repulsion type

Figure 3.123

force moving the pointer is, in each type, proportional to I^2. Because of this the direction of current does not matter and the moving-iron instrument can be used on d.c. or a.c. The scale, however, is non-linear.

The moving-coil rectifier instrument

A moving-coil instrument, which measures only d.c., may be used in conjunction with a bridge rectifier circuit, as shown in Figure 3.124 to provide an indication of alternating currents and voltages. The average value of the full wave rectified current is $0.637\,I_m$. However, a meter being used to measure a.c. is usually calibrated in r.m.s. values. For sinusoidal quantities the indication is

$$\frac{0.707\,I_m}{0.637\,I_m}$$

i.e. 1.11 times the mean value. Rectifier instruments have scales calibrated in r.m.s. quantities and it is assumed by the manufacturer that the a.c. is sinusoidal.

Comparison of moving-coil, moving-iron and moving-coil rectifier instruments (see Table 3.3)

Electronic voltmeter

An **electronic voltmeter** can be used to measure with accuracy e.m.f. or p.d. from millivolts to kilovolts by incorporating in its design amplifiers and attenuators. The loading effect of an electronic voltmeter is minimal.

Digital voltmeter

A **digital voltmeter (DVM)** has, like the electronic voltmeter, a high input resistance. For power frequencies and d.c. measurements a DVM will normally be preferable to an analogue instrument.

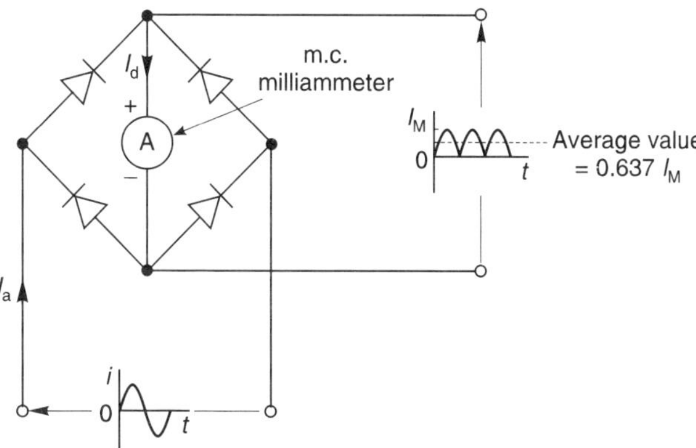

Figure 3.124

Table 3.3

Type of instrument	*Moving-coil*	*Moving-iron*	*Moving-coil rectifier*
Suitable for measuring	Direct current and voltages	Direct and alternating currents and voltages (reading in r.m.s. value)	Alternating current and voltage (reads average value but scale is adjusted to give r.m.s. value for sinusoidal waveforms)
Scale	Linear	Non-linear	Linear
Method of control	Hairsprings	Hairsprings	Hairsprings
Method of damping	Eddy current	Air	Eddy current
Frequency limits	–	20–200 Hz	20–100 kHz
Advantages	1 Linear scale 2 High sensitivity 3 Well shielded from stray magnetic fields 4 Lower power consumption	1 Robust construction 2 Relatively cheap 3 Measures d.c. and a.c. 4 In frequency range 20–100 Hz reads r.m.s. correctly regardless of supply waveform	1 Linear scale 2 High sensitivity 3 Well shielded from stray magnetic fields 4 Lower power consumption 5 Good frequency range
Disadvantages	1 Only suitable for d.c. 2 More expensive than moving-iron type 3 Easily damaged	1 Non-linear scale 2 Affected by stray magnetic fields 3 Hysteresis errors in d.c. circuits 4 Liable to temperature errors 5 Due to the inductance of the solenoid, readings can be affected by variation of frequency	1 More expensive than moving-iron type 2 Errors caused when supply is non-sinusoidal

The ohmmeter

An **ohmmeter** is an instrument for measuring electrical resistance. A simple ohmmeter circuit is shown in Figure 3.125(a). Unlike the ammeter or voltmeter, the ohmmeter circuit does not receive the energy necessary for its operation from the circuit under test. In the ohmmeter this energy is supplied by a self-contained source of voltage, such as a battery. Initially, terminals XX are short-circuited and R adjusted to give a f.s.d. on the milliammeter. If current I is at a maximum value and voltage E is constant, then resistance $R = E/I$ is at a minimum value. Thus the f.s.d. on the milliammeter is made zero on the resistance

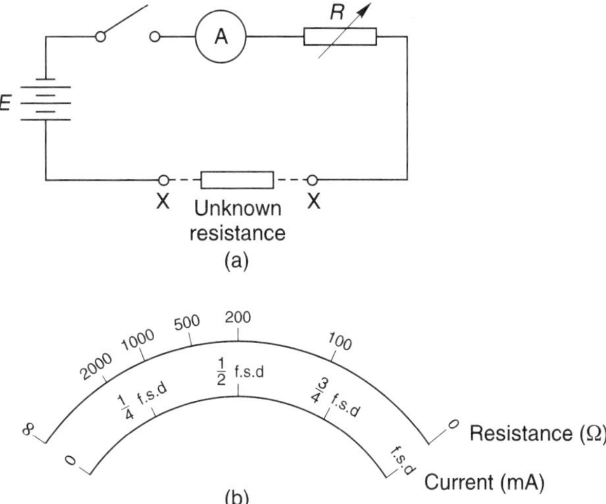

Figure 3.125

scale. When terminals XX are open-circuited no current flows and $R(=E/0)$ is infinity, ∞.

The milliammeter can thus be calibrated directly in ohms. A cramped (non-linear) scale results and is 'back to front', as shown in Figure 3.125(b). When calibrated, an unknown resistance is placed between terminals XX and its value determined from the position of the pointer on the scale. An ohmmeter designed for measuring low values of resistance is called a **continuity tester**. An ohmmeter designed for measuring high values of resistance (i.e. megohms) is called an **insulation resistance tester** (e.g. **megger**).

Multimeters

Instruments are manufactured that combine a moving-coil meter with a number of shunts and series multipliers, to provide a range of readings on a single scale graduated to read current and voltage. If a battery is incorporated then resistance can also be measured. Such instruments are called **multimeters** or **universal instruments** or **multirange instruments**. An 'Avometer' is a typical example. A particular range may be selected either by the use of separate terminals or by a selector switch. Only one measurement can be performed at one time. Often such instruments can be used in a.c. as well as d.c. circuits when a rectifier is incorporated in the instrument.

Wattmeters

A **wattmeter** is an instrument for measuring electrical power in a circuit. Figure 3.126 shows typical connections of a wattmeter used for measuring power supplied to a load. The instrument has two coils:

(i) a current coil, which is connected in series with the load, like an ammeter, and

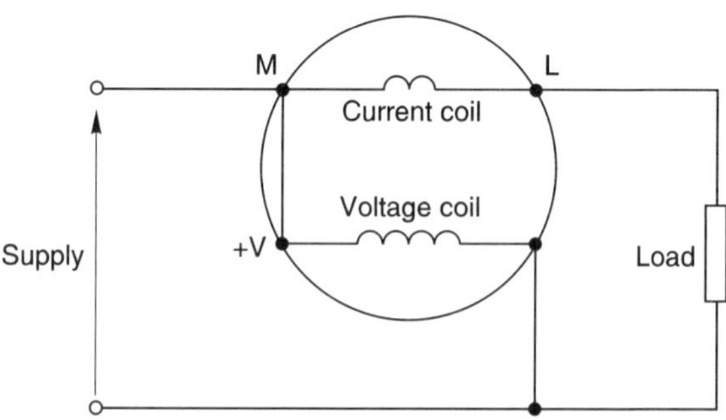

Figure 3.126

(ii) a voltage coil, which is connected in parallel with the load, like a voltmeter.

Cathode-ray oscilloscope

The cathode-ray oscilloscope (CRO) may be used in the observation of waveforms and for the measurement of voltage, current, frequency, phase and periodic time. For examining periodic waveforms the electron beam is deflected horizontally (i.e. in the X direction) by a sawtooth generator acting as a timebase. The signal to be examined is applied to the vertical deflection system (Y direction) usually after amplification.

Oscilloscopes normally have a transparent grid of 10 mm by 10 mm squares in front of the screen, called a graticule. Among the timebase controls is a 'variable' switch which gives the sweep speed as time per centimetre. This may be in s/cm, ms/cm or μs/cm, a large number of switch positions being available. Also on the front panel of a CRO is a Y amplifier switch marked in volts per centimetre, with a large number of available switch positions.

(i) With **direct voltage measurements**, only the Y amplifier 'volts/cm' switch on the CRO is used. With no voltage applied to the Y plates the position of the spot trace on the screen is noted. When a direct voltage is applied to the Y plates the new position of the spot trace is an indication of the magnitude of the voltage. For example, in Figure 3.127(a), with no voltage applied to the Y plates, the spot trace is in the centre of the screen (initial position) and then the spot trace moves 2.5 cm to the final position shown, on application of a d.c. voltage. With the 'volts/cm' switch on 10 V/cm the magnitude of the direct voltage is 2.5 cm × 10 V/cm, i.e. 25 V.

(ii) With **alternating voltage measurement**, let a sinusoidal waveform be displayed on a CRO screen as shown in Figure 3.127(b). If the s/cm switch is on, say, 5 ms/cm, then the **periodic time T** of the sinewave is 5 ms/cm × 4 cm i.e. **20 ms** or **0.02 s**. Since

$$\text{frequency } f = \frac{1}{T}, \ \textbf{frequency} = \frac{1}{0.02} = \mathbf{50\,Hz}$$

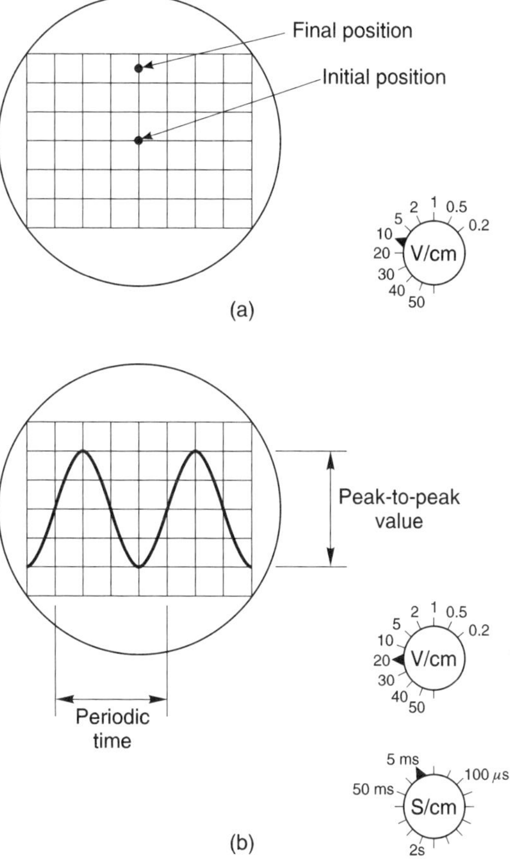

Figure 3.127

If the 'volts/cm' switch is on, say, 20 V/cm, then the **amplitude** or **peak value** of the sinewave shown is 20 V/cm × 2 cm, i.e. 40 V.

Since

$$\text{r.m.s. voltage} = \frac{\text{peak voltage}}{\sqrt{2}}$$

$$\text{r.m.s. voltage} = \frac{40}{\sqrt{2}} = \mathbf{28.28\,V}$$

Double beam oscilloscopes are useful whenever two signals are to be compared simultaneously. The CRO demands reasonable skill in adjustment and use. However its greatest advantage is in observing the shape of a waveform – a feature not possessed by other measuring instruments.

Problem 3.211 For the CRO square voltage waveform shown in Figure 3.128 determine (a) the periodic time, (b) the frequency and (c) the peak-to-peak voltage. The 'time/cm' (or timebase control) switch is on 100 μs/cm and the 'volts/cm' (or signal amplitude control) switch is on 20 V/cm.

(a) The width of one complete cycle is 5.2 cm, hence the periodic time, $T = 5.2\,\text{cm} \times 100 \times 10^{-6}\,\text{s/cm} = \mathbf{0.52\,ms}$.

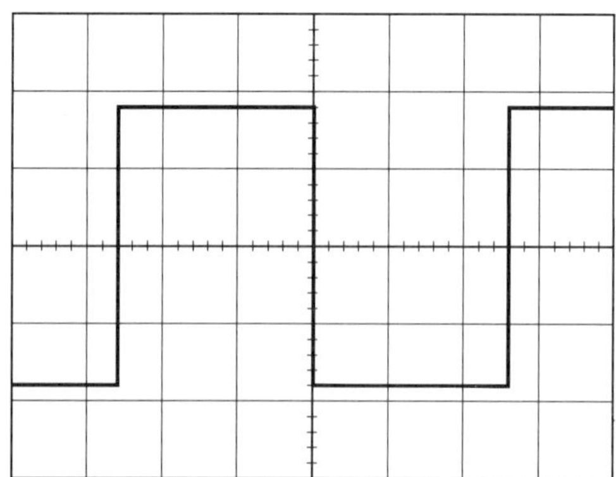

Figure 3.128

(b) Frequency, $f = \frac{1}{T} = \frac{1}{0.52 \times 10^{-3}} = \mathbf{1.92\,kHz}$

(c) The peak-to-peak height of the display is 3.6 cm, hence the peak-to-peak voltage = 3.6 cm × 20 V/cm = **72 V**.

Problem 3.212 For the double-beam oscilloscope displays shown in Figure 3.129 determine (a) their frequency, (b) their r.m.s. values, (c) their phase difference. The 'time/cm' switch is on 100 μs/cm and the 'volts/cm' switch is on 2 V/cm.

(a) The width of each complete cycle is 5 cm for both waveforms. Hence the periodic time, T, of each waveform is 5 cm × 100 μs/cm, i.e. 0.5 ms. Frequency of each waveform

$$f = \frac{1}{T} = \frac{1}{0.5 \times 10^{-3}} = \mathbf{2\,kHz}$$

(b) The peak value of waveform A is 2 cm × 2 V/cm = **4 V**.

Hence the r.m.s. value of waveform A $= \frac{4}{\sqrt{2}} = \mathbf{2.83\,V}$.

The peak value of waveform B is 2.5 cm × 2 V/cm = **5 V**.

Hence the r.m.s. value of waveform B $= \frac{5}{\sqrt{2}} = \mathbf{3.54\,V}$.

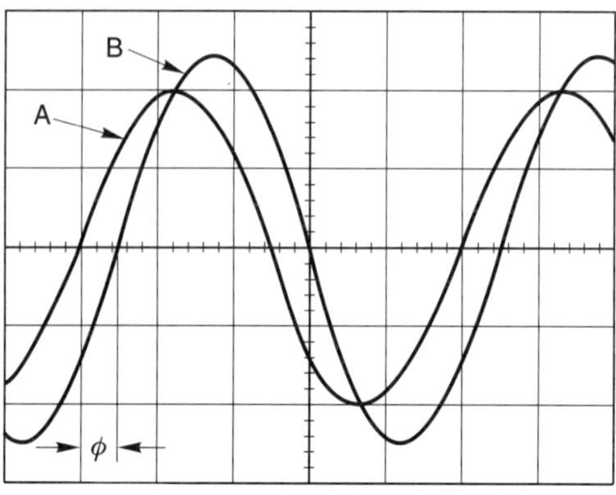

Figure 3.129

Test your knowledge 3.55

A sinusoidal voltage trace displayed by a CRO is shown in Figure 3.130. If the 'time/cm' switch is on 500 μs/cm and the 'volts/cm' switch is on 5 V/cm, find, for the waveform, (a) the frequency, (b) the peak-to-peak voltage, (c) the amplitude, (d) the r.m.s. value.

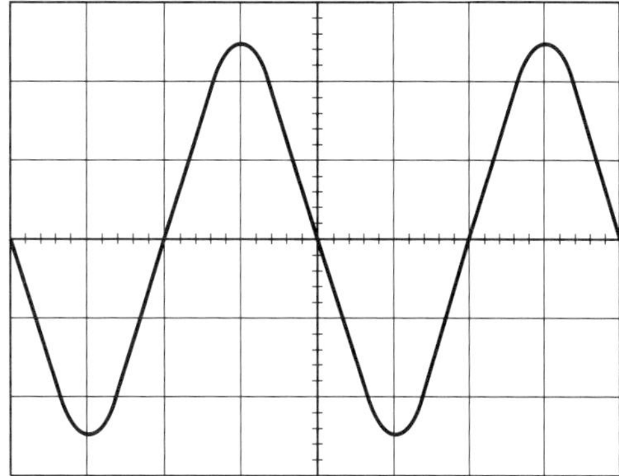

Figure 3.130

(c) Since 5 cm represents 1 cycle, then 5 cm represents 360°, i.e. 1 cm represents $(360/5) = 72°$.
The phase angle $\phi = 0.5\,\text{cm} = 0.5\,\text{cm} \times 72°/\text{cm} = 36°$.
Hence waveform A leads waveform B by 36°.

Test your knowledge: answers

3.1	1.	644	
	2.	2618	
	3.	83	
3.2	1.	−34	
	2.	−27	
3.3	1.	$\frac{7}{16}$	
	2.	$2\frac{7}{30}$	
	3.	$3\frac{1}{20}$	
3.4	1.	$\frac{1}{17}$	
	2.	17 g	
3.5	1.	(a) 30.739	(b) 157.0872
	2.	(a) 23.56	(b) 14.78
3.6	1.	(a) 46.2%	(b) 275%
	2.	(a) 2.50 g	(b) 7.64 t
3.7	1.	5^2	
	2.	3^4	
	3.	±16	

3.8	1.	(a) 4.76×10 (b) 3.2×10^{-3}
		(c) 5.126×10^4
	2.	(a) 1.54×10 (b) 5.71×10^{-1}
	3.	2.8×10^7
3.9	1.	7.40
	2.	0.170
	3.	(a) 381 mm (b) 56.35 km/h
		(c) 376 km (d) 52 lb 13 oz
		(e) 6.82 g (f) 54.55 l
		(g) 5.5 gallons
3.10	1.	$t = 3$
	2.	$x = 1$
	3.	$c = 7\frac{1}{2}$
	4.	$x = 2$
	5.	$a = \pm 6$
3.11	1.	$u = 23$
	2.	$P_2 = 2.6 \times 10^5$
3.12	1.	$t = 6.950$ s
	2.	$R = 19.6\,\Omega$
	3.	0.005 02 m or 5.02 mm
	4.	0.1440 J
3.13	1.	$R = \frac{E}{I}$
	2.	$C = \frac{5}{9}(F - 32)$
	3.	$u = \frac{2s}{t} - v$ or $u = \frac{2s - vt}{t}$
3.14	1.	$y = 14.5$
	2.	$(-1, 2)$
	3.	(a) $4, -2$ (b) $-1, 0$
		(c) $-3, -4$ (d) $0, 4$
3.15	1.	(a) 40°C
		(b) 127.5 Ω
	2.	(a) 0.25
		(b) 12
		(c) $F = 0.25\,L + 12$

(d) 89.5 N

(e) 592 N

(f) 212 N

3.16 1. (a) $120\,m^2$ (b) $1.2 \times 10^6\,cm^2$

(c) $0.12 \times 10^9\,mm^2$

2. (a) $4000\,cm^3$ (b) $0.004\,m^3$ (c) $4 \times 10^6\,mm^3$

3.17 1. $7000\,kg/m^3$

2. 50 l

3. $9000\,kg/m^3$

3.18 1. 4.3 kN at $-25°$

2. 26.7 N at $-82°$

3.19 1. 2.56 N at $-20.04°$

2. 53.5 kN at 37° to force 1
(i.e. 117° to the horizontal)

3.20 1. 10 min 40 s

2. (a) 72 km/h (b) 20 m/s

3.21 1. (a) 46 min 36 s (b) 42.86 km/h

(c) 45 km (d) 53.4 km/h

3.22 1. $9.26 \times 10^{-4}\,m/s^2$

2. 1.42 s

3. 39.2 km/h

3.23 1. (a) $\frac{5}{9}\,m/s^2$ (b) 750 N

2. $0.560\,m/s^2$

3.24 1. 200 mm

2. $\mathbf{F}_1 = 3.5\,kN$, $\mathbf{F}_2 = 1.5\,kN$

3.25 1. 2.625 kN, 5.375 kN

2. 36 kN

3.26 800 N

3.27 1. 490 J

2. (a) 2.5 J (b) 2.1 J

3. 14.72 kJ

3.28 1. 420 J

2. 4 m

3.29 1. (a) 80 N (b) 200 W

2. (a) 450 W (b) 600 W

3. (a) 1500 W (b) 510 mm/s

3.30 1. 250 MPa
2. 12.78 mm

3.31 1. 0.3 mm
2. (a) 127.3 MPa (b) 0.001 25 or 0.125%
3. (a) 157.1 kN (b) 80.03 MPa

3.32 1. 10 mm
2. (a) 15.92 MPa (b) 0.332 mm
3. (a) 140 MPa (b) 11.0 kN

3.33 1. 343 kPa
2. 300 MN
3. 758 mm

3.34 743 mm

3.35 (a) Bourdon gauge
(b) MacLeod gauge
(c) Mercury filled U-tube manometer
(d) Inclined manometer
(e) Fortin barometer

3.36 1. (a) 210 K (b) −48°C
2. 130 J/(kg °C)
3. 220°C

3.37 1. 18.08 MJ
2. 24.70 MJ

3.38 (a) Mercury-in-glass
(b) Copper–constantan thermocouple
(c) Chromal–alumel thermocouple
(d) Alcohol-in-glass thermometer

3.39 1. 50.0928 m
2. 0.0425%
3. (a) 4.1933 m (b) 4.2214 m

3.40 1. 16 A
2. 4 A

3.41 1. 20 kΩ, 125 mA
2. (a) 2400 Ω or 2.4 kΩ
(b) 600 000 Ω or 600 kΩ or 0.6 MΩ

3.42 1. 80 mW
2. (a) 500 V (b) 2.5 kW

3. 2 kW, 12 kWh, 84p
4. £10.29

3.43 (a) 5 A (b) 13 A

3.44 1. 1.44 Ω
2. 0.017 μΩ m
3. 0.180 Ω

3.45 1. 23.83 Ω

3.46 1. (a) 2.17 V (b) 1.6 V (c) 0.7 V
2. (a) 0.25 A (b) 14.5 V

3.47 0.5 A, 4.5 V, 2.75 W

3.48 1. (a) 2 Ω (b) 4 A
2. (a) all four in parallel
(b) two in series, in parallel with another two in series
(c) three in parallel, in series with one
(d) two parallel, in series with two in series
3. (a) 4 A (b) 44 A

3.49 (a) 50 V
(b) lamps A and B will not operate

3.50 1. 1.5 T
2. 250 mm

3.51 1. 21 N, 14.8 N
2. (a) upwards
(b) to the right
(c) out of the paper
(d) S on left, N on right

3.52 1. 200 m/s
2. (a) 3.75 mV (b) 3.25 mV (c) 1.875 mV

3.53 1. (a) 0.4 s (b) 10 ms (c) 25 μs
2. 625 Hz

3.54 1. (i) 62.5 Hz (ii) 10 A (iii) 10 A
2. 235.5 V, 166.5 V

3.55 (a) 500 Hz (b) 25 V
(c) 12.5 V (d) 8.84 V

Unit 4 Engineering in society and the environment

Summary

The main aspects of engineering technology of which this unit requires us to be aware are *information technology*, *automation*, and *materials*. In particular, we are expected to be conversant with the most recent 'high tech' applications of technology in the home, in our leisure activities, in industry and in conjunction with our health. We are also required to be familiar with the effects on the environment of material extraction and processing, of engineering manufacture and the subsequent disposal of unwanted waste.

The first section deals with all of the above but with none of it exhaustively. This is because engineering technology is continually evolving. In order to be aware of the latest developments and legislation, it is important that you regularly read the newspapers, trade journals and technical magazines. Situations can change appreciably almost overnight and can even be unrecognizable after a couple of years.

The application of engineering technology in society

Engineering technology in the home

In general the aim of new technology in the home is to reduce manual effort or otherwise improve the operation. Examples are as follows:

Domestic chores

Domestic toaster

The older models depend upon a bi-metal strip bending under the heat generated by the toaster heating element. The time taken for the bi-metal strip to bend to a pre-set position, after which time it switches off the current supply to the heating element, is a determining factor as to how brown the finished slice of toast is. We all know how unreliable this method of toast browning control can be for producing a series of slices of toast in quick succession at breakfast time. The problem has been largely reduced in the newer toasters where the crude bi-metal browning controller has been replaced by a pre-programmed microchip to give an accurate and consistent degree of browning.

Activity 4.1

Investigate the operation of a bi-metal strip used to control the temperature of a domestic appliance such as a toaster, electric iron, or heater. Write a brief illustrated report showing how this device works and give reasons for inconsistencies in its operation.

Automatic washing machines

Not only are the different washing routines automated by the use of pre-programmed microchips, there are now economies of saving the amount of water used for a given wash by using an atomized spray of water. Additional facilities also available include:

- automatic compensation for low mains water pressure
- automatic foam sensing control
- fault diagnosis for service engineers.

Upright vacuum cleaners

Some of the latest versions of these are now fitted with microporous filters and disposable bags for the control of pollen and house dust removal.

Frost-free fridge freezers

These are now available with automatic defrosting systems to alleviate the periodic chore of having to chip away large amounts of ice around the inside. They are also available with digital temperature displays and audio warning signals in the event of a malfunction occurring.

Food processors

Simple food mixers reduce the manual labour required to use a wooden spoon and mixing bowl. However, they do not use a variety of changeable tools to cut, chop, slice or blend mixtures, as do the new range of food processors. Not only has the manual effort of mixing been kept low, a whole new range of food processing has also been made available.

Stainless steel or plastic kitchen equipments

The use of stainless steel or plastic as an easy to clean material is not new. Stainless steel is a particularly hard wearing, rust proof, easy to clean metal. It is therefore very suitable for hygienic contact with hot or cold foods. Plastic kitchen equipments are similarly hygienic but can be damaged by hot, fatty foods.

UPVC window frames, doors and external house cladding
Until the mid-1980s, most of the above fittings were made from wood. They needed to be painted each two or three years to remain decorative and to provide a protective coating from the weather. The advent of the rigid plastic, UPVC (ultra high density polyvinyl chloride) enabled it to be used in the construction of houses in place of wood. If made from UPVC, window frames and doors are impervious to most weather conditions and never need to be painted.

Activity 4.2

1 Obtain an estimate of the cost of repainting (rub down, undercoat and top coat) the doors and window frames of the house or flat in which you live.

2 Now obtain a price for the fitting of replacement UPVC windows and doors.

3 If the replacement windows are to be double glazed units, estimate the saving in house heating costs.

4 Assuming that today's prices are maintained in the future, and that the house would need a full repaint each three years, estimate the time it will take for the cost of the replacement windows and doors to be recovered from the ensuing savings.

5 Write a brief, word-processed, report summarizing the results of 1–4.

PTFE (Polytetrafluoroethylene)
This is another plastic polymer material. It is noted for its very marked chemical inertness and heat resistance. It is widely used for a variety of engineering and chemical purposes, and as a 'non-stick' coating in kitchen cooking equipments.

Houses cooled by the sun
For some years householders have had the assistance of solar panels for heating water. Technology is taking another step; this time in the opposite direction. House cooling systems using solar panels are under development. Conventional air conditioning systems need to compress a gas, called a refrigerant, to a high pressure. This is done using an electrically driven compressor pump and it is an expensive operation. The compressed gas is allowed to expand rapidly, so cooling it, and it is this cold gas that is used in a heat exchanger to extract heat from the surrounding room.

The new system requires only one thousandth of the electricity of the conventional gas compressor type system. Figure 4.1 shows an annotated sketch outlining the method of operation of the new system.

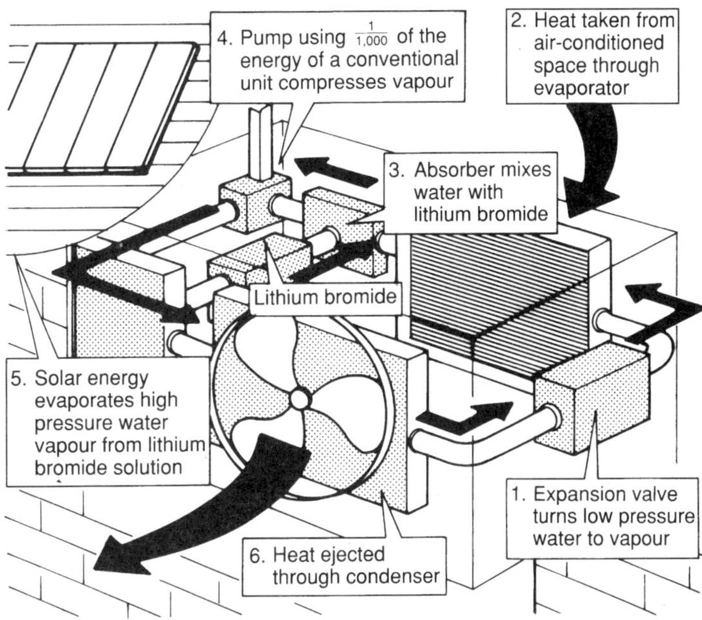

Figure 4.1 *Energy saving vapour absorption system*

> **Test your knowledge 4.1**
>
> Do solar panels provide energy on a hazy summer's day or do they only operate under direct sunlight? Explain your answer.

A water vapour refrigerant is used in the new system. Its pressure can be cheaply increased by a pump after being absorbed in lithium bromide liquid, from which it is then separated using heat from solar panels. The new system is claimed to be cheap, efficient and environment friendly. It does not damage the ozone layer as do conventional refrigerants.

The automated home

It is reported that the first automated homes are being tested in Scotland. The idea is for the householder to be able to dial their home telephone and then, through low voltage signalling lines, switch on the oven, draw the curtains or increase the central heating temperature. In addition to giving the householder remote control of space and water heating, he or she will be able to connect to burglar and smoke alarm detectors. These, if activated, will send a signal to a prearranged telephone number.

Home entertainment

This tends largely to be centred on the various television services that are ever expanding and improving technically.

The television set

This brings news programmes and entertainment directly into the home. When used with a video recorder, it can be used to play pre-recorded, hired video cassettes and to record television programmes. Further, the television can be switched from its normal picture mode into its text mode. The screen presentation can then paged round, using a remote hand-operated controller, giving the viewer the latest information on most things from the weather to stock market prices.

Satellite TV

This TV service is based on the use of geostationary satellites that broadcast programmes directly to our homes. It enables viewers to watch live sports and other programmes that are transmitted from overseas. Satellite TV requires special aerial systems and receiving equipment to interface signals to normal domestic TV receivers.

Test your knowledge 4.2

A TV aerial isn't required to receive cable TV signals. Does this mean that you don't need a licence to receive cable TV?

Cable TV

This is a system which provides a TV service without the need for an aerial. It therefore means that there is a possibility of reducing the amount of radio air traffic and freeing some frequency bands for other use. The cable TV user pays a fee to be connected to a TV cable that runs beneath the road outside his or her house.

Music centres

These become more sophisticated almost daily. Digital techniques using compact discs (CDs) have superseded the older record discs and high-quality stereophonic sound is standard.

Test your knowledge 4.3

What do the initials LCD and IC stand for?

Hand-held games

These are small, hand-held plastic packages fitted with LCD displays that use dedicated ICs to control sophisticated colour or monochrome displays. These graphic displays, or pictures, can sometimes be transferred to the domestic TV screen for increased picture size and clarity.

Engineering technology and leisure activities

Motor cars

Recent innovations include:

Catalytic converters to help in the reduction of toxic pollutants in the car's exhaust gases. For example, by the burning of poisonous carbon monoxide to the less harmful carbon dioxide.

Fuel injection systems are replacing the older carburettor method of mixing air and petrol for fuelling the engine. The carburettor method does not result in the correct burning of the fuel in the engine. This produces exhaust gases containing pollution levels which exceed those demanded by many countries – especially the USA. Computer managed, direct fuel injection is a major factor which makes this possible.

Re-cyclable plastics and aluminium for styling and lightness.

Air bags for both front and side impact driver and passenger protection.

Test your knowledge 4.4

Is it a legal requirement for air-bags to be fitted to all new cars?

Improved tyre technology for increased stability and safety.

Activity 4.3

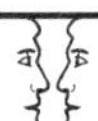

Find out just what the tyre improvements over the past ten years have been and how they affect stability and safety. Present your findings in the form of a brief article on 'Tyre Safety' that will appear in your local newspaper. You should limit the size of your article to between 750 and 1000 words.

Hint: Your local tyre fitting centre may be able to advise you and provide you with some manufacturers' literature.

Global positioning systems for navigation purposes, using an on-board computer and a map of the UK stored in computer memory.

Engines and automatic transmission systems can now be controlled by powerful on-board computers which also control the car's traction and braking. If a wheel spins because of excessive acceleration, the computer will either back-off the engine's power until the spinning stops, or if it decides that the wheel-spin is unlikely to cause a skid, will keep the power on and apply a brake, on whichever side is necessary, to stop the wheel-spin without cutting the power.

The same *microchip technology* will also prevent wheels locking under hard braking conditions and give automatic adjustment of the vehicle's suspension's shock absorbers immediately to suit any changing driving conditions.

A latent heat reservoir is reported as being an optional facility offered by BMW. Apparently, the system uses a salt-based chemical paste to absorb the heat generated by the engine and can retain this heat for several days. When the car is next started, the heat reservoir will have warm air blowing on the screen within half a minute, while the pre-warmed engine will be emitting less air pollution through its exhaust.

Test your knowledge 4.5

1 What is an alloy?
2 Name some examples of common alloys.
3 How are alloys made?

Bicycles

Magnesium alloy wheels for bicycles are very light but strong. The best frames for lightness and strength are made from carbon fibre section.

Music

The latest clavinovas are powerful digital pianos which will not only play prerecorded tunes but also provide a sophisticated musical backing for any tune being played on it. At the throw of a switch, a relatively unaccomplished piano player can enhance his or her performance by bringing in an accompaniment of percussion, wind or strings, etc., at will.

Personal communications

Mobile phones have become increasingly prolific since the mid-1990s. The digital communication techniques make it technically possible to have a complete transmitter/receiver in your pocket. This mobile telephone can be used just like a fixed installation. It makes communicating very convenient for the user, especially in genuine emergency use, but there are undesirable consequences:

- Their use in crowded public places is very annoying to bystanders.
- It has been shown that the electromagnetic emissions from electronic devices can interfere with the functioning of integrated chips in nearby equipments. It is suspected that the sophisticated electronics fitted to control the brakes on some modern motor cars may even have malfunctioned due to the effects of a nearby mobile phone. Modern cars have become increasingly vulnerable to electrical disruption. Not only brakes but fuel injection, ignition, windows and even seats of many cars are now controlled by microchips whose delicate components malfunction if exposed to even relatively weak radio waves.
- It is said that some mobile phones generate electromagnetic fields which are more than twice the safe level in a car. Perhaps the use of a mobile phone in a car will be forbidden – it already is in an aircraft. In the meantime, manufacturers of automobile electronic systems are beginning to incorporate sophisticated test procedures in order to ensure that their equipment is unaffected by stray radio frequency fields.

Personal radios and stereos have become even more prolific than the mobile phone. This is not surprising since they are much cheaper and can be used for entertainment purposes. However, since the use of a personal radio involves the wearing of headphones, there is an element of risk introduced by the user not being able to hear approaching traffic.

Engineering technology in industry and commerce

Manufacturing

In recent years, manufacturing technology has undergone considerable change due to the introduction of new technology. Two notable examples are:

- CNC machining centres can act as a lathe and vertical and horizontal mill in one. They can work 24 hours each day and many times faster than a manually controlled machine.
- PLC are programmed to automate many routine industrial processes. This ensures increased production and product repeatability.

Activity 4.4

1 What do the initials CNC and PLC stand for?
2 Investigate one example of the use of each of these devices.
3 Present your findings in the form of a brief article to be published in your school or college newsletter. Make sure that your article is suitable for the non-technical reader.

Information processing

Internet is the name given to the international communications system or *information superhighway* giving people access to many new services. Users of the system are interconnected through a personal computer and modem to the international telephone network. Also connected are sources of information such as universities, libraries, banks, stock markets, hospitals and the like. This internet system, which is expanding daily, enables users to use their PC keyboards and printers for such activities as:

- writing letters to each other (electronic mail or *e-mail*)
- accessing centres of information and then browsing through the different topics covered
- undertaking postal shopping from catalogues appearing on the PC screen
- making financial transactions
- taking educational courses.

Facsimile (or fax) transmissions of pictures and text over the telephone have been commercially available since the early 1980s. The latest fax machines are now sufficiently cheap for the householder to consider buying one. They can also double as copiers and, if properly interfaced to a PC, as a text scanner.

Computer aided design (or CAD) is not new but should not be forgotten as being an excellent tool for the design of engineering components or systems. The beauty of it is that the designs can be can be tried out by computer simulation before going to the expense of manufacturing a prototype.

Engineering technology in health and medicine

Engineered products are playing an increasingly important part in health care. Some typical examples are highlighted below.

Replacement joints

Figure 4.2 shows details of a replacement hip joint. More than 50 000 of these operations are carried out in the UK each year,

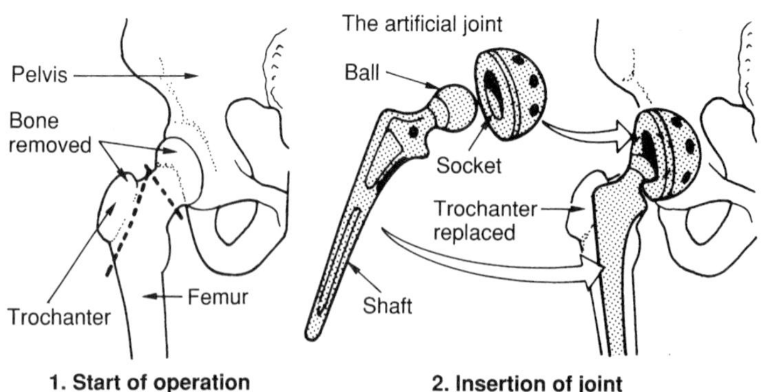

Figure 4.2 *A hip joint replacement*

mostly on people in their 70s or 80s. It has been the advances in anaesthesia which have made such operations possible for old people. Prior to this they would have been regarded as medically unfit to withstand the shock of such drastic surgery.

The operation is required usually because the head of the femur, or thigh bone, has worn away or broken. In the operation, the femur is cut through below the round, ball-like head of the bone that sits in its socket in the pelvis. The pelvic socket is reamed out and a new plastic socket is put in place. In older patients, it is usually cemented in.

Artificial hearts

For some years, real donor hearts have been transplanted into needy patients under the age of 60. Patients over 60 are usually judged as being too old to receive a heart transplant under the National Health Service. This restriction may not apply to a new artificial heart which has recently been developed. This is a battery powered heart and is designed to operate indefinitely. It assists the patient's real heart which hopefully will recover during the period of assistance. The artificial heart responds to physical activity and can increase the blood flow around the body by up to 10 litres per minute. Figure 4.3 shows the artificial heart in position.

Apparently, development work is continuing to produce an artificial heart the size of a man's thumb nail and which can be fitted inside a heart.

Internal body scanners

There is a wide range of sophisticated medical diagnostic equipment such as magnetic resonance imaging (MI) and computerized axial tomography (CAT) scanners which combine engineering and computer technologies. Some of these advanced techniques are used to detect not only the structure of part of a person's body but also whether it is functioning correctly. Basically, the techniques involve the examination of the electromagnetic and mechanical properties of the atoms of an organ within the body. It does this by looking at individual sectional slices of the organ, using a computer to analyse the electrical

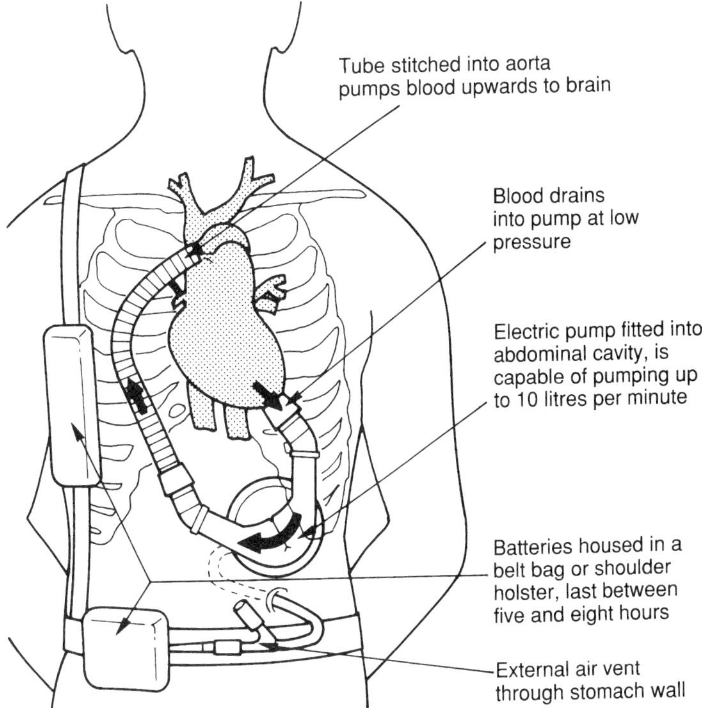

Figure 4.3 *How the electric heart works*

characteristics of the atoms in the slices and finally constructing an image of the whole organ.

IT assisted surgery

A further example of high technology engineering currently under development in the medical field is computer-aided surgery. Existing scanning techniques are to be used accurately to define the position of, say, a brain tumour and the associated pattern of local blood vessels. The operation can then be planned for minimum damage to the area surrounding the tumour. The operation on the tumour would be undertaken through a relatively small hole cut through the skull. When the technique is perfected, the idea is for the surgeon's implement to have three-dimensional positional guidance using computer-aided control. The operating surgeon will need to be specially trained in the new technique.

Telesurgery

Information technology is being used with a long-distance surgery system which has been tried successfully on pigs and is being developed for use on humans. It enables doctors to operate on patients thousands of miles away.

The system produces a three-dimensional picture of the patient which is displayed many miles away on a console and is viewed by the surgeon. The surgeon will slip his thumbs and fingers through control loops on two instruments connected to robotic arms that operate on the patient, wherever he or she is.

Telemedicine

This is not so dramatic as telesurgery and is already used by the American armed forces. In a reported example of its use, a digital coloured image of a patient and other medical data were sent by satellite from Bosnia to physicians in America. There they prescribed a drug which saved the patient's life.

Also available is a wide range of advanced monitoring and life support equipments for use in hospital operating theatres and intensive care units. Some of the monitors used to check the condition of patients are called *ambulatory biomonitors*. These can transmit body function details to a nurses' station while the patient is walking about – another way to save a bed!

Activity 4.5

Find out the range of specialist medical equipment available in your local health clinic for your GP's use. What are the names of the various equipments and for what are they used? Write a brief, word-processed report for use in a local information pack for families who may be thinking about moving into your area.

Hint: You may well find that your doctors now have a database of their patients which record each visit and the medication prescribed. There may even be a portable electrocardiograph available to your GP.

The effect of engineering activities on the environment

Extraction of raw materials

Economic and social effects stem from the regional wealth that is generated by the extraction of the raw material and its subsequent processing or manufacture into useful engineering materials. For example, the extraction of iron-ore in Cleveland and its processing into pure iron and steel has brought great benefit to the Middlesbrough region. The work has attracted people to live in the area and the money they earn tends to be spent locally. This benefits trade at the local shops and entertainment centres and local builders must provide more homes and schools and so on. The increased numbers of people produce a growth in local services which includes a wider choice of different amenities, better roads and communications and arguably, in general, a better quality of life.

On the debit side, the extraction of raw materials can leave the landscape untidy. Heaps of slag around coal mines and steelworks together with holes left by disused quarries are not a pretty sight. In recent years much thought and effort has been expended on improving these eyesores. Slag heaps have been remodelled to become part of golf courses and disused quarries filled with water to become centres for water sports or fishing. Disused mines and quarries can also be used for taking engineering waste, in what is known as a landfill operation, prior to the appropriate landscaping being undertaken.

Disposal of industrial waste materials

Engineering activities are a major source of *pollutants* causing many types of *pollution*. Air, soil, rivers, lakes and seas are all, somewhere or other, polluted by waste gases, liquids and solids discarded by the engineering industry. Because engineering enterprises tend to be concentrated in and around towns and other built up areas, these tend to be common sources of pollutants.

Electricity is a common source of energy and its generation very often involves the burning of the *fossil fuels*: coal, oil and natural gas. In so doing, each year, billions of tonnes of carbon dioxide, sulphur dioxide, smoke and toxic metals are released into the air to be distributed by the wind. The release of hot gases and hot liquids also produces another pollutant: heat. Some electricity generating stations use nuclear fuel which produces a highly radioactive solid waste rather than the above gases.

The generation of electricity is by no means the only source of toxic or biologically damaging pollutants. The exhaust gases from motor vehicles, oil refineries, chemical works and industrial furnaces are other problem areas. Also, not all pollutants are graded as *toxic*. For example, plastic and metal scrap dumped on waste tips, slag heaps around mining operations, old quarries, pits and derelict land are all *non-toxic*. Finally, pollutants can be further defined as *degradable* or *non-degradable*. These terms simply indicate whether the pollutant will decompose or disperse itself with time. For example, smoke is degradable but dumped plastic waste is not.

Carbon dioxide in the air absorbs some of the long-wave radiation emitted by the earth's surface and in so doing is heated. The more carbon dioxide there is in the air, the greater the heating or greenhouse effect. This is suspected as being a major cause of global warming, causing average seasonal temperatures to increase. In addition to causing undesirable heating effects, the increased quantity of carbon dioxide in the air, especially around large cities, may lead to people developing respiratory problems.

Oxides of nitrogen are produced in most exhaust gases and nitric oxide is prevalent near industrial furnaces. Fortunately, most oxides of nitrogen are soon washed out of the air by rain. But if there is no rain, the air becomes increasingly polluted and unpleasant.

Sulphur dioxide is produced by the burning of fuels that contain sulphur. Coal is perhaps the major culprit in this respect. High concentrations of this gas cause the air tubes in people's lungs to constrict and breathing becomes increasingly difficult. Sulphur dioxide also combines with rain droplets eventually to form sulphuric acid or *acid rain*. This is carried by the winds and can fall many hundreds of miles from the sulphur dioxide source. Acid rain deposits increase the normal weathering effect on buildings and soil, corrode metals and textiles and damage trees and other vegetation.

Smoke is caused by the incomplete burning of the fossil fuels. It is a health hazard on its own but even more dangerous if combined with fog. This poisonous combination, called *smog*, was prevalent in the early 1950s. It formed in its highest concentrations around the large cities where many domestic coal fires were then in use. Many deaths were recorded, especially among the elderly and those with respiratory diseases. This led to the first Clean Air Act which prohibited the burning of fuels that caused smoke in areas of high population. So-called smokeless zones were established.

Dust and grit (ash) are very fine particles of solid material that are formed by combustion and other industrial processes. These are released into the atmosphere where they are dispersed by the wind before falling to the ground. The lighter particles may be held in the air for many hours. They form a mist, which produces a weak, hazy sunshine and less light.

Toxic metals, such as lead and mercury are released into the air by some engineering processes and especially by motor vehicle exhaust gases. Once again, the lead and mercury can be carried over hundreds of miles before falling in rain water to contaminate the soil and the vegetation it grows. Motor vehicles are now encouraged to use lead-free petrol in an attempt to reduce the level of lead pollution.

Ozone is a gas that exists naturally in the upper layers of the earth's atmosphere. At that altitude it is one of the earth's great protectors but should it occur at ground level it is linked to pollution.

Stratospheric ozone shields us from some of the potentially harmful excessive ultraviolet radiation from the sun. In the 1980s it was discovered that emissions of gases from engineering activities were causing a 'hole' in the ozone layer. There is concern that this will increase the risk of skin cancer, eye cataracts and damage to crops and marine life.

At ground level, sunlight reacts with motor vehicle exhaust gases to produce ozone. Human lungs cannot easily extract oxygen (O_2) from ozone (O_3), so causing breathing difficulties and irritation to the respiratory channels. It can also damage plants.

This ground level or 'tropospheric' ozone is a key constituent of what is called photochemical smog or 'summer' smog. In the UK it has increased by about 60% in the last 40 years. There is a UK free-phone number, 0800 556677, that provides air quality information and gives details of the regional pollutant levels in the air.

Heat is a waste product of many engineering activities. A typical example being the dumping of hot coolant water from electricity generating stations into rivers or the sea. This is not so prevalent today as increasingly stringent energy saving measures are applied. However, where it does happen, river and sea temperatures can be raised sufficiently in the region of the heat outlet to destroy natural aquatic life.

Chemicals dumped directly into rivers and the sea, or on to land near water, can cause serious pollution which can wipe out aquatic life in affected areas. There is also the long term danger that chemicals dumped on soil will soak through the soil into the ground water which we use for drinking purposes and which will therefore require additional purification.

Radioactive waste from nuclear power stations or other engineering activities which use radioactive materials poses particular problems. Not only is it extremely dangerous to people – a powerful cause of cancer – its effects do not degrade rapidly with time and remain dangerous for scores of years. Present methods of disposing of radioactive waste – often very contentious however – include their encasement in lead and burial underground or at sea.

Test your knowledge 4.6

Name a pollutant that fits each of the following categories:

- toxic and degradable
- toxic and non-degradable
- non-toxic and degradable
- non-toxic and non-degradable.

Derelict land is an unfortunate effect of some engineering activities. The term derelict land may be taken to mean land so badly damaged that it cannot be used for other purposes without further treatment. This includes disused or abandoned land requiring restoration works to bring it into use or to improve its appearance. Land may be made derelict by mining and quarrying operations, the dumping of waste or by disused factories from by-gone engineering activities.

Pollution

Requirements for managing the environmental effects of engineering activities

Engineering activities can have harmful effects on the physical environment and therefore on people. In order to minimize these effects, there is a range of legislation – rules and regulations – which engineering companies must observe.

The appropriate United Kingdom Acts of Parliament include Deposit of Poisonous Wastes Act, Pollution of Rivers Act, Clean Air Act, Environmental Protection Act, Health and Safety at Work, etc., Act and the like. Additionally, not only are there local by-laws to be observed, there are also European Union (EU) directives that are activated and implemented either through existing UK legislation in the form of Acts of Parliament or mandatory instructions called Statutory Instruments (SIs).

New Acts and directives are introduced from time to time and industry needs to be alert to and keep abreast of these changes. Typical of these new initiatives is the European Electromagnetic Compatibility (EMC) legislation. This states that with effect from 1st January 1996 it is a requirement that all products marketed must conform with the new legislation. This new EMC legislation, at last, officially recognizes the well known problem of unwanted electromagnetic wave radiation that emanates from most pieces of electrical equipment. The unwanted radiation can interfere with adjacent electronic equipments causing them damage or to malfunction.

In the case of UK Acts of Parliament, the above legislation is implemented by judgement in UK Courts of Justice in the normal manner but based on EU legislation, if more appropriate, or by judgement of the European Court of Justice.

The purpose of this legislation is to provide the following functions:

- *prevent* the environment being damaged in the first place
- *regulate* the amount of damage by stating limits, for example, the maximum permitted amount of liquid pollutant that a factory may discharge into the sea
- *compensate* people for damage caused, for example, from a chemical store catching fire and spreading wind-borne poisonous fumes across the neighbourhood
- *impose sanctions* on those countries or other lesser parties that choose to ignore the legislation
- *define who is responsible* for compliance with legislation to persons who can be named and their precise area of responsibility documented.

For the purposes of showing competence in the above matters, you are not expected to have a detailed understanding of the various Acts but you should be aware of the general provisions of the legislation and what it is trying to achieve. Your school, college or local town library will most probably have details of the above Acts in the Reference Section.

The effects of the above legislation on engineering activities have, in general, made them more difficult and more expensive to implement. A few simple examples of this follow.

Chemical factories can no longer discharge their dangerous waste effluent straight into the river or sea without first passing it through some form of purification process.

Coal fuelled power stations must ensure that their chimney stacks do not pollute the neighbourhood with smoke containing illegal limits of grit, dust, toxic gases and other pollutants. A system of smoke filtration and purification must be, expensively, incorporated.

All *electrical equipment* including TVs, PCs, power hand tools, electromedical machines, lighting and the like, must be tested and certified that they comply with the relevant EMC (electromagntic compatibility) legislation. So, in addition to the cost of reducing any excessive radiation from the product itself, the purchase or hire of expensive EMC test equipment and the training of people to use it must also be taken into account. Further, because of delays in obtaining an official EMC examination and supporting certificate (EC), the introduction of new product designs can also be delayed and this may have adverse marketing effects.

Motor car exhaust gases must be sufficiently free of oxides of nitrogen, carbon monoxide and other toxic gases. This can only be achieved by, among other things, replacing the crude petrol carburettor with a more sophisticated petrol injection system and fitting a catalyser in the exhaust pipe. All this has added to the price of the motor car and has made it more difficult for the DIY motorist to service his vehicle.

Activity 4.6 

Two important ways of generating electricity involve the burning of fossil fuels or using nuclear fuel. You have been asked to take part in a discussion to be broadcast on local radio in which the case for and against each type of fuel will be aired. List the relative advantages and disadvantages of the two methods and decide which viewpoint you would take in the discussion and which points you would wish to make in support of your views. Prepare the notes that you will use during the broadcast and write them down on a single sheet of A4 paper.

Career options and pathways in engineering

This section starts by identifying the main sectors of the engineering industry. The type of companies involved, their products and the range of jobs they offer are then discussed. This is followed by a summary of the engineering qualifications and experience required by these jobs and the methods by which they are obtained. Sources of information on jobs available are listed and guidance, mainly for the young school or college leaver, as to how to apply for a first time job is given. This includes interpreting job adverts and responding to them in the way most likely to result in the applicant being called for interview. Letters applying for jobs and CVs are treated in depth, as are interview techniques.

The main sectors of the engineering industry

Mechanical

Of all the engineering disciplines, the one that probably covers the widest range of activities is that of mechanical engineering. It applies the theory and practice not only of mechanics, but additionally, pneumatics, hydraulics, thermodynamics and the properties of materials, to the design and manufacture of all types of transport vehicles, engines, power generation plant and all types of machines and their components.

The sectors of the engineering industry which employ specialist mechanical engineers include manufacturing, food and drink, petrochemicals, construction, aerospace, railway, motor, marine, local government, the Civil Service and the Armed Forces.

Electrical and electronic

The number of industrial, commercial and domestic equipments which depend upon electrical power, or electronic control, increases almost daily. The bulk of our domestic entertainment is provided for us through the radio or television and we communicate using telephones, modems, facsimile machines and computers as an everyday event. We travel in relative safety by land, sea and air because of the movement of transport vehicles being controlled by signalling equipment, radar and electronic visual aids. The operation of our manufacturing industry is similarly dependent upon electronic controls, computer assistance for design and manufacturing, machine processes and the like.

The people who ensure that all this electrical equipment is available, that it works properly and is regularly serviced and maintained, are our electrical engineers. Clearly, like their mechanical engineering counterparts, they are employed in virtually all sectors of industry and commerce.

Manufacturing

This is concerned with the production of the goods which we consume and export. Manufacturing or production engineering includes activities such as the selection and layout of plant and machinery, designing safe production systems to facilitate the efficient flow of work through the factory, monitoring and maintaining product quality and cost and, finally, controlling and encouraging the production staff, the people who do the manufacturing.

Because production engineers are involved with the manufacture of virtually all types of goods, they are to be found in all sectors of engineering from aircraft manufacture to the zinc plating of metal fabrications.

Chemical

Chemical engineers are involved with simple experimental laboratory preparations, using glass retorts and bottles, for the translation of these chemical processes for use with commercial manufacture units. These could be large process plants entailing the use of high-pressure steel containers, special pumps and sophisticated process control gear for quality and safety control.

Chemical engineers are involved with the manufacture of chemicals and pharmaceuticals; petroleum, petrochemicals and plastics; food and drink; paper and special metals; soaps, detergents and perfumes and the like.

Civil

This sector of engineering is concerned with the design, building and maintenance of roads, railways, airports, canals, harbours, jetties, bridges, tunnels, water supplies, dams, reservoirs, pipelines, hydroelectric power, sewers, sewage works, irrigation schemes and so on.

Choice of career

Broadly speaking, whichever of the above engineering disciplines you eventually choose for a career, they could all offer similar specialist functions such as research and development (R&D), production, marketing, sales, service, finance, personnel and general administration. For example, in a business producing its own range of motor cars, there would be a definite requirement for mechanical engineering expertise in all of the first five departments and possibly in the other three. Let's look at these in turn to see what is on offer.

Research and development

It is in this department of a motor business that new shapes for body shells, new designs for engines and transmissions, and so on, are sketched and, in conjunction with other departments, prototypes manufactured for testing and evaluation. It is difficult to imagine this department operating without several very experienced mechanical engineers.

It must be said at this point that this particular R&D department would also need the services of other than mechanical engineers. Electrical and production engineers would also be needed to provide their specialisms to the various motor car projects.

Production

This department decides how the motor car is to be made in the correct quantity, at the correct quality and at the correct price. Once again, engineers spanning several engineering disciplines will be required to work in the production planning team.

Marketing

This term involves several functions such as advertising, exhibitions, ascertaining product demand, etc. Few of these activities involve a permanent specialist engineering input and in most cases, if required, this would be provided by an engineer on loan from, say, the sales department.

Sales

This department, together with the service department, provides the motor company's chief contact with the public. The sales staff operate from well appointed sales rooms, demonstrating and discussing the different vehicles to potential buyers. The sales people need to be well turned out, patient and well mannered. Some of them, if not all, need to have a good technical knowledge about the cars they are selling. Nothing looks worse than a technically inclined customer asking simple technical questions about one of the company's products and there being nobody around who can give a sensible answer.

Service

In some companies, this forms part of the sales department because both functions have much contact with the public. In the case of the service engineer, it is with people who have previously bought a car and subsequently require it to be serviced or repaired. It is also common to offer a spare parts service for the DIY enthusiast who wishes to do his own repairs. Clearly, there is a definite need to employ trained service engineers for these activities.

Personnel

The personnel department may have engineers on its permanent staff if there is a lot of work involved with the selection and subsequent training of engineering staff belonging to the other departments within the company.

Finance

It would not be usual to have a trained engineer working in this department, which is principally concerned with the collection of money from the company's debtors and payment of money to its creditors and dealing with banks, shareholders, etc.

Administration

Any engineering staff with this department would not be directly concerned with the company's main product: motor cars. Rather, they would be the maintenance engineering staff responsible for the servicing and repair of the company's production machinery. This could involve many qualified engineers from several disciplines.

Qualifications and skills required

Each of the major engineering disciplines has its own institution. Most practising engineers join one or more of these institutions, as appropriate to their work, to help them in further developing their careers. This, in part, is achieved through monthly newsletters and magazines which are sent to all members. In addition, regular meetings and lectures are arranged, often on a regional basis, and these serve partly as a means of keeping up to date with engineering developments and partly as social occasions. One of the larger engineering institutions is the Institution of Electrical Engineers, known by the letters, IEE. The mechanical engineers have their IMechE and the civil engineers their ICE. There are more than ten different major engineering institutions that have formed a national organization in the UK called the Engineering Council. One of the purposes of this council is to achieve common high standards across a range of criteria throughout the engineering profession.

Perhaps the single main reason for an engineer to join an engineering institution is the recognition received from within the engineering profession, according to the grade of membership. This is because, for each grade of membership, the institutions demand certain levels of academic achievement, practical training and job responsibility. The grades of institution membership tend to be those of *student*, *associate member*, *member* and finally *fellow*.

A *Chartered Engineer (CEng)* will be a Corporate Member of one of the engineering institutions of the Engineering Council, e.g. MIMechE, MIEE, MICE and therefore typically will:

- have an approved honours degree in engineering or have passed a professional examination of similar standard
- have had two years post-graduate training
- be over 25 years of age and have successfully served in a responsible engineering appointment for at least two years
- have been recommended by his or her professional institution as being worthy of the title, CEng, which is an Engineering Council award.

Chartered Engineers are the leaders in their profession. They are employed in a rapidly changing world which requires them constantly to be updating and developing their engineering knowledge and skills. They are the managers of high-risk and capital-intensive engineering projects.

A *Corporate Engineer* will be a Fellow or Member of an engineering institution. He or she will have the same qualifications, training and job experience as the Chartered Engineer but has not yet applied or been recommended for a CEng award. Corporate Engineers are employed in much the same role as are Chartered Engineers.

An *Associate Member* (of some engineering institutions) is an engineer who is at least 23 years of age, is academically or professionally qualified, has completed his or her post-graduate training but cannot yet meet the institution's criteria for job responsibility as required for full corporate membership. An Associate Member will usually be a member of a team headed by a Corporate or Chartered Engineer.

An *Incorporated Engineer (IEng)* is a member of a recognized engineering institution and will:

- be qualified to BTEC Higher National Certificate or Diploma standard
- have had workplace training
- have had a professional review by his or her institution.

Incorporated Engineers can be very experienced and are often project leaders responsible to Chartered Engineers.

Test your knowledge 4.7

What do each of the following abbreviations mean?

1 CEng
2 IEng
3 Eng.Tech
4 IEE
5 IMechE

An *Engineering Technician (Eng.Tech.)* is a member of a recognized engineering institution and will:

- be qualified to BTEC National Certificate or Diploma standard (broadly equivalent to Advanced GNVQ)
- have had workplace training
- have had a professional review by his or her institution.

Technician Engineers will typically have been extensively trained by their employers and are expected to put their knowledge and practical skills to good use in one or other of the production, drawing office, out-inspection or service departments.

Acquisition of qualifications and skills

In the previous section we have seen brief details of the qualifications which engineers are expected to achieve for the different membership grades of their professional institutions. This section gives a brief insight as to the ways of obtaining these qualifications.

Academic

An engineering qualification begins in the secondary school. In an ever-changing educational world, the General Certificate of Secondary Education (GCSE) is, at the moment, the first meaningful achievement. From the wide choice of subjects available, the aspiring engineer should aim for six to eight GCSE passes and these should include mathematics, physics and English language; at Grade A to C minimum.

The more academic student may wish to stay at school to obtain A-levels in maths and physics plus one other before going to university to obtain a degree in engineering. It should be noted here, that some engineering institutions do not recognize all university degrees; only those on their approval list. Similarly, mathematics and physics degrees, by themselves, are unlikely to be given approval.

Vocational

The vocational education system in UK is undergoing traumatic changes and is unlikely to settle in its final format until about the year 2000.

For some years, the leading awarding body in vocational education has been the Business and Technology Education Council (BTEC). The awards are either a First, National or Higher National Certificate or Diploma. The certificates are intended for employed technicians attending a college on one day a week over two years (only one year for the First awards). The diplomas are intended for full-time college students, again taking one or two years. Both the National and Higher National Certificates and the Diplomas are accepted by most universities for entry to their engineering degree courses. The higher awards sometimes allow the student to join a university course in its second year.

The intention of the present government is to introduce an additional engineering vocational qualification, called the General National Vocational Qualification (GNVQ) in Engineering. It will be awarded, by BTEC, City and Guilds, or RSA, at three levels: Foundation, Intermediate and Advanced. The Foundation level is comparable to GCSE Grades E, F, G; the Intermediate to GCSE Grades A, B, C or a BTEC First award and the Advanced, to A-levels or a BTEC National award.

The traditional route for the achievement of these BTEC certificates, diplomas, and now GNVQs, has been through the local further education college. It may remain so, but the government is trying to open additional pathways through existing secondary schools and sixth form colleges. In general, the colleges of further education are already equipped to deliver vocational education and training. They have been doing it for years. They have the vocationally experienced lecturers to teach the subjects and expensively equipped workshops and laboratories. Without spending a great amount of time and money, most secondary schools and sixth forms will have neither of these. Figure 4.4 shows a chart of the different pathways currently available to students of engineering.

Main sources for careers information

Having decided the type of job that you would like and having listed what you have to offer an employer, the next step is to start your search for that job. There are a number of sources of information about jobs, the main ones are listed below.

Careers advisers

Careers advisers are usually based in a careers office, the location of which can be obtained from the local telephone directory – possibly under *Careers Services* or the like. These services were originally established to help school and college leavers get their first full-time job. However, careers advisers now also help with the task of finding jobs for some of the large number of unemployed adult people that exists today. It is also common practice in some schools and colleges to have a specialist school teacher or college lecturer trained to advise students about career prospects.

Employment agencies

These operate by having a pool of unemployed people on their books which they try to match against employers' requests for new staff. Invariably this operation needs to be supported by the employment agency advertising unfilled jobs in newspapers and on cards which they have in their own office windows.

The jobs offered by an employment agency can be low paid but quite suitable for a first time job, especially if the job is with a

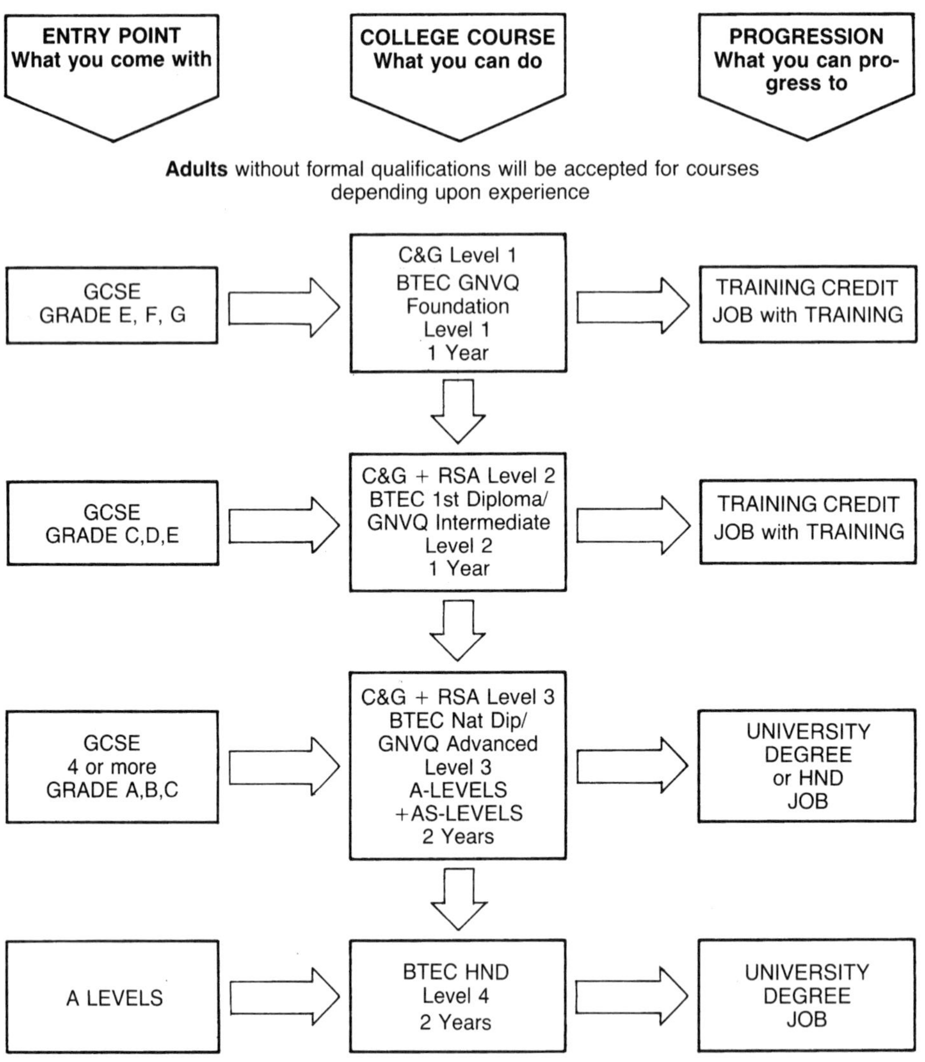

Figure 4.4 *Engineering courses and pathways available to full-time students*

well established firm. Remember, if you start with a job somewhat lower in status or pay than you originally planned, if there are promotion prospects – and there must be with a large firm – then consider taking that job and working your way up. How many times have you heard the story about the managing director who started his time at the firm as the office boy?

Employment agencies obtain their living by charging employers for their services. It saves the employer a lot of hard work and time in composing meaningful newspaper advertisements, reading lots of job application letters and then selecting the candidates with whom they wish to have a first interview. After the first or even second set of interviews are completed, the employment agency will usually offer the employer only three or four candidates from which a final choice can be made.

A word of warning about employment agencies. It is not legal for an employment agency to charge a person seeking a job an agency registration fee. The *Employment Agencies Act 1973* makes this illegal, unless the agency is acting for the entertainment business or the like.

Job centres

These are good sources of information about low paid jobs in the locality. The employers using this service often expect their job vacancies to be filled straight away and are often looking only for temporary or part-time workers. Of course, full-time students can use these temporary vacancies for holiday jobs or possibly for work experience. Nevertheless, job centres do have some full-time job opportunities for school and college leavers; usually in the form of advertisements from local large companies requiring apprentices or trainee professionals.

Shop windows and notice boards

While many of the advertisements on display are for used articles for sale, shop windows and notice boards can be useful sources of information for part-time jobs and possibly work experience. It is unlikely that a suitable full-time job will be advertised but, if you are in a supermarket or outside a newsagents shop, you have nothing to lose by having a quick look at the cards on display. You never know your luck; even if you do not see a job going there may be a bargain of second-hand computer or motor bike that you have been seeking!

Media advertisements

Radio

Although some large companies do use local radio stations to broadcast their staffing requirements, unless you or an interested acquaintance happens to be listening at the time, it is unlikely that you will hear them. If you think that you have missed something, you could always contact the offices of the local radio station to make enquiries.

National newspapers

These are another unlikely source of job information for young people because they tend to specialize in the more highly paid professional jobs for people in the age range 25 to 55.

Local newspapers

Perhaps one of the best job information sources for a young person is the local newspaper. The ones which are particularly useful in this respect are those which are delivered to your house free of charge. They cover their costs from the printed advertisements they run together with the inserted fly-posters.

Trade publications

These are directed at people already in the trade – that is why they buy the trade publication in the first place – and do not normally carry details of first jobs. But, if a copy is available, have a look in it anyway, you never know!

Applying for a job

The first true step in applying for a job is to decide just what sort of job is the one for you. Do you want to work as part of a team, say in a production department, or more on your own as an outside service or sales engineer? Do you want a job which involves much travel? Do you want to work in a large or small firm? Do you want to continue to live with your parents when you start your new job or would you welcome the independence of moving to a new part of the country? Which engineering discipline is of the most interest to you: electrical, electronic, aerospace, automobile, marine, mechanical or other?

Assuming that you have decided upon the type of job that you feel would best suit you, there are usually three hurdles that you must jump following your seeing an advertisement and the landing of the job itself. These are, the initial letter of application, the compilation of a CV – or filling in a Job Application Form – and finally the interview, that is, assuming that the other two stages have been successful.

Job advertisements

Suppose the following advertisement appeared in your local newspaper:

NEWFIELD GARAGES LTD
Head Office
16 Castle Street
Newfield
Loamshire

TRAINEE MOTOR TECHNICIAN WANTED

We are a private company, with three garages in the county specializing in the sales and service of MBW motor cars, for which we are a Main Dealership. Because we are expanding our operation at our Windlebury garage, we are planning to appoint there a trainee technician who will initially work on the repair and servicing of cars but who should have the will and ability to progress to become a Service Manager in due course.

The first year of the appointment will be largely involved with periods of in-house training together with four one-week attachments to the MBW training school in Hamburg.

We are inviting applications for this post from young people, with or without motor trade experience. However, we anticipate appointing a person who has obtained an engineering education/training qualification at school or college.

If you are interested please write a brief letter – not a CV – to Mr J.J.White, Service Director, at our Head Office stating why you consider yourself suitable for the post.

Understanding the advertisement

From the job advert you can 'read between the lines' to get a better idea of the nature of the job and whether you are really qualified to apply and whether it is your 'cup of tea'. Different people may read different things from the advert but some of the more obvious conclusions are as follows.

From the first paragraph

Because the company is described as 'private' it will most likely be a family concern with members of the family occupying most of the senior management positions.

The desirability of the successful candidate progressing through promotion is emphasized. This could mean an employer who likes his managers to have worked their way up from the shop floor and therefore well versed in the company procedures. If nothing else, it does indicate that the initial job is not a 'dead end' and that there are two other garages to give the necessary scope for promotion.

From the second paragraph

The company is prepared to 'invest in its people' by giving them specialist training in the repair of MBW cars. Since the MBW is a European car of some stature, the training will open other employment opportunities with other companies.

The person getting the job must be prepared to travel to and train in Germany. This pattern of training and familiarization with new MBW products could apply to all levels of employees. The question is whether you will like this overseas commitment.

If you are a school or college leaver your practical knowledge is bound to be limited. This offer of product training is almost an invitation for you to apply.

From the third paragraph

Another strong hint that ex-students – young people – may apply; the proviso being however, that they should have proven engineering qualifications. Perhaps this could apply to school leavers with GCSE or A Levels in maths, science or technology subjects. It would be 'right up the street' for a college leaver with a GNVQ in Engineering, especially if it included some motor vehicle optional subjects.

From the fourth paragraph

The Service Director himself is handling this appointment so it must be regarded as an important venture for Newfield Garages.

Since a CV is specifically ruled out, the brief letter requested assumes added importance. The letter will have say sufficient to get you through to the next stage of the selection procedure. This presumably will be an interview or perhaps the filling in of an Application Form sent to you – that is, if your first letter is held in sufficient regard by Mr White.

The applicant

John Short is 18 years old, lives at home with his parents and is seeking his first full-time job. He has had an interest in all things mechanical since he was 12 years old. His father is a self-employed driving instructor. He taught John to drive a motor bike and more recently a car. John started to tinker with the engine of his second-hand motor bike from the day he was given it as a 17th birthday present by his parents. He rapidly picked up information on how it worked by reading motor magazines and other technical literature.

John used his motor bike to travel to and from Axmead College of Technology where he studied for, and obtained, an Advanced GNVQ in Engineering with automotive subjects as his optional units.

John's work experience from college and holiday work for extra pocket money, amounts to a total of seven weeks. This time was roughly shared between three garages. His duties ranged from serving petrol from the manually operated pumps, in an older establishment, to helping with repair work in the vehicle service bays.

For his 18th birthday he bought, again with his parents' help, a 15 year old MBW316 which he has serviced and repaired himself. Any spare parts that he requires he buys from Newfield Garages. He is fortunate enough to have bought a MBW316 Workshop Maintenance Instruction Manual from a local charity shop for £1.

When he was not working on his motors or studying for his GNVQ assessments, John followed up his GCSE German Language B Grade – he obtained passes in seven further subjects including Maths and English Language – by following the BBC German language programmes 'Deutsch Direct' on the radio and TV.

Figure 4.5 shows an example of a very poor letter which John may have sent to Mr White. In addition to it having been written on lined paper, a cardinal sin against good taste, we can comment on this letter as follows:

1 The address block is incomplete. It has no 'county', 'postcode' nor 'telephone number'.

2 Mr White's name and full address should appear here.

3 It is usual to print the subject of the letter just after and below the salutation 'Dear . . . '.

4 The first paragraph has the word 'that' missing before the word 'you'. Also, it does not say for which job nor for which garage the letter of application applies. Mr White may have several vacant posts in his three garages advertised.

5 It is too vague of John to simply write that he has an 'engineering qualification'. He should say which one: 'an Advanced GNVQ in Engineering'.

6 Comments of this nature should never be aired!

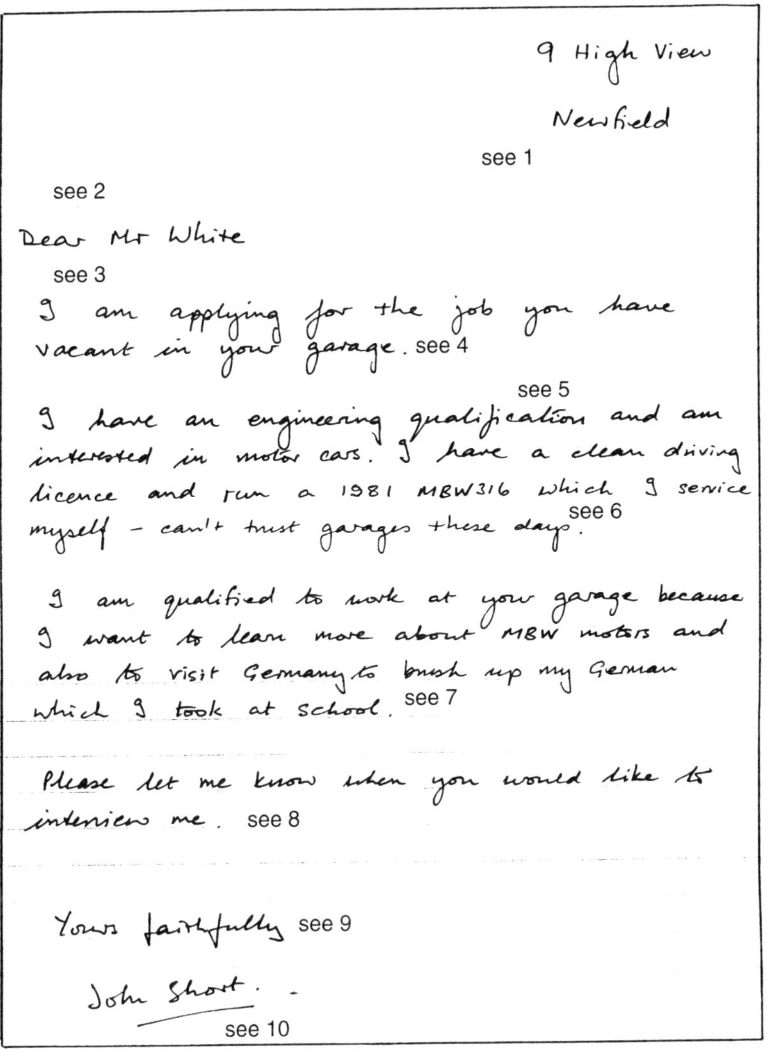
9 High View

Newfield

see 1

see 2

Dear Mr White

see 3

I am applying for the job you have vacant in your garage. see 4

see 5

I have an engineering qualification and am interested in motor cars. I have a clean driving licence and run a 1981 MBW316 which I service myself – can't trust garages these days. see 6

I am qualified to work at your garage because I want to learn more about MBW motors and also to visit Germany to brush up my German which I took at school. see 7

Please let me know when you would like to interview me. see 8

Yours faithfully see 9

John Short.

see 10

Figure 4.5 *Example of poor job application letter*

7 Mr White will not be interested, and could be offended, by being told what a potential employee intends to take out of his company. He will be more impressed with a list of proposed benefits to be given to the company.

8 This is a very arrogant way to end the letter. It implies that there is no doubt about John being offered an interview.

9 Letters which begin with 'Dear Sir' or 'Dear Madam' are closed with 'Yours faithfully'. Letters starting with 'Dear Mary' or 'Dear Mr Smith' require a 'Yours sincerely' ending.

10 You must always print or type your name after your closing signature. You may be able to read it, but others may not.

Overall, this first letter of John's is a shocker. Some employers are so busy that they would not give the time to reply to it. In any event, it would be most unlikely that John would be considered further for the job. John's much improved letter of application is shown in Figure 4.6.

Because it has been typed, John's second letter is much easier to read. However, if your handwriting is easy to read, there is no reason why you should not use it. In fact, some employers specifically ask that letters of application should be in the applicants' own handwriting.

Note how John has done a little research on Newfield Garages before replying to the advertisement; he has found the correct postcode. Mr White will like the interest John has shown by doing this; that is, if Mr White notices it at all.

In his second paragraph, John really tries to sell himself to Mr White and in his third, he says how his skills at German could be an asset to the company. Nowhere does he mention what he wants to take from the company.

The ending to John's letter is correct. He does not indicate that he is expecting an automatic success from his application.

Test your knowledge 4.8

If you were Mr White, based on this second letter, would you ask John to an interview? Give reasons.

9 High View
Newfield
Loamshire
NW10 5RX

Tel: 01444 869431

10th August 1996

Mr JJ White
Service Director
Head Office
Newfield Garages Ltd
16 Castle Street
Newfield
Loamshire
NW1 4TD

Dear Mr White

TRAINEE MOTOR TECHNICIAN - WINDLEBURY

I am responding to your advertisement in last Friday's local paper for the above appointment.

I consider myself suitable to apply for the job because I am a young person (18 years old), already hold an Advanced GNVQ in Automobile Engineering, have had several weeks relevant work experience in three local garages and am an enthusiastic hard worker of smart appearance.

In addition, I have a clean full driving licence and own a 1981 MBW316 which I service and repair myself. I also have a GCSE, Grade B, in German and am continuing with my studies by following the *Deutsch Direct* German language series on BBC radio and TV. I can 'get by' in German and feel that this could be an asset to your company should I need to travel for you in Germany.

Should you so wish, I would be pleased to meet you to discuss my application in more detail.

I look forward to hearing from you in due course.

Yours sincerely

John Short

John Short

Figure 4.6 *A much improved letter of application*

Curriculum vitae (CV)

In many respects a CV is similar to an employer's Job Application Form which is sent to job applicants. They both record personal data and history about the applicant. The main difference is that the applicant can choose the layout of his or her CV and what is written in it. It is the employer who decides the layout of the Job Application Form and, if the questions are answered correctly, what is written in it. We shall have a closer look at the CV a little later.

Interviews

Let's presuppose that you have applied for a job and to your delight receive a letter asking you to attend for an interview at a certain place and time. This could be your last hurdle before being offered the job and so it is important that you are prepared for it. Whatever the interview, the factors that need to be considered are similar. The following paragraphs are intended to highlight either what pre-interview work you can do or what you must look out for.

You have been invited to the interview

1 Note the name and job of the person who is to interview you.

2 Work out the time it will take you to travel to the place of interview. Arrange to arrive, calm and collected, five minutes before the interview is due to start. If you are too early you may be a nuisance; if you are too late you may have lost your interview.

3 Write a reply to the invitation saying that you will be there. Even if you should decide that job is not for you after all, you should still write a letter saying simply that 'I much regret that because of unforeseen circumstances I will not be able to attend your kind invitation for an interview'.

Preparation before you go to the interview

You will impress the interviewer if you know something about the firm's business. It shows that you have taken an interest in the firm and really do want the job:

1 Do a little research about the company to find out what it sells and to whom it sells.

2 Find out about the organizational structure of the company. If it is part of a group of companies, your promotion prospects could be better than if it is only a small family concern. Your local Job or Career Centre will be able to tell you something about local firms. For national companies, your school or college library will carry a reference copy of The Kompass Register. This lists thousands of companies and for each one states the

number of employees, its products, its sales offices, and the like.

3 Think how you would answer the following awkward questions:

'Why do you want to work for this company?'

Have some positive reason worked out otherwise the interviewer will think that you are only interested in the job and not the company itself. 'I have heard that it is a good employer'; 'It specializes in the manufacture of quality products with which I would be pleased to be associated'; 'It is part of a major group and therefore the chances of promotion must be good'; 'I have heard that your company policy is to promote from within', are the sort of responses the interviewer wants to hear.

'Why do you want this job?'

Typical answers might be: 'I like working with motor cars', 'The work interests me', 'I like to travel', 'I enjoy meeting new people' and 'Design work fascinates me'.

Each answer does not mean a lot on its own but if backed up with an example it can show what attracted you to the job. For example, you may have a weekend job helping behind the counter in a shop. You may find that you enjoy meeting the continually changing customers you have to serve. If you are also an enthusiastic car owner it may explain why you are interested in applying for a job as a trainee motor car salesman.

Your appearance at the interview

1 The first impression that you make on the interviewer(s) when you enter the room is very important. An interviewee is expected to arrive with neatly cut and groomed hair, a man to be wearing a collar and tie and, for best impressions, dressed in a suit and wearing polished shoes. Men should be properly shaved or be wearing a neatly trimmed beard. Ladies are advised not to wear trousers, if this is at all possible. Clean fingernails are a must; as is a clean handkerchief.

2 When you enter the interview do not rush. Walk calmly up to the interview desk with a disarming smile; if you can muster one. Be prepared to shake hands but do not make the first offer to do so.

3 Any handshaking should be brief and with a firm grip of the hands – not just the finger ends. Do not try to crush the offered hand – your physical strength is not under test – nor offer a limp hand which may give the interviewer the impression of shaking a dead chicken's foot.

4 Sit down only when invited to and not before. Once sitting, do not fidget; simply sit still, perhaps holding any papers you may have brought with you on your knees.

Above all, do not lean forward and rest on the interviewer's desk.

5 It will help you to speak more clearly if you talk more slowly than you would normally. Also, do not rush to answer questions – think before replying.

6 Never swear; it is neither clever nor helpful to your chances of success. Try not to use meaningless phrases such as 'you know' and 'didn't I'.

Curriculum vitae (CV)

What it is

'Curriculum vitae' is Latin and it means 'the course of life'. Advertisements for job vacancies frequently call for the submission of a 'CV', 'Biodata' or 'Career history' and in practice they all mean the same thing – a curriculum vitae.

A curriculum vitae is a form of biographical summary of your life to date as it applies to a particular situation; usually that of applying for a new job. It takes into account not only the history of jobs that you have had earlier but also your education, training, outside interests and transferable skills.

You must not confuse the CV with an application form for a job. The two may contain similar information. The major difference between them is that the application form is designed by the prospective employer whereas the CV is of your own design. The layout of your CV can be altered to give a different impact on different prospective employers to your advantage.

Basic criteria for a CV

Whatever the format of your CV, in order that it is most effective it must:

- be brief yet clear
- tell the employer why you are the best person for the job
- say what you can offer the employer.

There are many publications all of which give advice as to how to produce the 'ideal CV'. Much of this advice is contradictory and it may be that you should choose a style of CV which you consider right for you, that is, with which you feel comfortable. However, despite the varied advice available, there is general agreement on three main points regarding a completed CV. It must:

- be typed or printed and definitely not handwritten
- be as informative as possible, ideally limited to a single side of A4 paper, and definitely not more than two
- have a first-class appearance.

CVs can be written for you

Because of the time and effort required to produce an effective CV, some people have theirs written for them by specialist businesses. Whether you resort to this method depends upon whether you wish to incur the cost involved and your own confidence in producing your own CV. The commercially produced CV is sometimes criticized as being impersonal and therefore lacking that 'personal touch' that your own 'home-grown' CV can have.

General purpose CVs

If you are busy applying for lots of jobs it is sometimes worthwhile writing speculative letters to the firms that you know are likely to have suitable jobs. The purpose of a speculative letter is to find out if there are any suitable vacancies now or in the near future and it would be in order to attach to your letter a general all-purpose CV to tell the personnel officer – or human resources manager – something about your experience and capabilities. Hopefully, one or more of the firms will respond positively to your enquiry by sending you an application form for a particular job or by calling you for an initial interview.

Individual, tailor-made CVs for specific jobs

The all-purpose CV should never be used in connection with an application for a particular job. For this you should already know from the newspaper advertisement the general job specification and the qualities the prospective employer is expecting of the successful applicant. In this situation it is important that you produce a tailor-made CV. You can organize its layout and wording to highlight precisely those qualities that you have and for which the employer is looking.

The information to put in a CV

Below is a suggested list of headings under which your personal details and experience can be documented:

NAME
: You should always write your full name and, should there be any doubt, indicate your family name by underlining it.

DATE OF BIRTH
: In order to prevent any confusion, especially if applying to an overseas firm, write all dates in the form day, month, year, e.g. 21st October 1976.

ADDRESS
: Always use your permanent address. If you want a message sent to a temporary holiday address this, and any dates, should be included in the letter you send with the CV.

TELEPHONE NUMBER
: Include both day and evening numbers if there is a difference.

BRIEF CHARACTER PROFILE

Under this heading you have the opportunity to write a very brief pen-picture of yourself and how you match the job. You can show that you have the same skills as those called for in the advertisement. For example:

Hard working college leaver with an Advanced GNVQ in Engineering and work experience in the motor trade, seeks a first full-time post.

Enthusiastic, reliable college student with Intermediate GNVQ in Engineering looking for an engineering apprenticeship or other trainee post in an electronics engineering environment.

MARITAL STATUS; CHILDREN

These two headings need a little thought before being included. If there is no apparent need to include details of whether you are married or single or have children, then it may be sensible to leave these out.

INCAPACITY

If you are writing to an 'equal opportunities employer' any incapacity you may have could be of no consequence. On the other hand, if you suspect that the job for which you are applying could be difficult for you with your particular disability, then it would be correct to include details. Much depends upon the nature and severity of your disability and the type of the job for which you are applying. Perhaps the best advice is to do what you think is best for you.

EDUCATION AND QUALIFICATIONS

Under this heading you should start with your last school or college and state the highest qualification you achieved there. For example:

1990–95 Weyford Secondary School
GCSE: 5 subjects including Maths (D), English Language (C), Physics (D).

Make sure that you do not list any subjects which you have not passed. Also, only places of secondary education and above are of interest to an employer.

PROFESSIONAL QUALIFICATIONS

These are vocational qualifications and as such should be kept separate from your educational qualifications. For example:

1995–96 Weyford College GNVQ in Engineering (Intermediate). Optional units taken were Electrical Principles and Materials and Electrical Circuits and Instrumentation.

Once again, it it important to make no mention of any failed examinations.

TRAINING AND SHORT COURSES

Include only relevant information. For example, if you are applying for a job as a trainee electrical technician, there is nothing to be gained by giving details of a short course on landscape painting that you attended last year. Had the course been an introductory course to technical German and the electricians job entailed some work in Germany then it would be well worth a mention.

PREVIOUS JOBS

If you have had any previous jobs then these should be listed, starting with the last job first and giving dates and a brief note on what your responsibilities were.

PERSONAL SKILLS

If you are a school or college leaver trying to obtain your first full-time job you should find this heading particularly useful. You can list here your hobbies, sports, pastimes and holiday or weekend work that show you to have skills which can be transferred to some aspect of the advertised job.

For example, if your father is a qualified electrician and in your school holidays you frequently assisted him in his work, then mentioning this could help in getting you an interview for the trainee electrical technician job.

Similarly, had you been the captain of your school cricket team, it would indicate that you have leadership qualities and are prepared to accept responsibility. These are much sought after qualities for some jobs and should always be mentioned.

Why you want the job

This need not be given unless asked for but if it is, be very careful how you answer. Do not let the prospective employer think that you are planning to 'use' him or her. This will be the case if you use expressions such as 'I wish to gain further experience in . . . ' or 'I have not been able to obtain my first choice job in motor car sales . . . ' or 'I wish to work nearer my home town of . . . '. Even though all these may be perfectly true you must not use them.

It is better to write, 'I feel that my qualifications and recent work experience in a garage will make me an asset to you as a trainee motor technician' or 'My Advanced GNVQ in Engineering together with my enthusiasm and reliable time keeping can be put to good use in your company'.

In short, let the employer feel that you are primarily wanting to put something into the business rather than take something out.

Writing the CV

The way to start this is to read the specification of the job for which you have decided to apply and compare the qualities called for with a list of your own qualities – qualifications, experience, skills and the like. Then decide which of the headings shown above you wish to use in producing your CV and draft your first attempt. Make sure that you show that you have as many as possible of the qualities called for in the job specification. Try to imagine what the employer would like to see on your CV and try to match that as closely as possible. Also remember that if you have one piece of information which you consider to be by far most important then arrange for this to appear one third of the way down on the first page of your CV. This is known as the focal point of a page and is where the reader's eyes naturally focus first.

Another job advertisement

Now let's take a look at another job advertisement:

> JOB VACANCY
>
> LIGHTWATER ELECTRONICS – A leading manufacturer of electronic products offers a unique opportunity to train for a career in sales, ultimately culminating in management or marketing.
>
> We are looking to recruit a trainee sales engineer to embark on a thorough and comprehensive training programme.
>
> Applicants should possess a knowledge of electrical/electronic application and above all a determination to succeed in a competitive business.
>
> Those interested should apply by letter to Mr John Grass enclosing a full CV and should state why they consider themselves suitable for the post.
>
> Lightwater Electronics
> High End
> Surrey
> GU22 61ZX

The applicant

Let's suppose that an 18 year old called Brian Oldnall sees the advertisement and decides to apply for the post. Brian lives with his parents about five miles from Lightwater Electronics which he knows to be a reputable and expanding business. Brian attended Weyford Secondary School and obtained a reasonable set of GCSE pass grades in six subjects but was unplaced in history and art. Rather than stay on in the school's sixth form, Brian decided to follow a vocational rather than academic course. He applied to Weyford College to join their Advanced GNVQ in Engineering Course. He was accepted on the course and two years later obtained the GNVQ he wanted. His results on this course were very good; much better than those he obtained in his GCSE at school.

While he was at school Brian played centre forward for the first soccer team and in the summer months played cricket for the second eleven, which he captained. He also represented the school at swimming. The amount of time he spent on sporting activities may have accounted for his not having passed history and art at GCSE level.

Brian's work experience from school and college has all been with electrical or electronic businesses. He is also an electronics hobbyist. He has built his own portable radio from a kit he purchased and is 'into' radio controlled model cars and aeroplanes. If anything electrical around his parents' house becomes faulty, he likes to have a go at repairing it.

Brian's CV could look something like the following:

CURRICULUM VITAE

Brian Oldnall
24 Windy Lane
Stopford
Surrey
GU23 2XX
Tel: 01176 149245
Born: 10th May 1978

PERSONAL PROFILE

I am a college leaver qualified in engineering and have had work experience with electrical and electronic firms. I am also an enthusiastic electronics hobbyist and am now seeking full-time employment in the electronics industry.

QUALIFICATIONS

1994–96 Weyford College – Advanced GNVQ in Engineering
Optional subjects were:
Electrical Principles
Electrical Technology
Electronics
Engineering Instrumentation.

WORK EXPERIENCE

From Weyford Secondary School:

June 1993 for two weeks. Smiths Electrical of Stopford, a small private business (10 people) selling and repairing domestic electrical goods. I assisted in the repairs of electric irons, kettles, toasters, etc. I was shown how to diagnose faults and fit replacement parts.

June 1994 for two weeks. Johnson TV Hire Co, Stopford (15 people). I assisted in the TV repair workshop and also in the sales room by demonstrating the TVs to the public.

From Weyford College:

March 1995 for two weeks. Newton Electrical Panels Ltd, Weyford (100 people, part of national group of companies) making special to order control panels for process plants. I assisted in the test and out-inspection of completed panels.

August 1995 for three weeks. Paid holiday work again with Smiths Electrical of Stopford selling and repairing domestic electrical goods, which now includes radio and audio equipments.

March 1996 for two weeks. Weyford Electrical Ltd (12 people) specializing in domestic and industrial electrical installation work to 16th Edition IEE Regs. I assisted in the installation of an inductive internal communications loop in Weyford Hospital.

SKILLS AND INTERESTS

Since 1993 I have been a member of Weyford Model Aircraft Club where I fly my home-built radio controlled aeroplanes.

In 1994 I successfully built up a Disco console for Weyford Youth Fellowship. I designed the enclosure for the console and did all the wiring.

I played 1st team soccer and captained the cricket 2nd eleven at Weyford Secondary School. I now play cricket for Stopford village.

REFEREES

Mr John Rowley	Mr Paul Townley
Weyford College	President
Heath Lane	Weyford Model Aero Club
Weyford	14 Station Road
GU22 8TY	Weyford
Tel: 01438 345678	GU22 3TX
Ext 348	Tel: 01483 112245
(My course tutor)	

Some comments on Brian's CV

- He has managed to limit it to no more than two pages of A4 paper.
- He has not mentioned his academic performance at Weyford Secondary School; it was not as good as his GNVQ result which is a superior qualification anyway.
- Nor has he mentioned his swimming for the school; he felt that this was irrelevant to the advertised job.
- Because he is lacking in true full-time work experience he has made as much play as he dare about his work experience attachments from school and college and the fact that he has put his skills to good use in building up a disco console for his youth club.
- He has mentioned his soccer and cricket activities because this, he feels, shows that he can work as part of a team.
- Also, being a cricket captain shows his leadership qualities which will be required if he is to become a manager.
- He has not said why he considers himself to be suited for the job and he had better put this in the covering letter which he sends to Mr Grass with his CV.

The covering letter

This is the letter which accompanies the CV. It needs to be a properly written letter – word processed, if possible – referring to the advertisement or the job. A suggested letter that Brian could have written is as follows

24 Windy Lane
Stopford
Surrey
GU23 2XX

Tel: 01176 149245

1st July 1996

Mr John Grass
Lightwater Electronics
High End
Surrey
GU22 61ZX

Dear Mr Grass

TRAINEE SALESPERSON

I am replying to your advertisement in the last Friday's edition of the Weyford Gazette for the above post. As requested, I enclose my CV.

I have a GNVQ (Advanced) in electronics and, over the last three years, work experience – some in a sales showroom – with four electrical or electronic firms. I enjoy working as part of a team and meeting new people. Also, as an electronics enthusiast, a hard worker prepared to accept responsibility, I feel that I am a suitable candidate for the post. If you agree, I would be pleased to attend for an interview at your convenience.

Looking forward to hearing from you in due course.

Yours sincerely

Brian Oldnall

Brian Oldnall

Test your knowledge 4.9

Brian has finished his letter of application with a hand-written signature. Why is this important?

Test your knowledge 4.10

State three different methods of presenting information to a prospective employer. Give TWO advantages and TWO disadvantages of each method.

Activity 4.7

Find the names the names and addresses of FOUR local engineering companies and state in which one of the following engineering sectors they should be placed:

- mechanical engineering
- electrical and electronic engineering
- motor vehicle engineering
- aerospace or aeronautical engineering
- production engineering (e.g. manufacture of metal products)
- chemical engineering
- other.

Which of the companies that you have identified might be in a position to provide you with an opportunity to develop your career? Explain why.

Activity 4.8

Using a word processing package, produce your own CV using a single sheet of A4 paper. Pay careful attention to layout and clarity. Make sure that the document contains all of the essential information that an employer might require, together with details of your skills and interests.

Activity 4.9

The following advertisement appears in your local paper:

> TRAINEE ENGINEERING TECHNICIAN
>
> Axford Aerospace are seeking a Trainee Engineering Technician to work as a member of a team developing new products for the aerospace industry. Applicants should have previous experience in a design or manufacturing environment or should hold relevant ONC, BTEC or GNVQ qualifications.
>
> In the first instance, applicants should apply by letter for an Application Form. Your letter should include a brief personal profile and should state why you consider yourself suitable for the job.
>
> Letters of application should be addressed to:
>
> Mr D.E.Fisher
Personnel Officer
Axford Aerospace
Axford Aerodrome
Axford
BN13 2UB

Use a word processing package to write a letter of application to Axford Aerospace. Make sure that you fully comply with the requirements stated in the advertisement!

Multiple-choice questions

1 A metal commonly used for hygiene purposes in the kitchen is:

A iron
B copper
C stainless steel
D bronze

2 The main purpose of automating a process is to

A make workers redundant
B reduce costs and increase quality

C increase sales
D reduce the amount of energy consumed

3 The Internet is

A a railway network which uses the Channel Tunnel
B a satellite TV channel
C a mobile telephone network
D an international computer network

4 Electromagnetic emissions from mobile phones can cause:

A TV pictures to flicker
B radioactive burns on the user
C excessive AM radio interference
D microelectronic control systems to malfunction

5 Recent developments in computerized engine technology have led to:

A improved fuel economy
B pollution free diesel engines
C engines never needing an oil change
D more vehicles on the road

6 Modern motor car fuel systems use fuel injection rather than a carburettor because

A carburettors are very expensive
B it makes compliance with international pollution standards easier
C it produces a marked decrease in the engine fuel consumption
D lead-free petrol cannot be used with a carburettor

7 Which ONE of the following is a toxic waste material?

A sulphur dioxide
B steam
C furnace ash
D farmyard manure

8 A CV is a

A summary of personal data to be used in connection with an application for a job
B an application for a particular job
C another name for a job advertisement
D a confidential report concerning a person's behaviour

9 A letter starting with 'Dear Mr Smith', should be closed by the phrase

A Yours faithfully
B Yours truly
C Yours obediently
D Yours sincerely

10 A letter of application for a job should be written on:

A lined paper to ensure neatness
B perfumed paper to give an impression of cleanliness
C plain white paper because this is what is expected
D coloured, or fancy edged paper, to show that the writer has a modern approach to life

Index